GEOMETRIC FOLDING ALGORITHMS

Folding and unfolding problems have been implicit since Albrecht Dürer in the early 1500s but have only recently been studied in the mathematical literature. Over the past decade, there has been a surge of interest in these problems, with applications ranging from robotics to protein folding. With an emphasis on algorithmic or computational aspects, this comprehensive treatment of the geometry of folding and unfolding presents hundreds of results and more than 60 unsolved "open problems" to spur further research.

The authors cover one-dimensional (1D) objects (linkages), 2D objects (paper), and 3D objects (polyhedra). Among the results in Part I is that there is a planar linkage that can trace out any algebraic curve, even "sign your name." Part II features the "fold-and-cut" algorithm, establishing that any straight-line drawing on paper can be folded so that the complete drawing can be cut out with one straight scissors cut. In Part III, readers will see that the "Latin cross" unfolding of a cube can be refolded to 23 different convex polyhedra.

Aimed primarily at advanced undergraduate and graduate students in mathematics or computer science, this lavishly illustrated book will fascinate a broad audience, from high school students to researchers.

Erik D. Demaine is the Esther and Harold E. Edgerton Professor of Electrical Engineering and Computer Science at the Massachusetts Institute of Technology, where he joined the faculty in 2001. He is the recipient of several awards, including a MacArthur Fellowship, a Sloan Fellowship, the Harold E. Edgerton Faculty Achievement Award, the Ruth and Joel Spira Award for Distinguished Teaching, and the NSERC Doctoral Prize. He has published more than 150 papers with more than 150 collaborators and coedited the book *Tribute to a Mathemagician* in honor of the influential recreational mathematician Martin Gardner.

Joseph O'Rourke is the Olin Professor of Computer Science at Smith College and the founding Chair of the Computer Science Department. He has received several grants and awards, including a Presidential Young Investigator Award, a Guggenheim Fellowship, and the NSF Director's Award for Distinguished Teaching Scholars. His research is in the field of computational geometry, where he has published a monograph and a textbook, and coedited the *Handbook of Discrete and Computational Geometry*.

Geometric Folding Algorithms

Linkages, Origami, Polyhedra

ERIK D. DEMAINE

Massachusetts Institute of Technology

JOSEPH O'ROURKE

Smith College

CAMBRIDGE
UNIVERSITY PRESS

University Printing House, Cambridge CB2 8BS, United Kingdom

One Liberty Plaza, 20th Floor, New York, NY 10006, USA

477 Williamstown Road, Port Melbourne, VIC 3207, Australia

4843/24, 2nd Floor, Ansari Road, Daryaganj, Delhi - 110002, India

79 Anson Road, #06-04/06, Singapore 079906

Cambridge University Press is part of the University of Cambridge.

It furthers the University's mission by disseminating knowledge in the pursuit of education, learning and research at the highest international levels of excellence.

www.cambridge.org
Information on this title: www.cambridge.org/9780521715225

First published 2007

A catalogue record for this publication is available from the British Library

Library of Congress Cataloging in Publication data
Demaine, Erik D., 1981–
Geometric folding algorithms : linkages, origami, polyhedra / Erik D. Demaine, Joseph O'Rourke.
   p.   cm.
Includes index.
ISBN-13: 978-0-521-85757-4 (hardback)
ISBN-10: 0-521-85757-0 (hardback)
1. Polyhedra –Models.   2. Polyhedra – Data processing.
I. O'Rourke, Joseph.   II. Title.
QA491.D46   2007
516'.156 – dc22   2006038156

ISBN 978-0-521-85757-4 Hardback
ISBN 978-0-521-71522-5 Paperback

To my father, Martin Demaine

– Erik

To my mother, Eleanor O'Rourke

– Joe

Contents

Preface

At how many points must a tangled chain in space be cut to ensure that it can be completely unraveled? No one knows. Can every paper polyhedron be squashed flat without tearing the paper? No one knows. How can an unfolded, precreased rectangular map be refolded, respecting the creases, to its original flat state? Can a single piece of paper fold to two different Platonic solids, say to a cube and to a tetrahedron, without overlapping paper? Can every convex polyhedron be cut along edges and unfolded flat in one piece without overlap? No one knows the answer to any of these questions.

These are just five of the many unsolved problems in the area of geometric folding and unfolding, the topic of this book. These problems have the unusual characteristic of being easily comprehended but they are nevertheless deep. Many also have applications to other areas of science and engineering. For example, the first question above (chain cutting) is related to computing the folded state of a protein from its amino acid sequence, the venerable "protein folding problem." The second question (flattening) is relevant to the design of automobile airbags. A solution to the last question above (unfolding without overlap) would assist in manufacturing a three-dimensional (3D) part by cutting a metal sheet and folding it with a bending machine.

Our focus in this book is on geometric folding as it sits at the juncture between computer science and mathematics. The mathematics is mainly geometry and discrete mathematics; the computer science is mainly algorithms, more specifically, computational geometry. The objects we consider folding are 1D linkages, 2D paper, and the 2D surfaces of polyhedra in 3-space, and thus our title: *Geometric Folding Algorithms: Linkages, Origami, Polyhedra*. As we devote Chapter 0 to a foretaste of the topics covered, we restrict ourselves here to meta-issues.

Because the topics are tangible, physical intuition makes them accessible to those with a wide variety of preparations. Indeed, we have had success using this material to teach mathematical and computational concepts at the high-school, middle-school, and even grade-school levels. However, our presentation in this book is pitched more at the college/graduate-school levels, and we believe professional researchers will find much of interest as well. Because the topics are so variegated, a reader can sample and absorb as much from the book as their background allows. Although full appreciation of every nuance requires the reader's background to encompass the union of the authors' knowledge, there is much to be extracted by those with rather different training. In particular, high-school geometry, basic discrete mathematics, familiarity with big-Oh notation, and a dollop of that intangible "mathematical sophistication" will make accessible at least three-quarters of the book. The remaining quarter needs, in various combinations, concepts in calculus, linear algebra, differential equations, differential geometry, and graph theory; some algorithmic techniques (e.g., linear programming and dynamic programming); and aspects of complexity theory, especially the theory of NP-completeness. In general the material is elementary, though it is sometimes challengingly intricate.

This book is a not a textbook; there are no exercises, for example. It is more a "research monograph," but we hope one with wider appeal than the usual implication of that phrase. Both of us have taught courses based on material in this book, and we know from our experiences that it can easily form the core of an undergraduate or graduate course.

Both because the field of investigation is relatively new—most of what we describe has been discovered in the last decade—and because the problems are hard, there are many unsolved, specific problems, which are known as "open problems" in the field. These problems have the unusual characteristic of being at the same time shallow and deep: shallow in the sense that they may be stated without much technical jargon (like the five mentioned above), and deep in that at least some of them apparently require significant insights to resolve. We isolate many such open problems throughout the book, boxing them to stand out and attract attention (and all are listed in the index). We should warn the reader that the status of these problems varies considerably: some are old chestnuts that have resisted repeated cracking attempts, and others occurred to us as we were writing and are quite unexplored. (The distinction is usually evident from citations to the problem originators; no citation implies we posed it.) Highlighting these problems is an invitation to work on them, not a judgment of their depth or difficulty. In fact, several were solved as the book was being written. Please check the book's Web site[1] for the latest status.

A portion of what we describe in the book is based on published papers written with a large number of coauthors, and indeed we sometimes incorporate sections from these papers with only minor alterations. In this sense, all of our collaborators are coauthors of this book, and for the honor and thrill of working with them, we thank Pankaj K. Agarwal, Oswin Aichholzer, Rebecca Alexander, Greg Aloupis, Esther M. Arkin, Boris Aronov, Devin J. Balkcom, Nadia Benbernou, Michael A. Bender, Marshall Bern, Therese Biedl, David Bremner, Patricia Cahn, Jason H. Cantarella, Roxana Cocan, Robert Connelly, Mirela Damian, Martin L. Demaine, Satyan Devadoss, Melody Donoso, Heather Dyson, David Eppstein, Jeff Erickson, Robin Flatland, Blaise Gassend, Julie Glass, George W. Hart, Barry Hayes, Ferran Hurtado, Hayley N. Iben, Biliana Kaneva, Eric Kuo, Stefan Langermann, Sylvain Lazard, Bin Lu, Anna Lubiw, Andrea Mantler, Henk Meijer, Joseph S. B. Mitchell, Pat Morin,

---

[1] http://www.gfalop.org; see also http://cs.smith.edu/~orourke/TOPP for several of the most prominent open problems. For access to the passworded area of the site use the password clamfucg.

James F. O'Brien, Mark Overmars, Irena Pashchenko, Steve Robbins, Vera Sacristán, Catherine A. Schevon, Saurabh Sethia, Steven S. Skiena, Jack Snoeyink, Michael Soss, Ileana Streinu, Godfried T. Toussaint, Sue Whitesides, and Jianyuan K. Zhong.

A number of others contributed to the book by correcting errors, offering suggestions, drawing figures, and providing a variety of other advice, for which we are deeply grateful: Timothy Abbott, Reid Barton, Asten Buckles, Michael Burr, Helen Cameron, Matthew Chadwick, Michiko Charley, Beenish Chaudry, Sorina Chircu, Elise Huffman, Stephanie Jakus, Reva Kasman, Robert Lang, Dessislava Michaylova, Veronica Morales, Duc Nguyen, Sonya Nikolova, Katya Rykovanova, Wolfram Schlickenrieder, Don Shimamoto, Amanda Toop, and Emily Zaehring.

We owe a special debt to Günter Rote, who read a large portion of the manuscript while on vacation, sent us pages of corrections and suggestions, and caused one open problem to disappear from the draft. The staff at Cambridge has been very helpful, especially Pooja Jain and the ever sagacious Lauren Cowles. Finally, we thank Martin Demaine for creative help and extensive support throughout the project.

The first author was partially supported by NSF grant CCF-0347776, DOE grant DE-FG02-04ER25647, and a MacArthur Fellowship. The second author was partially supported by NSF Distinguished Teaching Scholars award DUE-0123154.

Erik D. Demaine   Joseph O'Rourke
Cambridge, Massachusetts   Northampton, Massachusetts

0 Introduction

The topic of this book is the geometry of folding and unfolding, with a specific emphasis on algorithmic or computational aspects. We have partitioned the material into three parts, depending on what is being folded or unfolded: linkages (Part I, p. 7–164), paper (Part II, p. 165–296), and polyhedra (Part III, p. 297–441). Very crudely, one can view these parts as focusing on one-dimensional (1D) objects (linkages), 2D objects (paper), or 3D objects (polyhedra). The 1D–2D–3D view is neither strictly accurate nor strictly followed in the book, but it serves to place related material nearby.

One might classify according to the process. Folding starts with some unorganized generic state and ends with a more structured terminal "folded state." Unfolding is the reverse process, but the distinction is not always so clear. Certainly we unfold polyhedra and we fold paper to create origami, but often it is more useful to view both processes as instances of "reconfiguration" between two states.

Another possible classification concentrates on the problems rather than the objects or the processes. A rough distinction may be drawn between *design* problems—given a specific folded state, design a way to fold to that state, and *foldability* questions—can this type of object fold to some general class of folded states. Although this classification is often a Procrustean bed, we follow it below to preview specific problem instances, providing two back-to-back minitours through the book's 1D–2D–3D organization. We make no attempt here to give precise definitions or state all the results. Our goal is to select nuggets characteristic of the material to be presented later in detail. The first pass through (design problems) emphasizes geometry, the second pass (foldability questions) emphasizes computational complexity.

0.1 DESIGN PROBLEMS

0.1.1 I: Kempe Universality

A planar *linkage* is a collection of fixed-length, one-dimensional segments lying in a plane, joined at their endpoints to form a connected graph. The joints permit full $360°$ rotation, and the rigid segments are permitted to pass through one another freely. With one or more joints pinned to the plane, the motion of any particular free joint J is constrained by the structure of the linkage. A specific question here is this:

Let S be an arbitrary algebraic curve in the plane. Is there some linkage so that the motion of some free joint J traces out precisely S?

Figure 0.1. There is a linkage that traces a thin version of this collection of curves.

The surprising answer is YES, even if the "curve" includes cusps and multiple pieces. Thus, there is linkage that "signs your name" (Figure 0.1).

The original idea for the construction of a linkage to trace a given curve is due to Kempe from 1876, but technical difficulties in the proof were not completely resolved until 2002. We discuss the history, sketch Kempe's beautiful but flawed proof, and the recent repairs in Section 3.2. An interesting open question here (Open Problem 3.2) is whether there is a linkage to follow any algebraic curve that does not self-intersect during the motion.

0.1.2   II: Origami Design

The epitome of a folding design question is provided by origami:

> Given a 3D shape (an origami final folded state), find a crease pattern and sequence of folds to create the origami (if possible) from a given square piece of paper.

Stated in this generality, this problem remains unsolved, which is one reason that origami is an art. However, in practice origami shapes are a subset of all possible 3D shapes. They are those shapes constructible in two steps: creating an origami *base*, and creasing and adjusting the remaining paper to achieve the desired design. An origami base can be considered a *metric tree*: a tree with lengths assigned to the edges. Creating the base is the hard part of origami design; the secondary adjusting steps are relatively easy (for origami masters).

In the past decade, Robert Lang has developed an algorithm to construct a crease pattern to achieve any given "uniaxial base," the most useful type of origami base. This represents a huge advance in origami design, which previously relied on just a handful of known origami bases. It has allowed Lang to design amazingly intricate origami, such as the mule deer shown in Figure 0.2.

We describe his algorithm, implemented in a program he calls TreeMaker, in Chapter 16. One issue remaining here is that although there is strong experimental evidence that the crease pattern output by TreeMaker leads to a non-self-intersecting folded state, this has yet to be formally proved.

0.1.3   III: Unfolding to Net

The oldest problem we discuss in this book goes back (in some sense) to Albrecht Dürer in the sixteenth century, who drew many convex polyhedra cut open along edges and unfolded flat to a single nonoverlapping piece, now called a *net*. See Figure 0.3 for an example from his *Painter's Manual* (1525).

It remains unresolved today (Open Problem 21.1) whether this is always possible:

> Can the surface of every convex polyhedron be cut along edges and unfolded to a net?

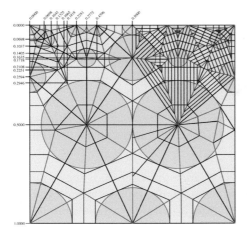

Figure 0.2. A "mule deer" folded by Robert Lang (opus 421): http://www.langorigami.com/art/gallery/gallery.php4?name=mule_deer The crease pattern was designed using TreeMaker.

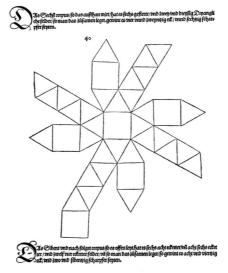

Figure 0.3. A net for the snub cube, drawn by Dürer.

Here the design target is not a specific shape, but rather any planar shape that avoids overlap. We discuss in Chapter 22 evidence for and against the conjecture that the answer to this question is YES, and spin off in a number of related directions. One such question is obtained by removing the *edge-unfolding* restriction: that the cuts must be along edges of the polyhedron. For unrestricted cuts, the answer to the posed question is known to be YES (Section 24.3).

0.2 FOLDABILITY QUESTIONS

We now make our second pass through the three parts of the book, this time concentrating on foldability.

0.2.1   I: Ruler Folding

A polygonal *chain* is a linkage whose graph is just a path. If one views it as a "carpenter's ruler," it is natural to seek to fold it up into as compact a package as possible. This is known as the "ruler folding problem":

> Given a polygonal chain with specific given (integer) lengths for its n links, and an integer L, can it be folded flat (each joint angle either 0 or 180°) so that its total length is $\leq L$?

Of course, sometimes the answer is YES and sometimes NO, depending on the link lengths and L. What is interesting here is the "computational complexity" of deciding which is the case. It was proved in 1985 that answering this question is difficult: NP-complete in the technical jargon. Effectively this means that for, say, $n = 100$, it is not feasible to decide. We present the (easy) proof in Section 2.2.2, one of several proofs throughout the book that establish similar NP-completeness results.

0.2.2   II: Map Folding

A *flat folding* of a piece of paper is a folding by creases into a multilayered but planar shape. The paper is permitted to touch but not penetrate itself. A fundamental question on flat folding is this:

> Given a (rectangular) piece of paper marked with creases, with each subsegment marked as either a mountain or valley crease, does it have a flat folded state?

It was proved in 1996 that answering this question is NP-hard, which means at least as intractable as an NP-complete problem. We present a (complex) proof in Section 13.2 for the easier case when the mountain—valley assignments are not given. When they are specified, the proof is even more difficult. An interesting variant remains open (Open Problem 14.1):

> Given a (rectangular) piece of paper marked by a regular square grid of unit-separated creases, with each subsegment marked as either a mountain or a valley crease, can it be folded into a single 1×1 square?

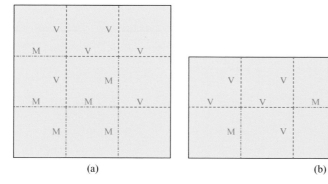

(a)   (b)

Figure 0.4. (a) A 3×3 map that can be folded (but not easily!) (cf. Figure 14.1); (b) a 2×5 map that cannot be folded (Justin 1994).

Figure 0.5. A polygon that folds to a convex polyhedron of 16 triangular faces.

See Figure 0.4 for examples. The added constraints on the creases in the map-folding problem may make this tractable even though the unconstrained problem is not.

0.2.3 III: Polygon Folding

The inverse of unfolding a convex polyhedron to a net is folding a polygon to a convex polyhedron:

Given a polygon of n vertices, can it fold to some convex polyhedron?

Here we assume that the folded polygon covers the surface of the polyhedron precisely once: there are neither gaps in coverage nor double coverage. Nor are we insisting that the polygon be an edge unfolding of the polyhedron; rather, the cuts that produce it are arbitrary.

Figure 0.5 shows an example that can fold. Not all polygons can—there are "unfoldable polygons"—so the question makes sense. It was first settled for a special case, when the folding is restricted to glue whole edges of the polygon to one another. Then an $O(n^3)$-time algorithm is known for a polygon with n vertices. Without the edge-to-edge restriction, only an exponential-time algorithm is available. Both algorithms are discussed in Section 25.1. It remains open (Open Problem 25.3) whether the question can be decided in polynomial time.

These quick tours give some sense of the material, but there is no substitute for plunging into the details, which we now proceed to do.

PART I

Linkages

1 Problem Classification and Examples

Our focus in this first part is on one-dimensional (1D) linkages, and mostly on especially simple linkages we call "chains." Linkages are useful models for robot arms and for folding proteins; these and other applications will be detailed in Section 1.2. After defining linkages and setting some terminology, we quickly review the contents of this first part.

Linkage definitions. A *linkage* is a collection of fixed-length 1D segments joined at their endpoints to form a graph. A segment endpoint is also called a *vertex*. The segments are often called *links* or *bars*, and the shared endpoints are called *joints* or *vertices*.[1] The bars correspond to graph edges and the joints to graph nodes. Some joints may be *pinned* to be fixed to specific locations. Although telescoping links and sliding joints are of considerable interest in mechanics, we only explore fixed-length links and joints fixed at endpoints. (We'll use the term *mechanism* to loosely indicate any collection of rigid bodies connected by joints, hinges, sliders, etc.) An example of a linkage is shown in Figure 1.1.

Overview. After classifying problems in this chapter, we turn to presenting some of the basic upper and lower complexity bounds obtained in the past 20 years in Chapter 2. We then explore in Chapter 3 classical mechanisms, particularly the pursuit of straight-line linkage motion. In contrast to these linkages, whose whole purpose is motion, we next study in Chapter 4 when a linkage is *rigid*, that is, cannot move at all. Most of the remainder of Part I concentrates on chains, starting with reconfiguring chains under various constraints in Chapter 5. We then reach in Chapter 6 what has been the driving concern in the community for at least a decade: deciding under what condition a chain is *locked*. In the language of configuration spaces (to be introduced shortly), rigidity corresponds to an isolated point in configuration space and lockedness to a disconnected component in configuration space. This leads naturally to interlocked collections of chains in Chapter 7. We close with a study of fixed-angle chains in Chapter 8, which leads directly to protein folding in Chapter 9.

We now turn to classifying the problems pursued in later chapters.

---

[1] Sometimes it is convenient to place an endpoint joint of one link in the interior of another rigid link. This structure can always be simulated by links that only share endpoint joints by adding extra links to ensure rigidity, as we will show later (Figure 3.10). So we sometimes will use interior joints in figures, but in our analysis assume all joints are mutual endpoints.

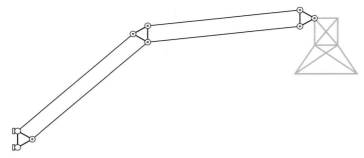

Figure 1.1. A linkage. The circles represent joints at which turning is possible; the two leftmost joints are pinned to the plane. The shaded "lamp" structure is rigid (because of the interior diagonals).

1.1   CLASSIFICATION

There are six features or "parameters" which together classify most of the material we discuss in Part I. These features fall into two groups: those classifying the linkages and those classifying the problems studied. We now quickly survey this classification, with many details deferred to later sections.

First, linkages can be distinguished according to their graph structure, the dimension, and intersection conditions. The graph structure may be a general graph (e.g., Figure 1.1), a tree, a single cycle, or a simple path. A linkage whose graph is a cycle is called a *polygon*; one whose graph is a path we will call a *polygonal chain* or just a *chain*. In the literature a chain is also called an *arc*, a *robot arm*, or an *arm*.

The second parameter is the dimension of the ambient space in which the linkage lives: two, three, four, or higher dimensions. We abbreviate these as 2D, 3D, 4D, and kD, and often use *planar* to mean 2D.

The third parameter classifies how the linkage may intersect itself or intersect obstacles in its environment. With no constraints and no obstacles, the linkage may freely pass through itself, with all joints permitting arbitrary rotation. For 2D chains, this model can be realized with the links at slightly different levels parallel to a plane, with joints realized as short pegs perpendicular to that plane. An intermediate situation of some interest is a linkage that can freely pass through itself, but which is forbidden to penetrate certain fixed obstacles in its environment. This is especially useful for modeling workspace restrictions for robot arms. Finally, much work has assumed an obstacle-free environment but insists that the linkage never self-intersect. A linkage which does not self-intersect is (confusingly) said to be *simple* in the literature, so often this restriction is phrased as demanding the *simplicity* of the linkage. For example, a polygon that does not self-intersect is a *simple polygon*. There is a somewhat subtle distinction between self-intersection and self-crossing, which we will revisit later (in Section 6.8.2).

The foregoing three parameters classify the linkages. We now turn to classifying the questions asked. There are again three primary features that distinguish the questions: the geometric issue, the answer desired, and the type of complexity bound sought.

The most specific geometric problem is "reconfiguration." A *configuration* of a linkage is a specification of the location of all the link endpoints (and therefore of the link orientations and joint angles).[2] A configuration respects the lengths and non-self-intersection of the linkage (if so specified), but may penetrate obstacles. A configuration which avoids all obstacles is said to be *free*, and one that touches but does not penetrate obstacles is *semifree*.

For example, a configuration of an n-vertex polygon in 3D may be specified by $3n$ numbers: n triples of vertex coordinates. The *configuration space* is the space of all configurations of a linkage.[3] In the polygon example, this space is a subset of $3n$-dimensional space, $\mathbb{R}^{3n}$. The *reconfiguration* problem asks, given an initial configuration $\mathcal{A}$ and a final configuration $\mathcal{B}$, can the linkage be continuously reconfigured from $\mathcal{A}$ to $\mathcal{B}$, keeping all links rigid (their original length), staying within the ambient space (e.g., a 2D plane), without violating any imposed intersection conditions? (When a linkage satisfies all the conditions it is said to be in a *legal* configuration.) A slightly less specific problem is to determine *reachability*: whether a particular point (usually a link endpoint) of a linkage can reach, that is, coincide with, a given point of the ambient space. Here the configuration that achieves the reaching is considered irrelevant. This is typical of robot arm applications, where, for example, a soldering robot arm must reach the soldering point, but whether the elbow is raised or lowered when it does is of less concern. The final and least specific problem we consider concerns what we will call *locking*: Are every two legal configurations of a linkage connected in the configuration space, or might a linkage be *locked* or "stuck" in one component of the space and thereby isolated from configurations in another component? There are several variations on this theme for different linkage structures, which will be detailed later (in Chapter 6).

The second problem parameter is the answer desired. *Decision problems* seek YES/NO answers: for example, can the arm reach this point? *Path planning problems* request more when the answer is YES: an explicit path through the configuration space that achieves the reconfiguration.

Finally, a third problem parameter is the complexity measure employed. For path planning problems, the combinatorial complexity of the path may be of interest, for example, the number of constant-degree piecewise-algebraic arcs composing the path. But in general, the algorithmic computational complexity is the primary measure: for example, $O(n^p)$, $\Omega(n^q)$, NP-complete, NP-hard, PSPACE-complete, PSPACE-hard, and so on.[4]

These parameters collectively map out a rich terrain to explore, as indicated in Table 1.1. Before embarking on this study, we quickly survey some of the applications that inspired the models.

1.2 APPLICATIONS

1.2.1 Robotics

It is sometimes useful to view an articulated robotic manipulator as a chain of links. For example, the robot arm shown in Figure 1.2, developed by Forward Thompson Ltd, is a six-axis (6 degrees of freedom) articulated linkage that is designed to apply adhesive

---

[2] In the literature, a configuration is sometimes called a placement, confirmation, or realization. We use the term "realization" for a different notion in Chapter 4.

[3] This is also called the *moduli space* of the linkage.

[4] Refer ahead to page 22 for a brief tutorial on complexity classes.

Table 1.1: Classification parameters for 1D linkage problems

Focus	Parameter	Values
Linkage	Graph structure	General, tree, polygon, chain
	Intersection constraints	None, obstacles, simple
	Dimension	2D, 3D, 4D, kD
Problem	Geometric issue	Reconfiguration, reachability, locked
	Answer desired	Decision, path planning
	Complexity measure	Combinatorial, computational bounds

Figure 1.2. A six-axis robot arm. [By permission, Forward Thompson Ltd.]

tape to the edges of plate glass units for protection. The settings of the four joints determine the exact position of this *robot arm*, and so its configuration may be specified by a point in a 4D *configuration space*. Construction of this configuration space for a specific robot arm and analysis of its geometric and combinatorial structure have proven to be a fruitful methodology (e.g., for "workspace clearance") since its introduction by Lozano-Pérez and others in the 1980s (e.g., Lozano-Pérez 1987; Lumelsky 1987).[5] We will make essential use of configuration spaces in Sections 2.1.1 and 5.1.1.2, and in fact throughout this book.

The complexity of the configuration-space approach for robots with a large number of degrees of freedom has led to probabilistic methods (e.g., Barraquand and Latombe 1990), an area which, a decade later, is dovetailing with protein folding, as we will see below.

Another concern of robot designers is *inverse kinematics*: given a desired tool position, compute joint angles which achieve that position. We will touch upon this topic ("reachability") in Section 5.1.1.2. There is currently great interest in the graphics community in inverse kinematics for animating articulated human models (e.g., Zhao and Badler 1994).

1.2.2   Mechanisms

The study of linkages, and more generally, mechanisms, has long been important to engineering. Often the kinematics of mechanisms is of central concern in practical applications (e.g., Hunt 1978), but for our purposes only the geometry of linkages will play a role. A *pantograph* is a typical useful linkage. Its essence is a parallelogram linkage, one of whose joints has its movement duplicated by an attached bar (see Figure 1.3; here the scale factor is 2). The pantograph has been used for centuries to copy and/or

[5]   The roots of this idea go back to Lozano-Pérez and Wesley (1979).

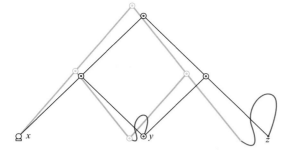

Figure 1.3. A pantograph. Joint x is pinned. The movement of joint y is duplicated and doubled by point z.

Figure 1.4. Thomas Jefferson's pantograph (1804). [Thomas Jefferson Polygraph (i.e., pantograph), 1804, Special Collections, University of Virginia Library, by permission.]

enlarge drawings. Thomas Jefferson used a pantograph to make copies of his extensive correspondence. Figure 1.4 shows the "polygraph" on display at Monticello, his home in Virginia.[6] The duplication property of pantographs finds myriad uses, even carving carousel ponies from a master model.[7] The parallel motion property of pantographs are used in certain truck windshield wipers,[8] in electric trolleys and trains to draw power from overhead cables, in the design of tool box drawers, and in many other applications. Their ability to translate a motion will be used in Section 3.2.

1.2.3   Bending Machines

Three manufacturing processes lead to interesting folding questions: sheet-metal bending, pipe bending, and box folding.

In sheet-metal bending, a 3D part is constructed from a single flat piece of metal by a bending machine. The machine is typically capable of creasing the part along one line at a time, to a given dihedral angle θ, often $\theta = \pi/2$. The part is positioned with the crease over a "die" of angle θ, and a "punch" quickly presses the crease into the die, causing the part to rotate to either side, as depicted in Figure 1.5. As the part nears completion, it becomes a more complex 3D shape, so that avoiding interference between the intermediate "workpiece" and both itself and the machine parts becomes a challenge. Engineering researchers automatically generate "process plans" via heuristic search to find creasing schedules that avoid intersection and minimize overall production time

---

[6] http://www.monticello.org/reports/interests/polygraph.html.
[7] http://www.carousels.com/articles.php.
[8] http://www.b-hepworth.com/pages/neum.html.

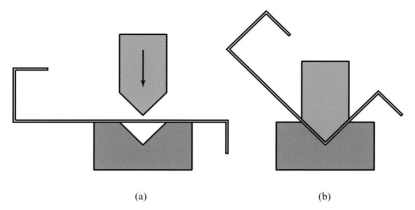

(a)   (b)

Figure 1.5. (a) The part is about to be creased by the punch moving vertically into the well of the die; (b) after creasing.

(including the considerable setup and handling times) (Gupta 1999; Kim et al. 1998; Wang 1997). These search algorithms are all worst-case exponential-time algorithms as a function of the number of bends n.

Pipe bending can be viewed as a 1D version of sheet-metal bending. Especially when bending long hydraulic tubes, used, for example, in jet aircraft, similar workspace clearance issues arise.

An investigation into the computational complexity of (idealized) versions of these problems justifies (in some sense) the engineers' heuristics. Arkin et al. (2003) proved that determining whether there exists any creasing sequence (let alone an optimal sequence) that avoids part self-intersection is an intractable problem[9] for both sheet-metal bending and pipe bending.

A third area is box or carton folding, folding a package carton from a flat cardboard blank. One implemented system uses a robot arm and specialized fixtures to carry out the folding sequence (Lu and Akella 2000). The sequence itself is generated using a motion planning algorithm that finds a piecewise-linear path in the configuration space, which is, for their examples, as large as 9D.

1.2.4   Protein Folding

A protein is composed of a chain of amino acid residues joined by peptide bonds.[10] It can be modeled for many purposes as a polygonal chain representing the protein's "backbone," with three atom vertices per residue, and adjacent atoms connected by bond links. A typical protein is constructed from between 100 and 1,000 amino acids, with some (e.g., the muscle protein titin) containing as many as 30,000 residues.[11] The covalent bond angles between two adjacent backbone atoms is largely fixed by the local chemistry, leaving one rotational degree of freedom (called a "dihedral motion" for reasons to be explained in Section 8.1) between each pair of incident links.

---

[9] Technically, weakly NP-complete.
[10] The "residue" is what is left of the amino acid when a water molecule is lost in the formation of the peptide bond.
[11] http://www.ks.uiuc.edu/Research/titinIg/.

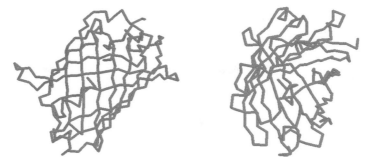

Figure 1.6. Two views of the backbone of a protein of 236 amino acids, the so-called green fluorescent protein (GFP), code 1EMB in the Protein Data Base. Images generated by Protein Explorer, http://proteinexplorer.org.

Natural proteins have two remarkable folding properties. First, they appear to have a unique minimum-energy folded state, determined solely by their "primary structure," that is, the sequence of specific amino acids along the backbone. Second, they curl up unerringly to this state in about 1 sec. The great[12] unsolved "protein folding problem" is to predict the unique folded state (the "tertiary structure" in biochemical terminology) from the primary structure, which will likely entail understanding how proteins accomplish the feat. Because the functionality of a protein is largely determined by its shape, the ability to predict the folding from the sequence of acids would enable quick, targeted drug design. IBM alone has committed US$100 million to an exploratory research effort on the protein folding problem, using their "Blue Gene" supercomputer, among the world's most powerful supercomputers.[13] Figure 1.6 shows a typical protein backbone.

Much of the work on protein folding can be classified as variations on energy minimization (e.g., Maranas et al. 1996), on which we will touch in another context (Section 6.7.3), but which will otherwise play little role here. Three research directions we will explore rely more thoroughly on geometry. The first is the "probabilistic roadmap technique," which uses sampling to find a folding path through the high-dimensional configuration space. The second is the HP model, which approximates the attractions and repulsions of amino acids on a (usually 2D) lattice. Both of these topics will be addressed in Chapter 9. The third is exploring the configuration space of "fixed-angled chains," for bond angles are nearly fixed (Section 8.1).

1.2.5 Mathematical Aesthetics

Although the movement of linkages has many practical applications, we would like to make clear that the research on which we report in this book is not primarily applications-driven. The problems that intrigue us are often suggested or inspired by certain applications, but what usually drives our own work (and that of our various collaborators) is curiosity and mathematical aesthetic considerations. For example, although robot arms are the epitome of practicality, questions concerning the NP-completeness of problems for n-link robot arms for arbitrarily large n (Section 2.1.2)

---

[12] "One of the great challenges of molecular biochemistry and biology" (Troyer et al. 1997).
[13] As of 2005. http://www.research.ibm.com/bluegene/.

are only inspired by real robot arms, which usually have $n \leq 6$. Fundamental computational complexity questions are natural to ask and to attempt to answer, even if they are only tenuously connected to the practical problems from which they arose. On the other hand, time and again theoretical advances have returned later to have practical applicability. A prime example of this is the roadmap algorithm (Section 2.2), which sat at the extreme end of theory in the mid-1980s when it was introduced, but which, as mentioned above, is now being applied (in a modified form) to advance on the protein folding problem. Although we will continue to note the spots at which folding and unfolding research makes contact with practical applications, we will make no further apology for addressing theoretical questions for their own sake.

2 Upper and Lower Bounds

2.1 GENERAL ALGORITHMS AND UPPER BOUNDS

2.1.1 Configuration Space Approach

One of the earliest achievements in the field is an algorithm by Schwartz and Sharir that can solve any "motion planning" problem, including essentially any linkage problem. The technique is to explicitly construct a representation of the free space for the mechanism and then answer all questions with this representation. For example, a reconfiguration decision question reduces to deciding if the initial configuration $\mathcal{A}$ is in the same connected component of the free space as the final configuration $\mathcal{B}$. An example of the configuration space for a 2-link arm in the presence of obstacles is shown in Figure 2.1. Because each configuration is mapped to a point in configuration space, the complex motion of the arm in (a) of the figure is reduced to the path of a point in configuration space (b). In this case the configuration space is a torus representing two angles θ_1 and θ_2, each with wrap-around ranges $[0, 2\pi]$ (see p. 61).

Their original algorithm, developed in the justly famous five "piano mover's" papers,[1] resulted in a doubly exponential algorithm. Subsequent research has improved this to singly exponential. Although we will not employ this algorithm subsequently, it is an important milestone and serves as a baseline for all subsequent work, and so it is worth stating the result more precisely.

The complexities are best expressed in terms of three parameters:

1. k, the number of degrees of freedom of the mechanism. This is the number of parameters necessary to fully specify the configuration of the mechanism, so that a configuration can be represented by a point in $\mathbb{R}^k$.
2. m, the number of "constraint surfaces," each recording some distance or non-penetration constraint, and each represented by a collection of polynomial equalities and inequalities.
3. d, the maximum algebraic (polynomial) degree of the constraint surfaces.

We will reserve n for the number of links in a linkage. Note that the dimension of the space in which the linkage is embedded is not independently relevant; rather, this dimension is absorbed into k. To repeat our earlier example, for a n-vertex polygon in 3D, $k = 3n$.

---

[1] Reprinted in Hopcroft et al. (1987).

θ_2

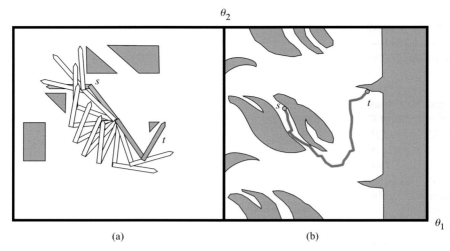

θ_1

(a)   (b)

Figure 2.1. (a) Arm and obstacles; (b) (θ_1, θ_2) configuration space. s and t mark initial and final positions. [Based on Figure 2 of Hwang and Ahuja 1992.]

Schwartz and Sharir used a technique called "cylindrical algebraic decomposition" to achieve an algorithm that can solve any motion planning problem in $O((md)^{3^k})$ time (see Box 2.1). This was subsequently improved by Canny with his "roadmap" algorithm (Canny 1987) to be singly exponential in k, which can even run in polynomial space (Canny 1988). The latest improvements to roadmap algorithms (Basu et al. 2000) yield a general motion planning algorithm with time complexity $O(m^{k+1}d^{O(k^2)})$ (see Box 2.2).

Study of configuration spaces continues. For example, the spaces representing the coordinated motion of several robots and/or mechanisms is especially rich topologically (Abrams and Ghrist 2002). However, we focus primarily on the simplest spaces.

2.1.1.1 Specialization to Planar Robot Arms

Let us specialize this complexity to the important case of a polygonal chain of n links, in an obstacle-free, planar environment, requiring non-self-intersection of the chain.[2] We simulate the non-self-intersection by considering the arm itself as an obstacle, so that the configuration space becomes the free space and the general motion planning results apply.

We need to compute m and k. One can represent the chain with $k = n+1$ parameters, two coordinates of one end (the "shoulder" of the arm) and one angle for each of the $n-1$ joints. However, the use of angles makes it more problematical to keep all constraints semialgebraic, which is the situation to which the theoretical results apply. So instead let us say that $k = 2n + 2$, using two Cartesian coordinates for each joint. Factoring out the trivial rigid motions by specifying the coordinates for one point reduces this to $k = 2n - 2$. The constraints come in two varieties: the link lengths must be constant and two links may not intersect. The former can be represented by n equations specifying the squared length of each link. The second constraint is provided by the obstacles, which are the links themselves. Each of the $\Theta(n^2)$ pairs must be specified to not intersect. This nonintersection condition can be written as inequalities involving

---

[2] Analyzed in Alt et al. (2004) and in Connelly and Demaine (2004).

Box 2.1: Cylindrical Algebraic Decomposition

A semialgebraic set is a subset of $\mathbb{R}^k$ defined by a Boolean combination of a collection $\mathcal{F}$ of polynomial equations and inequalities. A *cylindrical algebraic decomposition (CAD)* for $\mathcal{F}$ is a decomposition of $\mathbb{R}^k$ into finitely many *cells*, which have the property that each polynomial in $\mathcal{F}$ evaluates to the same sign $\{-, 0, +\}$ for every point in the cell. For univariate polynomials, a CAD is realized by a partition of $\mathbb{R}^1$ at the real roots of the polynomials. A CAD for multivariate polynomials is achieved by Collins' recursive algorithm by partitioning each cylinder over the lower-dimensional cells into sign-invariant sectors. The recursion leads to a doubly exponential number of cells: $O((md)^{3^k})$.

With cell adjacency stored in a connectivity graph, a motion planning problem can be solved by searching for a collision-free path through this graph between the cells containing the initial and final configurations of the mechanism, in "randomized expected time" proportional to the number of cells.[a]

---

[a] The randomization is needed only to choose a generic direction for coordinate axes.

Box 2.2: Roadmap Algorithm

Canny's algorithm constructs a network of piecewise algebraic curves (the *roadmap*) that (a) preserves the connectivity of the free space F in that every component of F contains a connected roadmap component and (b) is reachable from any configuration. The construction[a] sweeps a hyperplane H through the space and traces out all extremal points of the intersection of H with the obstacle surfaces, forming "silhouette" curves. These curves might not be connected within each component of F, violating (a). They are linked together at the "critical points" of the sweep, where new extrema appear or disappear. The connection is accomplished via a recursive call to the roadmap algorithm, this time within H and so one dimension lower. This recursion results in only singly exponential complexity. The initial and final configurations can be linked into the network by treating them as critical points. Motion planning is then reduced to searching the roadmap.

Subsequently Canny (1988) strengthened the result by showing that the roadmap algorithm and path planning can be made to run in polynomial space, and so the path planning problem is in PSPACE. Canny's work also relies on general position assumptions, which lead to an m^k term in the complexity. Recent work of Basu et al. (2000) removed this assumption at the cost of increasing that term to m^{k+1}, and leading to the quoted complexity, $O(m^{k+1}d^{O(k^2)})$.

---

[a] See LaValle (2006, Sec. 6.4.3), Sharir (2004, Thm. 47.1.2), and Latombe (1991, p. 191ff) for expositions of Canny (1987).

---

the coordinates of the four endpoints.[3] All equations have degree at most 2, so $d = 2$. Substituting into general motion planning result of $O(m^{k+1}d^{O(k^2)})$ yields a complexity of $O(n^{4n+2}2^{O(n^2)}) = 2^{O(n^2)}$.

For small n, this singly exponential complexity is manageable. And indeed, there are implementations that construct complex configuration spaces. For example,

---

[3] For example, see the formulation in terms of signed areas of triangles in (O'Rourke 1998, Ch. 1).

Halperin et al. (1998) implemented an "assembly planner" that has successfully solved a problem with more than $m = 1,000$ constraint surfaces in a configuration space of $k = 5$ dimensions. But for arbitrary n, this algorithm provides only an upper bound on the problem complexity. Although the complexity might appear to be large, it is in some sense close to optimal, for there are environments with m obstacles and mechanisms with k degrees of freedom whose configuration spaces have $\Omega(m^k)$ (or nearly this many) connected components. For example, if the mechanism is a single link in 3D, $k = 5$, and there is a construction with $\Omega(m^4)$ components (Ke and O'Rourke 1988). Thus the roadmap complexity of about $O(m^{k+1})$ is not far from the best possible among algorithms that explicitly construct a representation of the full configuration space. Furthermore, it is PSPACE-hard to decide whether there is a motion between two configurations of a planar tree linkage or of a 3D open chain (Alt et al. 2004), so it is unlikely that subexponential algorithms can solve these general problems. Canny's result (Canny 1988) establishes that general path planning is in PSPACE; so together these results determine the complexity of these problems as PSPACE-complete.

2.1.1.2 Solving Problems with Configuration Spaces

Let us briefly sketch how the various problems mentioned in Chapter 1 can be solved with a representation of the configuration space (more specifically, the free space) in hand.

1. Reconfiguration decision question. As mentioned previously, this reduces to locating $\mathcal{A}$ and $\mathcal{B}$ in the space and asking whether they fall within the same connected component.
2. Reconfiguration path. All the general algorithms permit the construction of a piecewise algebraic path in the space connecting two given points, which we will sketch below. This path specifies a continuous reconfiguration from $\mathcal{A}$ to $\mathcal{B}$.
3. Reachability decision question, and path. One can view a reachability question as only providing partial information about $\mathcal{B}$. For example, for a robot arm, usually the coordinates of the "hand" p_n are specified, say $p_n = (x_n, y_n, z_n)$. Any configuration $\mathcal{B} = (\ldots, x_n, y_n, z_n)$ reaches p_n, so the decision question reduces to asking if the component of space containing the initial configuration intersects this subspace of configurations. A path requires one point $\mathcal{B}_0$ in this intersection and then can be constructed as above.
4. Locked questions. Here the issue is simply this: Does the configuration space consist of more than one connected component?

With any reasonable representation of the configuration space, all of these questions can be answered in time proportional to the complexity of that space, perhaps with a small additional overhead. For example, paths are often found as follows. The configuration space is partitioned into cells such that all points within the cell are surrounded by the same set of constraint surfaces. Two adjacent cells share a surface patch that represents a "criticality" of some sort. One then constructs a graph, with a node for each cell and arcs for adjacency, and searches this graph, say using depth-first search. Connectivity in the graph leads to a path in configuration space.

The upshot is a generic exponential upper bound, roughly m^k, for answering any specific question involving obstacles of complexity m and a mechanism of k degrees of freedom. We should note that this algorithm does not help answering the more general questions we will consider in Chapter 6, such as "Can *any* 2D chain lock?" because they

Figure 2.2. Wooden "Hedgehog in a Cage" puzzle. http://www.cleverwood.com/.

involve an arbitrary number n of links. Nor does the exponential upper bound show that the problems are truly hard; it only shows that the problems are not *too* hard. In fact we will see that many instances can be solved efficiently. We consider one more example before turning to lower bounds.

2.1.2 Example: Separation Puzzles

There is a class of puzzles, called "separation" or "take-apart" puzzles, which consist of an arrangement of a collection of distinct, rigid parts, with the goal to separate the parts, to take apart the object.[4] Some of these puzzles are extremely difficult to solve, even those which have no hidden parts. Nevertheless all of these puzzles could in theory be solved by a general motion planning algorithm in "constant time," because for any specific puzzle, the parameters are constant.

Consider the clever "Hedgehog in a Cage" puzzle shown in Figure 2.2. It consists of just two rigid parts. The cage could be represented by two squares for the top and bottom and four segments for the bars. The hedgehog could be represented by, say, a dodecahedron and 14 segment spikes (8 short, and 6 long). In terms of the key complexity parameters, $k = 6$, because the hedgehog has 6 degrees of freedom, and $m \approx 50^2$, because there are about 50 vertices, segments, and faces that pairwise interact to provide constraint surfaces. The combinatorial portion of the complexity of the roadmap algorithm (p. 19) is dominated by the term m^{k+1}; in this case, $50^{14} > 10^{23}$. Regardless of the magnitude of this term, with sufficient time, the roadmap algorithm could create a map for the hedgehog to escape from the cage, as well as solving any other similar take-apart puzzle.

We should note that one can generalize separation puzzles to have parameters that increases without bound. For example, the Towers of Hanoi puzzle with n disks, or n disks and k pegs, is such a parameterized puzzle. For such puzzles it does make sense to establish complexity bounds (see, for example, O'Rourke 1998, Section 8.7).

---

[4] See, e.g., http://www.puzzlesolver.com/list.php?cat=2.

2.2   LOWER BOUNDS

2.2.1   Introduction

In this section we review the early flurry of results that established that several key versions of linkage reachability questions are intractable in the technical, computational complexity sense, and report on the latest status. Four results were published in a period of 6 years, which together settled several basic lower bounds by 1985. Each successive result can be viewed as a strengthening of the one that preceded it. We first sketch the history before describing the constructions.

History. Reif obtained the first result: that reachability for a tree-like object of polyhedra connected at point joints, in the presence of polyhedral obstacles, in 3D, is PSPACE-hard (Reif 1979, 1987). Together with Canny's roadmap algorithm described in Section 2.1.1 and Box 2.2, which establishes that these motion planning problems are in PSPACE, the problem is PSPACE-complete.

As a minitutorial, the four complexity classes we are concerned with in this book may be sorted as

$$P \subseteq NP \subseteq PSPACE \subseteq EXP.$$

Here P, *polynomial time*, is considered efficient or tractable, and *EXP*, *exponential time*, is intractable. Without resolution of the famous $P \overset{?}{=} NP$ question, it remains unknown whether the intermediate classes are truly difficult, but it is widely presumed that they are. The suffix -*hard* is used to indicate "at least as hard as," while -*complete* means "among the hardest questions in that class." This leads to the following difficulty spectrum:

$$P \subseteq NP\text{-complete} \subseteq NP\text{-hard} \subseteq PSPACE\text{-complete} \subseteq PSPACE\text{-hard} \subseteq EXP.$$

We will follow the usual conventions in calling all but P *intractable*. Reif's result is well out on this spectrum, and means that the complexity of his reachability question grows more than polynomially (and likely exponentially) with the degrees of freedom of the object, unless P = PSPACE (and in particular P = NP).

Reif's result left two important possibilities unresolved: it might be that a chain (rather than a tree) is easier to reconfigure; and it might be that reachability in 2D is easier than in 3D. Reif's proof uses both the tree structure and 3D in essential ways. The second question was resolved by Hopcroft, Joseph, and Whitesides (henceforth, HJW), but at the cost of increasing the geometric complexity. They proved that the reachability question for a planar graph-linkage, without obstacles and permitting self-intersections, is PSPACE-hard (Hopcroft et al. 1984). In a second paper (Hopcroft et al. 1985) they reduced the geometric complexity from a graph to a path/chain, and established a slightly weaker NP-hardness bound for reachability, this time with a few obstacles, and again permitting self-intersections of the chain. The final early result, by Joseph and Plantinga (1985), strengthens this in a slightly altered form: they prove that reachability, in the presence of many obstacles, is PSPACE-complete.

Following this work nearly two decades later, Alt et al. (2004) extended the techniques of Joseph and Plantinga to prove two hardness results without obstacles. First,

Table 2.1: Lower bounds[a]

Dim	Link graph	Linkage intersection	Obstacles	Question	Complexity
3D	Tree	Simple	Polyhedra	Reachability	PSPACE-complete (Reif 1987)
3D	Chain	Simple	None	Reconfiguration	PSPACE-complete (Alt et al. 2004)
2D	Graph	Permitted	None	Reachability	PSPACE-complete (Hopcroft et al. 1984)
2D	Chain	Permitted	4 segments	Reachability	NP-hard (Hopcroft et al. 1985)
2D	Chain	Permitted	Polygons	Reachability	PSPACE-complete (Joseph and Plantinga 1985)
2D	Tree	Simple	None	Reconfiguration	PSPACE-complete (Alt et al. 2004)

[a] "Simple" means non-self-intersecting.

deciding reconfigurability of a tree linkage in 2D between two given configurations is PSPACE-complete. Second, the same problem for a chain in 3D is PSPACE-complete.

The six results mentioned above are summarized in Table 2.1, including some further details to be explained shortly.

Missing from this table is what might be considered the most natural problem: deciding whether a 2D chain can be reconfigured without self-intersection between two specified configurations in the absence of obstacles. The reason for omitting this problem is that here the answer is always YES! This lockability issue is the topic of Chapter 6.

2.2.2 Three Constructions Sketched

Reif's proof reduces acceptance by a particular type of Turing machine, an abstract "automaton" used to prove results in the theory of computation, to the solution of a reachability problem of a tree-linkage P in 3D. The machine M is a space-bounded symmetric Turing machine, whose technical definition need not detain us.[5] M "accepts" a word w (that is, M reads w and eventually reaches a special "accepting" state) if and only if an initial configuration of P that depends on w can move through an obstacle course and reach the "exit." See Figure 2.3 for one (small) section of the obstacle course. The motion of P simulates the Turing machine's state transitions. Because the acceptance problem for these machines is known to be PSPACE-complete and because Reif's reduction is not too expensive (technically, "log-space"), the reachability problem is established to be PSPACE-hard.

The first HJW proof has a similar high-level structure, but its underlying details are different, and more interesting for our purposes. They reduce the acceptance question

---

[5] See Lewis and Papadimitriou (1997). The number of work-tape cells used is bounded as a function of the input length ("space-bounded"), and the state transition graph is undirected ("symmetric").

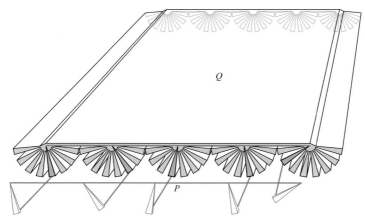

Figure 2.3. The movable object P consists of a series of identical tetrahedra linked by segments. It is poised to enter a "tunnel" Q (at the darkly shaded portals). Each cylindrical fan of triangular tunnels represents one state of the machine, with one triangle per possible tape symbol. Connecting several such tunnels together forces P to change states in a determined way. (After Figure 10 of Reif 1979).

for another type of machine, a "linear bounded automaton" (or LBA), which again is known to be PSPACE-complete, to a reachability question for a planar linkage. Because the construction is polynomial, again the reachability problem is established to be PSPACE-hard. Without obstacles, however, the complexity has to be embodied in the linkage itself. They require linkages that form a general graph structure. They use joint positions to encode variables and arrange the linkage to restrict the variables to have essentially binary values. At any one time, only one variable can "flip its bit," so to speak (by moving continuously between two extreme positions). A key linkage used in their proof is one designed by Kempe in 1876 that constructs linkages to "solve" polynomial equations (Kempe 1876). Kempe was concerned with making a linkage to follow any algebraic curve; HJW extend his method to multivariate polynomials. This permits them to implement an AND-gate as $x_1 x_2 = 0$, and in fact to build linkages to implement any Boolean function. This forms the basis of their LBA reduction. We will explain Kempe's work, as much for its intrinsic interest as for its application to lower bounds, in Section 3.2.

Deviating from historical order, the construction of Joseph and Plantinga is similar to Reif's in its manner of simulation, and similar to HJW's in that it simulates an LBA. But it accomplishes this with a planar polygonal chain, albeit in the presence of complex obstacles. See Figure 2.4 for a portion of their construction.

As mentioned earlier, this thread of research was resumed most recently by Alt et al. (2004), who strengthened some intractability results to hold without obstacles. Their result on tree linkages in 2D follows the Joseph–Plantingua construction, but uses one rigidified portion of the tree to act like the obstacles (such as that in Figure 2.4) needed to construct the gadgets. The tree rigidification is obtained with a locked tree along the lines of Section 6.5 (p. 94). This enables them to prove PSPACE-completeness for trees in 2D. They also extend their 2D construction into a 3D "maze," using interlocked 3D chains from Chapter 7 (p. 123), which leads to the natural endpoint of this line of research: deciding the reconfigurability of an open chain in 3D is PSPACE-complete.

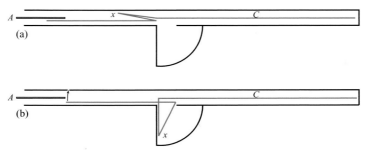

Figure 2.4. The moving chain is C. The backward and forward position of the "switch" whose tip is x encodes the bit 0 and 1. The gadget shown checks the bit in the sense that if C starts below the obstacle A with the switch backward as in (a), it cannot move to above A. If, however, the switch is initially forward as in (b), x can enter the circular well and open enough to shorten the chain, permitting passage to above A. (After Figure 5 of Joseph and Plantinga 1985.)

Ruler folding. The earlier proof of HJW that reachability for a planar polygonal chain is NP-hard is a simple and beautiful argument that has served as a model for several later proofs. It is the only one we present in detail.

They start with the *ruler folding problem*:

> **Ruler Folding:** Given a polygonal chain with links of integer lengths $\ell_0, \ldots, \ell_{n-1}$ and an integer L, can the chain be *folded flat*—reconfigured so that each joint angle is either 0 or π—so that its total folded length is $\leq L$?

Here the chain may self-cross, and there are no obstacles. So any move is legal. And yet they prove the problem is difficult:

Theorem 2.2.1 (Hopcroft et al. 1985). *The ruler folding problem is NP-complete.*

Proof: The proof is by reduction from Set Partition (henceforth: PARTITION), a known NP-complete problem:[6]

> PARTITION: Given a set of n positive integers, $S = \{x_1, \ldots, x_n\}$, does there exist a partition of S into $A \subseteq S$ and $B \subseteq S$ so that
>
> $$\sum_{x_i \in A} x_i = \sum_{x_j \in B} x_j? \tag{2.1}$$

From a given instance of PARTITION, a ruler is constructed as follows. Let $s = \sum_{i=1}^{n} x_i$ be the sum of all the elements in S. The ruler R consists of links of lengths

$$(\ell_0, \ell_1, \ell_2, \ldots, \ell_{n+1}, \ell_{n+2}, \ell_{n+3}) = (2s, s, x_1, \ldots, x_n, s, 2s).$$

The claim is that R can be folded into a length of at most $2s$ if and only if the instance of PARTITION has the answer YES.

---

[6] This problem is *weakly NP-complete*, meaning that it is NP-complete as a function of n and the binary encoding of the input numbers, but if the integers were written in unary notation (thus inflating the input size), the problem might become polynomial, and so admit a "pseudopolynomial-time algorithm."

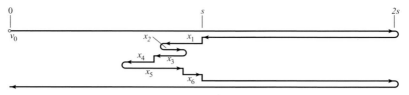

Figure 2.5. Ruler folding reduction. Here $x_1 + x_3 + x_4 = x_2 + x_5 + x_6$.

Consider the links directed from v_i to v_{i+1}, and fold the ruler along the real line, with $v_0 = 0$ and $v_1 = 2s$. The vertices v_2 and v_{n+2} must both lie at s if all is to fit within $[0, 2s]$. This forces the links between to consume zero total displacement: so the sum of the leftward-pointing links must equal the sum of the rightward-pointing links (see Figure 2.5). Therefore R folds into $2s$ if and only if $\{x_1, \ldots, x_n\}$ can be partitioned into two equal-length halves. It is clear that it cannot fold to $< 2s$, because there are links of that length. So we have established the claim for $L = 2s$.   □

Recently a "fully polynomial-time approximation scheme"[7] has been found for finding the minimum folding length of a ruler (Călinescu and Dumitrescu 2005).

This folding problem does not fit one of the "standard" varieties we outlined earlier (in Section 1.1), but it can be used to establish the intractability of planar arm reachability:

Theorem 2.2.2 (Hopcroft et al. 1985). *Whether a planar polygonal chain can reach a given point p, in the presence of obstacles but permitting the chain to self-intersect, is NP-hard.*

Proof: Let a ruler folding problem specify a ruler R of total stretched length L. We construct a reachability problem as follows. A chain C, anchored at the origin $v_0 = (0, 0)$, consists of $100L + 1$ unit-length links, followed by links whose lengths are 100 times those in R. Call this latter portion of the chain R_{100}. The obstacles consist of four segments, as illustrated in Figure 2.6: three sides of a $100(L + k) \times (2 + \varepsilon)$ rectangle, forming a "tunnel," and one long segment s of length $201L$ nearly touching v_0. The tunnel width is $2 + \varepsilon$ for some small $\varepsilon > 0$ to permit unit-length segments to turn around in the tunnel avoiding contact with the tunnel walls. We now argue that the hand endpoint at the end of R_{100}, call it v, can reach $p = (200L, 1)$ if and only if R can be folded to length $\leq k$.

The construction ensures that p can only be reached if the ruler enters the tunnel and passes through the $100k$ "gap" to the left of the origin. The shortest link of R_{100} has length 100 (because the links of R are integers); so once inside the thin tunnel, it is unable to rotate more than slightly. Thus only if R_{100} folds to at most $100k$ might p be reached. On the other hand, the unit-length links of C have enough room to rotate inside the tunnel. Thus, if R_{100} does indeed fold to gap length, then p may be reached by the following sequence: first fold R_{100} outside the tunnel, draw it inside the tunnel by crumpling the $100L$ unit links, unfurl $100k$ of them to position the folded R_{100} as illustrated in (b) of the figure, pass the folded ruler through the unit links into the upper half of the tunnel, and finally extend out toward p. The distance from v_0 to p is $\sqrt{(200L)^2 + 1}$, and C has length $200L + 1$, more than enough to reach p.   □

---

[7] This finds a solution within $(1 + \varepsilon)$ of the optimum, in time polynomial in $1/\varepsilon$, and in the input and output sizes of the problem.

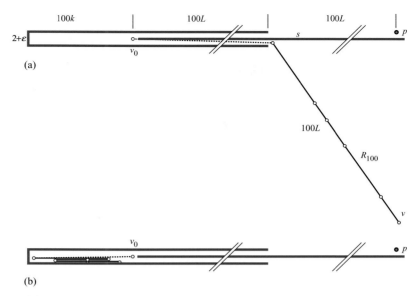

Figure 2.6. (a) Initial configuration. The dashed segment starting from v_0 represents a large collection of unit links. The slashes indicate interruptions. Note the tunnel is only $2 + \varepsilon$ units high. (b) Intermediate configuration, when R_{100} is folded and about to pass through the gap.

Path planning. The three intractability results that simulate automata with linkage problems all also establish an exponential lower bound on path planning versions of the problems, because they each require an exponential number of "elementary moves" of the linkage, under any reasonable definition of "elementary." The ruler folding proof is fundamentally different: the number of moves needed to reach p is polynomial (in fact, linear) in n; what is difficult is finding the moves.

Although the results in Table 2.1 establish that sufficiently rich linkage problems are intractable, room remains for efficient (polynomial-time) algorithms in at least two directions. First, all these results rely on arbitrarily long or complex linkages. As soon as the number of degrees of freedom is fixed, the Roadmap Algorithm (Box 2.2) shows there are polynomial algorithms. Second, the various assumptions on the intersection conditions and the number and type of obstacles are often important to the proofs, and with different assumptions the complexity could change. For example, we will see in Section 5.1 that reducing the number of obstacles from four segments to none in Theorem 2.2.2 (line 3 of Table 2.1) drops the complexity from NP-hard to $O(n)$!

Universality theorem. It remains true, however, that the most general linkage problems are intractable. This was recently given a strong form via a general "universality theorem," established in slightly different forms by Jordan and Steiner (1999) and by Kapovich and Millson (2002).

Theorem 2.2.3. *Let $X \subseteq \mathbb{R}^d$ be a compact real algebraic variety, i.e., the locus of zeros of a polynomial in d variables. Then X is homeomorphic[8] to some components of the configuration space of a planar linkage.*

---

[8] Roughly speaking, two geometric objects are *homeomorphic* if one can be continuously stretched and deformed into the other.

Intuitively this says that the configuration space of a linkage can be as complicated as can be specified algebraically. See Connelly and Demaine (2004), O'Hara (2005), and Shimamoto and Vanderwaart (2005) for discussions of this complex topic. We will return to universality in a particularly appealing form (there is a linkage with a joint whose "orbit" follows any given algebraic curve) in Section 3.2.

3 Planar Linkage Mechanisms

In this chapter, we discuss aspects of the movements of linkage mechanisms embedded in the plane, permitting links to pass through one another, without obstacles. Our goal is to explain Kempe's universality theorem (Theorem 3.2.1).

3.1 STRAIGHT-LINE LINKAGES

In preparation for presenting this universality result, we discuss the fascinating history of the design of linkages to draw straight lines, which led directly to Kempe's work.

3.1.1 Degrees of Freedom

Linkages that have a point that follows a particular curve have 1 degree of freedom, that is, the configuration space is one-dimensional. We consider two types of joints: pinned and unpinned or free. Both are universal joints, but a pinned joint is fixed to a particular point on the plane. Let a linkage have j free joints and r rods (or links). Pinned joints add no freedom, a free joint adds 2 degrees of freedom, and a rod removes 1 degree of freedom if it is not redundant (Graver 2001, p. 135). Thus, if all rods are "nonredundant" (a common situation), a linkage has $2j - r$ degrees of freedom. Thus, to follow a curve, a linkage should have $r = 2j - 1$ rods; in particular, it must have an odd number of rods.

3.1.2 Watt Parallel Motion

It was a question of considerable practical importance to design a linkage so that one point moves along a straight line, a motion needed in many machines, for example, to drive the piston rod of a steam engine. James Watt invented a simple linkage in 1784 that nearly, but not quite, achieves this.[1] Watt's "parallel motion" linkage (something of a misnomer that has stuck) is illustrated in Figure 3.1. Two pinned joints x and y are centers of equal-radii circles traced by joints a and b respectively. Here $j = 2$ and $r = 3$, so it has $2j - r = 1$ degree of freedom. The midpoint c of ab moves approximately on a vertical straight line. Joint a is pulled leftward and b rightward on their circles, with

---

[1] We rely here primarily on Kempe (1877).

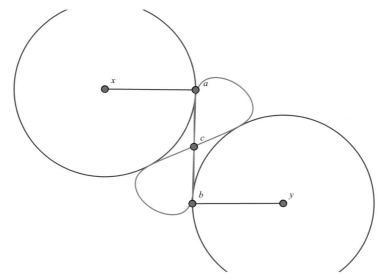

Figure 3.1. A Watt linkage. Joints x and y are pinned; joints a and b are free. The locus of point c is the figure-8 curve. [Construction from Cinderella.]

c staying approximately balanced between. Point c actually follows a "lemniscate," a figure-8 shape with one long, nearly straight section. Variations on this theme are now called "Watt linkages." Watt wrote, "I am more proud of the parallel motion than of any other mechanical invention I have ever made."[2]

3.1.3   Peaucellier Linkage

It was 80 more years before exact straight-line motion was achieved in 1864, by Charles-Nicolas Peaucellier, a captain in the French army. He received a prize, the Prix Montyon, from the Institute of France for his discovery.[3] His remarkable linkage is illustrated in Figure 3.2. Joints x and y are pinned, and rods xa and xb keep joints a and b on circle C_x and rod yc keeps joint c on circle C_y, which also includes x. There are $j = 4$ free joints and $r = 7$ rods, so the linkage has $2j - r = 1$ degree of freedom. The rhombus $abcd$ expands and narrows symmetrically about the diagonal ab. As the cell $abcd$ is rotated clockwise, the gap between the circles C_x and C_y narrows, squeezing the cell, remarkably just enough so that joint d stays on a vertical line L. It is clear that x and the tips c and d of the rhombus are collinear. It can be shown that the product of the distances from x to c and to d is a constant, and that this implies that d follows a straight line.

When Lord Kelvin worked a model of the device, he is reputed to have remarked that "it was the most beautiful thing he had ever seen" (Fergusson 1962).

Peaucellier's linkage was discussed by J. J. Sylvester in a lecture in 1874, and soon improved by Hart to a linkage of only five links, as we will describe in Section 3.3. It was

---

[2]   Hunt (1978, p. 134), quoting Reuleaux from 1875.

[3]   The history is tangled. A student of Chebyshev, Lippman Lipkin, independently discovered the same mechanism in 1871 (contra: Coxeter and Greitzer 1967, p. 109), as Peaucellier acknowledged in 1873 (Cabillon 1999). Peaucellier's result was first published in 1864 (Peaucellier 1864). See also Koetsier (1999) and Kempe (1877, p. 12).

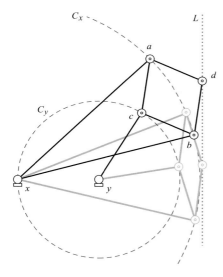

Figure 3.2. Peaucellier linkage. The dark lines show it in one position, the light lines in another.

this lecture that initiated Alfred Bray Kempe[4] to the intricacies of linkages and inspired his work toward his universality theorem.[5]

3.2   KEMPE'S UNIVERSALITY THEOREM

An attractive version of Kempe's result was formulated by Thurston in the 1970s (King 1999, Cor. 2.1):

There is a linkage that signs your name.

As mentioned, Kempe's proof of Theorem 3.2.1 was flawed, and the subsequent history rather complex. We will not detail this history (see King 1999 or Connelly and Demaine 2004), but content ourselves by mentioning that the first complete, detailed proof is generally acknowledged to be by Kapovich and Millson (2002). In King (1999), the universality theorem is credited to Thurston and Kapovich-Millson. Because Kempe is acknowledged by all to be the source of the main ideas, we will continue to call it "Kempe's universality theorem":

Theorem 3.2.1 (Kempe 1876). *Let C be a bounded portion of an algebraic curve in the plane, that is, the intersection of zero-set of a real-coefficients polynomial $\phi(x, y) = 0$ with a closed disk. Then there exists a planar linkage such that the orbit of one joint is precisely C.*

We now describe Kempe's proof, tracking his own presentation (Kempe 1876), followed by a discussion of the flaw and one route to repair.

---

[4] Kempe was a barrister by profession, but an excellent amateur mathematician. He is best known for his false proof of the four-color theorem in 1879, which stood for 11 years before Heawood found a flaw. Several ideas in his "proof" were, however, valid, and survive in the form of "Kempe chains" in the computer-aided proof of Appel and Haken a century later (Appel and Haken 1977). The story with linkages is similar: his proof was flawed, but the overall conception was brilliant.

[5] There has since been an extension to 3D (Zindler 1931).

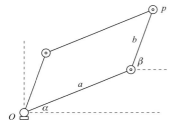

Figure 3.3. Linkage parallelogram.

3.2.1 Kempe's Proof

Let $\phi(x, y) = 0$ be the algebraic curve to be followed by some joint p. The construction begins by placing p at the apex of a parallelogram as shown in Figure 3.3. This parallelogram constrains the range of p's motion, but if we choose a and b equal and large, p is free to move anywhere within distance $2a$ of the pinned vertex O. The feature of this construction is that, if we view O as the origin $(0, 0)$, then the coordinates of $p = (x, y)$ may be expressed in terms of cosines of angles α and β:

$$x = a \cos\alpha + b \cos\beta,$$
$$y = a \cos(\alpha - \pi/2) + b \cos(\beta - \pi/2).$$

(This linkage has $j = 3$ and $r = 4$ and so $2j - r = 2$ degrees of freedom.) Substituting into $\phi(x, y)$ and repeatedly applying the trigonometric product-to-sum identity

$$\cos A \cos B = \frac{1}{2} \cos(A + B) + \frac{1}{2} \cos(A - B) \tag{3.1}$$

converts powers and products of x and y into cosines of sums of angles, resulting in the following general form for the polynomial $\phi(x, y)$:

$$\phi(x, y) = c + \sum_i c_i \cos(r_i\alpha + s_i\beta + \delta_i), \tag{3.2}$$

where c and c_i are constants, r_i and s_i are integers, and $\delta_i \in \{0, \pm\pi/2\}$. For example, the term xy^2 expands and reduces to

$$\frac{1}{4}\big[a^3 \cos\alpha + 2ab^2 \cos\alpha - a^3 \cos(3\alpha) + ab^2 \cos(\alpha - 2\beta) + a^2b \cos(2\alpha - \beta)$$
$$+ 2a^2b \cos\beta + b^3 \cos\beta - b^3 \cos(3\beta) - 3a^2b \cos(2\alpha + \beta) - 3ab^2 \cos(\alpha + 2\beta)\big].$$

Each term of the sum in Equation (3.2) is achieved by a link of length c_i, at a suitable angle, with the whole shifted by c. Thus the problem reduces to constructing an angle of the form $r_i\alpha + s_i\beta + \delta_i$ from α and β.

Kempe shows that such a construction can be accomplished with three "gadgets": a "translator" for translating a motion, an "additor" for adding any two angles, and a "multiplicator" for multiplying an angle by a positive integer. The essence of the translator is the action of a pantograph, touched on earlier in Section 1.2.2. Note that while a multiplicator could naturally be implemented using a collection of additors (adding an angle repeatedly to itself), in fact the additor construction uses a multiplicator, and thus both are necessary. Hence, we describe these gadgets in reverse order.

Multiplicator. First, a *contraparallelogram* is a four-bar linkage formed by flipping two adjacent sides of a parallelogram across the diagonal they determine (see Figure 3.4).

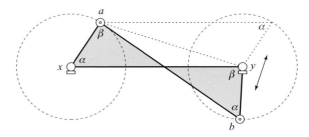

Figure 3.4. A contraparallelogram. $|xa| = |yb|$ and $|xy| = |ab|$. (Note: There is no joint at the point at which the two rods xy and ab cross.)

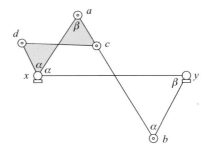

Figure 3.5. Kempe's "multiplicator": two nested, similar contraparallelograms.

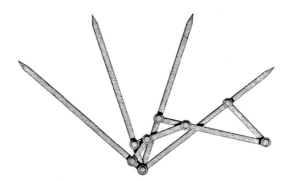

Figure 3.6. Kempe's angle trisector (Kempe 1877, p. 42).

(This structure plays a key role in Hart's linkage, as we will see in Section 3.3.) The opposite angles α and β are equal as in a parallelogram; the two shaded triangles are congruent. The linkage has 1 degree of freedom with x and y pinned[6] and so fixing either one of α or β determines the other. Kempe cleverly joined two similar contra-parallelograms as shown in Figure 3.5 so that the angle β at joint a is shared. The two have their lengths in proportion: $|xy|/|xa| = |xa|/|xd|$. Because fixing β determines α in a contraparallelogram, and because of their similarity, the angle α is duplicated at x in the smaller, shaded contraparallelogram, effectively multiplying the angle α by 2. This is why Kempe called this linkage a "multiplicator." Its dynamic action is quite pleasing.[7]

This same process may be repeated, as illustrated in Figure 3.6 for $r_i = 3$, to multiply an angle by a positive integer r_i. Kempe notes that this linkage of eight bars exactly

---

[6]   In this case, the xy rod is redundant. So $j = 2$ and $r = 3$, and $2j - r = 1$.
[7]   See http://www.gfalop.org/I.

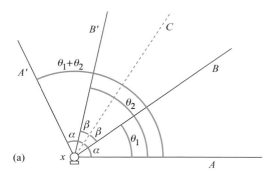

(a)

Figure 3.7. (a) A reflects to A' about C, as does B to B'; (b) the construction with superimposed reversors.

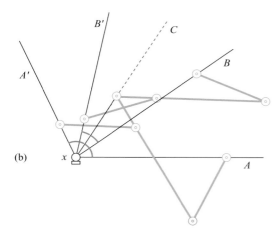

(b)

trisects an angle, a construction not possible by straight edge and compass. We will return to this point in Part II, Chapter 19 (p. 285).

Additor. Kempe's "additor" adds two angles together. This relies on the multiplicator, but only in its times-2 version in Figure 3.5, which he also called a "reversor" because it can be viewed as reflecting the angle α about the line xa. The construction for adding two angles θ_1 and θ_2 is illustrated in Figure 3.7. It consists of five links, all joined at x, as illustrated in Figure 3.7(a). A makes an angle α with C and reflects across C to A'. Similarly, B makes an angle β with C and reflects across C to B'. If these reflections are maintained, then A and A' are placed symmetrically α around C and B and B' are placed symmetrically β around C. Then the angle $\theta_1 = \alpha - \beta$ and $\theta_2 = \alpha + \beta$ so that $\theta_1 + \theta_2 = 2\alpha$, which is the angle between A and A'. Thus, with A the "base" link, we can "input" the angle θ_1 with link B and θ_2 with link B', and then A' reads off the "output" $\theta_1 + \theta_2$.

The reflections about C can be maintained by Kempe's "reversor" in Figure 3.5. This requires erecting an appropriate contraparallelogram that maintains the reflected angle between A and C, and another to maintain the reflection of B about C. These extra links and joints are shown shaded in Figure 3.7(b).

Translator. Kempe's simplest gadget is the translator, whose purpose is to copy a link at its same orientation elsewhere. His design is shown in Figure 3.8(a). Link A, making an angle α with the horizontal, is displaced to the parallel link C. It makes sense to choose $|D| = |E|$ so that C is free to be placed anywhere within a disk centered on x,

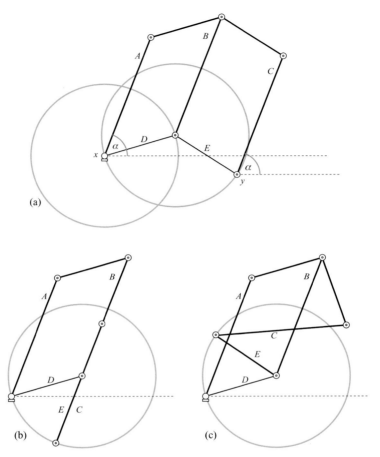

(a)

(b)   (c)

Figure 3.8. (a) Kempe's translator; (b) alignment of C with B; (c) second parallelogram flips to a contraparallelogram. [See also Figure 2.3.2 in Hopcroft et al. 1984.]

that is, the annulus of reachability for the point y (according to Theorem 5.1.2) is a disk. There is, however, a flaw in this design, to which we will return below.

Overall design. Returning to Equation (3.2), the "multiplicator" permits multiplying any angle by an integer r_i. The "additor" permits adding angles, and the "translator" permits connecting together the terms of the sum. Combining these gadgets in an appropriate arrangement yields a joint p' that stays at a horizontal distance $\phi(x, y) - c$ from the y-axis. Using a Peaucellier linkage to force p' to stay on the vertical line $x + c = 0$ in turn forces p to stay on the curve $\phi(x, y) = 0$. Thus, this complex linkage of 1 degree of freedom has the joint p follow the equation $\phi(x, y) = 0$ within the range of x values provided by the parallelogram linkage in Figure 3.3 (see Figure 3.9).

3.2.2 The Flaw and Repairs

Concerning the translator in Figure 3.8, Kempe (1876, p. 214) says: "it is clear that the link $[C]$ is equal and parallel to $[A]$, but is otherwise free to assume any position." What he missed is that when link C aligns with B ((b) of the figure), the second parallelogram is free to flip to a contraparallelogram, as in (c), in which case C is no longer parallel

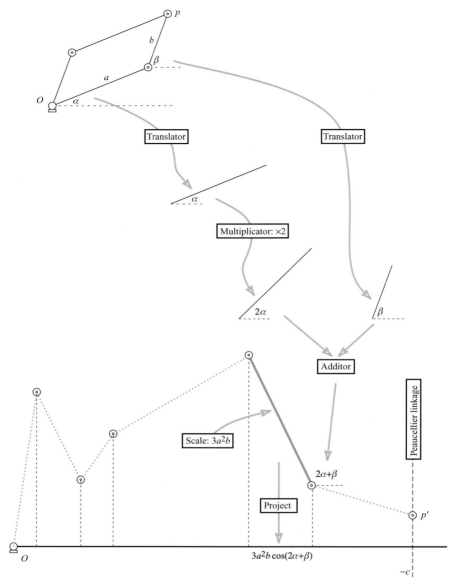

Figure 3.9. A schematic of the overall structure of Kempe's construction. Each term of Equation (3.2) is simulated by a link whose projection on the horizontal axis is that term. The point p' is constrained to follow a particular vertical line by a Peaucellier linkage. The joints labeled O are the same joints.

to A! This unwanted passage through a degeneracy to another part of the configuration space is a generic problem throughout Kempe's construction, noted by Thurston, and in Hopcroft et al. (1984), and perhaps by others. This is how the problem is phrased by Kapovich and Millson (2002):

> The main problem with Kempe's proof is that it works well only for a certain subset of the moduli space,[8] however, near certain "degenerate" configurations the moduli

---

[8] Another term for the configuration space of the linkage.

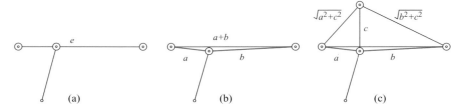

Figure 3.10. Simulating a joint in the interior of a link.

space splits into several components and the linkage fails to describe the desired polynomial function. We use the "rigidified parallelograms" to resolve this problem and get rid of undesirable configurations (p. 3).

To repair Kempe's proof, we need to "brace" both the parallelogram and the contra-parallelogram to prevent flipping from one to the other. For the first bracing, we follow King's presentation (King 1999), of Kapovich and Millson's resolution. The latter authors did not need to brace the contraparallelogram, as their proof deviates from Kempe's and does not employ that mechanism. For the second bracing, we follow a recent idea of Abbott and Barton (2004).

3.2.2.1 Bracing the Parallelogram

It is convenient to be able to locate a joint in the interior of a rigid bar (as in Figure 3.5). One could either permit this as part of the definition of a linkage or retain the restriction that joints are only at endpoints of bars and simulate an interior joint as follows. Refer to Figure 3.10. The joint in the interior of edge e in (a) of the figure can be simulated with links whose length sums the length of e, as shown in (b). This linkage is not "infinitesimally rigid" (as defined in Section 4.4, p. 49), so one might want to add the "stiffening truss" shown in (c).

Now we show how to "rigidify" parallelograms. Consider a rhombus linkage (see Figure 3.11). With its two bottom vertices a and b pinned, it has the natural circular motion shown in (a). However, as Kapovich and Millson pointed out, it also has the degenerate motions shown in (b) and (c). So the configuration space of this linkage is three circles, each pair intersecting in a single point. Blocking of the degenerate motions is accomplished by adding a bar between the midpoints of opposite sides, as in (d). These extra links are needed at many points throughout Kempe's construction, for example, in the parallelogram that is part of the Peaucellier linkage (Figure 3.2) employed in Kempe's construction.

3.2.2.2 Bracing the Contraparallelogram

Bracing the contraparallelogram is a bit more complicated. Here we describe a bracing due to Abbott and Barton (2004).

Let the contraparallelogram be $abcd$, and let p, q, r, and s be the midpoints of its links. The linkage is braced by four extra bars to a common point x on the perpendicular bisector of the line containing those midpoints (see Figure 3.12(a)). As long as these four extra bars are chosen long enough, the contraparallelogram is prevented from flipping to the parallelogram configuration. The reason why is hinted at in Figure 3.12(b). When $abcd$ is a parallelogram, so is $pqrs$. Now, returning to (a) of the figure, we know that x is on the perpendicular bisector of sq and on the perpendicular bisector of pr. This

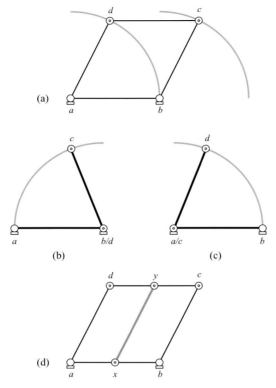

Figure 3.11. (a–c) Three circular motions of a rhombus; (d) rhombus rigidified.

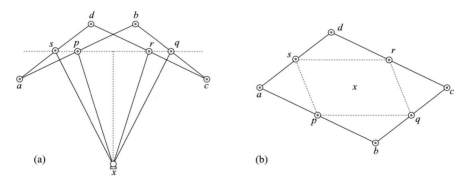

Figure 3.12. (a) Braced contraparallelogram; (b) flipping to the corresponding parallelogram is not possible.

would place x at the center of the parallelogram $pqrs$. But if the extra bars are chosen long enough, this is impossible. See Box 3.1 for details.

3.2.3   Recap

We have chosen to present Kempe's original proof with repairs, but as mentioned, the first full proof is due to Kapovich and Millson. They follow Kempe's construction in spirit but not detail. Their proof employs modified versions of many of his linkages, but avoids

Box 3.1: Braced Contraparallelogram

Let $r_1 = |xp| = |xr|$ and $r_2 = |xs| = |xq|$ be the two lengths of the bracing bars. These lengths are chosen so that $r_2^2 - r_1^2 = \frac{1}{4}(|ab|^2 - |ad|^2)$. A few facts:

1. p, q, r, and s are collinear.
2. $|sp| = |rq|$.
3. x is on the perpendicular bisector of sq and on the perpendicular bisector of pr.
4. $|sp| \cdot |sr| = \frac{1}{4}(|ab|^2 - |ad|^2)$.

This last claim is not obvious, but we defer a proof to the discussion of Hart's inversor (Box 3.2). Let L be the linkage composed of just the closed 4-chain $abcd$, and let L_b be the braced version. Using the facts above, Abbott and Barton (2004) prove that any degenerate (flat) or contraparallelogram configuration of L can be extended to a configuration of L_b, and, conversely, the bracing of L_b restricts the configurations of $L \subset L_b$ to be just the degenerate and contraparallelogram configurations.

his multiplicator in favor of a more complex multiplier mechanism (incorporating three translators, a pantograph, three inversors (modified Peaucellier linkages), and an adder) whose properties are more easily controlled. The collection of elementary linkages allow a nearly straightforward composition of a linkage for complex polynomial maps, $f : \mathbb{C}^n \to \mathbb{C}$, without recourse to the trigonometric expansions employed by Kempe. This complex linkage is then modified further to obtain a linkage realizing a real polynomial map $f : \mathbb{R}^n \to \mathbb{R}$, achieving a version of the universality theorem (Theorem 2.2.3): for any compact real algebraic set X (e.g., a bounded portion of a curve), there is a linkage L such that the map is an "analytically trivial covering"[9] of X. Thus, the formidable full proof finally firmly establishes that, indeed, there is a linkage that signs your name, even if (as King 1999, notes) your signature includes cusps and multiple strokes! And, as previously noted, this beautiful generalization of the task of creating straight-line mechanisms has led to a thorough understanding of the rich topology of the configuration space of linkages.

It is natural to ask how many links are required to trace a curve defined by a polynomial of degree n. This was first addressed in Gao et al. (2002), where an $O(n^4)$ bound was obtained on the number of links. This was subsequently improved by Abbott and Barton (2004) to $O(n^2)$, and extended to arbitrary dimension d to $O(n^d)$ bars. They also offer strong evidence that these bounds are optimal.

Two interesting open questions were recently posed on this topic:

Open Problem 3.1: Continuous Kempe Motion. [a] Given a (connected) curve C in the plane, is there is a linkage that traces exactly that curve via a continuous motion/reconfiguration?

---

[a] Abbott and Barton (2004).

---

[9] The map is an *analytic isomorphism*, with both f and f^{-1} restrictions of analytic maps (i.e., locally, power series).

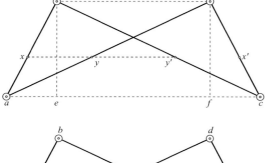

Figure 3.13. Hart's inversor. $|ab| = |cd|$ and $|bc| = |da|$, so that $abcd$ is a contra-parallelogram.

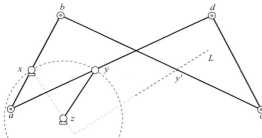

Figure 3.14. Hart's mechanism to convert circular motion of y to linear motion of y'. Joint x is pinned and the center of inversion. L is perpendicular to the line containing xz.

The point here is that although the proofs construct a linkage such that the positions of one joint are exactly C, it is possible that the whole of C cannot be traced by a continuous reconfiguration of the linkage. Perhaps it is necessary to jump to a different configuration to trace different sections of C.

The next open problem ties into linkages forbidden from self-intersection, a topic discussed at great length later on in Part I:

Open Problem 3.2: Noncrossing Linkage to Sign Your Name.[a] Is there a linkage that traces any given algebraic curve, and moves without any self-intersection throughout the tracing? If not, what is the class of curves that can be traced by a noncrossing linkage?

---

[a] Don Shimamoto, April 2004.

3.3   HART'S INVERSOR

We complete the story of linear-motion linkages by sketching Hart's 1875 5-link mechanism,[10] which "improves" upon Peaucellier's 7-link arrangement at the cost of a loss of some symmetry and the use of one more complex joint.

Start with a contraparallelogram $abcd$ as in Figure 3.13. The symmetry of the contraparallelogram guarantees that ac and bd remain parallel during movement. Hart takes four points x, y, y', x' dividing the four links in the same proportion. It is then not hard to see that these four points are collinear, and remain collinear and parallel to ac and bd throughout the free motion of the linkage. The key property—not at all obvious—is

---

[10] Hart (1875), as described in Kempe (1877, p. 18ff). See also Hunt (1978, p. 15) and http://poncelet.math.nthu.edu.tw/disk3/cabrijava/hart2.html.

Box 3.2: Geometry of Hart Inversor

To simplify the notation, we will use, for example, ab to mean both the segment with endpoints a and b and the length $|ab|$ of the segment. Thus, the contraparallelogram length condition is $ab = cd$ and $bc = da$. Our goal is to prove that $xy \cdot xy'$ is a constant. We prove this in two steps. We first reduce the goal to the constancy of $ac \cdot bd$. $\triangle axy$ is similar to $\triangle abd$ (sharing the angle at a) and so

$$xy/bd = ax/ab,$$
$$xy = bd \cdot ax/ab.$$

$\triangle bxy'$ is similar to $\triangle bac$ (sharing the angle at b) and so

$$xy'/ac = bx/ab,$$
$$xy' = ac \cdot bx/ab.$$

Thus

$$xy \cdot xy' = (bd \cdot ac)(ax \cdot bx)/ab^2. \tag{3.3}$$

Note that ab is a constant link length, and ax and bx are constant proportions of ab. Thus, if $bd \cdot ac$ is a constant, so is $xy \cdot xy'$.

Now we prove that $bd \cdot ac$ is a constant by analysis of right triangles in Figure 3.13. Drop perpendiculars from b and d to points e and f on ac. The right triangle $\triangle afd$ yields

$$af^2 + fd^2 = ad^2,$$
$$af^2 = ad^2 - fd^2,$$
$$af^2 = ad^2 - eb^2, \tag{3.4}$$

where the last step is justified because $fd = eb$ due to symmetry. The right triangle $\triangle aeb$ yields

$$ae^2 + eb^2 = ab^2,$$
$$ae^2 = ab^2 - eb^2. \tag{3.5}$$

Now we express ac and bd in terms of ae and af, and substitute using Equations (3.4) and (3.5):

$$bd \cdot ac = (af - ae)(af + ae)$$
$$= af^2 - ae^2$$
$$= (ad^2 - eb^2) - (ab^2 - eb^2)$$
$$= ad^2 - ab^2. \tag{3.6}$$

Because both terms in Equation (3.6) are constant link lengths, $bd \cdot ac$ is a constant, and by Equation (3.3), so is $xy \cdot xy'$.

that $|xy| \cdot |xy'|$ is a constant, which we prove in Box 3.2. This means that the points y and y' are inverses of one another with respect to a circle centered on x. Because circles through x invert to lines, adding one more link zy to constrain y to a circle that passes through x, as in Figure 3.14, forces y' to move on a line L (just as the link yc in Figure 3.2

forces d to move on a line). Note that joint y is "on top of" ad, which remains rigid underneath, a type of joint not present in Peaucellier's linkage.[11] Thus circular motion of y is converted to exact linear motion with only five links.

---

[11] We could retain y as a universal joint for the three links ya, yd, and yz, but at the cost of bracing ad with another link.

4 Rigid Frameworks

The area of *rigidity theory* studies a special class of problems about linkages: can the linkage move at all? This seemingly simple question has a deep theory behind it, with many variations and characterizations.

This chapter gives an overview of some of the key results in 2D rigidity theory, with a focus on the results that we need elsewhere in the book, in particular, in our study of locked linkages in Section 6.6 (p. 96). A full survey is beyond our scope; see Connelly (1993), Graver (2001), Graver et al. (1993), and Whiteley (2004) for more information.

4.1 BRIEF HISTORY

While rigidity theory fell out of prominence for the first half of the twentieth century, since then there has been increasing interest and connections made to applications. Perhaps the first result that could be identified as in this area was Augustin-Louis Cauchy's theorem from 1813 on the rigidity of convex polyhedra, settling part of a problem posed by Leonhard Euler in 1766. This and related results on polyhedral rigidity are covered in Section 8.2 (p. 143) and in Part III, Section 23.1 (p. 341). Another early result in the area is James Clerk Maxwell's work from 1864 on planar frameworks, and it is here that the development in this chapter culminates. In some sense, we follow backward in time, building on the new infrastructure and deep understanding of the area, to make it easier to reach some older and more difficult results.

4.2 RIGIDITY

We begin with some basic terminology. A configuration of a linkage is *flexible* if there is some nontrivial motion of the linkage starting at that configuration. (Rigid motions and zero-duration motions are *trivial*.) An inflexible configuration of a linkage is *rigid*. Figure 4.1 shows some examples. Equivalently, a rigid configuration corresponds to an isolated point in the configuration space of the linkage (after factoring out rigid motion), while a flexible configuration corresponds to a point in the configuration space surrounded by a neighborhood of dimension at least 1 (again ignoring rigid motion).

Surprisingly, the computational complexity of deciding this most basic form of rigidity does not seem to have been stated explicitly before,[1] though it seems likely to be intractable in general, as we detail now. An exponential-time algorithm follows from the

---

[1] Personal communication with Robert Connelly and Walter Whiteley, August 2003.

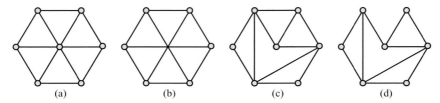

Figure 4.1. (a–c) Rigid; (d) flexible. In (b), the bars are not joined at the hexagon center.

general motion planning algorithms in Section 2.1 (p. 17). As hinted at above, testing rigidity of a configuration is equivalent to testing whether a given point is isolated (has "local dimension" 0) in the configuration space (pinning a bar to the plane to prevent rigid motions). The configuration space of a linkage is an *algebraic set*, meaning that it can be defined by a system of polynomial equations (the distance constraints between the endpoints of each bar).[2] In a field called algebraic complexity (cf. Blum et al. 1998), Koiran (2000, Thm. 6) proved that testing whether a given point is isolated in a given algebraic set is coNP-hard. (He also gives a relatively simple exponential-time algorithm for this problem (Sec. 3), and mentions that it is not known whether the problem "belongs to the standard polynomial hierarchy" (Sec. 5).) It seems likely that testing rigidity would therefore also be coNP-hard, because of the universality results about linkages simulating arbitrary algebraic sets—see Theorem 2.2.3 in Section 2.2.2 (p. 27)—but such a reduction would need some care, for example, making sure that the conversion can be computed efficiently.

4.3   GENERIC RIGIDITY

"Rigidity," without adjectival modification, is the most basic and intuitive form of rigidity, but many variations have been proposed. For example, one property of rigidity is that it depends on the particular configuration of the linkage, meaning that it depends on the specific coordinates assigned to the joints, in addition to the bar lengths (see Figure 4.2).

In contrast, "generic rigidity" is a form of rigidity that depends only on the graph structure of the linkage, meaning that it does not depend on the specific edge lengths nor joint coordinates. In this section, a *graph* acts as a linkage without preassigned link lengths; we still use the terms "joint" and "link" for its vertices and edges.

A *realization* of a graph is an assignment of coordinates to joints; this notion is similar to a configuration of a linkage, except that there are no restrictions on the edge lengths. A graph is *generically rigid* if almost every realization of the graph is rigid. The phrase "almost every" is a technical notion meaning that, if we chose a realization of the graph at random, it would be rigid with probability 1 ("almost always"). Thus, the only flexible realizations are those chosen with very specific and degenerate alignments: parallel edges, matching edge lengths, etc. They form a lower-dimensional subset of the full space of realizations (see Figures 4.3(a) and 4.4(a–b)).

On the other hand, if a graph is not generically rigid, then almost every realization of the graph must be flexible and we call the graph *generically flexible* (see Figures 4.3(b)

---

[2] Section 2.1.1.1 gives a more explicit description of how to express linkage configuration spaces with polynomials, but that setup considered linkages forbidden from self-intersection, causing the configuration space to be a semialgebraic set (involving polynomial inequalities). Here linkages may self-intersect, simplifying to an algebraic structure.

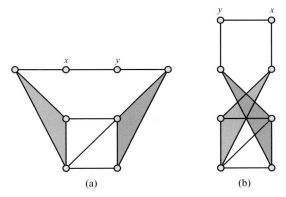

Figure 4.2. Dependence of rigidity on configuration. (a) Rigid embedding: the shaded triangles pull the chain containing x and y taut; (b) Flexible embedding: the link yx can move left or right (see also Figure 4.8).

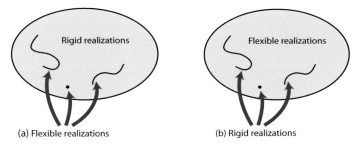

Figure 4.3. Schematic definition of generic rigidity and generic flexibility. (a) Space of realizations of a generically rigid graph; (b) space of realizations of a generically flexible graph.

and 4.4(c–d)). The nice property here is that generic realizations are either all rigid or all flexible (Graver 2001, Thm. 3.13). Thus, generic rigidity/flexibility is a well-defined combinatorial notion, depending only on the graph, which describes the behavior of almost all of its (geometric) realizations.

4.3.1   Laman's Theorem

Because generic rigidity applies to a graph without being specific to a particular realization, it becomes possible to characterize the notion combinatorially: which graphs are generically rigid? In 1970, Gerard Laman characterized the generically rigid graphs in 2D (Laman 1970):

Theorem 4.3.1 (Laman's theorem). *A graph of n joints is generically rigid in 2D precisely when it falls into one of the following two categories:*

1. *It has exactly $2n - 3$ links, and every induced subgraph[3] on k joints has at most $2k - 3$ links.*
2. *It has more than $2n - 3$ links, and it contains a generically rigid subgraph with $2n - 3$ links.*

If the graph has fewer than $2n - 3$ links, then it is never generically rigid in 2D.

---

[3] The subgraph H *induced* by a set S of vertices in G is the graph whose vertex set is S and whose edges are all those edges in G that connect two vertices in S. In other words, H is the maximal subgraph of G with vertex set S.

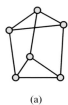 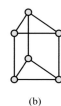 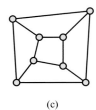 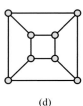

(a)   (b)   (c)   (d)

Figure 4.4. (a) Generically rigid; (b) rarely flexible; (c) generically flexible; (d) rarely rigid.

Intuitively, a linkage requires at least $2n - 3$ links to be rigid, but these links cannot be too concentrated in any induced subgraph, because then another subgraph would have too few edges to stay rigid. Generically rigid graphs with exactly $2n - 3$ links are called *minimally generically rigid* because the removal of any link results in a generically flexible graph; any additional edges constitute *overbracing*. Viewed another way, any generically rigid graph is an edge-superset of some minimally generically rigid graph.

In addition to this structural characterization, there is an algorithmic characterization of generically rigid graphs. A direct implementation of Laman's theorem would require testing all 2^n-induced subgraphs. Sugihara (1980) developed the first polynomial-time algorithm. Imai (1985) improved the running time to $O(n^2)$, a bound subsequently matched by other algorithms (Gabow and Westermann 1992; Hendrickson 1992). A practical variation of these algorithms is a "pebble game" by Jacobs and Hendrickson (1997), which runs in $O(mn)$ time, where m is the number of links. (This time bound is never better than $O(n^2)$, assuming the linkage is connected, and is worse when the linkage is significantly overbraced, with $m \gg n$.) Their algorithm also decomposes the graph into rigid and overbraced regions, and computes the generic number of degrees of freedom. Further experimental speedups of this algorithm have been studied (Moukarzel 1996). Recently, Lee et al. (2005) showed how to improve the running time of this algorithm to $O(n^2)$. An intriguing open question is whether the worst-case complexity can be improved below $O(n^2)$:

Open Problem 4.1: Faster Generic Rigidity in 2D. [a] Is there a $o(n^2)$-time algorithm to test generic rigidity of a given graph with n vertices?

---

[a] Implicit in work cited above.

Finally we mention a recent result characterizing the planar generically rigid graphs: a planar graph is generically rigid in 2D precisely if it can be embedded as a "pseudotriangulation" (Orden et al. to appear), and a planar graph is minimally generically rigid in 2D precisely if it can be embedded as a "pointed pseudotriangulation" (Haas et al. 2005). We describe these two notions of "pseudotriangulation" in Section 6.7.2 (p. 108).

4.3.2   Higher Dimensions

Laman's theorem provides necessary and sufficient conditions for minimal generic rigidity in 2D. The d-dimensional generalization of his conditions are still necessary, but no longer sufficient. A minimally generically rigid graph in d dimensions has exactly

$$dn - \binom{d+1}{2} \tag{4.1}$$

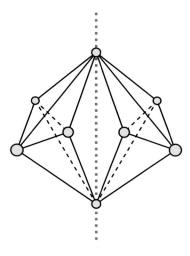

Figure 4.5. The "double banana" example which satisfies Laman's combinatorial condition yet is generically flexible: the bananas can pivot around the dotted line.

links. Note that, for $d = 2$, this formula gives the $2n - 3$ of Theorem 4.3.1. For $d = 3$, the formula gives $3n - 6$. The smallest example that satisfies Laman's conditions in 3D yet is not generically rigid is the "double banana" shown in Figure 4.5. Here there are $n = 8$ joints and $3n - 6 = 18$ links. Moreover, all induced subgraphs on k joints have at most $3k - 6$ edges. Yet the two "bananas" can spin around the axis through their common points.

Despite attracting the attention of significant study, characterizing generically rigid graphs in 3D (or higher dimensions) remains one of the central open problems in rigidity theory:

Open Problem 4.2: Generic Rigidity in 3D.[a] Characterize the generically rigid graphs in 3D. In particular, can these graphs be recognized in polynomial time?

---

[a] See, e.g., Graver (1993, Ch. 5).

Because the double-banana graph is only 2-connected,[4] it is natural to wonder whether vertex connectivity can be used to formulate sufficient conditions. A recent result answers this question in the negative (Mantler and Snoeyink 2004a) by constructing flexible linkages in 3D that meet Laman's condition and yet have the highest possible connectivity.

4.3.3 Henneberg Constructions

Returning to 2D, another characterization of minimally generically rigid graphs was given much earlier, by Ernst Lebrecht Henneberg in 1911 (Henneberg 1911). He defined two types of constructions, starting from a single link connecting two joints, for incrementally constructing minimally generically rigid graphs:

---

[4]   That is, the removal of only two vertices suffices to disconnect the graph.

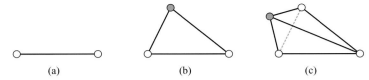

Figure 4.6. A Henneberg sequence. (a) A single edge; (b) construction 1 adds a new vertex; (c) construction 2 deletes an edge and adds a new vertex.

1. Add a new joint, and connect it to two other joints by two new links (Figure 4.6(b)).
2. Remove an existing link, add a new joint, and connect it to the endpoints of the removed link as well as one other joint, for a total of three new links (Figure 4.6(c)).

These straightforward operations turn out to be complete, giving an intuitive characterization of generic rigidity:

Theorem 4.3.2 (Henneberg's theorem). *A graph is minimally generically rigid in 2D precisely if it can be constructed by a sequence of Henneberg constructions. A graph is generically rigid only if some subgraph is minimally generically rigid.*

Laman's proof of his Theorem 4.3.1 uses Henneberg's theorem 4.3.2. So while Laman's characterization is more useful for constructing algorithms, Henneberg's characterization is useful for proving mathematical results such as Laman's characterization. We refer the reader to Graver's book (Graver 2001) for readable proofs of both theorems.

4.3.4   Global Rigidity

A stronger form of rigidity demands that not only is there no motion starting from the configuration, but there is simply no other configuration of the linkage. Formally, a configuration of a linkage is *globally rigid* if every other configuration of the linkage is equivalent via rigid motion and/or reflection. We can also call the whole linkage globally rigid if it has only one configuration up to rigid motion and/or reflection. As with rigidity, the complexity of deciding global rigidity does not seem to have been considered, and here it seems even more difficult to apply a result from algebraic complexity because the universality Theorem 2.2.3 may make a configuration space with duplicate copies of an algebraic set.

The global rigidity property is particularly interesting because it means that the configuration of the linkage can, in principle, be reconstructed uniquely (up to rigid motion and reflection) from just the linkage (graph and link lengths). Such reconstructions are of interest for computing the global geometry of sensor networks and proteins given just information about distances between pairs of sensors or atoms (see Bădoiu et al. 2006; Berger et al. 1999; Cabello et al. 2004). Unfortunately, the reconstruction problem is NP-hard (Saxe 1979; Yemini 1979), even in the case of "range graphs" (Bădoiu et al. 2006) and for infinitesimally rigid planar embeddings (Cabello et al. 2004), although the complexity seems to be open in the globally rigid case.

The generic case is potentially easier. A graph is *generically globally rigid* if almost every realization of the graph is globally rigid. Connelly (2005) and Jackson and Jordán (2005) recently characterized generic global rigidity in the plane in terms of generic rigidity, leading to a polynomial-time algorithm to test generic global rigidity in the plane:

Theorem 4.3.3 (Connelly 2005; Jackson and Jordán 2005). *A graph on at least four vertices is generically globally rigid precisely if it is vertex 3-connected and removal of any edge leaves a generically rigid graph ("redundant generic rigidity").*

The easy half of this theorem—that both of these conditions are necessary for generic global rigidity—was proved earlier by Hendrickson (1992). He also conjectured that these conditions were sufficient, in any dimension d with vertex $(d+1)$-connectivity. The theorem above resolves this conjecture in 2D. Unfortunately, Connelly (1991) showed that the conjecture is false in 3D.

Because the proof of Jackson and Jordán (2005) does not explicitly construct the unique realization, we are left with the following open problem on reconstructing these graphs:

> **Open Problem 4.3: Realizing Generically Globally Rigid Graphs.** [a] What is the complexity of finding the unique embedding of a generically globally rigid graph when assigned a generic realizable set of bar lengths?
>
> ---
>
> [a] Bǎdoiu et al. (2006).

4.4 INFINITESIMAL RIGIDITY

Suppose we have a motion of a linkage starting at some configuration at "time 0." Assuming that this motion is differentiable, the first derivative of the motion at time 0 is called a *first-order motion* or *infinitesimal motion* of the linkage; it specifies a way to "get started moving." More generally, an infinitesimal motion assigns a velocity vector to each of the vertices, each specifying the first-order direction that the vertex moves and its relative speed. Such an infinitesimal motion might exist even when it does not come from the derivative of an actual motion.

4.4.1 Infinitesimal Length Constraint

To be valid, an infinitesimal motion must preserve the length of each link to the first order. To see the form of this condition, we need some notation. Suppose that the initial configuration is given by points $\mathbf{p}_i$ for each vertex i, and the velocity vectors are $\mathbf{v}_i$ for each vertex i. Consider any two vertices i and j, and for now assume both lie on the same horizontal line (see Figure 4.7). Let $\mathbf{v}_i = (x_i, y_i)$, and let $L = \|\mathbf{p}_i - \mathbf{p}_j\|$ be the length of the link. Its new link length is the square root of

$$(L - x_i + x_j)^2 + (-y_i + y_j)^2.$$

This expression expands to

$$L^2 - 2L(x_i - x_j) + x_i^2 + x_j^2 + y_i^2 + y_j^2 - 2(x_i x_j + y_i y_j).$$

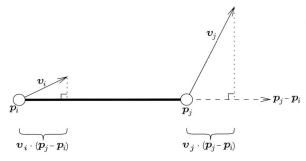

Figure 4.7. Measuring the first-order change in the distance between vertices $\mathbf{p}_i$ and $\mathbf{p}_j$ when subjected to the infinitesimal motion given by velocity vectors $\mathbf{v}_i$ and $\mathbf{v}_j$. [Figure 7 of Connelly et al. 2003.]

Now most terms in this expression are products of two infinitesimals, that is, second-order terms. To the first order, the squared length is $L^2 - 2L(x_i - x_j)$, which is L^2 (constant) only if $x_i = x_j$.

More generally, we can remove the assumption that $\mathbf{p}_i$ and $\mathbf{p}_j$ lie on the same horizontal line by writing x_i and x_j as the lengths of the projections of the velocity vectors onto the line through $\mathbf{p}_i$ and $\mathbf{p}_j$:

$$x_i = \mathbf{v}_i \cdot (\mathbf{p}_j - \mathbf{p}_i),$$
$$x_j = \mathbf{v}_j \cdot (\mathbf{p}_j - \mathbf{p}_i).$$

The difference between the two projected lengths is given by

$$\mathbf{v}_j \cdot (\mathbf{p}_j - \mathbf{p}_i) - \mathbf{v}_i \cdot (\mathbf{p}_j - \mathbf{p}_i)$$
$$= (\mathbf{v}_j - \mathbf{v}_i) \cdot (\mathbf{p}_j - \mathbf{p}_i)$$
$$= (\mathbf{v}_i - \mathbf{v}_j) \cdot (\mathbf{p}_i - \mathbf{p}_j).$$

When this difference is zero, the distance between $\mathbf{p}_i$ and $\mathbf{p}_j$ stays constant to the first order. Therefore, the contraints on an infinitesimal motion are the following:

$$(\mathbf{v}_i - \mathbf{v}_j) \cdot (\mathbf{p}_i - \mathbf{p}_j) = 0 \quad \text{for each link } \{i, j\}. \tag{4.2}$$

4.4.2 Rigidity Matrix

Suppose that we want to compute an infinitesimal motion (the $\mathbf{v}_i$'s) given a particular configuration of a linkage (the $\mathbf{p}_i$'s). We need to satisfy all the constraints given in Equation (4.2). In each equation, the points $\mathbf{p}_i$ and $\mathbf{p}_j$ are known and the vectors $\mathbf{v}_i$ and $\mathbf{v}_j$ are unknown. The equation is in fact linear in its unknowns—the known vector $\mathbf{p}_i - \mathbf{p}_j$ specifies coefficients on each of the unknown components of the vector $\mathbf{v}_i - \mathbf{v}_j$:

$$\left(p_i^x - p_j^x\right)v_i^x - \left(p_i^x - p_j^x\right)v_j^x + \left(p_i^y - p_j^y\right)v_i^y - \left(p_i^y - p_j^y\right)v_j^y = 0.$$

The system of constraints in Equation (4.2) is therefore a linear system of equations. We can rewrite this system as a matrix equation

$$R\mathbf{v} = 0,$$

Box 4.1: Rigidity Matrix

Consider the square shown below, whose vertices have coordinates

$$\mathbf{p}_1 = (0, 0),$$
$$\mathbf{p}_2 = (5, 0),$$
$$\mathbf{p}_3 = (5, 5),$$
$$\mathbf{p}_4 = (0, 5).$$

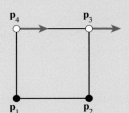

The rigidity matrix is

$$R = \begin{pmatrix} -5 & 0 & 5 & 0 & 0 & 0 & 0 & 0 \\ 0 & 0 & 0 & -5 & 0 & 5 & 0 & 0 \\ 0 & 0 & 0 & 0 & 5 & 0 & -5 & 0 \\ 0 & -5 & 0 & 0 & 0 & 0 & 0 & 5 \end{pmatrix}.$$ (4.3)

Solving $R\mathbf{v} = 0$, with $\mathbf{v} = (v_1^x, v_1^y, v_2^x, v_2^y, v_3^x, v_3^y, v_4^x, v_4^y)$, yields

$$-5v_1^x + 5v_2^x = 0,$$
$$-5v_2^y + 5v_3^y = 0,$$
$$5v_3^x - 5v_4^x = 0,$$
$$-5v_1^y + 5v_4^y = 0,$$

which simplifies to

$$v_1^x = v_2^x,$$
$$v_2^y = v_3^y,$$
$$v_3^x = v_4^x,$$
$$v_1^y = v_4^y.$$

If we pin $\mathbf{p}_1$ and $\mathbf{p}_2$ by setting their velocities to zero, $\mathbf{v}_1 = (0, 0)$ and $\mathbf{v}_2 = (0, 0)$, then we obtain these equations:

$$v_3^x = v_4^x,$$
$$v_3^y = v_4^y = 0.$$

This solution describes an infinitesimal motion which translates to $\mathbf{p}_3$ and $\mathbf{p}_4$ moving horizontally in concert, representing the familiar flexibility of a rhombus.

where R is called the *rigidity matrix*. The rigidity matrix has one row per bar (corresponding to an equation) and two columns per vertex (corresponding to the coordinates of a velocity vector). Most entries of the rigidity matrix are 0, and the other entries are differences between coordinates of the $\mathbf{p}_i$'s. Specifically, for each row (bar), the two coordinates of $\mathbf{p}_i - \mathbf{p}_j$ appear in the two columns corresponding to the first endpoint i of the bar, and two coordinates of $\mathbf{p}_j - \mathbf{p}_i$ appear in the two columns corresponding to the second endpoint j; see Box 4.1 for an example.

The rigidity matrix provides a direct connection between infinitesimal rigidity and linear algebra (see, e.g., Graver 2001, p. 101). First, the space of infinitesimal motions is the "null-space" or "kernel" of R: the space of vectors mapped to zero when left-multiplied

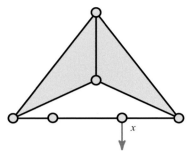

Figure 4.8. The movement of x illustrated projects so that the incident bars do not lengthen. So the linkage is infinitesimally flexible. But the linkage is in fact rigid because the three collinear bars are stretched taut by the rigid shaded quadrilateral. [Based on Figure 3.8 of Graver 2001.]

by R. Then the rank-nullity theorem of linear algebra tells us that the dimensionality of infinitesimal motions plus the rank of R equals the number of columns, which is dn in dimension d. A linkage is *infinitesimally rigid* precisely if the dimensionality of infinitesimal motions equals the dimensionality of rigid motions, which is 3 in 2D (1 for rotation and 2 for translation), 6 in 3D (3 for rotation and 3 for translation), and $\binom{d+1}{2}$ in general. Thus, a linkage is infinitesimally rigid in 2D precisely if the rank of R is $2|V| - 3$, and in 3D precisely if the rank is $3|V| - 6$. The general form is $dn - \binom{d+1}{2}$, the same expression we encountered for minimal generic rigidity earlier in Equation (4.1). (The reason for this "coincidence" is that, at generic configurations, generic rigidity is equivalent to infinitesimal rigidity, a point which will be discussed in the next section.) Perhaps most importantly, this yields a polynomial-time algorithm to test whether a configuration is infinitesimally rigid, using standard Gaussian elimination.

4.4.3   Connection Between Rigidity and Infinitesimal Rigidity

As we hinted at above, a motion of a linkage can be converted into an infinitesimal motion of the starting configuration. This fact is not obvious, however, because it would seem to require that the motion has a derivative at time 0. Fortunately, it turns out the initial configuration being flexible implies the existence not only of some motion, but also of an analytic motion, meaning in particular that it has infinitely many derivatives. Thus, we can take the derivative of this motion and obtain an infinitesimal motion. This argument proves the following:

Theorem 4.4.1. *Flexibility implies infinitesimal flexibility. Equivalently, infinitesimal rigidity implies rigidity.*

Thus, infinitesimal rigidity is a stronger form of rigidity. In contrast, the reverse implication does not hold: some linkages are infinitesimally flexible but still rigid because of higher-order derivatives not working out. Figure 4.8 shows an example.

While our study of different notions of rigidity ends here, several others exist. For example, Connelly and Whiteley (1996) have studied *second-order rigidity*, which considers the existence of the second derivative of a motion in addition to the first derivative. This notion more precisely captures rigidity in the sense that infinitesimal (first-order) rigidity implies second-order rigidity, which implies rigidity. This notion hints at an infinite hierarchy of various levels of rigidity, and as their complements, various levels of flexibility (see Figure 4.9). However, this hierarchy has not yet been settled, even in terms of definitions, beyond second order. See Connelly and Servatius (1994) for a discussion of this issue, and some failed attempts.

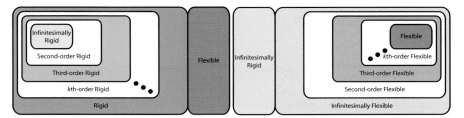

Figure 4.9. Rigidity and flexibility hierarchies of all configurations (of all linkages). The hierarchies are hypothetical for kth order, $k \geq 3$. [Based loosely on Figure 2 of Connelly and Whiteley 1996.]

4.5 TENSEGRITIES

One branch of rigidity theory considers more general objects than linkages called *tensegrities*. A tensegrity[5] can consist of the usual *bars* (links) as well as new types of edges called *struts* and *cables*. In contrast to bars, whose lengths must remain fixed throughout any motion, struts can grow but not shrink, and cables can shrink but not grow. The increased flexibility of these edges makes the notions of rigidity and flexibility more interesting (see Figure 4.10 for examples). While tensegrities may seem to take us a little astray from linkages, they in fact will become crucial for understanding locked chains in Section 6.6 (p. 96).

As with regular linkages, we can define global rigidity for tensegrities, and in fact the example in Figure 4.10(d) is globally rigid. However, the notion of generic rigidity is no longer meaningful: most tensegrity graphs have many realizations in which they are rigid as well as many realizations in which they are flexible. For example, the tensegrity in Figure 4.10(a) becomes rigid when the bars form a convex polygon, because then the struts force every vertex angle to increase, which is impossible (as we will see in Lemma 8.2.1, p. 143).

4.5.1 Infinitesimal Rigidity of Tensegrities

The notion of infinitesimal rigidity generalizes fairly straightforwardly to tensegrities. Considering our computation from before, the dot product from Equation (4.2),

$$(\mathbf{v}_i - \mathbf{v}_j) \cdot (\mathbf{p}_i - \mathbf{p}_j),$$

is positive when the distance between $\mathbf{p}_i$ and $\mathbf{p}_j$ increases, and is negative when the distance decreases. Thus the constraints on an infinitesimal motion become

$$(\mathbf{v}_i - \mathbf{v}_j) \cdot (\mathbf{p}_i - \mathbf{p}_j) = 0 \quad \text{for each bar } \{i, j\},$$
$$(\mathbf{v}_i - \mathbf{v}_j) \cdot (\mathbf{p}_i - \mathbf{p}_j) \geq 0 \quad \text{for each strut } \{i, j\}, \tag{4.4}$$

and

$$(\mathbf{v}_i - \mathbf{v}_j) \cdot (\mathbf{p}_i - \mathbf{p}_j) \leq 0 \quad \text{for each cable } \{i, j\}. \tag{4.5}$$

---

[5] The term was coined by the architect and polymath R. Buckminster Fuller, who used 3D tensegrities in much of his work. The word "tensegrity" comes from the phrase "tens(ional int)egrity."

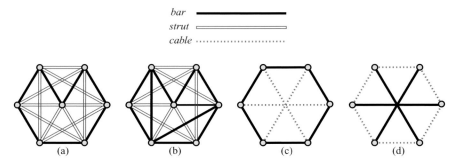

Figure 4.10. Tensegrities. (a–c) are flexible; (d) is rigid, even globally. (a–b) follow from Theorems 6.6.1 and 6.7.2, respectively; (c–d) are based on Figure 3 of Connelly and Whiteley (1996) and Figure 4.2 of Connelly (1993).

4.5.2   Connection to Linear Programming

These constraints are no longer just linear equations; some are linear inequalities. Such a problem is called a *linear program*. There are well-known polynomial-time algorithms to solve linear programs, except that now the polynomial running time of the algorithm depends additionally on the number of desired bits of accuracy in the output (see, e.g., Todd 2004). A long-standing open problem is whether there is a *strongly polynomial-time* algorithm for linear programming whose polynomial running time depends just on the number of variables and constraints.

However, before we can simply apply such a linear-programming algorithm to find infinitesimal motions, we need to prevent trivial motions from being valid solutions, or else the algorithm might find a trivial motion when nontrivial motions exist. We can prevent rigid motions by first rotating the tensegrity so that any particular edge $\{i, j\}$ becomes horizontal, and then setting $v_i^x = v_i^y = v_j^y = 0$, pinning vertex i to the plane (preventing translation) and forcing vertex j to stay on the horizontal line (preventing rotation). To prevent the zero motion ($\mathbf{v}_k = 0$ for all k), we add an *objective* to the linear program: among all solutions to the constraints, we desire one that maximizes a particular function $f(\mathbf{v}_1, \mathbf{v}_2, \ldots, \mathbf{v}_n)$. We sequentially try to maximize v_k^x, then $-v_k^x$, then v_k^y, then $-v_k^y$, for $k = 1, 2, \ldots, n$, each time calling the linear-programming algorithm. At each step, we find a nonzero infinitesimal motion if there is some motion in which the objective can be made positive. If the algorithm returns a nonzero motion at any step, then we have a nontrivial infinitesimal motion and the tensegrity is flexible. On the other hand, if the algorithm always returns the zero motion, then the tensegrity must be rigid, because no v_k^x or v_k^y can be made nonzero. Therefore, in polynomial time (an extra $O(n)$ factor over linear programming), we can determine whether a given tensegrity is infinitesimally rigid or flexible, and if flexible, find a nontrivial infinitesimal motion.

4.5.3   Stress

An important notion that arises with tensegrities is the idea of an "equilibrium stress."

A *stress* in a tensegrity is an assignment of real numbers, called *forces* or *stresses*, one to each edge (bar, cable, or strut). A bar can hold arbitrary stress, a strut can hold

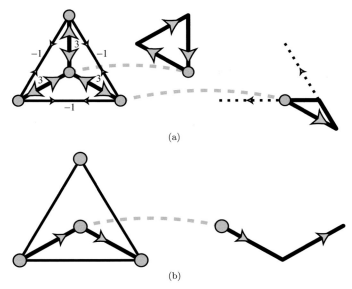

(a)

(b)

Figure 4.11. Examples of equilibrium stresses in tensegrities. Here we suppose that all edges are bars, but edges with negative stresses could be struts and edges with positive stresses could be cables. (a) An equilibrium stress; (b) a tensegrity with no equilibrium stress other than the everywhere-zero stress.

only negative stress (or zero stress), and a cable can hold only positive stress (or zero stress). We think of a negative stress as the edge "pushing" on its two endpoints by the same amount given by the stress itself. Similarly, a positive stress corresponds to the edge "pulling" on its two endpoints by the same amount given by the absolute value of the stress. And a zero stress corresponds to no force at all. A stress is *in equilibrium* if these forces balance out at each vertex, and therefore nothing moves. To write this as a formula, let $\omega_{i,j}$ denote the stress along the edge $\{i, j\}$. The direction of the force lies along the edge, so a pushing force felt by vertex i has direction $\mathbf{p}_i - \mathbf{p}_j$. We can view the magnitude of the stress $\omega_{i,j}$ as measured relative to the length of the edge, $\|\mathbf{p}_i - \mathbf{p}_j\|$. Thus, the force felt by vertex i is $\omega_{i,j}(\mathbf{p}_i - \mathbf{p}_j)$. The equilibrium constraint, that the total force at vertex i balances out, is then

$$\sum_j \omega_{i,j}(\mathbf{p}_i - \mathbf{p}_j) = 0. \tag{4.6}$$

In particular, the *everywhere-zero stress*, $\omega_{i,j} = 0$ for all $i,\ j$, always satisfies this property.

Figure 4.11(a) shows an example of an equilibrium stress. The ω values on the outer cycle are -1, while the ω values on the other edges are 3. Thus the edges of the outer cycle could be either bars or struts, and the other edges could be either bars or cables. The figure shows equilibrium graphically by adding the force vectors and confirming that they return to the origin, that is, the force polygons close.

On the other hand, the tensegrity in Figure 4.11(b) has no equilibrium stress except the everywhere-zero stress, even if all edges are bars. This property holds because neither the central vertex nor the top vertex (both of which have degree 2) could be in equilibrium with nonzero stresses because the two nonparallel force vectors could never sum to zero.

4.5.4   Duality

Stresses have a powerful connection to infinitesimal rigidity. Roughly speaking, if it is possible to stress a strut or a cable, then it acts effectively like a bar; otherwise, it is possible to stretch the strut or shrink the cable. Intuitively, the strut tries to stretch (pushing on its endpoints) or the cable tries to shrink (pulling on its endpoints), but if these forces can somehow be resolved into equilibrium by the rest of the tensegrity pushing or pulling, then the strut or cable gets stuck. Formally, we can state the result in two equivalent ways:

Theorem 4.5.1 (Connelly and Whiteley 1996, Thm. 2.3.2). *There is an equilibrium stress on a tensegrity that is nonzero on a particular strut or cable if and only if every infinitesimal motion holds that strut or cable fixed in length.*

Equivalently, a tensegrity has an infinitesimal motion that changes the length of a particular strut or cable if and only if every equilibrium stress is zero on that strut or cable.

We refer to this result as the *duality* between stresses and infinitesimal motions. If you are familiar with duality in linear programming (Farkas's lemma), these notions correspond directly by way of the connection to linear programming described above. In particular, the stress equilibrium constraints (4.6) form the dual linear program of the infinitesimal motion constraints (4.4).

One useful consequence of Theorem 4.5.1 is that, for a tensegrity to be infinitesimally rigid, all of its struts and cables must be fixed in their length. By the theorem, there must be an equilibrium stress for each strut and cable that is nonzero on that strut or cable. If we add these equilibrium stresses together, edge by edge, we obtain the following result:

Theorem 4.5.2 (Roth and Whiteley 1981, Thm. 5.2). *Any infinitesimally rigid tensegrity has an equilibrium stress that is nonzero on every strut and cable.*

The converse of this theorem is not true. An equilibrium stress that is nonzero on every strut and cable only forces the struts and cables to act like bars. To determine whether the tensegrity is infinitesimally rigid, we need to determine whether the *corresponding linkage*, where we replace every strut and cable by a bar, is infinitesimally rigid. Note in particular that Theorem 4.5.2 tells us nothing about tensegrities that happen to be linkages (i.e., when every edge is a bar): the everywhere-zero stress is in equilibrium and, in this case, nonzero on every (nonexistent) strut and cable. The following result lets us apply our knowledge about infinitesimal rigidity of linkages—in particular, the rigidity matrix of Section 4.4.2—to the infinitesimal rigidity of tensegrities, by way of equilibrium stresses:

Theorem 4.5.3 (Roth and Whiteley 1981, Thm. 5.2). *A tensegrity is infinitesimally rigid if and only if*

1. *it has an equilibrium stress that is nonzero on every strut and cable, and*
2. *the corresponding linkage (obtained by replacing every strut and cable by a bar) is infinitesimally rigid.*

4.5.5   Linkages and Tensegrities

The last theorem connected infinitesimal rigidity of tensegrities to infinitesimal rigidity of linkages by way of equilibrium stresses. We conclude our discussion on this topic by showing that equilibrium stresses can be used in pure linkages as well, to detect overbracing:

Theorem 4.5.4. *An infinitesimally rigid linkage has an equilibrium stress that is nonzero on a particular bar if and only if the linkage remains infinitesimally rigid after removing that bar.*

Proof: Define the two *corresponding tensegrities* of the linkage by replacing the particular bar by either a strut or a cable. If the linkage has an equilibrium stress that is nonzero on the particular bar, then one of the corresponding tensegrities can carry the same stress (choosing the strut version if the stress is negative and the cable version if the stress is positive). Also, the corresponding linkage of either corresponding tensegrity is the original linkage, which the theorem presupposes is infinitesimally rigid. Thus, Theorem 4.5.3 says that one of the corresponding tensegrities is infinitesimally rigid if and only if the linkage has an equilibrium stress that is nonzero on the particular bar. Now Roth and Whiteley (1981, Cor. 5.3) states that the linkage resulting from an infinitesimally rigid tensegrity by deleting a strut or cable that can carry a nonzero stress in an equilibrium stress, and replacing every other strut and cable by a bar, is also infinitesimally rigid. Hence, if one of the corresponding tensegrities is infinitesimally rigid, then the linkage resulting from deleting the particular bar is also infinitesimally rigid; and the converse is also trivially true. Putting the results together, the linkage has an equilibrium stress that is nonzero on the particular bar if and only if the linkage remains infinitesimally rigid after deleting the particular bar.   □

4.6   POLYHEDRAL LIFTINGS

An interesting characterization of equilibrium stresses from Section 4.5.3 comes from a connection to 3D. Imagine a tensegrity (or linkage) as lying in the $z = 0$ plane. Suppose that the tensegrity is *planar* in the sense that no edges cross, thereby defining the notion of a *face*, a connected region outlined by edges. A *polyhedral lifting* is an assignment of z coordinates to the vertices of the tensegrity so that each face remains planar. While we use the term "lifting," the z coordinates can also be negative, so some vertices may be dropped instead of lifted.[6] By applying a suitable vertical shear to the polyhedral lifting, we can assume that the exterior face lies in the $z = 0$ plane. See Figure 4.12 for an example of a polyhedral lifting, where the linkage is a triangulation.

One natural distinction between different types of edges in a polyhedral lifting is based on their dihedral angle (see Figure 4.13). Depending on whether the dihedral angle is convex ($< 180°$), reflex ($> 180°$), or flat ($= 180°$), we call the edge a *valley*, *mountain*, or *flat edge*.

---

[6]   An appropriate and often-used term for this notion is "polyhedral graph," as in the graph of a function $z = z(x, y)$, but this term is easily confused with the notion of an abstract graph with vertices and edges.

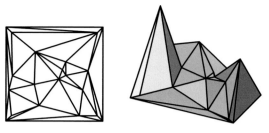

Figure 4.12. A triangulated planar graph, and a polyhedral lifting.

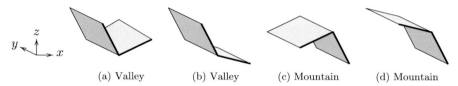

       (a) Valley   (b) Valley   (c) Mountain   (d) Mountain

Figure 4.13. Characterizing edges in a polyhedral lifting as valleys or mountains. The thick edges denote the intersection with a vertical plane. [Based on Figure 8 of Connelly et al. 2002b.]

It turns out that these polyhedral liftings are a geometric view of equilibrium stresses:

Theorem 4.6.1 (Maxwell-Cremona Theorem (Cremona 1890; Maxwell 1864; Whiteley 1982)). *A planar tensegrity has an equilibrium stress precisely if it has a polyhedral lifting. Furthermore, there is a one-to-one correspondence between equilibrium stresses and polyhedral liftings (modulo vertical translation and shear) satisfying these properties:*

1. *A valley edge corresponds to a negative stress.*
2. *A mountain edge corresponds to a positive stress.*
3. *A flat edge corresponds to a zero stress.*

In particular, the everywhere-zero stress corresponds to the everywhere-flat lifting.

This chapter on rigidity will play a crucial role when we study locked chains in Chapter 6, particularly in 2D (Section 6.6, p. 96). In particular, we use struts in tensegrities to capture a notion of "expanding" a linkage, we use duality to relate infinitesimal motions to equilibrium stresses, and we use the polyhedral lifting viewpoint to look at the vertex of maximum z coordinate (a notion difficult to capture with stresses).

5 Reconfiguration of Chains

After seeing that many fundamental linkage reconfiguration problems are intractable, we specialize in this chapter to the simplest situations, where polynomial-time—sometimes even linear-time—algorithms are possible.

We start in Section 5.1 by exploring from Table 1.1 the simpler end of each spectrum: the simplest linkage graph stucture, an open chain; the simplest intersection constraints, none; the simplest dimension, 2D; and the simplest problem, reachability. In Section 5.2 we continue to permit the chain to cross itself, but not a surrounding boundary. Finally, we initiate the exploration of reconfigurations of chains that avoid self-intersection in Section 5.3, saving the richest topic (locked chains) for Chapter 6.

5.1 RECONFIGURATION PERMITTING INTERSECTION

5.1.1 Chain Reachability

Let C be an open chain of n edges/links e_i, with lengths $|e_i| = \ell_i$, $i = 1, 2, \ldots, n$, and with $n + 1$ vertices/joints $v_0, \ldots, v_n$ so that $e_i = v_{i-1}v_i$ (see Figure 5.1). Such a pinned chain is often called an *arm* (or "robot arm"), with v_0 the *shoulder* joint and v_n the *hand*. Fix v_0 at the origin of a coordinate system.

5.1.1.1 Connectivity of Configuration Space

The configuration space of this arm C is certainly connected. From any configuration with $n \geq 2$ links, we can rotate the last link e_n to become a collinear extension of the next-to-last link e_{n-1}. Then the arm can be viewed as an arm with one fewer link, treating the last two links e_{n-1} and e_n as fused together into a single "virtual link" of length $\ell_{n-1} + \ell_n$. By repeating this process, we eventually reduce the arm down to a single virtual link of length $\ell_1 + \ell_2 + \cdots + \ell_n$, which corresponds to a straightened out configuration of the arm. Because any configuration can be brought to this configuration, any two configurations can be brought to each other via this intermediate state.

5.1.1.2 Annulus

The next question we investigate here is this: What is the set of points reachable by the hand of this arm C? The answer is simple (and perhaps obvious): an annulus centered on v_0.

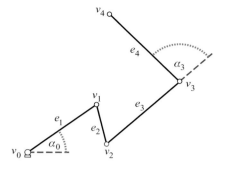

Figure 5.1. An arm/pinned open chain. The angle α_i is the turn angle between adjacent links.

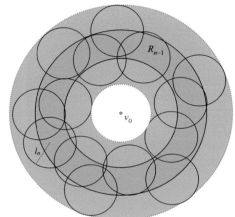

Figure 5.2. Annulus sum.

Define an *annulus* as the closed set of points bounded by (between) two concentric circles. If the two circles have the same radius, then the annulus is a circle; if the inner circle has radius zero, then the annulus is the closed disk determined by the outer circle. We first look at the shape of the reachability region, and then its radii.

Lemma 5.1.1. *The set of points reachable by the hand v_n of an arm C of n links is an annulus centered on the shoulder joint v_0.*

Proof: The proof is by induction. An arm of $n = 1$ link can reach the points on a circle of radius ℓ_1 centered on v_0, which is an annulus by our definition. Suppose now the lemma holds for all arms of up to $n - 1$ links. Let $C' = [\ell_1, \ell_2, \ldots, \ell_{n-1}]$ be the arm C with the last link removed. By the induction hypothesis, the C' reachability region R_{n-1} for v_{n-1} is an annulus centered on v_0. Let $S(r)$ be the circle of radius r centered on the origin. Then the reachability region for C is the set

$$\bigcup_{p \in R_{n-1}} p + S(\ell_n)$$

that is, the union of circles of radius ℓ_n centered on every point of R_{n-1} (see Figure 5.2). This is again an annulus: with outer radius larger by ℓ_n, and inner radius either reduced by ℓ_n or to zero (if the origin can be reached by v_n). $\qquad\square$

If one defines the pointwise set-sum $A \oplus B$ of two sets A and B to be

$$\{x + y \mid x \in A, \ y \in B\},$$

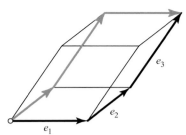

Figure 5.3. The vector sum $e_1 + e_2 + e_3$ is independent of the order of the terms.

where $x + y$ is the coordinate-by-coordinate vector sum of the points x and y, then this lemma follows from the fact that the set-sum of an annulus and a circle is an annulus.

Before computing the radii of the annulus, we should note that this annulus should not be confused with configuration space (p. 11) of the robot arm. Only in the case of a 1-link arm do the two coincide. A 2-link arm's configuration can be parameterized by two angles (at v_0 and v_1), so the configuration space is 2D. Because the angles 0 and 2π represent the same position for each link, the space is a torus (a "2-torus"), obtained by identifying the left and right, and top and bottom sides of a $[0, 2\pi] \times [0, 2\pi]$ square. See again Figure 2.1 (p. 18). In general, the configuration space of an n-link arm is a n-torus (see Thurston and Weeks 1984).

The outer radius of the annulus is obviously the sum of the link lengths, but the inner radius is perhaps less self-evident. We incorporate the claim of Lemma 5.1.1 into this theorem for later reference:

Theorem 5.1.2 (Hopcroft et al. 1985). *The set of points reachable by the hand v_n of an arm C of n links is an annulus centered on the shoulder joint v_0. The outer radius of the annulus is*

$$r_o = \ell_1 + \cdots + \ell_n.$$

Let ℓ_M be the length of a longest link of C, and let s be the total length of all the other links: $s = r_o - \ell_M$. Then, if $\ell_M \leq s$, $r_i = 0$ (i.e., the annulus is a disk); else, $r_i = \ell_M - s$.

Proof: First note that the region of reachability R of an arm C is independent of the order of the links in the arm. This can be seen as follows. Suppose $v_n = p \in R$, and look at a particular configuration of C that realizes this placement of v_n. View each link e_i as a vector displacement from v_{i-1} to v_i. Then, with v_0 at the origin, we have

$$e_1 + \cdots + e_n = v_n = p,$$

that is, p is reached by that vector sum. Because vector addition is commutative, any reordering of the links in C can reach the same point p (see Figure 5.3).

Now, reorder the links of C by moving a longest link e_M to the front, incident to the shoulder. This does not alter the reachability region, but helps intuition. Now it should be clear that if $\ell_M > s$, the inner radius is $r_i = \ell_M - s$, and when $\ell_M \leq s$, the inner radius diminishes to zero (see Figure 5.4). $\qquad\square$

Corollary 5.1.3. *Reachability for an n-link arm can be decided in $O(n)$ time.*

Proof: Compute r_o and r_i in $O(n)$ time. p is reachable precisely if $r_i \leq |p| \leq r_o$. $\quad\square$

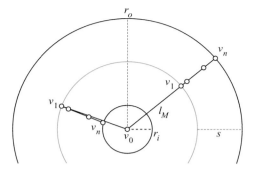

Figure 5.4. Inner radius $r_i = \ell_M - s$.

Recall from Theorem 2.2.2 that the addition of obstacles renders this problem intractable.

5.1.1.3 Two-Kinks Theorem

We now move up a notch in complexity among the options in Table 1.1, and consider the same geometric situation, but ask a more difficult question: find a configuration of an arm that reaches a given point p with its hand v_n. Note that the previous algorithm gives no hint how to configure the arm to reach the point.

There is a beautiful theorem here, not deep but nevertheless surprising, that helps a good deal: the "two-kinks theorem" (Kutcher 1992; Lenhart and Whitesides 1991). Say an *internal joint* v_i, $i = 1, \ldots, n-1$, is *kinked* if its two incident links e_i and e_{i+1} are not collinear, that is, the angle α_i between them is not 0 or π (or a multiple of π). The two-kinks Theorem not only says that at most two joints need be kinked, it also identifies which two. Define the *median* link of an arm as the link e_m that covers the midpoint of the arm when all links are stretched out straight in a line (i.e., $\alpha_i = \pi$ for all internal joints), or either link incident to the midpoint if it falls at a joint. (We will continue to say "the" median link even in the latter case.)

Theorem 5.1.4. *If an n-link arm C can reach a point p, it can reach p with at most two internal joints kinked. The two kinked joints may be chosen to be at either end of the median link.*

Proof: The proof proceeds by "freezing" all but the two indicated joints, and showing that the reachability region of the new arm C' is the same as that for C. By Theorem 5.1.2, the reachability region depends only on r_i and r_o. Freezing joints does not change the sum of the link lengths, so r_o is not altered. So the onus of the proof is to show that r_i is also unaltered by freezing. We consider two cases:

1. $r_i > 0$ (Figure 5.5(a)). The inner radius is positive only when the longest link length ℓ_M strictly exceeds the sum of the other lengths: $\ell_M > s$. This means that the longest link e_M covers the midpoint regardless of where it appears in the sequence of links. Thus $e_M = e_m$, the median link. Now, freezing the joints except for those incident to e_M does not change the fact that e_M is the longest link, for the chains to either side of e_M can be at most s long, and $s < \ell_M$. Thus both ℓ_M and s are the same for C' as they are for C, and therefore r_i is unchanged.
2. $r_i = 0$ (Figure 5.5(b)). Theorem 5.1.2 tells us that in this case, $\ell_M < s$. Let e_m be the median link, and freeze the joints before and after e_m to form C'. Now we

(a)

(b)

Figure 5.5. (a) ($\ell_M = 6$) > ($s = 5$); (b) ($\ell_M = 6$) < ($s = 11$); ($\ell_{M'} = 8$) ≤ ($s' = 9$).

could have created a new longest link through this freezing (as illustrated in the figure), so the argument used in the previous case does not apply. However, the new longest link cannot be longer than half the total length of all the links, for it must be to one side of e_m. Therefore $\ell_{M'} \leq s'$, where $\ell_{M'}$ is the length of the new longest link and s' is the sum of the lengths of the remaining links. Theorem 5.1.2 then shows that $r_i = 0$, so again it has not changed. $\square$

This theorem leads to a linear-time algorithm for finding a configuration that reaches a given point: Identify the median link in $O(n)$ time, freeze the joints to produce a 3-link arm, and solve the 3-link problem in $O(1)$ time.

5.1.1.4 Solving a 3-Link Problem

A 1-link reachability problem has a unique solution determined by the location of v_1 on the circle of radius ℓ_1 centered on v_0. A 2-link reachability problem has (in general) two solutions determined by the intersections of the two circles centered on v_0 and v_2 of radii ℓ_1 and ℓ_2 respectively. A 3-link problem has (again, in general) an infinite number of solutions, and the only challenge is to design a procedure to select a finite subset that guarantees reaching a given v_3 position. It is proved in O'Rourke (1998, Lem. 8.6.4) that if a 3-link arm can reach a given v_3, then either it can reach with one of the two internal joints v_1 or v_2 frozen at angle 0 to align the two incident links, or it can reach for any orientation α_0 of the first link. These cases all lead to 2-link problems, and therefore to a constant-time algorithm to solve any 3-link problem. Together with the two-kinks theorem, this implies:

Theorem 5.1.5. *A configuration for an n-link arm to reach a given point can be computed in $O(n)$ time.*

Finally we mention a recent result on motions of n-link arms: any "motion planner" controlling an n-link robot arm will have "degree of instability of at least $2n - 1$" (Farber 2002, Thm. 9.1). Although it would take us far afield to describe this result in detail, essentially it means that reconfiguration in general will need to pass through several movement discontinuities. Thus "instabilities . . . are inevitable in most practically interesting cases."

5.1.2   Turning a Polygon Inside-Out

We have seen in the previous section that an open polygon chain that is permitted to self-intersect is easily understood, with both reachability and reconfiguration computable in linear time. In this section we explore closed polygonal chains (polygons) in two dimensions, again permitting self-intersection. Although in the end everything

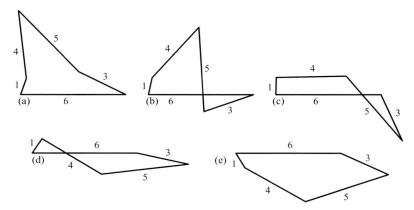

Figure 5.6. Inversion of a closed polygonal chain. The links are marked with their lengths. The sequence is inverted from $(6, 3, 5, 4, 1)$ to $(1, 4, 5, 3, 6)$.

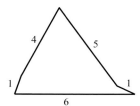

Figure 5.7. A noninvertible polygon, with link lengths $(6, 1, 5, 4, 1)$.

is still linear-time computable, the closure of the chain introduces some interesting complications. In particular, the configuration space of a closed chain might not be connected. This possibility is most easily understood by asking a question first formulated (and solved) by Lenhart and Whitesides (1991, 1995) (and independently by Kapovich and Millson 1995): When can a polygon be turned inside-out?

Figure 5.6 illustrates just such an "inside-out" reconfiguration of a polygon. Let the links be labeled $L_1, L_2, \ldots, L_n$ in some consecutive order around the chain C. We say that C is *invertible* if it may be reconfigured so that the labeling order is reversed to $L_n, \ldots, L_2, L_1$, that is, "turned inside-out." Figure 5.7 shows a polygon that is not invertible. Lenhart and Whitesides proved this theorem characterizing invertability (phrased differently but equivalently by Kapovich and Millson):

Theorem 5.1.6 (Kapovich and Millson 1995; Lenhart and Whitesides 1995). *A polygon is invertible if and only if the second and third longest links together are not longer than the total length of the remaining links.*

Their proof then leads to the conclusion that the configuration space of a polygon has exactly one connected component if the chain is invertible and exactly two components if not. Here we prove half of Theorem 5.1.6,[1] and only sketch the other, algorithmic half.

---

[1] Our proof is suggested as an "alternative proof" in their paper, but not detailed there.

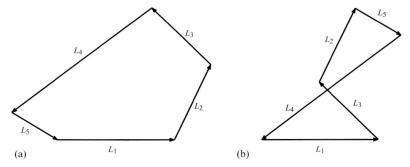

Figure 5.8. Under the permutation $\pi = (1, 3, 2, 5, 4)$, (a) is altered to (b).

The observation that reachability of an open chain (a robot arm) is independent of the ordering of the links considerably simplified the proof of Theorem 5.1.2. Here we seek a similar simplification, but phrased in terms of connectivity rather than reachability.

Let $\pi = (i_1, i_2, \ldots, i_n)$ be a permutation of $(1, 2, \ldots, n)$; so $\pi(k) = i_k$. Let $\mathcal{C} = (\ell_1, \ell_2, \ldots, \ell_n)$ be an abstract[2] open chain, and let $\pi(\mathcal{C}) = (\ell_{\pi(1)}, \ell_{\pi(2)}, \ldots, \ell_{\pi(n)})$ be the open chain composed of the same links but permuted by π. Similarly, for a configuration C of $\mathcal{C}$, let $\pi(C)$ be the configuration obtained by permuting the link vectors of C according to π (see Figure 5.8).

Lemma 5.1.7. *Let $\mathcal{C}_1$ and $\mathcal{C}_2 = \pi(\mathcal{C}_1)$ be two closed chains composed of the same link lengths in a different permutation π. Two configurations C_1 and C_1' of $\mathcal{C}_1$ are in the same component of the configuration space of $\mathcal{C}_1$ if and only if $\pi(C_1)$ and $\pi(C_1')$ are in the same component of $\mathcal{C}_2$'s configuration space.*

Proof: Because vector addition is commutative, if C_1 is closed, so is $\pi(C_1)$, for it simply reorders the link vectors, whose sum remains zero. Thus every configuration C''_1 intermediate between C_1 and C_1' has a counterpart configuration $\pi(C''_1)$. Thus the reconfiguration of $\mathcal{C}_1$ is paralleled (literally!) by a reconfiguration of $\mathcal{C}_2$, and vice versa. □

Corollary 5.1.8. *$\mathcal{C}_1$ is invertible if and only if $\mathcal{C}_2 = \pi(\mathcal{C}_1)$ is invertible, for any permutation π.*

This corollary permits us to reorder the links in a way that will render the seemingly strange claim of Theorem 5.1.6 nearly obvious. In particular, order and relabel the links of $\mathcal{C}$ so that

$$\ell_1 \geq \ell_2 \geq \cdots \geq \ell_n.$$

Then Theorem 5.1.6 says that $\mathcal{C}$ is invertible if and only if

$$\ell_2 + \ell_3 \leq \ell_1 + (\ell_4 + \cdots + \ell_n). \qquad (5.1)$$

Fix $L_1 = (v_0, v_1)$ to lie on a horizontal line H. For $L_2 = (v_1, v_2)$ to invert, it must pass through H from above to below. Consider the moment when v_2 is first on H. The

---

[2] An *abstract* chain is just this list of lengths, without any specific embedding in the plane, i.e., without specifying a configuration of the chain.

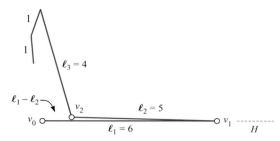

Figure 5.9. $\ell_1 - \ell_2$ and the remainder of the chain are insufficient to span ℓ_3. (The link lengths here are those used in Figure 5.7: $5 + 4 > 6 + 1 + 1$.)

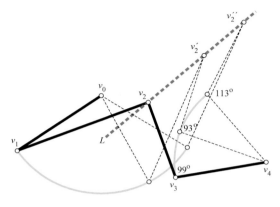

Figure 5.10. A line-tracking motion. The initial position of the chain is shown with solid lines, with v_2 tracking line L to positions v_2' and v''_2. This is not a simple line-tracking motion because the angle at v_3 changes nonmonotonically, as indicated by the angle measurements.

remaining chain must connect v_0 to v_2, two points separated by $\ell_1 - \ell_2$ (see Figure 5.9). This is only possible if

$$\ell_3 \le (\ell_1 - \ell_2) + (\ell_4 + \cdots + \ell_n) \tag{5.2}$$

or

$$(\ell_1 + \ell_2) \le \ell_3 + \cdots + \ell_n \tag{5.3}$$

for the gap and the rest of the chain must be able to stretch the length of L_3. But Equation (5.2) is just a rearrangement of Equation (5.1), and Equation (5.3) follows easily from Equation (5.1), which proves its necessity. Sufficiency was established in Lenhart and Whitesides (1995) by an algorithm that performs the inversion. Here we only describe their basic "move," which will have occasion to use later in Sections 5.3.1 and 6.4.

Let v_0, v_1, v_2, v_3, v_4 be five consecutive joints of a chain. A motion in which

(a) v_0 and v_4 remain fixed and
(b) v_2 moves (in one direction) along a line

is called a *line-tracking motion* (see Figure 5.10). If, in addition, all five joint angles change monotonically (i.e., none reverse direction during the motion), it is called a *simple line-tracking motion*. They proved that any two configurations of an n-link closed chain in the same connected component of configuration space can be reached from the other by a sequence of $O(n)$ simple line-tracking motions. In particular, the reconfiguration can be achieved by moving at most five joints at once. Their proof reconfigures any given chain to a standard triangular "normal form" by iterating line-tracking motions to straighten one joint (v_1 or v_3 in the above notation) at a time, until the configuration is geometrically a triangle, with all but its corner joints straightened.

Finally, we briefly touch on extensions to higher dimensions. Because the line-tracking motion works in 3D, it is nearly a straightforward observation that, if motion is permitted in $\mathbb{R}^3$, every closed polygonal chain has a connected configuration space.

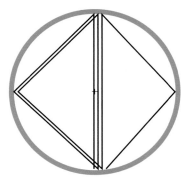

Figure 5.11. The long links must remain near a diameter. The consecutive pairs of short links each can be placed left or right of that diameter. This yields $2^{n/3}$ disconnected configurations.

Given such a chain, reconfigure it to the (planar) triangular normal form, rotate it in 3D to flip it over to its mirror image, and from there it may be moved to any other "inside-out" configuration. Thus, the Lenhart and Whitesides algorithm provides a way to reconfigure a closed chain in 3D (permitting self-intersections) with $O(n)$ 5-joint moves.

Sallee (1973) had earlier designed a way to accomplish this reconfiguration with 4-joint moves, shown by Toussaint (2005) to use $O(n)$ moves. See Toussaint (2003) for a careful analysis of 4-bar line tracking. These results also apply to chains in dimensions beyond 3, in particular establishing the following result:

Theorem 5.1.9. *For any $d \geq 3$, the configuration space of any closed chain in d dimensions is connected via motions in d dimensions.*

5.2 **RECONFIGURATION IN CONFINED REGIONS**

We have seen in Section 5.1.1.2 that reachability for a chain that is permitted to self-intersect is easily understood, but Theorem 2.2.2 showed that with the addition of just four segment obstacles, reachability can become NP-hard. This latter negative result prompted a search for tractable intermediate situations. Three have been studied, all leading to polynomial-time—indeed, linear-time—algorithms for specific problems: the chain is confined inside a circle, inside a square, or inside an equilateral triangle. The chain is permitted to self-cross itself, but not penetrate the boundary of the confining region. In the first two cases, the shoulder joint is pinned; in the latter case, results were obtained without any joint fixed. We will proceed chronologically, describing each result at a high level, but leaving the detailed proofs to the original papers.

5.2.1 Chains Confined to Circles

Hopcroft, Joseph, and Whitesides explored circles to avoid "the complexities that arise in situations where a link can jam in a corner" (Hopcroft et al. 1985, p. 308). They succeeded in developing polynomial algorithms for both reconfiguration and reachability. These results were by no means a foregone conclusion. For example, the configuration space of an n-link chain in a circle can have $2^{\Omega(n)}$ connected components (see Figure 5.11).

Their key lemma depends on the notion of "link orientation." Say link $L_i = (v_{i-1}, v_i)$ has *left orientation* if the center of the circle is left or on the line containing L_i; *right orientation* is defined similarly. A link collinear with a circle diameter has both left and

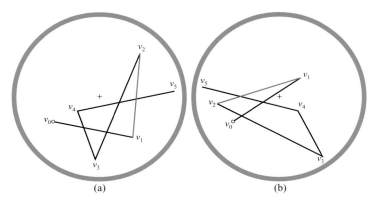

(a)   (b)

Figure 5.12. The five links have (a) orientations LLRRL and (b) orientations RLLLR. Thus all but L_2 are reoriented.

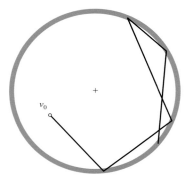

Figure 5.13. Normal form for the chain shown in Figure 5.12(a).

right orientation (see Figure 5.12). It is an obvious necessary condition that if C_1 can be reconfigured to C_2, then it must be possible to reorient each link whose orientation differs in C_1 and C_2. They proved this condition is also sufficient. Moreover, whether a link can be reoriented can be decided from knowing the maximum and minimum distance its endpoints can be moved from the circle, distances that can be computed in polynomial time. This provides a decision algorithm for testing if two configurations are in the same connected component in configuration space. To actually find a sequence of "simple moves" (in this case, monotonic change of at most four joints at once) to achieve reconfiguration, they employ an intermediate "normal form" where as many joints as possible are on the circle (see Figure 5.13). Finally, to solve reachability, they proved the surprising result that the boundary of the reachability region for a joint can be covered by a constant number of circular arcs, a constant independent of n. They proved a bound of 148 on this constant, which they admit is "probably very generous." A closer analysis by Kantabutra and Kosaraju (1986) improved their $O(n^3)$ algorithm complexity to $O(n)$, and likely implies a tighter constant. We summarize in a theorem:

Theorem 5.2.1 (Hopcroft et al. 1985; Kantabutra and Kosaraju 1986). *For an n-link chain confined to a circle, with one end pinned, the decision, motion planning, and reachability questions can all be answered within $O(n)$ computation time.*

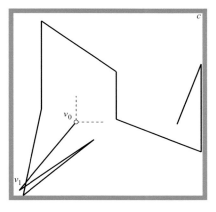

Figure 5.14. Link $L_1 = (v_0, v_1)$ is "stuck" because it cannot reorient to "aim" toward corner c.

5.2.2 Chains Confined to Squares

A result on chains confined to squares was obtained by Kantabutra (1997): if every link is at most half the side length of the square, then whether two pinned configurations are reachable from another can be decided in $O(n)$ time. We will only mention his key idea. Say a link $L_i = (v_{i-1}, v_i)$ is *stuck* if it cannot be reoriented so that both v_i and the corner c of the square furthest from v_{i-1} lie in the quadrant whose origin is v_{i-1} (see Figure 5.14). He proves that every "short" linkage configuration (satisfying the half-side-length condition) has at most one stuck link and that two configurations are reachable from one another if and only if the same link (if any) is stuck in both. Finally, the stuck link can be identified in linear time.

There seem to be no results for chains with links of arbitrary lengths confined to a square.

5.2.3 Chains Confined to Equilateral Triangles

Acute angles in the confining region seem to create considerable complexity in the possible reconfigurations. The only result available is highly specialized, but nevertheless intriguing: van Kreveld et al. (1996) restricted attention to (a) equilateral triangle regions, containing (b) *equilateral chains* (each link the same length),[3] with (c) neither endpoint pinned, and with the goal (d) deciding whether a configuration can reach a *folded state*, where all links are collinear, collapsed on top of one another like a folded carpenter's ruler. With these constraints, and with the equilateral triangle having unit edge length, they proved the following alternation property: there are three link lengths, $x_1 \approx 0.483$, $x_2 = 0.5$, and $x_3 \approx 0.866$, such that all chains with link lengths in the ranges $[0, x_1]$ or $(x_2, x_3]$ fold, whereas there are chains with lengths in the complementary ranges $(x_1, x_2]$ and $(x_3, 1]$ that cannot fold (see Figure 5.15). For chains in the two folding ranges, they found linear-time algorithms to perform the folding.

Figure 5.16 shows three critical "stuck examples." The first and most complicated example has nine links slightly longer than

$$x_1 \approx 0.483 = \frac{1}{4}\left(12 + 7\sqrt{3} - (6 + 3\sqrt{3})\sqrt{4\sqrt{3} - 3}\right).$$

---

[3] In later contexts (p. 153) these are called *unit chains*.

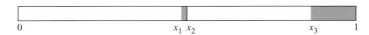

Figure 5.15. There are chains with link lengths in the shaded regions $(0.483, 0.500]$ and $(0.866, 1.000]$ that cannot fold.

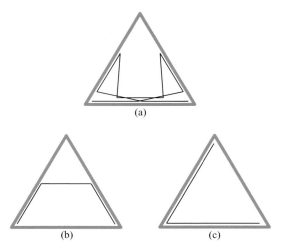

Figure 5.16. "Stuck" chains: (a) $x_1 > 0.483$; (b) $x_2 = 0.5$; (c) $x_3 > 0.866$.

The second shows an example with $x_2 = 0.5$, and the third an example with lengths slightly greater than the triangle altitude $x_3 = \sqrt{3}/2 \approx 0.866$. The complexity of Figure 5.16(a) (and the expression for x_1) hints at the difficulty of the analysis. It remains an open question whether a similar alternation property holds for other confining regions.

5.3   RECONFIGURATION WITHOUT SELF-CROSSING

For the remainder of Part I of this book, we explore chains with the physically natural constraint that the chain cannot cross itself. The somewhat subtle notion of "self-cross" is discussed in Box 5.1.

Perhaps the most fundamental question for noncrossing chains is this: Is the configuration space of a polygonal chain, when confined to a plane, connected? The answer is YES, but the only known proof is not at all straightforward. We defer this topic and related questions about connectivity to Chapter 6. Here we concentrate on several easier results concerning reconfiguration of chains within one component of configuration space. We start in 2D with convex polygons (Section 5.3.1), then consider polygons in 2D but permitting movement into 3D (Section 5.3.2), and finally we look at a particular case of open chains in 3D (Section 5.3.3).

5.3.1   2D: Convex Polygons

A convex polygon of more than three vertices has flexibility and can be reconfigured in the plane to other polygons having the same counterclockwise sequence of edges. A particularly simple problem is to reconfigure one convex polygon P to another convex polygon P' realizing the same sequence of edge lengths. The algorithm of Lenhart

Box 5.1: Self-Crossing

Whenever two noncollinear segments share a point in the relative interior of each, they are said to *cross*. It does not suffice to rely on this to define self-crossing of a chain, however, because there might be crossings involving vertices and/or collinearity. This motivates the following definition.

Parameterize a chain C by arc length $t \in [0, 1]$; we shall use the notation $C(t)$ for the point of the curve at t and $C[t_1, t_2]$ for the portion of the curve between $C(t_1)$ and $C(t_2)$. The curve has a left and right side in the neighborhood of each point, determined by the tangent for points interior to an edge and, at a vertex, consistent with this information in a neighborhood of a vertex (Figure 5.17(a)). We say that C *self-crosses* if there are two disjoint intervals (which may degenerate to points) $[t_1, t_2]$ and $[t_3, t_4]$ such that

(a) the curves match over the intervals: $C[t_1, t_2] = C[t_3, t_4]$; and
(b) $C(t_1')$ and $C(t_2')$ are on opposite sides of $C(t_3)$ and $C(t_4)$ for t_1' and t_2' in a neighborhood of t_1 and t_2 respectively.

The usual situation is that the intervals are points, and this definition merely ensures that one part of the curve crosses another portion from one side to the other. The awkward definition is needed to handle situations like that illustrated in Figure 5.17(b), where the crossing is not determined locally, but rather is "deferred" to the ends of the matching interval.

Usually it is this notion of self-crossing that is physically relevant, rather than *self-intersection*, which occurs whenever two nonadjacent links of the chain share a point. Some proofs are simpler if noncrossing grazing contact is permitted, and we will take advantage of this when convenient.

These issues are readdressed in Section 6.8.

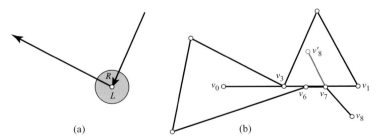

(a)   (b)

Figure 5.17. (a) Left and right in neighborhood of a vertex; (b) $C = (v_0, \ldots, v_8)$ does not self-cross, for it stays on the same side of $v_0 v_1$ both in a neighborhood of v_3 and in neighborhoods of v_6 and v_7. But if v_8 is replaced by v_8', then $C' = (v_0, \ldots, v_8')$ does self-cross.

and Whitesides (Section 5.1.2) achieves this when edges are permitted to cross, by reconfiguring to a triangular normal form. Although they did not design their algorithm to avoid self-crossings, in the convex-to-convex situation, it is not difficult to see that it does. Their algorithm repeatedly performs line-tracking motions (Figure 5.10, p. 66), moving v_3 outward along a line perpendicular to the diagonal $v_1 v_5$ until either v_2 or v_4 straightens. In the convex case, this alters the shape of a convex pentagon, maintaining convexity. If their algorithm is modified to stop a line-tracking motion earlier if either v_1 or v_5 straightens, then it preserves convexity also at the joints between the pentagon and the remainder of the polygon, and therefore avoids self-intersection.

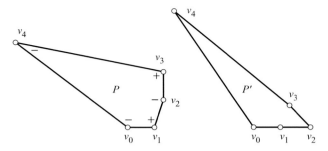

Figure 5.18. The sign sequence $(-, +, -, +, -)$ alternates four times.

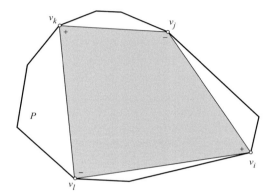

Figure 5.19. The inscribed quadrangle (v_i, v_j, v_k, v_l).

On the other hand, Aichholzer et al. (2001) designed an algorithm that avoids an intermediate normal form: it has the property that each vertex angle changes monotonically and thus reconfigures directly from P to P'. We now sketch this algorithm, which employs several tools we will use later.

Their key idea is to use a lemma Cauchy employed in his celebrated theorem on the rigidity of convex polyhedra. Let P and P' be two convex planar configurations of the same closed chain. Let α_i and α_i' be corresponding angles in the two polygons. Label the angle at v_i in P as one of $\{-, 0, +\}$, depending on whether α_i' $\{<, =, >\}$ α_i respectively. Thus '$+$' means that α_i must increase to reach α_i' (see Figure 5.18).

Cauchy's lemma (now carrying Steinitz's name as well after he corrected an error in Cauchy's proof) is as follows:

Lemma 5.3.1 (Cauchy–Steinitz). *In any $\{-, 0, +\}$ labeling derived from distinct convex configurations of a polygon, there are at least four sign alternations in a complete circuit of the angles.*

We will prove this lemma in Part III, Section 23.1.1.

This lemma permits identifying vertices v_i, v_j, v_k, v_l in cyclic order around the polygon, whose angles are labeled $+$, $-$, $+$, $-$ in that order (see Figure 5.19). Now the convex polygon is treated as the inscribed quadrangle (v_i, v_j, v_k, v_l), holding rigid the subchains of the convex polygon connecting those four vertices: $[v_i, v_j], [v_j, v_k], [v_k, v_l]$, and $[v_l, v_i]$. The quadrangle is flexed according to the $+, -, +, -$ angular labels, stopping once one

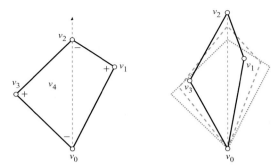

Figure 5.20. Flexing a quadrangle. [After Figure 2 of Aichholzer et al. 2001.]

angle matches the desired value in P', that is, a label changes from $+$ or $-$ to 0. That this flexing is possible is guaranteed by this lemma:

Lemma 5.3.2. *For any convex quadrangle $Q = (v_0, v_1, v_2, v_3)$, moving v_2 along the directed line from v_0 through v_2 decreases the angles at v_0 and v_2, and increases the angles at v_1 and v_3. The motion can continue until one of the increasing angles reaches π. Neither of the decreasing angles can reach 0 before an increasing angle reaches π.*

Proof: Because the distance $|v_2 - v_0|$ increases, the angles at both v_1 and v_3, which are apexes of triangles $\triangle_1 = (v_0, v_1, v_2)$ and $\triangle_3 = (v_0, v_3, v_2)$ based on v_0v_2, increase throughout the motion (see Figure 5.20). Because no angle goes past 0 or π, Q remains convex throughout the motion; so by Lemma 5.3.1, there must be four sign alternations. This proves that the angles at v_0 and v_2 decrease throughout the motion. Finally, neither decreasing angle can reach 0 until both $\triangle_1$ and $\triangle_3$ collapse, which requires v_1 and v_3 to reach π, respectively. □

An attractive alternate formulation of the essence of this lemma is provided by Toussaint (2003): For a convex 4-bar linkage with no three vertices collinear, one diagonal increases in length if and only if the other decreases in length.

This Lemma 5.3.2 supports the following algorithm to reconfigure from P to P'. Identify a quadrangle $Q = (v_i, v_j, v_k, v_l) \subseteq P$ whose angles are labeled $(+, -, +, -)$ by Lemma 5.3.1. Flex according to Lemma 5.3.2 until one of the four angles $\alpha_i, \alpha_j, \alpha_k, \alpha_l$ in P reaches its target angle in P'. We now argue that at least one target angle will be reached at or before the flex straightens an angle of Q. Lemma 5.3.2 establishes that no angle of Q can collapse to 0 before one straightens to π. Because the angle in P is at least as large as the corresponding angle in Q, at or before an angle straightens in Q, it must reach its target convex angle in P'. Thus, the flexing motion cannot terminate before a target angle is reached.

Repeating this process leads to a sequence of at most n flexings, each changing four angles at once, that reconfigures P to P'. Because each step changes the label of an angle from $+$ or $-$ to 0, and that label persists, the motion is angle-monotone. Finally, it is shown in Aichholzer et al. (2001) that these operations can be computed in $O(n)$ time on a pointer machine with real numbers. These authors have also shown that it is not possible in general to find an *expansive* or "distance-monotone" reconfiguration, in which the distance between every pair of vertices increases monotonically. Such expansive motions will prove important in Section 6.6.

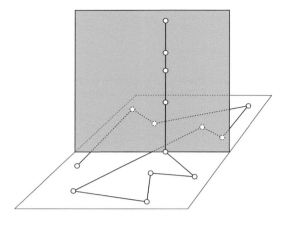

Figure 5.21. A planar arc partially lifted into a vertical line.

5.3.2   2D/3D: Arbitrary Polygons

The natural next topic is reconfiguring arbitrary polygons in a plane. As mentioned previously, we defer this complex topic to Chapter 6, and instead allow ourselves to move into 3D. Our goal will be to reconfigure the chain into a *canonical form*: for open chains, a *straight configuration*—all vertex angles equal to 180°; for closed chains, a *convex configuration*—a planar polygon, that is, one having all interior vertex angles less than or equal to 180°. The result of the previous section shows that convex configurations are indeed "canonical" in the sense that any one can be reconfigured to any other. Given a non-self-intersecting chain in the plane, we would like to know (a) if it may be reconfigured in 3D, avoiding self-crossings, to the canonical form, and (b) if so, how to plan the reconfiguring motion. Both questions are easily answered for open chains (Biedl et al. 2001a): simply pick up the chain into 3D link-by-link, at all times maintaining a prefix of the chain in a vertical line orthogonal to the plane containing the arc (see Figure 5.21). Each lifting motion causes two joint angles to rotate so that the lifted vertical line remains vertical, while the rest of the chain remains in its original plane. The $O(n)$ moves of the motion can be planned in $O(n)$ time.

For closed chains, the answer to question (a) is also YES and the next section discusses three different algorithms that answer question (b).

5.3.2.1 Pocket Flipping

The roots of this question go back to a problem Erdős posed in 1935 in the *American Mathematical Monthly* (Erdős 1935). Define a *pocket* of a polygon to be a region exterior to the polygon but interior to its convex hull, bounded by a subchain of the polygon edges and the pocket *lid*, the edge of the convex hull connecting the endpoints of that subchain. Every nonconvex polygon has at least one pocket. Erdős defined a *flip* as a rotation of a pocket's chain of edges into 3D about the pocket lid by 180°, landing the subchain back in the plane of the polygon, such that the polygon remains simple (i.e., non-self-intersecting). He asked whether every polygon may be convexified by a finite number of simultaneous pocket flips. The answer was provided in a later issue of the *Monthly* by de Sz. Nagy (1939). First, Nagy observed that flipping several pockets at once could lead to self-crossing (see Figure 5.22). However, restricting to one flip at a time, Nagy proved that a finite number of flips suffice to convexify any polygon. See Figure 5.23 for a three-step example. Applets are available to perform

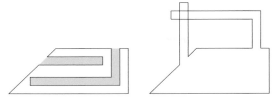

Figure 5.22. Flipping several pockets simultaneously can lead to crossings (de Sz. Nagy 1939).

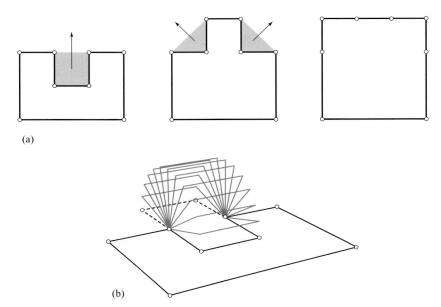

(a)

(b)

Figure 5.23. (a) Flipping a polygon until it is convex. Pockets are shaded. (b) The first flip shown in 3D.

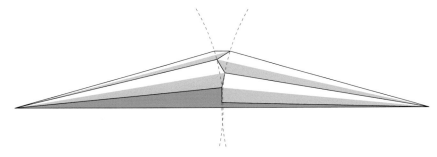

Figure 5.24. Quadrangles can require arbitrarily many flips to convexify.

pocket flipping for arbitrary polygons.[4] This beautiful result has been rediscovered and reproved several times, as uncovered by Grünbaum and Toussaint, and detailed in their histories of the problem (Grünbaum 1995; Toussaint 2005).

Before we come to Nagy's finiteness proof, it is algorithmically important to mention that the number of required flips can be arbitrarily large in terms of the number of vertices, even for a quadrangle, a fact first proved by Joss and Shannon in 1973. Figure 5.24

---

[4] http://www.cs.mcgill.ca/~cs507/projects/1998/mas/.

shows the construction (Grünbaum 1995). By making the vertical edge of the quadrangle very short and close to the horizontal edge, the angles after the first flip approach the mirror image of the original quadrangle, and hence the number of required flips approaches infinity. It remains open to bound the number of flips in terms of some natural shape measure or to determine the complexity of finding the shortest convexifying flip sequence.

Open Problem 5.1: Pocket Flip Bounds. [a] Can the number of flips be bounded in terms of natural measures of geometric closeness, such as the diameter (maximum distance between two vertices), sharpest angle, or the minimum feature size (minimum distance between two nonincident edges)?

---

[a] Mark Overmars, February 1998.

Open Problem 5.2: Shortest Pocket-Flip Sequence. [a] Determine the complexity of finding the shortest, or longest, sequence of flips to convexify a given polygon. Weak NP-hardness has been established for the related problem of finding the longest sequence of "flipturns."

---

[a] Aichholzer et al. (2002).

We now offer a proof of the finiteness property first established by Nagy. Despite the independent rediscoveries of this result, it has been difficult to reach a proof acceptable to all. Indeed Nagy's original proof, although brilliant in overall design, is flawed. The proof we offer below is based on that in Demaine et al. (2006b). We shunt into a side box discussion of some of the pitfalls not explicitly addressed in Nagy's proof.

Theorem 5.3.3 (Nagy). *Every polygon P can be convexified with a finite number of pocket flips.*

Proof: Let $P = (v_0, \ldots, v_{n-1})$. The superscript k will indicate the "descendant" polygon after k flips: $P^k = (v_0^k, \ldots, v_{n-1}^k)$. Let x be any point in $P = P^0$. Clearly we have $P^0 \subset P^1 \subset \cdots \subset P^k$, so x is inside all descendants of P. Let C^k be the convex hull of P^k; so $P^k \subseteq C^k$ (see Box 5.2(1)).

We first offer an outline of this long and subtle proof.

1. The distance from each vertex v_i^k to a fixed point $x \in P$ is a monotonically non-decreasing function of k.
2. The sequence P^k approaches a limit polygon P^*.
3. The angle θ_i^k at vertex v_i^k converges.
4. Any vertex v_i^k that moves infinitely many times converges to a flat vertex v_i^*.
5. The infinite sequence P^k cannot exist.

Box 5.2: Proof Pitfalls

1. Nagy says, "Each polygon in the sequence $P^0, C^0, P^1, C^1, P^2, C^2 \ldots$ contains obviously the forgoing ones in its interior." This is FALSE, as Figure 5.26 illustrates. This mistaken claim is used by Nagy to conclude that P^* is convex (see also point 3 below).
2. The danger here is that it is possible for distances to be monotonically nondecreasing and converging, but for v_i^k to "hop around" points on a circle centered on x, never approaching an accumulation point. The argument in Step 2 of our proof handles this by intersecting three circles.
3. Some proofs conclude that the limit polygon is convex because otherwise it could be flipped again. But it is conceivable that one intricate pocket would require, on its own, an infinite number of flips to converge, leaving other pockets untouched unless there were a careful "round-robin" limit process. These complications are avoided in our proof by analyzing nonflat Step 4 and flat Step 5 vertices separately.

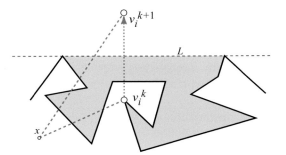

Figure 5.25. The distance from x to v_i is increased by a flip.

Step 1. First we prove that the distance from x to any particular vertex v_i is monotonically nondecreasing with k. Let $d(x, v_i^k)$ be this distance at step k. If the $(k+1)$st flip does not move v_i, then the distance remains the same. If v_i is flipped, then it flips over the pocket's line of support, L, which is the perpendicular bisector of $v_i^k v_i^{k+1}$ (see Figure 5.25). Because L supports the hull of P^k and x is inside P^k, x is on the same side of L as v_i^k. Thus $d(x, v_i^{k+1}) > d(x, v_i^k)$. This establishes that the distance from x to each vertex is a monotonically nondecreasing function of k.

Step 2. Next we argue that the sequence P^k approaches a limit polygon P^*. The perimeter of P^k is independent of k, for it is just the sum of the fixed edge lengths. The distance from x to v_i is bounded above by half the perimeter (because the polygon has to wrap around both x and v_i). Thus each distance sequence $d(x, v_i^k)$ has a limit. If we look at the distance sequences to v_i from three noncollinear points x_1, x_2, and x_3 inside P, their limits determine three circles (centered at x_1, x_2, and x_3) whose unique intersection point yields a limit position v_i^* (see Box 5.2(2)). Then $P^* = (v_0^*, \ldots, v_{n-1}^*)$ (see Box 5.2(3)).

Step 3. Let $\theta_i^k \in [0, 2\pi)$ be the directed angle $\angle v_{i-1}^k v_i^k v_{i+1}^k$. We observe that $\theta_i^k \in [\varepsilon_i, 2\pi - \varepsilon_i]$, where $\varepsilon_i = \min_k\{\theta_i^k, 2\pi - \theta_i^k\}$. Indeed this relation holds for θ_i^0, and for θ_i^k to get closer to 0 or 2π, $d(v_{i-1}^k, v_{i+1}^k)$ would have to decrease, which is impossible by the distance argument detailed in Step 1. Because θ_i^k stays away from 0 and 2π, and the edge lengths of P^k are fixed, and therefore cannot approach 0, θ_i^k is a continuous function of the coordinates of the three vertices in P^k that define

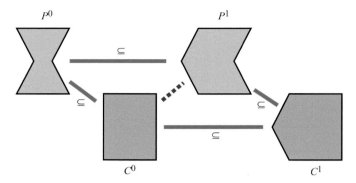

Figure 5.26. Red lines indicate the subset relation. Here $P^0 \subseteq C^0 \not\subseteq P^1$.

the angle. These vertices converge, so θ_i^k must also converge, and its limit is $\theta_i^* = \angle v_{i-1}^* v_i^* v_{i+1}^*$.

Step 4. We now distinguish between flat and nonflat vertices of P^*. A *flat vertex* v_i^* is one for which $\theta_i^* = \pi$. A *nonflat vertex* has $\theta_i^* \neq \pi$; it could be convex or reflex. Consider a vertex for which v_i^k moves an infinite number of times. We show that this vertex converges to a flat vertex v_i^* of P^*. Indeed, when v_i^k moves as a result of a pocket flip, $\theta_i^{k+1} = 2\pi - \theta_i^k$, as befalls any directed angle which is reflected. Consequently, there are infinitely many k for which $\theta_i^k \geq \pi$, and infinitely many for which $\theta_i^k \leq \pi$. Thus the limit θ_i^* can only be π.

Step 5. All that remains is to force a contradiction by showing that once the nonflat vertices of P^* have been reached, no further flips are possible. Here we use the convex hull C^k of P^k, and the hull C^* of P^*. Of course, $P^k \subseteq C^k$ and $P^* \subseteq C^*$. We will obtain a contradiction to Fact A: for any k, $P^{k+1} \not\subseteq C^k$. The reason this fact holds is that, at every flip, the mirror image of the pocket area previously inside C^k is outside C^k (e.g., cf. Figure 5.26: $P^1 \not\subseteq C^0$.)

Let $\bar{k}$ be a value of k for which only flat vertices of P^* have yet to converge. $P^{\bar{k}}$ includes all nonflat vertices in their final positions. Of course, P^* also includes all nonflat vertices in their final positions. Now, because the flat vertices of P^* cannot alter its hull beyond what the nonflat vertices already contribute, we know that $C^* \subseteq C^{\bar{k}}$. ($C^{\bar{k}}$ is conceivably a proper superset because of the vertices of $P^{\bar{k}}$ that are destined to be, but are not yet, flat vertices in P^*.)

Now consider $P^{\bar{k}+1}$. It is contained in all subsequent polygons and so in $P^* \subseteq C^*$. So we have reached Fact B: $P^{\bar{k}+1} \subseteq C^* \subseteq C^{\bar{k}}$. Fact B contradicts Fact A, so there cannot be an infinite sequence P^k. ☐

5.3.2.2 Deflations

Wegner (1993) proposed an interesting variant of a pocket flip, which he called a *deflation* operation. It is in a sense the reverse of a flip. Let v_i and v_j be two nonadjacent vertices of a polygon P, and L the line containing the segment $v_i v_j$. A deflation is possible if (a) L intersects the boundary only at v_i and v_j, and (b) the chain $v_i, v_{i+1}, \ldots, v_j$ reflects about L to the interior of P, resulting in a new simple polygon P'. These two conditions guarantee that L is a supporting line for P', so that a pocket flip would restore P. This relationship to flips led Wegner to conjecture that every polygon can be deflated only a finite number of times. This was disproved by Fevens et al. (2001). Their

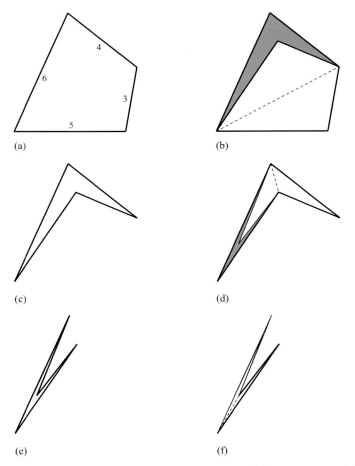

Figure 5.27. (a) Original polygon P_0, with lengths $3 + 6 = 4 + 5$; (b) deflation step; (c) P_1; (d) deflation step; (e) P_2; (f) deflation step.

family of counterexamples consists of any quadrilateral that satisfies two conditions on the lengths of its edges:

1.   The sum of opposite edge lengths are the same: $|ab| + |cd| = |bc| + |da|$.
2.   No two adjacent edges have the same length.

They prove that any quadrilateral in this class Q leads to an infinite sequence of deflations. Note that condition (1) cannot suffice on its own, because a square satisfies it but has no deflations. An example of a quadrilateral in their class is shown in Figure 5.27. Here the edge lengths are 5, 3, 4, 6, satisfying both conditions. The initial polygon P_0 (Figure 5.27(a)) is convex with two possible deflations. Selecting one (Figure 5.27(b)) yields a nonconvex P_1 (Figure 5.27(c)), which has only one possible deflation. All remaining polygons in the sequence are nonconvex. P_2 (Figure 5.27(d)) is thinner, and each folding over the diagonal deflates further. P_3 is already invisibly thin in Figure 5.27(f).

We now prove that any quadrilateral $P = (a, b, c, d)$ in class Q deflates forever. At any step, there must be at least one internal diagonal of P, say bd, and so condition (a) of a deflation holds. The deflation reflects a to a' (or equivalently, c to c') about the line L containing bd. The only way for this reflection to fail to produce a simple polygon

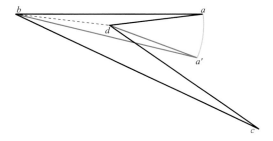

Figure 5.28. An attempted deflation produces a self-crossing polygon.

is if two opposite edges cross, as in Figure 5.28. (Of course two adjacent edges cannot cross.) If two opposite edges cross, then the convex hull of the crossing segments is a quadrilateral. It is a consequence of the triangle inequality that the sum of the lengths of the diagonals of a convex quadrilateral are strictly longer than the sum of the lengths of a pair of opposite sides. Thus, in the figure, we must have $|ab| + |cd| > |bc| + |da|$. But this violates condition (1) of the class Q. This inequality could only degenerate to an equality if one vertex reflected on top of another, for example, $a' = c$, turning the convex hull into a triangle. But in such a case, two adjacent edges of P must have the same length, violating condition (2) of Q. Therefore deflation is always possible for any quadrilateral in Q.

5.3.2.3 Variations on Flipping

Before turning to other ways to convexify a planar polygon in 3D, we mention several extensions of and variations on pocket flipping:

1. *Flips in nonsimple polygons.* Flips can be generalized to apply to nonsimple (i.e., possibly self-crossing) polygons, by flipping a subchain connecting two adjacent convex hull vertices across that hull edge. Ignoring self-crossing, the question is this: Can every nonsimple polygon can be convexified by a finite sequence of such flips? Grünbaum and Zaks (1998) proved that the answer is YES, Toussaint (2005) found a more easily computed sequence of flips, and Biedl and Demaine conjecture that every sequence of flips convexifies a nonsimple polygon.
2. *Pivots.* A rotation of a subchain about the line connecting its endpoints (not necessarily on the hull) is called a *pivot* in knot theory and in physics (Toussaint 2005). Following work by Millett (1994), Toussaint proved that any polygon can be convexified in 3D with a finite number of pivots, permitting self-crossing. Recall from Section 5.1.2 that convexification can be accomplished with $O(n)$ 5- or 4-joint line-tracking motions.
3. *Flipturns.* Motivated by their example needing an unbounded number of pocket flips (Figure 5.24), Joss and Shannon defined a new operation they could bound. A *flipturn* rotates by 180° a subchain bounding a pocket of the polygon, not in 3D about the pocket lid, but in 2D around the midpoint of the lid. Unlike flips, flipturns are not linkage motions because they reorder the edges. They proved that at most $(n-1)! - 1$ flipturns can be made before the polygon becomes convex, and conjectured that just $n^2/4$ flipturns should suffice. This was settled in Aichholzer et al. (2002) with a proof that the length of the longest flipturn sequence is at most $n^2/4 - O(1)$, improving on results in Ahn et al. (2000). Although there was some hope that the shortest flipturn sequence might have

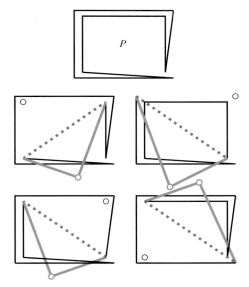

Figure 5.29. A polygon P such that every pop produces a self-crossing. Four of the 8 pops are illustrated.

a better bound, recently Biedl (2006) constructed polygons whose shortest sequence needs $(n-1)^2/8$ flipturns.

4. *Pops.* This notion was first defined by Millett (1994), named *mouth flip* by Toussaint (2005), and independently named *pops* by Ballinger and Thurston.[5] A *pop* is a pivot by 180° of one vertex v about the line connecting the endpoints of its two incident edges. The endpoints stay fixed while v rotates into 3D and back to the plane. Millet showed that "star-shaped" equilateral polygons can be convexified, and Toussaint improved his proof. Thurston gave an example of a simple polygon that becomes self-crossing with any pop (see Figure 5.29). Ballinger and Thurston proved that almost every (nonsimple) polygon can be convexified by pops (permitting self-crossing), but it remains open whether the "almost" in this claim can be dropped.

Open Problem 5.3: Pops. [a] Can every (nonsimple) polygon in the plane be convexified by a finite sequence of pops?

---

[a] B. Ballinger and W. Thurston, November 2001.

5.3.2.4 Arch Algorithms

Returning to the problem of convexifying a planar polygon in 3D without self-crossings, the unboundedness of the pocket-flipping algorithm leads to the question of convexifying within a polynomial number of moves. This was first accomplished by an algorithm inspired by the straightforward algorithm for lifting an open chain (Figure 5.21), and subsequently simplified. We now sketch just the main ideas of these two algorithms. Both algorithms lift the polygon, link by link, at all times maintaining a convex chain (or *arch*) lying in a plane orthogonal to the plane Π containing the polygon. The algorithms

---

[5] Personal communication from W. Thurston, November 2001.

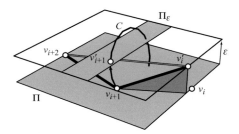

Figure 5.30. v_{i+1} rotates up the circle C until it hits Π_ε.

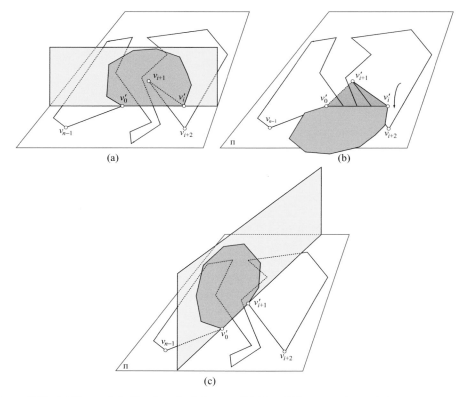

(a)

(b)

(c)

Figure 5.31. (a) The arch in Π_z after the i-th step; (b) After lifting v_{i+1}, the arch is rotated down to the plane Π_ε; (c) After reconvexification in Π_ε "absorbs" v_i', the arch can be safely rotated about its new base edge to a vertical plane again.

differ in the invariant on the form of the arch, how additional links are added to the arch, and how the invariant is restored.

Convex arch algorithm. The first algorithm (Biedl et al. 1999, 2001a) avoids crossings by using a plane Π_ε parallel to and slightly ($z = \varepsilon > 0$) above the plane Π of the polygon in which to convexify at each step. At a generic step, the polygon from v_0 to v_i has been pulled up into a convex arch A_i in a vertical plane Π_z, whose base $v_0' v_i'$ rests in Π_ε (see Figure 5.31(a)). Then v_{i+1} is rotated up to Π_ε, via a 2-link chain movement with v_i' fixed in Π_ε, and v_{i+2} fixed in Π (see Figure 5.30). Once v_{i+1} is in Π_ε, the arch is rotated down to lie wholly in Π_ε (see Figure 5.31(b)). The arch is then reconvexified within Π_ε, safe

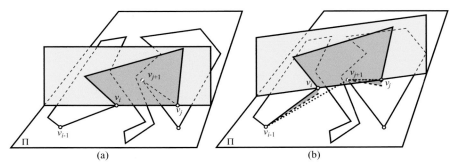

Figure 5.32. (a) The arch at a generic step; (b) v_i and v_j have been lifted slightly, producing a twisted trapezoid $v_i v_{i-1} v_{j+1} v_j$.

from interfering with the polygon in Π. Then A_{i+1} with base $v_0' v_{i+1}'$ is rotated up to Π_z ready for the next step, with convexity guaranteeing no penetration down to Π (see Figure 5.31(c)). Although the overall idea of this algorithm is simple, computation of a $\varepsilon > 0$ to ensure that each step can be accomplished without self-crossings is somewhat delicate. In any case, the end result is an algorithm that convexifies a planar polygon in $O(n)$ moves, each of which moves at most four joints at once.

Quadrilateral arch algorithm. An alternative lifting was presented in Aronov et al. (2002) that permits convexification in vertical planes, avoiding the use of Π_ε, substituting instead a nonplanar intermediate configuration. At a generic step of their algorithm, the arch A is a convex quadrilateral, with base $v_i v_j \in \Pi$. They lift two edges at once, $v_{i-1} v_i$ and $v_j v_{j+1}$, into the half-space above Π. During this lifting, $v_i v_j$ is kept parallel to Π and $v_{i-1} v_{j+1}$ never leaves Π. The result is, in general, a "twisted trapezoid" (Figure 5.32(b)), which then must be incorporated into the arch, made planar, and reduced to a quadrilateral. All this can be accomplished strictly above Π, again in $O(n)$ 4-joint moves.

We now sketch the critical lifting step, without, however, proving that it works. To ease notation, let $a = v_i$ and $a' = v_{i-1}$, and $b = v_j$ and $b' = v_{j+1}$. The vertices a' and b' stay fixed in Π, while the "virtual link" ab lifts parallel to Π. ab is "virtual" because it is not an edge of the polygon, but it is a link in that its length $r = |ab|$ is preserved throughout the motion. ($a'b'$ can also be viewed as a virtual link, for its length $r' = |a'b'|$ also does not change, because its endpoints are fixed.) Vertex a moves on a sphere S_a of radius $\alpha = |a'a|$ centered on a', and b moves on a sphere S_b of radius $\beta = |b'b|$ centered on b'. Assume without loss of generality that $\alpha \leq \beta$. Then the situation is as illustrated in Figure 5.33. The segment ab is lifted until it has reached maximum height above Π. This final position can take one of two forms:

1. $a'a$ becomes vertical and b lies on the circle that is the intersection of the plane $z = \alpha$ with S_b (dashed in Figure 5.33); or
2. ab lies in the vertical half plane V bounded by the line through $a'b'$.

The first termination possibility occurs whenever the virtual link ab is long enough. As the circle $\{z = \alpha\} \cap S_b$ (dashed in the figure) has radius $\sqrt{\beta^2 - \alpha^2}$, the condition is

$$r + \sqrt{\beta^2 - \alpha^2} \geq r'. \tag{5.4}$$

When $r = |ab|$ is not long enough to satisfy this equation, then the second termination condition holds, in which case (a, b, b', a') forms a plane trapezoid.

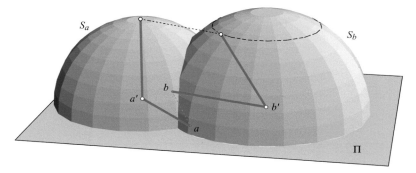

Figure 5.33. The centers of the spheres S_a and S_b are a' and b' respectively. The initial and final positions of $a'a$ and $b'b$ are shown red.

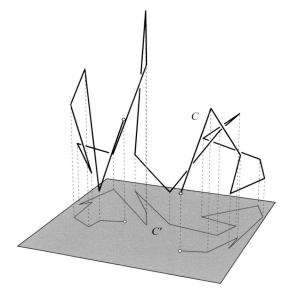

Figure 5.34. A 3D open chain that has a simple projection in the plane below.

Condition 1, in contrast, represents (in general) a nonplanar "twisted trapezoid." There are many paths for ab to follow on its way to this terminal state, and it requires a delicate argument, which we will not detail, to establish that there always is a lifting motion that is guaranteed to avoiding crossings.

5.3.3   3D: Simple Projection

We now move one more step away from planarity by considering an open chain C in 3D that projects on some plane as a non-self-intersecting chain C', as in Figure 5.34. In this circumstance, there is a simple algorithm that straightens C, working from one end of the chain (Biedl et al. 1999, 2001a). Let C' lie in the xy-plane. First, we claim that one can compute a radius r such that vertical cylinders (axes parallel to z) of that radius around each vertex of C' are (a) disjoint from one another and (b) intersected only by the two edges incident to the vertex of C on the cylinder axis. The radius r can easily be computed from the minimum distance between a vertex and edge of C'. Let

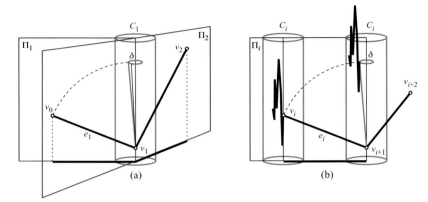

Figure 5.35. (a) Step 0: e_0 is rotated within Π_1 and then into Π_2; (b) Step i: the chain translates within Π_i.

C_i be the cylinder around v_i. The initial step of the algorithm rotates the first edge e_1 until its endpoint v_0 is within $\delta = r/n$ of the axis of C_1, and then twists to align with the next move (see Figure 5.35(a)). A generic step of the algorithm squeezes the previous links like an accordion into the next cylinder C_{i+1} (staying within the projected shadow C'), and then rotates within that cylinder. Thus, after Step i, all the links $e_1, \ldots, e_i$ are packed into C_{i+1} and ready for the next step. This can be done so that the chain remains monotone within the relevant vertical plane Π_i, as illustrated in (b) of the figure. It is then easy to unravel the final monotone chain by straightening one joint at a time, that is, rotating each joint angle to be straight, starting at either end of the chain. All this can be accomplished in $O(n)$ moves, each involving at most two joints at once, and computable in $O(n)$ time.

An easy generalization of this algorithm shows that any open chain embedded on the surface of a convex polyhedron can be straightened without self-crossing. However, extensions to closed chains with simple projections proved more difficult, and awaited the resolution of the 2D problem discussed in Section 6.6.

6 Locked Chains

6.1 INTRODUCTION

We know from Section 2.2.2 that configuration spaces of general linkages that are permitted to self-intersect, even in $\mathbb{R}^2$, can have exponentially many connected components. On the other hand, we know from Section 5.1.1.1 (p. 59) and Section 5.1.2 (p. 66) that configuration spaces of open and closed 3D chains that are permitted to self-intersect have just one connected component. We also know from Section 5.3 (p. 70) that configuration spaces of planar chains have just one connected component when permitted to move into $\mathbb{R}^3$ but forbidden to self-cross. We have until now avoided the most natural questions, which concern chains embedded in $\mathbb{R}^d$, with motion confined to the same space $\mathbb{R}^d$, without self-crossings. These questions avoid the generality of linkages on the one hand, and the special assumptions of planar embeddings or projections on the other hand.

The main question addressed in this context is which types of linkages always have connected configuration spaces. A linkage with a connected configuration space is *unlocked*: no two configurations are prevented from reaching each other. If a linkage in 3D or higher dimensions has a disconnected configuration space, it is *locked*. But for connectivity of the configuration space to be possible in 2D, we need to place an additional constraint, because planar closed chains cannot be turned "inside-out" as they could in Section 5.1.2 when we permitted the chain to self-intersect. A linkage is thus *locked* in 2D if there are two configurations in which every cycle is oriented the same (clockwise/counterclockwise) yet they cannot reach each other in configuration space. Otherwise, the linkage is *unlocked*. This condition corresponds to connectivity of each of finitely many disjoint subspaces of the configuration space, each of which makes a particular choice of the orientations of the cycles in the linkage. For example, a planar closed chain is unlocked if both of its orientations have connected subspaces of configuration space, and an open chain or tree linkage is unlocked simply if its configuration space is connected.

Because motions can be reversed and concatenated, deciding whether every linkage in some family is unlocked is equivalent to asking whether every linkage in that family can be reconfigured to some canonical form. Section 5.3 identified canonical forms of chains: our goal here is to determine whether every open chain can be straightened and whether every closed chain can be convexified. A chain is locked precisely if some configuration cannot be straightened or convexified without self-crossing. Another type of linkage of considerable study in this context are linkages whose graphs are trees. A canonical form for a tree is a *flat configuration*: all vertices lie on a horizontal line, and

Table 6.1: Summary of locked and unlocked results

Dimension	Can chains lock?	Can trees lock?
2	NO	YES
3	YES	YES
$d \geq 4$	NO	NO

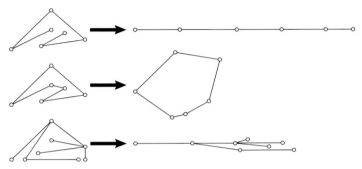

Figure 6.1. Three common linkages (open and closed chains, and trees) and their associated canonical configurations (straight, convex, and flat).

all links point "rightward" from a common root. It is easy to fold any flat configuration of a tree into any other. The three canonical configurations are shown in Figure 6.1.

The fundamental locked and unlocked questions—whether a chain or tree can lock in $\mathbb{R}^2$, $\mathbb{R}^3$, and $\mathbb{R}^4$ and beyond—have all been settled recently. The answers are summarized in Table 6.1. One immediate surprise is that the answers for open and closed chains are the same, despite significantly less freedom for closed chains. Not all the answers in the table are independent (e.g., if a chain can lock, then so can a tree), and not all are equally difficult. We will present the results in historical order of discovery, which roughly corresponds (not surprisingly) to increasing mathematical difficulty: 3D, 4D, and 2D, in Sections 6.3, 6.4, and 6.5–6.8 respectively. First we present an overview of the history of the topic.

6.2 HISTORY

The question of whether chains can lock in 2D was an open problem for more than 25 years before it was resolved. We sketch the history here; see Connelly et al. (2003) for more details. To our knowledge, the first person to pose the problem of convexifying closed chains in 2D was Stephen Schanuel, as a homework problem in algebraic topology. George Bergman learned of the problem from Schanuel in the early 1970s, and suggested the simpler question of straightening open chains. This line of interest resulted in the problems being included in Kirby's *Problems in Low-Dimensional Topology* (Kirby 1997, Prob. 5.18). Ulf Grenander's work in pattern theory led him, independently, to pose the same problems in a 1987 seminar.

In the discrete and computational geometry community, the problems were independently posed by William Lenhart and Sue Whitesides in 1991 and by Joseph Mitchell in 1992. The problems were published in Lenhart and Whitesides (1995).

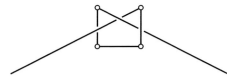

Figure 6.2. The "knitting needles" locked chain. The standard knot theory convention is followed to denote "over" and "under" relations.

In the mid-1990s, a growing group of people from this community were involved in trying to construct a 2D locked chain, or prove that every 2D chain is unlocked. Typically, someone in the group would distribute an example that she or he constructed or was given by a colleague. We would try various motions that did not work, and we would often try proving that the example was locked because it appeared so. For some examples, it took several months before we found an unlocking motion. This activity was given some focus in 1998 at a workshop organized by Anna Lubiw and Sue Whitesides.[1] Here the participants explored variations on the 2D problem, solving the problem for star-shaped polygons in 2D (Everett et al. 1998), developing the algorithm to convexify in 3D (Biedl et al. 1999) described in Section 5.3.2, and made several other advances. The most important for this accounting was settling three entries in Table 6.1: trees can lock in 2D (Biedl et al. 1998a), and both open and closed chains (and therefore trees) can lock in 3D (Biedl et al. 1999). This later result was independently established within the topology community by Cantarella and Johnston (1998) a year earlier.

With this progress reported in subsequent paper presentations (and more rapidly through the scientific "grapevine"), the locked/unlocked problems gained intense worldwide attention. The effort to find unlocking motions for seemingly locked examples continued. It was Günter Rote who noticed in June 1999 that all the unlocking motions found seem to "blow up" the chains in the sense that the distance between any two vertices never decreased—the motions were *expansive*. That fall, the key remaining entries of Table 6.1 were settled.

The second author and Roxana Cocan proved that neither open nor closed chains can lock in any dimension $d \geq 4$ (Cocan and O'Rourke 1999) (the same conclusion for trees came later; Cocan and O'Rourke 2001). Meanwhile, the more difficult 2D problem was yielding under a collaboration between the first author and Robert Connelly and Rote. Verifying Rote's hunch that expansive motions sufficed to straighten/convexify, they finally resolved the problem in early 2000: neither open nor closed chains can lock in 2D (Connelly et al. 2000a,b). Since then, alternative approaches with better algorithmic guarantees have been developed by Ileana Streinu (2000) and by Jason Cantarella, the first author, Hayley Iben, and James O'Brien (2004).

We now proceed to describe these results, again in roughly historical order.

6.3   LOCKED CHAINS IN 3D

6.3.1   Locked Open Chains

It is of course easy to lock a closed chain in 3D by tying it into a knot. But it is less immediate for open chains, or for closed but unknotted chains. The same example of a locked open chain of five links was found, described first by Cantarella and Johnston (1998), and later in Biedl et al. (1999) (see Figure 6.2). The latter group called this the

---

[1] The *International Workshop on Wrapping and Folding*, at the Bellairs Research Institute of McGill University, February 1998.

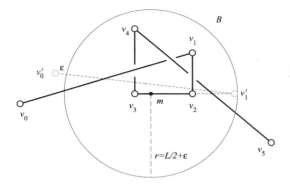

Figure 6.3. B is centered on the midpoint m along (v_1, v_2, v_3, v_4). A reconfiguration of v_1 to v_1' and v_0 to v_0' illustrates that v_1 remains interior and v_0 remains exterior to B.

"knitting needles" example, viewing the two long end links as needles and the three short middle links as a flexible cord connecting them. Although there is a sense in which it is obvious that the chain is locked, the claim still needs a proof, especially in light of the number of false starts made in the general area of locked and unlocked chains. We present a strengthening of the proof from Biedl et al. (1999), which will serve as a model for several later proofs (Chapter 7), and then remark on additional results obtained in Cantarella and Johnston (1998).

Theorem 6.3.1. *The chain $K = (v_0, \ldots, v_5)$ in Figure 6.3 cannot be straightened when the end links are longer than the lengths of the middle links combined.*

   Proof: Let $\ell_i = |v_{i-1} v_i|$ be the length of the ith link. Let $L = \ell_2 + \ell_3 + \ell_4$ be the total length of the short central links. We set ℓ_1 and ℓ_5 below. Let $r = L/2 + \varepsilon$, for small $\varepsilon > 0$, and center a ball B of radius r on the midpoint m of the three central links, i.e., m is $L/2$ from both v_1 and v_4 along the chain (see Figure 6.3). By construction, we have $\{v_1, v_2, v_3, v_4\} \subset B$ during any reconfiguration of the chain. Now choose ℓ_1 and ℓ_5 to be at least $2r + \varepsilon = L + 3\varepsilon$. Because this length is greater than the diameter of B, the vertices v_0 and v_5 both are necessarily exterior to B during any reconfiguration. Assume now that the chain K can be straightened by some motion. We will reach a contradiction. First we recall some terms from knot theory.[2] The *trivial knot* is an unknotted closed curve homeomorphic to a circle. An open polygonal chain is unknotted if, when closed by a last link, it becomes the trivial knot. The *trefoil knot* is the simplest knot, the only knot that may be drawn with three crossings.

   Because of the separation maintained between $\{v_0, v_5\}$ and $\{v_1, v_2, v_3, v_4\}$ by the boundary of B, we could have attached a sufficiently long unknotted string s from v_0 to v_5 exterior to B that would not have hindered the unfolding of K. But this would imply that $K \cup s$ is the trivial knot; but it is clearly a trefoil knot. We have reached a contradiction; therefore, K cannot be straightened. $\qquad\square$

   Later it will be useful to note that the ratio between the lengths of the longest and shortest links for which this theorem holds is

$$\frac{\ell_2 + \ell_3 + \ell_4 + 3\varepsilon}{\min\{\ell_2, \ell_3, \ell_4\}}.$$

---

[2] See, e.g., Livingston (1993) or Adams (1994).

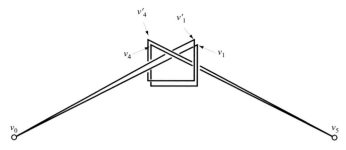

Figure 6.4. K^2 (K doubled): a locked but unknotted chain.

This ratio can be as small as $3 + \varepsilon'$, for any $\varepsilon' > 0$, achieved when all three central links have the same length. Cantarella and Johnston proved that, when the length condition of the theorem holds, the configuration space of the chain has three connected components: the straightened chain, and left- and right-handed versions of Figure 6.2. If the needles are not long enough, the space has just one component—the chain cannot lock. They also proved that all chains of fewer than five links are unlocked, regardless of link lengths.

Finally we should note that the trefoil knot is only the simplest knot on which to base a locked chain construction. Any knot that has a "stick realization" with an "exterior edge" e (on the convex hull; Adams et al. 1997) can be converted to a locked chain by deleting e and extending the incident edges.[3]

6.3.2   Locked Closed Chains

It was observed by Biedl et al. (1999) that a closed, unknotted chain can be locked by "doubling" K: adding vertices v'_i near v_i for $i = 1, 2, 3, 4$, and connecting the whole into a chain $K^2 = (v_0, \ldots, v_5, v'_4, \ldots, v'_1)$ (see Figure 6.4). Because $K \subset K^2$, the preceding argument applies when the second copy of K is ignored: any convexifying motion will have the property that v_0 and v_5 remain exterior to B, and $\{v_1, v_2, v_3, v_4\}$ remain interior to B throughout the motion. Thus the extra copy of K provides no additional freedom of motion to v_5 with respect to B. Consequently, we can argue as before: if K^2 is somehow convexified, this motion could be used to unknot $K \cup s$, where s is an unknotted chain exterior to B connecting v_0 to v_5. This is impossible, and therefore K^2 is locked.

Although it is easy to see that the example in Figure 6.4 is locked, it has the disadvantage of using 10 links. Cantarella and Johnston (1998) found a closed hexagon they proved is locked. Its topological structure is illustrated in Figure 6.5(a); lengths that lock it are shown in (c). Very roughly, the small loop in the chain encircling the long horizontal link cannot be freed. Their proof is similar to that in Theorem 6.3.1 in relying on knot theory. Toussaint (2001) subsequently found another class of locked hexagons, shown in (b) and (d) of the figure. He conjectures that "there are no more than five classes of nontrivial embeddings of hexagons in 3D."

Finally, it is not difficult to see that creating a closed chain as a "connected sum" of copies of, e.g., the configuration in Figure 6.5(a) leads to the conclusion that the number

---

[3]   Personal communication from Julie Glass, July 2004.

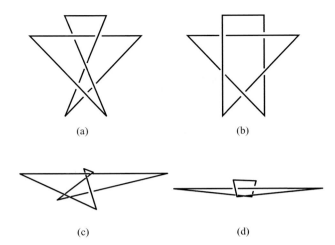

Figure 6.5. (a) Cantarella and Johnson locked hexagon; (b) Toussaint locked hexagon; (c) lengths from Cantarella and Johnston (1998) for (a); (d) lengths from Toussaint (2001) for (b). The latter two are perspective projections from 3D.

of connected components of the configuration space of a chain of n links can become arbitrarily large as $n \to \infty$ (Cantarella and Johnston 1998). We will return to this idea in Chapter 7, where we will consider the question of how many chain cuts are needed to untangle such a chain.

We will revisit the topic of locked chains in Chapter 9 (p. 148), posing some open problems arising in the context of protein folding.

6.3.3   Chains of Fat Links

We mentioned above that chains of five equal-length links cannot lock. It is unclear whether this remains true if the links are "fat" in some sense. One possible model for such links is as follows. Start with a chain of segment links. Let $s = ab$ be one segment of the chain with endpoints a and b. Define fat(s) to be the union of unit spheres centered at a and at b and a unit-radius cylinder with axis s and length the same as the length of s. A *chain of fat links* is then the union of fat(s) for all s in a chain of segments. Finally, we say that a chain of fat links is non-self-intersecting (or *simple*) if no two nonadjacent fat links intersect. Note that adjacent links are permitted to interpenetrate. The notion of locked remains the same as with segment chains. This leads to this problem:

> **Open Problem 6.1: Equilateral Fat 5-Chain.** Can an equilateral chain (all link lengths the same) of five "fat" links (in the above sense) be locked? If not, what is the smallest link length ratio that permits locking?

Establishing whether the configuration shown in Figure 6.6 is locked seems to require rather different techniques than employed for segment chains.

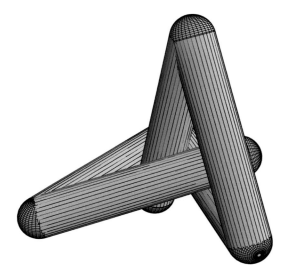

Figure 6.6. A "fat" chain of five links. Is it locked? [Figure created in collaboration with Sorina Chircu.]

6.4   NO LOCKED CHAINS IN 4D

In this section we discuss the results from Cocan and O'Rourke (1999, 2001), which can be summarized as follows:

Theorem 6.4.1. *Every open chain, closed chain, or tree of n vertices can be straightened/convexified (without self-crossing) in $\mathbb{R}^d$ for $d \geq 4$, with a number of moves polynomial in n, computable in polynomial time.*

Our aim here is to provide the intuition behind the proof techniques, rather than the details.

The proof (Theorem 6.3.1) that the knitting-needles chain is locked shows that there is a close connection in 3D between unknotted, locked chains and knots. However, it is well known that no 1D closed, "tame," non-self-intersecting curve C is knotted in $\mathbb{R}^4$ (e.g., Adams 1994, pp. 270–1). Because proofs of this theorem employ topological deformations, it seems they are not easily modified to help settle questions about chains in 4D. The rigidity of the links prevents any easy translation of the knot proof techniques to polygonal chains. However, it does suggest that it would be difficult to construct a locked chain by extending the methods of Section 6.3 used in 3D.

A second consideration lends support to this intuition. This is the inability to confine one segment in a "cage" composed of other segments in 4D. Consider the first link $e_1 = v_0 v_1$ of the knitting-needles chain in Figure 6.3. It is surrounded by other segments in the sense that it cannot be rotated freely about one endpoint (say v_1) without colliding with the other segments, taking them to be frozen in their initial positions. Let S be the 2-sphere in $\mathbb{R}^3$ of radius ℓ_1 centered at v_1. Each point on S is a possible location for v_0. Segment e_1 is confined in the sense that there are points of S that cannot be reached from e_1's initial position without collision with the other segments. This can be seen by centrally projecting the segments from v_1 onto S, producing an "obstruction diagram." It should be clear that v_0 is confined to a cell of this diagram. Although this by no means implies that the chain in Figure 6.3 is locked, it is at least part of the reason that the chain might be locked.

We will argue below that such confinement is not possible in 4D.

Open chains in 4D. This intuition leads to an algorithm that straightens open chains in 4D one link at a time: the first joint v_1 is straightened, "frozen," and the process repeated until the entire chain has been straightened. This is a procedure which, of course, could not be carried out in 3D because of the above confinement possibility. But there is much more room for maneuvering in 4D. In particular, the set of positions for v_0 with v_1 held fixed has the topology $\mathbb{S}^3$: it is a 3-sphere $S \subset \mathbb{R}^4$. It is proved in Cocan and O'Rourke (2001) that the *obstruction diagram* $\mathrm{Ob}(v_0)$ for v_0—the set of all positions for v_0 that cause intersection between $e_1 = v_0 v_1$ and the remainder of the fixed chain— is a collection of $O(n)$ arcs and points in S. Then the task of straightening the joint v_1 reduces to finding a path for v_0 through S that avoids the obstructions and reaches the particular spot (the "goal position") corresponding to a straightened angle at v_1. "Flying" through the 3D manifold S avoiding 1D obstructions is not difficult—it is this sense in which there can be no "cages" in 4D (whereas in 3D, the obstruction curves can confine v_1). One important detail is that the goal position itself might lie on the obstruction diagram. In this case, v_1 is moved slightly to break the degeneracy; an argument is needed to show this is always possible.

The sketched algorithm is inefficient, but a less-intuitive algorithm achieves straightening in $3n$ moves (straightening each joint in three moves, moving no more than three joints at once), which can be planned by an $O(n^2)$-time algorithm. This algorithm was implemented for chains in "general position." For a chain whose vertex coordinates are chosen randomly, the program straightens it with probability 1, for then the degenerate cases that cause the most theoretical difficulty are unlikely to occur. Figure 6.7 shows output for a chain whose $n = 100$ vertices were chosen randomly and uniformly in $[0, 1]^4$.

Closed chains in 4D. The algorithm for convexifying closed chains in 4D is more difficult. It employs the line-tracking motions used in Section 5.1.2 to straighten one joint at a time (see Figure 5.10, p. 66). These motions are applied repeatedly until a triangle is obtained (which is a planar convex polygon, the desired canonical configuration). Although the overall design of the algorithm is identical to that of

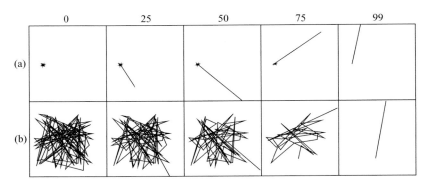

Figure 6.7. Snapshots of the algorithm straightening a chain of $n = 100$ vertices, initially (0), and after 25, 50, 75, and all 99 joints have been straightened (left to right). (a) Scale approx. 50:1; the entire chain is visible in each frame; (b) scale approx. 1:1; the straightened tail is "off-screen." (The apparent link length changes are an artifact of the orthographic projection of the 4D chain down to 2D.) [Figure 7 of Cocan and O'Rourke 2001, by permission, Elsevier.]

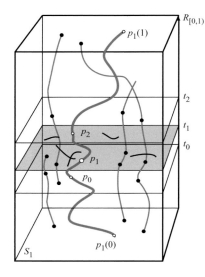

Figure 6.8. v_1 moves through its configuration space along path π_1 (red), connecting $p_1(0)$ to $p_1(1)$. Ob(v_1) includes curves over time $t \in [0, 1)$ and a collection of arcs at one special $t = t_1$. [Figure 15 of Cocan and O'Rourke 2001, by permission, Elsevier.]

Lenhart and Whitesides (1995) for convexifying polygons permitting self-crossing, avoiding self-crossings greatly complicates the task. We will say little about this algorithm except to connect it to the basic intuition for the open chains case.

Let $(v_0, v_1, v_2, v_3, v_4)$ be five consecutive vertices of a closed polygonal chain. Recall a line-tracking motion fixes v_0 and v_4 and moves v_2 along a line L, bending the elbow joints v_1 and v_3 to accommodate the movement of v_2. Let $\mathbb{R}_{[0,1)}$ be the interval $[0, 1)$ on the real line, open at 1. In $\mathbb{R}^3$, the configuration space for v_1 has the topology of $\mathbb{S}^1 \times \mathbb{R}_{[0,1)}$: $\mathbb{S}^1$ for the circle on which v_1 moves, and $\mathbb{R}_{[0,1)}$ for the line L on which v_2 rides. In $\mathbb{R}^4$, this space becomes $\mathbb{S}^2 \times \mathbb{R}_{[0,1)}$, for the elbow vertex v_1 can move on a 2-sphere. Defining the obstruction diagram Ob(v_1) in a manner similar to Ob(v_0) used for open chains, it can be shown that Ob(v_1) is a set of points and 1D curves in the configuration space. This permits v_1 to "fly through" the configuration space avoiding the obstacles, as depicted in Figure 6.8. At the same time that v_1 is moving through its configuration space, v_3, the other elbow joint, moves in a coordinated way through its space, both motions proceeding until some joint straightens. As all obstacles are piecewise algebraic, motion can be planned in polynomial time.

Finally we just mention that trees can be straightened in a manner similar to that used for open chains, and, not surprisingly, in dimension $d > 4$ there is even more freedom. Performing the straightening/convexification in appropriate 4-flats within $\mathbb{R}^d$ yields Theorem 6.4.1.

6.5   LOCKED TREES IN 2D

Motivated by the question about locked chains in 2D and by an application in computational origami (Demaine and Demaine 1997), a group of researchers (Biedl et al. 1998a, 2002) discovered a provably locked tree in 2D. One version of the tree is shown in Figure 6.9(a). It consists of one central high-degree vertex, connected to several three-bar *petals* whose endpoint wedges into the center. While Figure 6.9(a) has eight petals, the tree is locked for five or more petals.

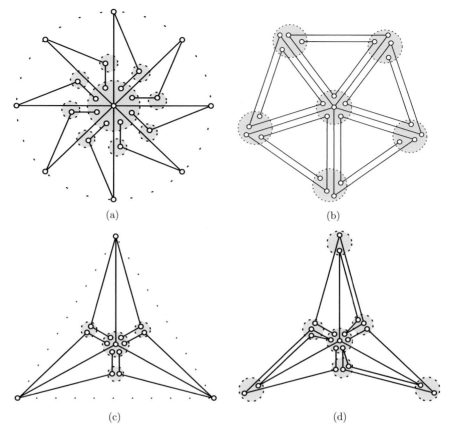

(a)   (b)

(c)   (d)

Figure 6.9. Locked planar polygonal trees. Points in shaded circles are closer than they appear. (a) Based on Figure 1 of Biedl et al. (1998a); (b) based on Figure 6 of Biedl et al. (1998a); (c) from Connelly et al. (2002a); (d) from Connelly et al. (2002a).

Intuitively, because of the cyclic structure of the petals, no petal can be angularly collapsed from its original state and no petal can be straightened out without additional angular room, so each petal can barely move at all. The key property is that each petal occupies an angle strictly less than $90°$, preventing it from straightening. By symmetry, this property requires that the tree must have at least five petals.

Despite it striking nearly everyone as intuitively clear that the tree in Figure 6.9(a) is locked, a proof as clear as the intuition has not been easy to find. The original proof that the tree is locked (Biedl et al. 1998a) is based on proving a cyclic dependence on critical events in any unlocking motion. Unfortunately, the precise formalization of this approach is tedious. A more recent approach based on rigidity theory (Connelly et al. 2002a) leads to a simpler proof, which we present in Section 6.8 once we have covered the necessary background.

Another, more intuitively locked, tree is shown in Figure 6.9(b). This tree has the additional property that the edges are nearly equilateral. Here the petals act as rigid triangles, pinned together at the central vertex. Again the tree is locked for five or more petals, and we will outline a proof in Section 6.8.

An interesting problem considered in Connelly et al. (2002a) is to what extent we need high-degree vertices to lock a tree. The trees in Figure 6.9(a–b) can use as little as one degree-5 vertex, which can be split into three slightly separated degree-3 vertices while keeping the same behavior (Biedl et al. 1998a). In contrast, Figure 6.9(d) shows a different tree with just a single degree-3 vertex, and all other vertices having degree 1 or 2. The underlying structure is another (locked) tree shown in Figure 6.9(c); this tree uses four degree-3 vertices to form six less-symmetric "petals." Then Figure 6.9(d) joins pairs of petals together into common chains to remove all degree-3 vertices except the center. The intuition is similar: no petal can be unfolded without penetrating an adjacent petal, and no petal can be collapsed. The proof that the tree is locked follows the same approach based on rigidity theory to be outlined in Section 6.8.

Recent work has moved further toward a characterization of which trees can lock. Kusakari et al. (2002) establish that no *x-monotone* tree can lock. Here each path from the root is a chain monotone with respect to the *x*-axis.[4] On the other hand, Kusakari (2003) shows a natural generalization to fail: it is possible to lock *radially monotone trees*, those such that points on each path from the root monotonically increase in distance from the root. Poon (2006) shows there are trees of diameter[5] 4 that are locked, but every *unit tree* of diameter 4, one whose link lengths are all 1, can be straightened. He also makes the bold conjecture that no unit tree can lock, in 2D or even 3D. We will return to such "equilateral" or "unit" chains and trees in Section 9.1.

6.6   NO LOCKED CHAINS IN 2D

Before it was resolved that there are no locked chains in 2D, various researchers posed several examples of "possibly locked" chains. To gain some intuition for the problem, let us consider one such example (independently posed by several people). The idea is to *double* each edge of a locked tree, splitting each degree-*k* vertex into *k* degree-2 vertices, and produce a polygon that can be made arbitrarily close to the original tree. Intuitively, because the tree is locked, these slightly modified polygons should also be locked. But this intuition is wrong! In contrast, this approach worked when we doubled the "knitting needles" locked 3D chain in Section 6.3. The key difference for trees is that the vertices of degree 3 or more behave very differently once split: the pieces can now separate. This new type of motion permits eventual convexification of the polygon. An example is shown in Figure 6.10. Here we see one tricky aspect of unfolding 2D chains: to make significant progress, the motion requires moving all joints at once.

Expansiveness. The rest of this subsection describes the main ideas behind the proof in Connelly et al. (2000b, 2003) that no polygonal chains are locked in the plane. As mentioned in Section 6.1, the unlocking motions are *expansive* in the sense that the distance between every pair of vertices never decreases. In fact, the motions are *strictly expansive* in the sense that the distance strictly increases between every pair of vertices except in the obvious impossible situations: the distance must remain constant between two vertices connected by a bar or a straight subchain of bars.

---

[4] A curve is *monotone with respect to a line L* if it meets every line orthogonal to *L* in at most one point, i.e., the curve projects orthogonally to *L* without overlapping itself.

[5] The *diameter* of a tree is the maximum number of edges along any node-to-node path. The tree in Figure 6.9 has diameter 6.

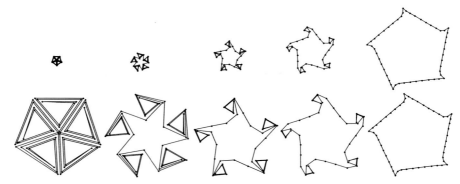

Figure 6.10. Two views of convexifying a polygon that comes from doubling each edge in the locked tree from Figure 6.9(b). The top snapshots are all scaled the same, and the bottom snapshots are scaled differently to improve visibility. [Figure 1 of Connelly et al. 2003, by permission, Springer.]

Multiple chains. More generally, the theorem applies to a disjoint collection of chains in the plane. In this context, however, there are some additional restrictions. Specifically, if one chain is nested in another closed chain, then in general we cannot expect the enclosed chain to straighten or convexify if there is insufficient room (see Figure 6.11). In the setting where self-crossings are forbidden, we are led to the following problem:

Open Problem 6.2: Unlocking Nested Chains. What is the complexity of deciding whether an open chain can be straightened while avoiding self-intersection when confined within a convex closed chain?

There are examples where the enclosing polygon must reshape itself several times to accommodate the unraveling of the confined open chain; so this problem has a rather different flavor from straightening within a rigid convex polygon. Moreover, as Section 5.2 (p. 67ff.) indicates, this problem is intricate even when the chain is permitted to self-cross.

The theorem of Connelly et al. (2003), on the other hand, guarantees that the *outermost* enclosing chains convexify without intersecting the enclosed chains, as in Figure 6.11. But once an enclosing chain becomes convex, the expansiveness property forces the enclosing chain and any enclosed chains to move rigidly in unison. Thus, from that moment on, the enclosed chains simply "come along for the ride." In particular, "strict expansiveness" permits (and indeed requires) the distance to remain constant between two vertices on the boundary of or interior to a common convex closed chain.

Theorem. To summarize the above discussion and add additional detail, the general theorem is as follows:

Theorem 6.6.1 (Connelly et al. 2003). *Every collection of disjoint polygonal chains in the plane has a strictly expansive motion to a configuration in which every outermost*

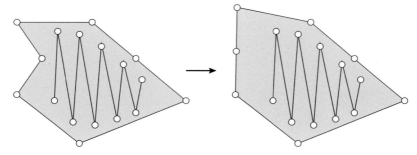

Figure 6.11. The nested open chain cannot be straightened inside the closed chain because of the lack of space. [Figure 2 of Connelly et al. 2003, by permission, Springer.]

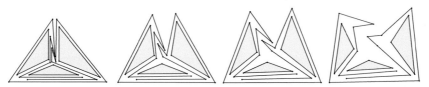

Figure 6.12. Four frames in the unlocking of four polygonal chains. The initial disjoint collection is shown in the leftmost frame.

Figure 6.13. Three frames toward the end of the unlocking motion, continuing Figure 6.12 at a different scale. [Figures 1 and 2 of O'Rourke 2000a.]

chain is either straight or convex. Furthermore, the motion is piecewise-differentiable (piecewise-C^1) and preserves any symmetry already present in the entire configuration, and any expansive motion increases the area of every closed chain.

Figures 6.12 and 6.13 illustrate the solution so obtained for a collection of three closed chains (triangles) entangled with one open chain.

We now turn to the proof of this theorem, concentrating on the case of a single chain. Multiple chains complicate the proof, especially if the motion must be expansive throughout, but most of the ideas remain the same. Here we just mention these differences and hint at how they are overcome.

6.6.1   Expansiveness and the Connection to Tensegrities

Expansiveness turns out to relate the problem of unlocking chains to the notion of tensegrities in rigidity theory described in Section 4.5 (p. 53).

The first useful property of expansiveness is that any expansive motion cannot cause crossings, because for any vertex to cross a bar, that vertex must first approach the bar and get closer to one of its endpoints. This property allows us to focus on the

$\Theta(n^2)$ expansiveness constraints—the distance between any pair of vertices cannot decrease—and ignore the noncrossing constraints. In some sense, this focus makes the problem more combinatorial: crossings might happen between a vertex and any point along a bar, which is inherently continuous (there are infinitely many such pairs of points), while expansiveness is just a requirement between pairs of vertices. Indeed, if a motion is expansive between all pairs of vertices, then it is also expansive between any two points along the chain. (This fact is Lemma 1 of Connelly et al. 2003.)

We can view a linkage that permits only expansive motions (and hence avoid crossings) as having *struts* between every pair of vertices not already connected by a bar. This augmentation turns the problem into a tensegrity flexibility problem: is there a motion that preserves the length of each bar and does not decrease the length of each strut? It is here that we will draw on the tools described in Section 4.5. We will argue that when the bars of a tensegrity form a collection of disjoint chains, and some of the outermost chains are not straight or convex, the tensegrity must be flexible.

There are two main pieces to this argument. First, we show that every such tensegrity is *infinitesimally* flexible. This part of the proof is the most combinatorial and most interesting, and it is here we will spend most of our time. As described in Section 4.4 (p. 49), infinitesimal flexibility is necessary for flexibility—in some sense specifying how to "get started" moving—but insufficient to guarantee flexibility. However, an infinitesimal motion can be obtained for every configuration of the linkage.

These infinitesimal motions define a vector field on the configuration space, and the second part of the proof is to show that flowing along this vector field causes the infinitesimal motions to be pieced together into an entire motion. This aspect of the proof is more continuous, and relies mainly on establishing the applicability of theorems from differential equations and convex programming. We will omit most of these details and simply describe how the pieces fit together.

6.6.2 Combinatorial Argument

Initial simplifications. As mentioned above, we add a strut between every pair of vertices not already connected by a bar. But some of these struts cannot be strictly expanded, for example, a strut that is completely covered by two or more collinear bars. To prevent this complication, we can conceptually "fuse" any vertex of the chain with angle 180°, and join the two incident bars into one (see Figure 6.14(a–b)). The result is a linkage with fewer vertices, and in which no strut is completely covered by bars.

For multiple chains, there are additional struts that cannot be strictly increased in length, those that connect two vertices on the boundary of or interior to a common convex closed chain. In this case, we would fuse any chains to any enclosing convex closed chain, allowing us to remove enclosed chains (see Figure 6.14(c–d)). Then we would remove the complete graph of struts between vertices on a common convex closed chain, and replace them with a planar triangulation of bars which rigidifies the convex closed chain. These additional bars cause some complication later on.

Duality. The first tensegrity tool we apply is the duality described in Section 4.5.4 (p. 56). Theorem 4.5.2 tells us that to prove the tensegrity is infinitesimally flexible, it is enough to show that the only equilibrium stress is everywhere-zero. Furthermore, Theorem 4.5.1 tells us that it is even enough to show that every equilibrium stress is zero on all the struts.

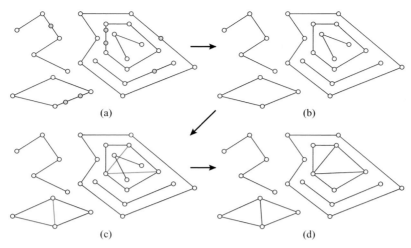

Figure 6.14. The first three modifications to the collection of chains. (a) Original collection; (b) with straight vertices removed; (c) with convex cycles rigidified; (d) with components nested within convex cycles removed.

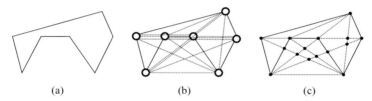

Figure 6.15. Planarizing the tensegrity of the chain's bars and all other struts. [Based on Figure 6 of Connelly et al. 2003.]

For the case of a single chain, we will show that the equilibrium stress must be everywhere-zero unless the chain is already straight or convex, in which case we are already finished. For multiple chains, the already straight and convex chains and their contents can have some stress on their bars, but still we would show that the struts have no stress on all the struts, unless every outermost chain is straight or convex.

Planarization. The next tensegrity tool we would like to apply is the Maxwell–Cremona theorem, Theorem 4.6.1 (p. 58), but for it to apply we need a planar tensegrity. To make our tensegrity planar, we apply the obvious transformation shown in Figure 6.15: wherever two edges (bars or struts) cross, introduce a vertex if there is not one already; and wherever an edge crosses over a vertex, subdivide the edge. Now there may be several copies of any one edge, in which case we remove all copies except one, leaving the most restrictive type of edge: a bar if there was one, and a strut otherwise.

A worry here is that we have introduced new vertices of articulation and thus potentially new degrees of freedom. But it turns out that the two frameworks are equivalent in terms of equilibrium stresses, which is all that we need:

Lemma 6.6.2. *If the original tensegrity (from the bars of the linkage and all other struts) has an equilibrium stress that is not everywhere-zero, then so does the planar tensegrity.*

Proof: We made two types of modifications to the tensegrity to make it planar: subdividing an edge at a new or old vertex, and removing duplicate edges. We show how to make corresponding changes to a supposed equilibrium stress.

First, when we subdivide an edge, we simply apply an equal force to the two sub-divided edges. Because we have defined in Section 4.5.3 (p. 54) the stress $\omega_{\{i,j\}}$ as a coefficient multiplier for the vector $\mathbf{p}_i - \mathbf{p}_j$, the new stresses for the subdivided edges $\{i, x\}$ and $\{x, j\}$ need to be rescaled appropriately for the new edge lengths:

$$\omega_{i,x} = \omega_{i,j} \frac{|\mathbf{p}_i - \mathbf{p}_j|}{|\mathbf{p}_i - \mathbf{p}_x|},$$

$$\omega_{x,j} = \omega_{i,j} \frac{|\mathbf{p}_i - \mathbf{p}_j|}{|\mathbf{p}_x - \mathbf{p}_j|}.$$

The result is an equilibrium stress on the subdivided tensegrity, because the original endpoints of the edge feel the same force as before, and the subdivision vertex feels equal opposing forces from the two subdivided edges which cancel out.

Second, when we coalesce duplicate edges into a single edge, we simply add up all the stresses. The resulting stress can be positive only if one of the original edges was a bar, and hence when the coalesced edge is a bar. Thus, we maintain the property that struts carry only negative stress. And the endpoints feel the same force as before, so the stress remains in equilibrium.

The only potential difficulty is that the stresses might have canceled out to become everywhere-zero even though the original stress was not everywhere-zero. But from our initial simplifications (Figure 6.14), we know that no strut is entirely covered by bars. Consequently, the portion of each strut that is not covered by bars cannot have its stress canceled out by summation. Thus, if one strut carried a nonzero stress, then some subdivided portion of it will continue to carry a nonzero stress. On the other hand, if no strut carried nonzero stress, then all the stress is on the bars, but no degree-1 or degree-2 vertex of bars (whose angle is not 180°) can satisfy equilibrium with nonzero incident stress. ☐

A consequence of this lemma is that to prove the only equilibrium stress of the original tensegrity is everywhere-zero, it is enough to prove the only equilibrium stress of the planar tensegrity is everywhere-zero.

This lemma also holds for multiple chains, but more generally not only is the planar stress not everywhere-zero, but it is also nonzero on some strut exterior to all convex closed chains. The idea is to argue that every strut has a portion exterior to all convex closed chains that is uncovered by bars. This stronger property is necessary later on, because it permits stress on struts interior to convex closed chains.

Maxwell–Cremona lift. Now that we have a planar tensegrity, we can apply the Maxwell–Cremona theorem (Theorem 4.6.1, p. 58). In particular, this theorem says that our goal of the only equilibrium stress being everywhere-zero is equivalent to the only polyhedral lifting being everywhere-flat. To prove this goal, we consider an arbitrary polyhedral lifting, and argue that the only possibility is everywhere-flat unless the chain is already straight or convex. This final argument is the heart of the (combinatorial half of) the proof.

Intuition. The intuition behind the final argument is as follows. Because struts can carry only negative stress, they must (by Theorem 4.6.1) lift to valleys (or flat edges). Thus, any mountains in the polyhedral lifting must project to bars in the tensegrity. But there are few bars, the subdivided bars from the original chain and at most two bars incident to a common vertex. In the lifting, each vertex has at most two incident mountains and the remainder valleys. Such a lifting can exist locally at a vertex, but if you think about the possible configurations with just one or two mountains, at least one of the mountains must point upward, ascending in z-coordinate. If we follow this path that ascends in z-coordinate, we reach another vertex which again has at most two bars, so as before there must be a second bar with ascending z-coordinate. We can continue in this way, visiting new vertices and always ascending in z-coordinate, seemingly forever. The only way this process could properly terminate is by visiting some vertices in a cycle, in which case the z-coordinate cannot increase, but must stay the same throughout. In this case, some more thinking reveals that the cycle of bars must always turn in the same direction (left or right), meaning that the cycle of bars is in fact a convex closed chain.

Maximum z. To formalize and simplify the argument above, it turns out to be enough to look at the top of the polyhedral lifting, all the points achieving maximum z-coordinate. Here it is enough to argue locally at each vertex of this maximum surface. From such a vertex at the maximum, any outgoing path must be locally descending or level.

Consider first the special case of a vertex at a local peak, surrounded by a descending surface. This case is in some sense the generic case we would expect, is particularly simple, and representative of the general case. Imagine slicing the polyhedral lifting along a plane slightly below the peak vertex and parallel to the base $z = 0$ plane containing the tensegrity. Because the vertex is a peak, the slice results in an entire small, planar polygon around the vertex (see Figure 6.16). Convex vertices of this polygon correspond to mountains in the lifting, and reflex vertices correspond to valleys. Because the peak was incident to at most two mountains, the polygon has at most two convex vertices. But this apparent property contradicts the following basic geometric fact:

Lemma 6.6.3. *A planar simple polygon has at least three convex vertices.*

Proof: Measure the *signed turn angle* at each vertex of the polygon, that is, the angle of deviation from being straight, signed positive for left turns and negative for right

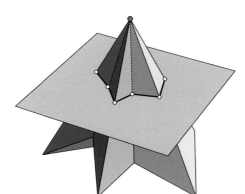

Figure 6.16. Slicing below a vertex local peak.

turns. Orienting the polygon counterclockwise, a convex vertex corresponds to a left turn, with positive turn angle, and the total turn angle in any polygon is 360°. Any turn has absolute value less than 180°: a full turn of 180° (or more) would create self-intersection. To add up to 360°, we must have at least three positive numbers less than 180°, so there must be at least three convex vertices. □

The conclusion is that this special case of a peak vertex cannot arise. Of course, some other cases can arise, and they are characterized by the following general lemma:

Lemma 6.6.4. *A vertex at maximum z-coordinate in the polyhedral lifting is either entirely flat or must have two incident mountains. In the latter case, in projection, if the two mountain ridges are not collinear, then the reflex wedge between them must be locally at maximum z as well.*

Proof: By definition, the reflex wedge locally contains no mountains. Intuitively, valleys are not enough to "reach around" the reflex angle, and the only possibility is for it to be entirely flat.

Imagine slicing the polyhedral lifting along a plane slightly below the vertex under consideration and parallel to the $z = 0$ plane. If the desired lemma were false, the slice results in a small, planar chain around the reflex wedge of the vertex, connecting the two bounding mountains in projection. (If there is only one mountain, we view it as a double boundary of a 360° reflex wedge.) Again, convex vertices of this chain correspond to mountains in the lifting, and reflex vertices correspond to valleys. Because there are no mountains in the reflex wedge, there are no convex vertices along the chain. But the chain cannot possibly reach around the reflex angle with only reflex and straight angles. Thus the lemma must be true. □

This lemma is the most we can say locally at each vertex. The remainder of the proof is just a bit of global analysis. The key property is that each edge of the maximum z surface must be a mountain, and hence projects to a bar. Thus, if we follow along the edges of the maximum z surface, the lemma tells us that the reflex side is locally at maximum z as well. For the vertices to not be entirely flat, the chain of edges must always turn in the same direction. The lemma tells us that the chain cannot end with a degree-1 vertex, so it must cycle around to form a convex closed chain. Otherwise, the entire maximum z surface is flat, encompassing the entire polyhedral lifting.

Now that we know the only polyhedral lifting is everywhere-flat, we conclude from Theorem 4.6.1 (p. 58) that the only equilibrium stress of the planar tensegrity is everywhere-zero, then we conclude from Lemma 6.6.2 (p. 100) that the only equilibrium stress of the original tensegrity is everywhere-zero, and finally we conclude from Theorem 4.5.2 that the original tensegrity is infinitesimally flexible. Therefore, any nonstraight open chain or nonconvex closed chain has an infinitesimal expansive motion.

For multiple components, we only need the weaker result that every edge exterior to all convex cycles lifts to a flat edge. From a similarly short global analysis, this result follows a generalization of Lemma 6.6.4. The generalization is only slightly more complicated. The difference is that convex cycles are triangulated with bars, so some vertices may have degree more than 2, but those vertices have all their bars enclosed in a convex wedge. Hence, except for two collinear bars (which might arise from subdivision), there is still a reflex wedge containing no bars, and the same proof of Lemma 6.6.4 applies.

6.6.3   Continuous Argument

The basic overview of the continuous argument is as follows.[6]

First, we define an ordinary differential equation on configurations, where the derivative is with respect to time. The differential equation at each configuration is given by a particular infinitesimal motion. There can be (infinitely) many different infinitesimal motions from any given configuration, which increase various struts at different relative rates. So to define the differential equation uniquely, we define a convex *objective function*, and at each configuration choose the infinitesimal motion that minimizes that objective function. This objective function turns the problem into a convex optimization with linear constraints (see Section 4.5.2, p. 54). By "stability" results in optimization theory, it can be proved that the resulting differential equation (or more precisely, its right-hand side) is differentiable.

By a general result in differential-equation theory, differentiability is enough to guarantee the existence of a solution from any initial configuration. The solution either moves the linkage forever (infinite time) or it reaches a limit point at which the differential equation is no longer defined. Infinitesimal motions fail to exist, and hence the differential equation fails to be defined, in three cases, which form the *boundary* of the differential equation:

1. When the chain is already straight or convex. In this case, we are finished.
2. When a vertex straightens out to an angle of 180°. In this case, as described previously, we fuse the vertex and reduce the linkage. This situation can arise only n times for a chain with n joints.
3. When the chain is self-touching or self-crossing. Because any motion (solution) that arises from the differential equation is expansive, we cannot reach such a configuration.

To summarize, as long as we reach the boundary of the differential equation after a finite amount of time, we are guaranteed to make discrete progress and ultimately straighten or convexify the chain. The one worry that remains is that the solution to the differential equation may never reach a boundary point, instead moving the linkage for an infinite amount of time. But this problem is easy to avoid by *rescaling* the differential equation so that every strut increases by at least some fixed rate, say 1. Each strut can expand in length to at most the diameter of the straight or convex configuration, so after finite time it must stop growing, and the motion terminates, meaning that we must hit the boundary of the differential equation.

Another general result of differential equation theory says that if the differential equation is continuously differentiable (C^1), then so are the solutions. We may piece together as many as n solutions to differential equations, so the total motion is only piecewise-C^1.

For multiple components, the proof gets significantly more complicated. The new issue that arises is that the various chains might fly apart from each other faster than they individually straighten and convexify. Our rescaling to increase every strut length at a rate of 1 may cause the chains to fly away to infinity at a growing speed without bound, so that the chains in some sense "reach infinity" before any of them straighten

---

[6]   See Connelly et al. (2002b, 2003) for full details.

or convexify. To argue that this situation cannot arise, we need to prove that the speed at which the chains fly to infinity is bounded from above by a fixed amount which depends on the initial configuration of the linkage. The bound may be very large; it only needs to be finite to prove that the chains straighten and convexify before flying away. Proving such a bound, though, is tedious.

On the other hand, there is an easier way to handle multiple components if we do not need to establish Theorem 6.6.1 exactly, and are instead satisfied with a motion that avoids self-intersection but may not be expansive. The idea is that once the components are separated sufficiently far (possibly on their way to infinity), they are effectively isolated from each other, so a motion of one by itself cannot interfere with (cross) the motion of another. Thus, after separation, we can apply the one-component result to each component in isolation and still avoid self-intersection.

This concludes our description of the proof of Theorem 6.6.1.

6.7   ALGORITHMS FOR UNLOCKING 2D CHAINS

Now that we know why there are no locked chains in 2D, we of course also want to find efficient algorithms for unlocking such chains. The proof described in the previous section leads to one approach, based on convex optimization and differential equations, which we describe below. Two other approaches have been developed since that achieve better running times, based on pseudotriangulations and energy, respectively. It is these more efficient approaches that we highlight.

6.7.1   Unlocking 2D Chains Using Convex Programming

The motion described in Section 6.6.3 (p. 104) is given by the solution to an ordinary differential equation, or more precisely an initial-value problem,

$$\mathbf{p}'(t) = \mathbf{v}(t), \quad \mathbf{p}(0) = \mathbf{p},$$

where $\mathbf{p}$ is the initial configuration, and the desired derivative $\mathbf{v}$ is the solution to a convex optimization problem with linear constraints. Such a description is far from "closed form," but because each of the components is fairly well-understood algorithmically, we can obtain an approximation to the true motion in finite time.

Convex optimization. First, for any configuration, we can solve the convex optimization problem in polynomial time. More precisely, we can find an approximate solution within any desired error tolerance ε_1 in time depending on $1/\varepsilon_1$. The classic approach to solving such a problem is the *ellipsoid method* (see, e.g., Grötschel et al. 1993). Its running time is $O(n^2/\varepsilon_1)$. Recently, Bertsimas and Vempala (2004) developed a simpler and more efficient algorithm, whose running time is $O(n/\varepsilon_1)$ with high probability.

Ordinary differential equation. Next, Connelly et al. (2003, Lem. 8) guarantees that the solution $\mathbf{v}$ to this optimization problem varies over all configurations in a continuously differentiable way. If we consider the motion until just ε_2 shy of the end of time (when $\mathbf{v}$ is undefined and its limit may be infinite), then $|\mathbf{v}|$ must be bounded from above by some bound D dependent only on $1/\varepsilon_2$ and the initial configuration of the linkage, **p**. Also, Connelly et al. (2003, Lem. 10) guarantees that we need to follow the motion for only a finite amount of time before completion, and that we can upper-bound this time by a number T that depends only on the combinatorial structure of the linkage and on the initial configuration **p** (the smallest separation, longest edge, etc.). These results establish in particular that the desired first derivative in the ordinary differential equation is "globally Lipschitz-continuous"[7] over the region of interest.

Forward Euler algorithm. Now we apply a standard approximation to solving ordinary differential equations. The *forward Euler* method repeatedly computes the desired derivative at a particular step and advances along that derivative for a short time h:

$$\hat{\mathbf{p}}(t + h) = \hat{\mathbf{p}}(t) + h\mathbf{v}(t); \quad \hat{\mathbf{p}}(0) = \hat{\mathbf{p}},$$

where $\hat{\mathbf{p}}(t)$ denotes the estimated configuration at time t. We thus need to make at most $\lceil T/h \rceil$ steps before completion.

Global error bound. By choosing h small enough, we obtain arbitrarily good approximations to the true motion. To state the error bound precisely, let L denote a bound on the Lipschitz constant on **v** over the time range of interest. We suppose that each iteration uses approximate arithmetic which may introduce an additive error of $Kh = \varepsilon_1$. (Thus, as h gets smaller, we need to increase the arithmetic precision.) Then we have the following bound on final absolute error E (Lambert 1992, p. 60):

$$E \le |\hat{\mathbf{p}}(T) - \mathbf{p}(T)| \le h\frac{1}{2}(D + K)Te^{LT}.$$

Because the terms after the h are independent of h, this bound approaches 0 as h approaches 0. If we want the error E to be at most ε_3, then according to this bound it suffices to choose

$$h \le \varepsilon_3 \Big/ \frac{1}{2}(D + K)Te^{LT},$$

which is large but nonetheless finite.

Putting the pieces together. The running time of this algorithm is $O(n/\varepsilon_1) = O(n/h)$ for each step, and hence $O(Tn/h^2)$ in total, where h is bounded by the previous equation. The resulting motion brings us to within ε_2 of the final time with an absolute accuracy of at most ε_3, where ε_2 influences D, and both D and ε_3 influence the bound on h.

Summary. In theory, forward Euler gives a finite algorithm for computing arbitrarily many discrete time snapshots of an arbitrarily close approximation to the motion satisfying Theorem 6.6.1 (p. 97). The bound is far from useful, but in practice we can use larger h at the sacrifice of losing an error guarantee but still usually producing a reasonable motion. This practical approach was used to produce the animations in Figures 6.12 and 6.13.

---

[7] A function f is *Lipschitz-continuous* if there is a constant $c > 0$ such that $|f(x) - f(y)| \le c|x - y|$ for all x and y.

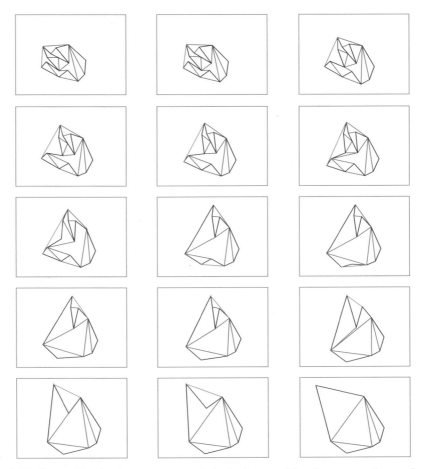

Figure 6.17. Convexification by a sequence of single-degree-of-freedom expansive mechanisms. [Figure courtesy of Ileana Streinu and Elif Tosun.]

6.7.2   Unlocking 2D Chains Using Pseudotriangulations

Streinu (2000, 2005) developed an algorithm for unlocking chains by a sequence of motions, each given by a linkage with additional bars that restrict motions to a single degree of freedom. Such single-degree-of-freedom linkages have two advantages. First, they are physically easy to implement: build a structure with the bars, and apply force to a vertex to push it along its single path of motion in the correct direction. This property makes the motions particularly natural. Second, such motions are *algebraic* (defined by polynomial equalities), so they are computationally more efficient than, for example, the differential equations from the previous section; algebraic motions can be computed up to a desired error tolerance ε in exponential time and polynomial space. These algebraic motions, however, do not capture the "canonical" motions described in the previous section, which for example preserve symmetries in the linkage.

See Figure 6.17 for an example of a motion arising from this approach.[8]

---

[8]   See also http://cs.smith.edu/~streinu/Research/Motion/motion.html.

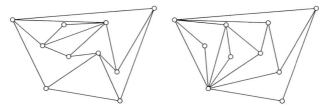

Figure 6.18. Two pointed pseudotriangulations of the same set of 10 points.

Pseudotriangulations. Streinu's approach is based on the notion of "pointed pseudo-triangulations." A *pseudotriangle* is a planar polygon with exactly three convex vertices. Each pair of convex vertices is connected by a reflex chain, which may be just one segment. In particular, a triangle is a pseudotriangle. A *pseudotriangulation* of a set of points in the plane is a collection of edges between the points that form a partition of the convex hull into pseudotriangles. Under the name of "geodesic triangulations," these structures found use in ray shooting (Chazelle et al. 1994), and then for visibility algorithms by Pocchiola and Vegter (1995), who named them (after a dual relationship to pseudoline arrangements) and studied many of their properties (Pocchiola and Vegter 1996). Recently the identification of *minimum pseudotriangulations* by Streinu (2000) has generated a flurry of new applications.

Minimum pseudotriangulations have the fewest possible number of edges for a given set of points. The term *pointed pseudotriangulation* is gaining prominence for the same notion because of the following property[9] (see Figure 6.18):

Lemma 6.7.1. *A pseudotriangulation of a point set is minimum precisely if every point p in the set is* pointed *in the sense that its incident edges span an angle less than π (i.e., they fall within a cone with apex at p and aperture angle less than π).*

Pointed pseudotriangulations on n points have remarkable regularity: they have exactly $n-2$ pseudotriangles. This property is shared by triangulations of the interiors of simple polygons, but not by usual triangulations of point sets which may have triangles counts from $n-2$ to $3(n-2)$. Another indication of their well-behavedness is the recent result of Kettner et al. (2003) that every set of n points in general position has a pointed pseudotriangulation with maximum vertex degree 5, a property not shared by triangulations of polygons or point sets.

Unlocking chains. In the context of unlocking chains,[10] we construct a *pseudotriangulation of the chains*, meaning a pseudotriangulation of the point set given by the joints of the chains, with the property that the pseudotriangulation has edges that match the bars of the chains. Such pseudotriangulations always exist: the chains themselves have the pointed property, and we can greedily add edges that maintain the pointed property, and in the end we have a pseudotriangulation by Lemma 6.7.1.

---

[9] Appears as Streinu (2000, Thm. 3.1(2)) (where the property is called "acyclicity") and as Kettner et al. (2003, Lem. 1).

[10] Although Streinu (2000) concentrates on unlocking a single chain, open or closed, the arguments extend to multiple chains as indicated below.

Mechanisms. The key property of a pseudotriangulation of chains is that it is minimally infinitesimally rigid: if the edges are viewed as bars in a linkage, this linkage is infinitesimally rigid and hence rigid, and the removal of any bar produces a flexible (nonrigid) linkage (Streinu 2005). Such a flexible linkage that is only one bar away from being rigid is called a *one-degree-of-freedom mechanism* or simply a *mechanism*. The configuration space of a mechanism is locally a 1D path, corresponding to changing the length of the removed bar. So there are precisely two directions (motions) to follow along this path, corresponding to increasing or decreasing the length of the removed bar. Moreover, and important for avoiding self-intersection, the removal of a bar along the convex hull of a pseudotriangulation produces a mechanism such that one direction of motion is expansive (and the other direction is the opposite—*contractive*):

Theorem 6.7.2 (Streinu 2005, Thm. 4.1). *The linkage of bars obtained from any pseudotriangulation of chains, minus any convex-hull edge, is a mechanism such that one direction of motion is expansive.*

We provide a long proof sketch of this result, which employs several of the key concepts used in the proof of Theorem 6.6.1.

Proof: *(Sketch)* The linkage being a mechanism follows from minimal rigidity of pseudotriangulations, whose proof we do not cover here. Thus we only need to establish that the linkage has an expansive motion. One proof of this result ties into the proof that there are no locked chains in 2D from Section 6.6 (p. 96). As before, we fuse any vertices of the linkage with a straight angle of $180°$. Once such vertices are removed, we argue that there is an infinitesimal motion of the tensegrity arising from the bars of the pseudotriangulation and additional struts between every pair of vertices not connected by a bar. Thus, the pseudotriangulation linkage has an expansive infinitesimal motion. Then we argue that such infinitesimal motions can be combined into a full expansive motion until the bars no longer form a pseudotriangulation.

To prove that the tensegrity is infinitesimally flexible, we apply Lemma 6.7.1 above together with (a generalization of) Lemma 6.6.4. Lemma 6.7.1 says that each vertex of the tensegrity has an incident reflex angle that contains no bars. The proof of Lemma 6.6.4 establishes the more general result (Connelly et al. 2002b, Lem. 6) that the reflex pie wedge around any such vertex lifts to the maximum z coordinate in any polyhedral lifting resulting from the Maxwell–Cremona theorem (Theorem 4.6.1, p. 58).

Might there still be parts of the lifting that are below the maximum z coordinate? We distinguish two types of faces of the linkage: triangles and nonconvex pseudotriangles. Nonconvex pseudotriangles must have their interiors lift to the maximum z-coordinate, to be consistent with (the generalization of) Lemma 6.6.4 at the reflex vertex. So only the interiors of triangles might lift below the maximum z-coordinate.

Consider the triangulations that remain after removing all nonconvex pseudotriangles. A triangulation can be a pseudotriangulation only if all its vertices are on the convex hull (any interior vertex could not be pointed). Thus, the polyhedral lifting looks like the following picture: triangulated convex polygons might lift to below the maximum z coordinate, but they are always surrounded by plateaus at maximum z. But if the boundary of a triangulated convex polygon lifts to the maximum z-coordinate, we claim that the interior must also be at the maximum z-coordinate. Any triangulation has a triangle with two edges on the boundary (an "ear"), and such a triangle must be wholly at the maximum z coordinate by planarity. Then we remove the triangle from the triangulated convex polygon of consideration, and repeat by induction.

We conclude that any polyhedral lifting of the planar tensegrity must be completely flat. But we have not yet used the fact that a bar was removed from the convex hull of the pseudotriangulation, so we must use this property in our connection between polyhedral liftings and infinitesimal flexibility. Theorem 4.6.1 (p. 58) says that every polyhedral lifting being flat implies that every equilibrium stress of the planar tensegrity must be everywhere-zero. The important remaining step is Lemma 6.6.2 (p. 100) and the extension described afterward, which requires the existence of a portion of some strut that is exterior to all convex cycles of bars. If we had not removed a convex-hull edge, every strut would be interior to a convex bar cycle, but with the removal, the strut that replaces the removed bar is partly (indeed, entirely) exterior to all convex cycles of bars.

Through this extension to Lemma 6.6.2, we obtain that any equilibrium stress of the original tensegrity is everywhere-zero. The duality between stresses and motions (Theorem 4.5.2) implies that the original tensegrity has an infinitesimal motion, which means that the pseudotriangulation linkage has an expansive infinitesimal motion.

No piecing-together of infinitesimal motions is necessary in this context: we are guaranteed a unique expansive infinitesimal motion. For multiple chains, we must be careful that the motion terminates after finite time, but the boundedness of any motion follows quite easily for these mechanisms because of their limited reach.   ☐

Pseudotriangulations have many other connections to rigidity (Chapter 4). In addition to being minimally infinitesimally rigid, they admit Henneberg-like constructions (Streinu 2000, Thm. 3.1(6)) similar to those described in Section 4.3.3 (p. 47).

Flipping. Theorem 6.7.2 expands a pseudotriangulation as long as it remains a pointed pseudotriangulation. The pointed-pseudotriangulation property would be violated by an angle switching from convex to reflex, or vice versa. Just before such a violation, two incident bars of the pseudotriangulation become collinear. If two incident bars of the original chain become collinear, then we can fuse the straight joint. Otherwise, we need to modify the pseudotriangulation in order to continue. Streinu (2000) showed how to modify the pseudotriangulation using *flips* at such collinearities, and showed that at most $O(n^3)$ flips will be made throughout the motion (Streinu 2005). An example of a flip is shown in Figure 6.19. In the end, we straighten or convexify the chain.

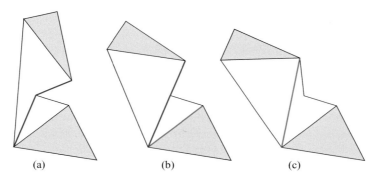

(a)   (b)   (c)

Figure 6.19. This flip in pseudotriangulation of single-degree-of-freedom mechanism is triggered by the collinearity in (b). The shaded triangles remain rigid throughout the motion.

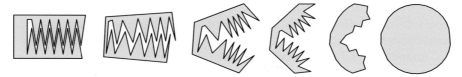

Figure 6.20. Convexifying a 29-link closed chain via gradient descent of energy. [From Cantarella et al. 2004.]

Figure 6.21. Convexifying a 380-link closed chain via gradient descent of energy. [From Cantarella et al. 2004.]

Algorithms. Each of the motions in between the $O(n^3)$ flips are single-degree-of-freedom mechanisms. Each such mechanism is defined by a system of polynomial equations specifying that each bar length remains fixed. The solution to such a system is called an *algebraic* motion. Algebraic motions are well-studied objects, so there are several algorithms, albeit all inefficient, to manipulate them. In the context of Streinu's method, we need to manipulate algebraic motions in two ways: to find when a motion must terminate because two incident bars become collinear, and to apply each motion to such termination times so that we can compute the next motion. Of course, to animate the motion, we also need the desired number of intermediate snapshots of the motions. All three of these operations can be performed, up to a desired accuracy tolerance ε, in time exponential in n (Mishra 2004). See Streinu (2003; 2005, page 41) for discussion of these numerical computational issues.

Extensions. The single-degree-of-freedom expansive mechanisms that arise from pseudotriangulations have a rich structure that is only recently being elucidated. In particular, it is shown in Rote et al. (2003) that these motions are the bounding rays of the cone of all expansive motions, as well as providing a new proof that there are no locked chains.

6.7.3 Unlocking 2D Chains Using Energy

Cantarella et al. (2004) developed an algorithm for unlocking chains based on following the downhill gradient of appropriate energy function, corresponding roughly to the intuition of filling the polygon with air. See Figures 6.20 and 6.21 for examples; a Java applet is also available.[11]

The resulting motion is not expansive; it essentially averages out the forces imposed by the various strut constraints. On the other hand, the existence of the downhill gradient relies on the existence of expansive motions from Theorem 6.6.1 (p. 97), by showing that the latter decrease energy. The motion avoids self-intersection not through

---

[11] http://www.cs.berkeley.edu/b-cam/Papers/Cantarella-2004-AED/.

expansiveness but by designing the energy function to approach $+\infty$ near (and only near) self-intersecting configurations; any downhill flow from an initially finite energy avoids such infinite spikes.

Cantarella et al. (2004) show that any energy function will suffice for the method to unlock a chain provided it is sufficiently smooth (C^2), decreases under expansive motion, and shoots to $+\infty$ as configurations approach self-intersection. A class of such energy functions arises simply by summing the reciprocal of the (squared) distance between every vertex and every nonincident edge in the configuration:

$$\sum_{\substack{v,e \\ \text{nonincident}}} \frac{1}{d(v, e)^2}.$$

(The square is for smoothness.) Any expansive motion increases all vertex-edge distances (Connelly et al. 2003, Cor. 1), and so decreases energy. If a configuration approaches a self-crossing configuration, then some vertex will approach some bar, and so the corresponding distance approaches zero and the reciprocal energy term shoots to $+\infty$. The exact definition of the distance between a vertex and an edge is flexible. Cantarella et al. (2004) use a C^∞ distance measure called *elliptic distance*, which measures the size[12] of an ellipse passing through the vertex v and with foci at the endpoints e_1, e_2 of the edge e:

$$d(v, e) = |v - e_1| + |v - e_2| - |e_1 - e_2|.$$

The mathematical consequence of this method is a C^∞ unlocking motion. In comparison, the convex-programming method of Section 6.7.1 (p. 105) produces piecewise-C^1 motions, and the pseudotriangulation method of Section 6.7.2 (p. 107) produces piecewise-C^∞ motions.

The algorithmic consequence is an easy-to-compute motion of pseudopolynomial complexity as described below. The motion is piecewise-linear in angle space: it decomposes into a finite sequence of moves and, during each move, every angle changes linearly from one value to another. Each move corresponds to a single step in the direction of steepest descent, which can be computed in $O(n^2)$ time.[13] From such a piecewise-linear representation of the motion, it is easy to extract an exact snapshot of the linkage at any time during the motion in linear time. Furthermore, the number of steps in the piecewise-linear motion is polynomial in the number n of links and the ratio r between the maximum edge length and the minimum elliptic distance between a vertex and an edge in the initial configuration. The established (conservative) polynomial bound is rather large—$O(n^{123} r^{41})$—but nonetheless this is the only known explicit bound of this type for any chain-unlocking algorithm. As a consequence, the total running time of the algorithm is $O(n^{125} r^{41})$.

Figure 6.22 compares all three convexifying algorithms—convex programming (Section 6.7.1), pseudotriangulations (Section 6.7.2), and energy—on the same example.

One interesting open problem arises in the context of this method and its pseudopolynomial bound dependent on both the combinatorial complexity n and the geometric complexity r of the linkage. Can the dependence be purely combinatorial by carefully chosen moves?

---

[12]   Precisely, the major axis minus the distance between the foci.
[13]   On a real RAM supporting arithmetic, $\sqrt{}$, sin, and arcsin.

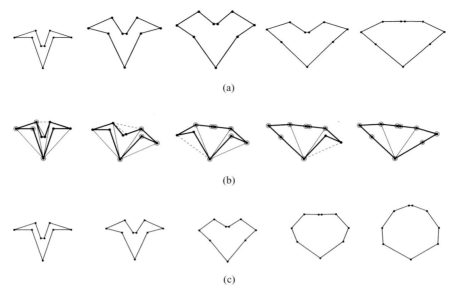

(a)

(b)

(c)

Figure 6.22. Convexifying a common polygon via all three convexification methods. (a) Via convex programming (Connelly et al. 2003); (b) Via pseudotriangulations (Streinu 2000). Pinned vertices are circled; (c) Via energy minimization (Cantarella et al. 2004). [Figure 9.3.3 of Connelly and Demaine 2004, by permission, CRC.]

> **Open Problem 6.3: Polynomial Number of Moves.** Can a planar polygonal chain be straightened or convexified in 2D by a sequence of moves, each of which is piecewise-linear in angle space, with the number of moves polynomial in the number n of links?

The energy approach is also applicable to the *linkage refolding* problem: to find the "shortest" motion connecting two given configurations of the same linkage. Here "shortest" might be defined by a geodesic path in the appropriate space of motions (see Iben et al. 2006).

6.8 INFINITESIMALLY LOCKED LINKAGES IN 2D

We have seen examples of 2D locked trees in Section 6.5 (p. 94), and showed that there are no 2D locked chains (Section 6.6, p. 96). A natural next problem, largely unsolved, is to characterize which linkages are locked. One version of this characterization problem is to understand the complexity of the decision question:

> **Open Problem 6.4: Characterize Locked Linkages.** [a] What is the complexity of deciding whether a given linkage is locked?
>
> ___
>
> [a] Posed for 3D chains in Biedl et al. (1999) and 2D trees in Connelly et al. (2002a).

As described in Section 2.1.1.1 (p. 18), the roadmap algorithm for general motion planning enables testing whether a linkage is locked in $2^{O(n^2)}$ time. A depth-first search in the roadmap representation of the configuration space determines all the connected components, and we can check whether any pair of connected components have the same clockwise/counterclockwise orientation for every cycle, in which case the linkage is locked.

A related, more difficult problem is deciding whether there is a motion between two given configurations of a linkage. As discussed in Section 2.1.2, this problem is PSPACE-hard, both for trees in 2D and for open chains in 3D. However, the examples used in the proof constructions (Alt et al. 2004) are locked: locked 2D trees and interlocked 3D chains are used to build a "scaffold" for the reduction. Thus it remains open whether a similar reduction can be made in which the reconfigurable (YES) instances are also unlocked. Alternatively, perhaps there are more efficient algorithms for deciding whether a linkage is locked, at least for the special cases of 2D trees or 3D chains. The interlocking of two or more chains, an interesting topic in its own right, will be explored in Chapter 7.

Proving linkages locked. Until recently, the few linkages that have been proved locked—the petal tree in Figure 6.9(a) (p. 95) and the knitting needles in Figure 6.2 (p. 88)—have complicated or ad hoc proofs. The general approach is to define a collection of "events," at least one of which must occur before the linkage unlocks, and argue that no event can occur. Typically the latter argument is cyclic, showing that no event can occur before some other event occurs, and hence no events can occur at all. This approach is powerful with the right ingenuity, but, at least for 2D trees, it seems to require complicated manipulation of error tolerances.

Here we consider a related, stronger notion of *infinitesimally locked linkages* in 2D, some forms of which can be tested efficiently. This approach, introduced in Connelly et al. (2002a), matches the intuition in many 2D locked linkages, leading to short proofs that several such linkages are locked. In particular, we will see a proof that the tree in Figure 6.9(b) (p. 95) is locked.

6.8.1   Stronger Definitions of Locked

As described above, the general definition of "locked" permits a linkage to be quite flexible, so long as some (perhaps "distant") configuration cannot be reached. In contrast, the trees shown in Section 6.5 (p. 94) are locked in a much stronger sense: they can barely move at all. In fact, the tighter we draw the examples, the less they can move.

To capture this notion, we introduce a metric on the configuration space; that is, we define the *distance* between any two configurations of a linkage. There are many ways to define this distance metric; one simple way is to measure the Euclidean distance between the two configurations if we consider them points in $3n$-dimensional space, where n is the number of joints.

We call a linkage configuration *locked within ε* if the only configurations it can reach by motions are at most distance ε away. In other words, the connected component of reachable configurations from this configuration is contained in a ball of radius ε centered at this configuration. Of course, the strength of this definition depends on the relative magnitude of ε, but for ε sufficiently small, being locked within ε implies being locked.

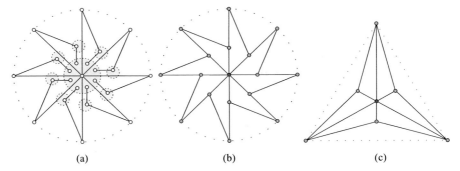

(a)   (b)   (c)

Figure 6.23. Self-touching configurations resulting from Figure 6.9 by contracting the dotted circles down to points. Only the geometry is illustrated; additional combinatorial information is specified by the insides of the dotted circles in Figure 6.9, and by the number of links along each edge.

The interesting property of the locked trees from Section 6.5 is that, in some sense, they are locked within ε for all $\varepsilon > 0$. More precisely, for any desired ε, there is a drawing of each tree that is locked within ε, where the small separations in the tightly drawn tree depend on the desired lock tolerance ε. But how do we capture the "drawings" of a tree? In what sense are two linkage configurations "drawings" of the same tree?

One way to capture a continuous family of "tightly drawn" linkages is to let the small separations shrink to zero, resulting in a *self-touching configuration* as shown in Figure 6.23, and consider small perturbations of that self-touching linkage (turning distances of zero into positive distances). The idea is that the smaller the perturbation, the more locked the configuration becomes (the smaller the ε). We call this the *strongly locked* property.

It remains to define perturbations precisely. A δ-*perturbation* of a linkage configuration modifies the configuration as well as the linkage—the link lengths may change—so that each joint moves within a radius-δ disk of its original position, and the combinatorial embedding of the linkage remains the same. Thus the distance between any two joints (and so the length of any link) changes by at most 2δ.

Now the definition of a configuration being strongly locked follows the usual definition of limits: for every ε, there is a δ such that every δ-perturbation is locked within ε. In this way we capture the arbitrarily locked aspect of the trees by looking at their self-touching limits.

6.8.2   Self-Touching Configurations

Self-touching configurations[14] behave somewhat differently from normal configurations, because it is no longer clear just from the geometry whether a vertex is on one side or the other side of an edge. These and other combinatorial properties must be specified separately from the geometry, as part of the configuration. To be precise about this notion requires some effort, so we leave most of the details to Connelly et al. (2002a). The basic idea, however, is simple: specify how many links lie along each edge and how many joints lie at each vertex, and at each vertex, specify a planar graph of the connections between links and joints. Any perturbation of a self-touching configuration must preserve the embedding determined by this combinatorial information.

---

[14]   See also the discussion of self-crossing (p. 71) and the definition of flat origami in Part II, Chapter 11.

6.8.3   Connection to Rigidity

In contrast to standard configurations of linkages which are always slightly flexible, self-touching configurations can be completely *rigid* in the sense that any motion causes bars to cross immediately. In comparison to the notions described earlier (Chapter 4), rigidity and flexibility need to be defined to capture the noncrossing constraints in addition to the bar-length constraints.

We also obtain the notion of *infinitesimal rigidity*, where the noncrossing constraints only need to specify that currently touching vertices and bars remain on the correct sides of one another. For the most part, these constraints are of the form "this vertex remains to the left of this bar"; such constraints can be viewed as infinitesimally short (zero-length) *sliding struts* between a vertex and a bar, where the strut slides along the bar to keep perpendicular to the bar. Another type of noncrossing constraint can arise when two vertices touch on their reflex sides, so that a vertex is on no particular side of any one bar. Here we focus on when such complex constraints do not arise, in which case we obtain the dual notion of *equilibrium stress* as well.

These notions have the following analogs to theorems in rigidity from Chapter 4:

Theorem 6.8.1 (Connelly et al. 2002a). *In the context of self-touching configurations of linkages with sliding struts (simple noncrossing constraints), the following hold:*

1. *Flexibility implies infinitesimal flexibility.*
2. *There is an equilibrium stress on a tensegrity that is nonzero on a particular strut or cable if and only if no infinitesimal motion can change the length of that strut or cable.*
3. *A tensegrity is infinitesimally rigid if and only if it has an equilibrium stress that is nonzero on every strut and cable and the corresponding linkage is rigid. (Here the "corresponding linkage" is obtained by replacing the sliding struts with sliding bars, which pin a vertex to anywhere along a bar.)*

More surprising is that rigidity is closely related to the notion of strongly locked defined in the previous section:

Theorem 6.8.2 (Connelly et al. 2002a). *If a self-touching configuration is rigid, then it is strongly locked. In particular, if a self-touching configuration is infinitesimally rigid, then it is strongly locked.*

The idea behind the proof of this theorem is to consider "sloppy rigidity" introduced in Connelly (1982, Thm. 1) (although not with that name). Sloppy flexibility allows edge lengths to change slightly, up to some specified constant $\varepsilon > 0$; this notion corresponds to $\varepsilon/2$-perturbations. The idea is that by turning the each bar-length equality constraint into a thin range of inequalities, the configuration space only slightly "thickens," leaving a small volume.

6.8.4   Proving Trees Locked

Theorem 6.8.2 is powerful in that it provides a simple way to prove a linkage to be strongly locked. Theorem 6.8.1(3) gives a particular method for proving (infinitesimal)

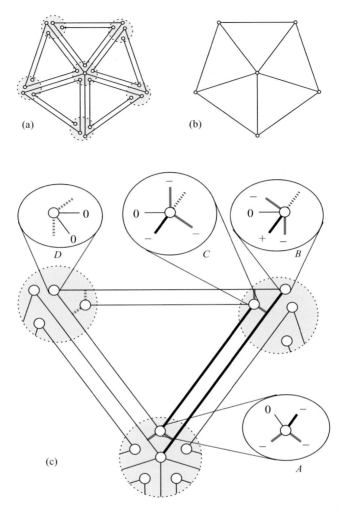

Figure 6.24. (a) Locked tree; (b) self-touching version; (c) one petal and adjacent pieces. Short light edges denote zero-length sliding struts. Blow-ups show the signs of stresses on edges to achieve equilibrium; thick edges carry stress; and dotted edges show negations of + edges.

rigidity: construct an equilibrium stress that is nonzero on all sliding struts, and check that the corresponding linkage is locked. This method can be used to prove that all of the trees in Figure 6.9 (p. 95) are strongly locked. In particular, Connelly et al. (2002a, Sec. 10) describes such proofs for the petal tree of Figure 6.9(a) and the single degree-3 tree of Figure 6.9(d). Here we consider the triangle tree of Figure 6.9(b), repeated for convenience in Figure 6.24(a), which has a somewhat simpler proof, while still illustrating the three-step process of the other proofs.

Step 1: Model as self-touching configuration. The first step is to model the linkage as a self-touching configuration. Shrinking the small separations shown by shaded circles in Figure 6.24(a) produces the self-touching geometry in Figure 6.24(b). Figure 6.24(c) focuses on just one "petal," relying on the rotational symmetry of the tree.

The vertices on the convex hull of the tree touch each other in pairs on their reflex sides. This situation is the case not captured directly by the infinitesimal theory and sliding struts. Thus, in particular, there are no sliding struts connecting these pairs of vertices. It turns out that the remaining sliding struts suffice to make the linkage infinitesimally rigid, and hence strongly locked. There are alternate ways to handle these abutting reflex vertices, when necessary, although none are completely satisfactory (Connelly et al. 2002a).

We also remove the dotted sliding struts in the upper-left. This modification is necessary for Step 3 because we would not be able to give these struts strictly negative stress. Fortunately, their removal does not make Step 2 harder in this case. The delicate balance between Steps 2 and 3 can be difficult to achieve.

Step 2: Prove corresponding linkage infinitesimally rigid. The second step is to prove infinitesimally rigid the corresponding linkage resulting from replacing sliding struts with sliding bars. This step is usually easy, and in this case it matches the intuition behind the tree. In particular, wherever we have a vertex wedged into a convex angle between two bars of the linkage, the two sliding bars force that vertex to stay along both linkage bars, and therefore stay pinned against the convex angle. As a consequence, each petal acts as a rigid triangle (with a link inside), and all the petals are glued together at the central vertex. Because the central angles of the petals sum to 360°, the corresponding linkage of the tensegrity is infinitesimally rigid.

The property that each petal "acts" as a rigid triangle is intuitively why the original tree is locked. With separations in the linkage, this property holds only approximately, because some motion is possible. With the sliding bars, however, the property holds precisely.

Step 3: Find equilibrium stress. The third step is to construct an equilibrium stress in the tensegrity that is nonzero (in fact, negative) on all sliding struts. We could simply apply linear programming to construct such a stress, but in this example we can avoid explicit numerical stresses.

The main intuition comes from examining stress on a degree-3 vertex, which has exactly one local equilibrium stress up to scaling (refer to Figure 6.25). First, the three incident edges can have stresses with the same sign precisely if the edges do not lie in a common half plane, so that all angles are less than 180°. Intuitively, the first two edges can apply different amounts of force so that they add up to a force in the direction of the third edge, and this third edge can apply an equal opposing force. Second, if the three incident edges are in a common half plane, then we must negate the sign of the stress on the middle edge, effectively reversing the edge's direction and effectively leaving the edges with angles less than 180°.

Now we construct an equilibrium stress as follows. The five petals are handled symmetrically so that the central vertex is guaranteed to be in equilibrium. For

(a)   (b)

Figure 6.25. Possible sign patterns for an equilibrium stress at a degree-3 vertex (when no two of the edges are collinear). (a) incident edges not in a half plane; (b) 3 incident edges in a half plane. [Figure 10 of Connelly et al. 2002a, by permission, American Mathematical Society.]

the vertex nearest the central vertex (blow-up A), we scale the balancing stresses so that they are all negative, which is possible because the incident edges do not lie in a common half plane. The upward-pointing edge is incident to another vertex (blow-up C), whose stresses we scale to match the stresses assigned so far; again the resulting stresses are all negative. The final vertex (blow-up B) has two negative stresses already assigned, which sum to a negative force along the direction of the remaining edge, because these stresses correspond to negations of the previous vertex. By assigning a matching positive stress to the remaining edge, equilibrium is obtained.

Finale of the argument. To complete the argument, we apply all the tools described above. By Theorem 6.8.1 (p. 116), Steps 2 and 3 imply that the self-touching linkage is infinitesimally rigid and hence rigid. By Theorem 6.8.2, this infinitesimal rigidity implies that the self-touching linkage is strongly locked, and hence that the original linkage is locked with sufficient small separation.

6.8.5   Other Problems on Self-Touching Linkages

It is natural to extend Theorem 6.6.1 (all chains can be straightened/convexified) to self-touching chains, but this has yet to be explored fully:

> **Open Problem 6.5: Self-touching Chains.** Can all 2D self-touching chains be straightened?

Most recently, the techniques described here have been used to establish locked chains of planar shapes (Jordan regions) (Connelly et al. 2006).

6.9   3D POLYGONS WITH A SIMPLE PROJECTION

In Section 5.3.3 (p. 84), we proved that an open chain in 3D that has a *simple projection*—an orthogonal/orthographic projection onto some plane as a non-self-intersecting chain—can always be straightened. We promised to revisit the same question for closed chains, and we do so now. In particular, we describe a proof that any 3D polygon with a simple projection (Figure 6.26) can be convexified: reconfigured to a planar convex polygon. So neither open nor closed chains with a simple projection can be locked.

The proof takes the form of an algorithmic procedure detailed in Calvo et al. (2001). The key idea is as follows. Let P be the 3D polygon and P' its simple projection onto (without loss of generality) the xy-plane. Let $e = (a, b)$ be an edge of P, and $e' = (a', b')$ the corresponding edge of P'. Any reconfiguration of P that preserves the z-coordinates of a and b preserves the length of the projection $|e'|$. Therefore, one can reconfigure P' in the xy-plane, and track that motion in $\mathbb{R}^3$ by maintaining the height of each vertex. If the reconfiguration of P' avoids self-intersection in the xy-plane, so does the corresponding reconfiguration of P in $\mathbb{R}^3$.

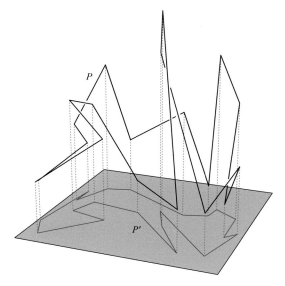

Figure 6.26. A 3D polygon P that has a simple projection P' in the plane below.

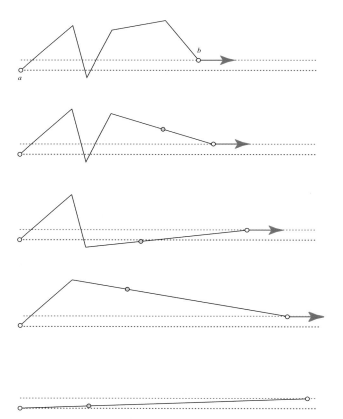

Figure 6.27. Pulling vertex b rightward straightens the planar, monotone chain to a segment while maintaining the heights of both endpoints a and b.

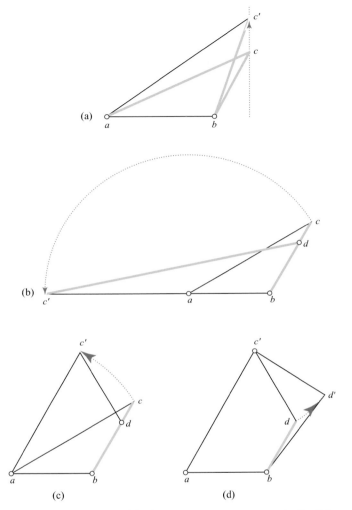

Figure 6.28. Blue edges are "springs." (a) ab is fixed and c moves vertically; (b) ac rotates about a; (c) dc' has become taut; (d) $c'd$ rotates about c'.

This leads to the following algorithm:

1. Reconfigure P' to a convex polygon, using the expansive motions of Section 6.6 (p. 96), and track this motion by reconfiguring P correspondingly.
2. Reconfigure the convex polygon in the xy-plane to a triangle, using the algorithm of Section 5.3.1 (p. 70), again tracking the movements in $\mathbb{R}^3$.
3. "Stretch" the chains of P in 3D that project to each triangle edge in the xy-plane.

We now explain this last step.

Once the projection P' is a triangle, P lies entirely in three vertical planes, each plane containing a monotone chain and projecting to an edge of the triangle. Steps 1 and 2 of the algorithm have not altered z-heights of vertices, but Step 3 will. The primitive motion is illustrated in Figure 6.27. One endpoint of a monotone chain is pulled horizontally, a process that can be continued until the chain becomes a segment. Note that the endpoints a and b remain at their original z-heights. Calvo et al. (2001) call a monotone chain that is more than one segment a *spring*, for it has the potential to stretch. Call a polygon with k chains that are springs a *k-spring*. So at the beginning of Step 3, P is a 3-spring.

From here we only sketch the remainder of the algorithm, which requires some care to ensure correctness. If P is a 3-spring or a 2-spring, it is reduced to a 2-spring or a 1-spring (respectively) as follows. Let $P = \triangle abc$, with two springs incident to c. Move c perpendicular to the line containing ab until on spring stretches taut at c' (see Figure 6.28(a)). If P is a 1-spring, the stretched length of that spring may not form a triangle with the other two sides. So a more complex procedure is needed.

Let bc be the one spring of $\triangle abc$ and choose labeling so that $\angle c$ is acute. Select an interior vertex d on the bc spring and "crack" the spring there. Fixing a, b, and d, rotate the segment ac about a until an event happens. Note that because the angle at c is acute, this motion retains convexity at c. The event could be that dc stretches enough to permit c to reach collinearity with ab at c', as in Figure 6.28(b), in which case we are back to a 2-spring $\triangle bdc'$. On the other hand, the event may be that dc is first stretched taut, forming the convex quadrilateral $abdc'$ as shown in Figure 6.28(c). In that case, fix a, b, and c' and rotate $c'd$ about c' until either $\{a, c', d'\}$ become collinear, or the spring bd straightens (as in Figure 6.28(d)). In this last case we are left with a springless convex quadrilateral.

It is established in Calvo et al. (2001) that Step 3 requires no more than $O(n)$ moves, each of which rotates at most seven joints at once. Whether a chain has a simple orthogonal projection can be determined in $O(n^4)$ time (Bose et al. 1996).

It remains an interesting challenge to find additional natural conditions on a chain (open or closed) that ensure that it is not locked. Next we consider several entangled polygonal chains.

7 Interlocked Chains

Having seen in the previous sections that chains can only lock in 3D, it is natural to investigate the conditions that permit chains to lock in 3D. Sections 5.3.3 and 6.9 showed that chains with non-self-intersecting projections cannot lock. In this section we report on the beginnings of an exploration of when chains can lock and, in particular, when pairs of chains can "interlock." This line of investigation was prompted by a question posed by Anna Lubiw (Demaine and O'Rourke 2000): Into how many pieces must a chain be cut (at vertices) so that the pieces can be separated and straightened? In a sense, this question crystallizes the vague issue of the degree of "lockedness" of a chain—how tangled it is—in the form of a precise problem. The "knitting needles" example (Figure 6.2, p. 88) is only slightly locked, in that removal of one vertex suffices to unlock it (recall that Cantarella and Johnson proved that no chain of fewer than 5 links can be locked). Concatenating many copies of the knitting needles leads to a lower bound of $\lfloor (n-1)/4 \rfloor$ on the number of cuts needed to separate a chain of n links, as illustrated in Figure 7.1: each copy of the 5-link chain must have one of its four interior vertices cut. An upper bound of $\lfloor (n-3)/2 \rfloor$ has been obtained, but otherwise Lubiw's problem remains unsolved.

Open Problem 7.1: Unlocking Chains by Cutting. [a] Deletion of how many vertices of an n-link chain in 3D suffice to separate and straighten the pieces? There are lower and upper bounds of (roughly) $\frac{1}{4}n$ and $\frac{1}{2}n$, but perhaps $\frac{1}{3}n$ suffices.

---

[a] Posed by Anna Lubiw (Demaine and O'Rourke 2000).

One of the impediments to answering Lubiw's question is that, even if each piece into which a chain has been cut is individually unlocked, the tangled collection still might be "interlocked." Say that a collection of disjoint, noncrossing chains can be *separated* if, for any distance d, there is a motion (avoiding self-crossing) that results in every pair of points on different chains being separated by at least d. If a collection cannot be separated, we say that its chains are *interlocked*. Which collections (usually, pairs) of short chains can interlock was investigated in several papers (Demaine et al. 2001c, 2002b, 2003b), which we now sample.

Table 7.1: Interlocking pairs of chains

	Open			Closed		
	2f	**3f**	**4f**	△	□	**Pentagon**
2f	−	−	−	−	−	−
3f	−	−	+	−	+	+
4f	−	+	+	+	+	+

(+) = can lock; (−) = cannot lock. kf = open, flexible k-chain; △ = closed (rigid) 3-chain; □ = closed, flexible 4-chain; pentagon = closed, flexible 5-chain.

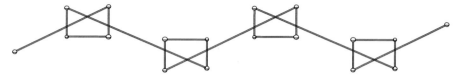

Figure 7.1. An $n = 17$ link chain that requires cutting at least $\lfloor (n-1)/4 \rfloor = 4$ vertices to separate. [Figure 1 of Demaine et al. 2003b, by permission, Elsevier.]

We should note that, because we will be investigating short k-chains (e.g., $k \leq 5$), whether any specific configuration of a finite collection of such chains is interlocked could be answered in principle by the any of the general motion planning algorithms, for example, the roadmap algorithm (Section 2.2, page 19). Aside from the impracticality of this approach, it is of no help in discovering a configuration that might be interlocked, nor in proving that no configuration in a class could be interlocked. As these issues are the focus of this section, we will introduce new, ad hoc techniques.

The work in the cited papers explored a variety of constraints on the chains to gain an understanding of what conditions are needed for interlocking. Four classes of restrictions on the freedom of the chains have been considered:

1. Open versus closed chains. Closed chains are more constrained.
2. *Flexible* chains: those without any constraints on joint motion (each joint is a "universal joint").
3. *Rigid* chains: each joint is frozen, and the entire chain is rigid.
4. *Fixed-angle* chains: chains that maintain the angle between links incident to each joint, but permit revolution of the links that keeps each angle fixed. These are intermediate in freedom between flexible and rigid chains.

Fixed-angled chains will be the topic of the next chapter, and will not be further discussed here. This still leaves a large array of possibilities, from which we select a few threads to illustrate the flavor of the results and the main proof techniques. We concentrate on the flexible chain results summarized in Table 7.1, with some remarks on rigid chains. We consider what type of chain is needed to interlock with open, flexible 2-, 3-, and 4-chains, the three rows of the table. The columns of the table cover both open and closed flexible chains out to $k = 4 - 5$.

The selected results we discuss rely on three different proof techniques: a scaling argument, a topological method, and geometric proofs based on orientation determinants.

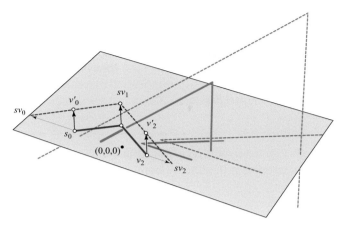

Figure 7.2. Three 2-chains (solid) scaled (dashed) by $s = 2$. The clipping for one chain (blue, lying in a vertical plane) is detailed.

7.1   2-CHAINS

We start with the easiest argument, which establishes the 2f/2f cell of the table as a corollary:

Theorem 7.1.1. *No finite collection of rigid 2-chains can be interlocked.*

    Proof: We will provide an explicit movement that separates the chains from one another. Choose a point not on any chain and make that the origin of a coordinate system. Consider the uniform scaling, or dilation, defined as the affine transformation $S : p \to sp$ for each $p \in \mathbb{R}^3$ and some real $s \geq 1$. Because affine transformations preserve incidence relations among lines, the scaling does not cause any pair of chains to intersect. Let $C = (v_0, v_1, v_2)$ be one 2-chain with link lengths ℓ_1 and ℓ_2, transformed by S to $C' = (sv_0, sv_1, sv_2)$. Again because S is affine, the angle at v_1 is preserved, but the lengths of the links are scaled by s. Imagine now clipping the links to (v'_0, sv_1, v'_2) so that v'_0 lies on (sv_0, sv_1) at a distance ℓ_1 from sv_1, and similarly for v'_2 (see Figure 7.2). This defines a rigid motion for C that, as s becomes large, separates C arbitrarily far from the other chains without intersecting them. Applying the dilation and clipping to all chains proves that the collection is not interlocked.   □

    This beautiful proof technique has a long pedigree, dating back at least to de Bruijn in 1954 (de Bruijn 1954),[1] and has been used in various guises in several papers on separating objects by rigid movements (Dawson 1984; Snoeyink and Stolfi 1994; Toussaint 1985). We will employ it again in Theorem 7.2.1. Of course, if collections of rigid chains cannot interlock, neither can two flexible chains, so this result is stronger than the 2f/2f entry in Table 7.1. Moreover, the exact same proof technique establishes that no collection of rigid k-stars interlocks, where a k-star is k segments sharing a vertex. So, for example, no number of "Cartesian axes" can interlock.

    It is established in Demaine et al. (2002b) that a flexible 2-chain cannot interlock with an open, rigid 4-chain (and so with an open, flexible 4-chain—the 4f/2f table entry), or

[1] We thank Jack Snoeyink for this reference.

with a closed, flexible 5-chain, but can interlock with an open, rigid 5-chain. It remains unclear which flexible chains can interlock with a flexible 2-chain. It may not be difficult to establish that a flexible 2-chain cannot interlock with a flexible quadrilateral, but this has not been explored as of this writing (and thus is marked with "?" in the table). Concerning open chains, it was recently established that a flexible 2-chain can interlock with a flexible, open k-chain (Glass et al. 2004), but the smallest k is unknown:

Open Problem 7.2: Interlocking a Flexible 2-Chain.[a] What is the smallest value of k that permits a flexible 2-chain to interlock with an open, flexible k-chain?

---

[a] Demaine et al. (2002b).

We mentioned that a rigid 5-chain suffices; presumably $k > 5$ is needed for a flexible chain, and $k \leq 11$ has recently been established (Glass et al. 2006).

7.2   3-CHAINS

For 3-chains we describe several $+$ and $-$ results from Table 7.1. The first, that two open, flexible 3-chains cannot interlock, is proved by a variation on the scaling-and-clipping proof technique used in Theorem 7.1.1.

Theorem 7.2.1. *Two flexible 3-chains cannot interlock.*

Proof: Let the two chains be $A = (a_0, a_1, a_2, a_3)$ and $B = (b_0, b_1, b_2, b_3)$. We assume the chains are in general position in the sense that no two nonincident links are coplanar and no three joints are collinear. Because the chains do not intersect themselves or each other, there is always room to perturb the joint positions so that general position holds. Let Π be a plane between and parallel to the middle links $a_1 a_2$ and $b_1 b_2$ of the chains. By the general position assumption, this plane exists and does not contain either middle link (for that would make them coplanar). Now arrange the coordinate system so that Π is the xy-plane. Finally, perturb again if necessary so that no two of the endpoint vertices a_0, a_3, b_0, and b_3 have the same z-coordinate.

We apply a nonuniform scaling transformation to each joint's z-coordinate, $S : z \rightarrow sz$, for $s > 1$. The middle links remain parallel to the xy-plane and retain their original length (see Figure 7.3). Because all endpoint vertices have different z-coordinates, the distance between them increases under S. Thus the end links of each chain lengthen and the chains grow further apart. We clip the end links to maintain their original length, just as in the proof of Theorem 7.1.1. Because affine transformations preserve incidences, the motion cannot cause any links to touch. As s gets large, the chains separate arbitrarily far and so are not interlocked.   □

It is not difficult to extend this theorem to the claim that no pair of flexible 3-chains and any (finite) number of flexible 2-chains cannot interlock. In contrast to these result, three 3-chains can interlock (Demaine et al. 2003b, Thm. 9).

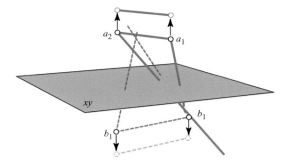

Figure 7.3. Nonuniform scaling keeps the middle links parallel to the xy-plane. (Scaled versions of the end links are not shown.)

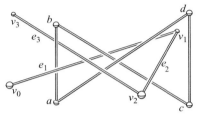

Figure 7.4. A quadrilateral and a 3-chain can interlock. [Figure 6 of Demaine et al. 2003b, by permission, Elsevier.]

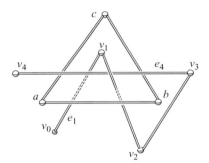

Figure 7.5. A triangle and a 4-chain can interlock. [Figure 2 of Demaine et al. 2003b, by permission, Elsevier.]

The following result is more difficult to prove:

Theorem 7.2.2. *A flexible 3-chain can interlock with a closed, flexible 4-chain.*

The example is illustrated in Figure 7.4. It should at least be plausible that if e_1 and e_3 are long enough, the chains are indeed interlocked. The proof technique here is topological. We will return to this theorem briefly after introducing the technique below.

<h2>7.3   4-CHAINS</h2>

For 4-chains we offer one proof sketch, an example of the topological method of Demaine et al. (2003b) that can be viewed as an extension of the topological proof used to show that the knitting needles are locked (Theorem 6.3.1).

Theorem 7.3.1. *A triangle can interlock with an open, flexible 4-chain.*

Proof: *(Sketch)* The interlocked chains, a triangle (a, b, c) and a 4-chain $C = (v_0, v_1, v_2, v_3, v_4)$, are shown in Figure 7.5. Let H be the plane containing $\triangle abc$, and

Box 7.1: Topological Links

A topological link is a multicomponent knot. Just as two inequivalent knots cannot be deformed into one another without self-crossing, two inequivalent links cannot be deformed into one another without one passing through the other. Topologists name a link by the minimum number of crossings in a drawing, with a superscript indicating the number of components, and a subscript an arbitrary table index. Thus 5_1^2 can be drawn with 5 crossings, it has 2 components, and it is the 1st (and, in this case, only) such in the standard tables. Although Figure 7.5 is drawn with 6 crossings, the e_1/e_4 crossing could be eliminated. A drawing of the link with 5 crossings is shown in Figure 7.6 (see, e.g., Adams 1994, p. 287, or Weisstein 1999, p. 1086).

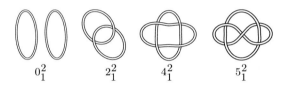

$$0_1^2 \qquad 2_1^2 \qquad 4_1^2 \qquad 5_1^2$$

Figure 7.6. The first few two-component links. [Images produced by knotplot: http://www.knotplot.com/.]

denote the halfspaces above and below by H^+ and H^-, respectively. Let the circumcircle of $\triangle abc$ has center o and radius r. We will view $\triangle abc$ and o as fixed.

Chain C is situated so that $\{v_0, v_2\} \subset H^-$ and $\{v_1, v_3, v_4\} \subset H^+$.

The plan of the proof is analogous to that used in Theorem 6.3.1: define a large ball, and connect v_0 to v_4 with a "rope" γ outside the ball, arguing that the topology of the resulting "links" cannot change. Let R be a distance representing the "short" links, $R = r + |e_2| + |e_3|$, and set the length of e_1 and e_4 to be much longer, $20R$. Define two open balls centered on o: B of radius $15R$, and B' of radius $4R$ (refer to Figure 7.7).

The initial placement of the chains has v_0 and v_4 outside of B (because the end links of C are so long), while a, b, c, v_1, v_2, and v_3 are all inside $B' \subset B$. As long as v_0 and v_4 stay outside B, we can attach a sufficiently long unknotted rope between v_0 and v_4 that remains outside B, and our configuration is equivalent to the topological link known as 5_1^2 (see Figure 7.6 and Box 7.1).

The non-interlocked configuration of the two closed chains corresponds to two separable unknots, 0_1^2 (Figure 7.6), so any motion separating this configuration requires v_0 or v_4 to enter the ball B'. Before that happens, v_0 or v_4 will have to enter the ball B.

Consider the first event when v_0 or v_4 touches the boundary of B, say v_0 without loss of generality. When v_0 touches B, point v_1 must be exterior to B' by at least R, because $20R > 15R + 4R$ (see Figure 7.7). Then the vertices v_1, v_2, and v_3 all must be out of B' (because $R > |e_2| + |e_3|$) but still inside B (because $10R > |e_2| + |e_3|$). Thus, the only elements inside B' are $\triangle abc$ and possibly the two links e_1 and e_4.

The remainder of the proof, which we do not repeat here, analyzes all the configurations possible in this constrained situation and concludes that they are either geometrically impossible (e.g., Figure 7.8) or topologically equivalent to 0_1^2, 2_1^2, or 4_1^2 (Figure 7.9). Because none of these are topologically equivalent to the starting configuration, and because the analysis concerns the situation at the first event of v_0 entering B, which must precede any topological change, this first event can never happen.   □

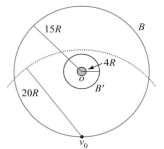

Figure 7.7. When v_0 touches B, point v_1 must be exterior to B' by at least R, and therefore v_2 and v_3 are also exterior to B'. [Figure 3 of Demaine et al. 2003b, by permission, Elsevier.]

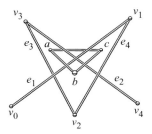

Figure 7.8. A configuration incompatible with the fact that v_0 or v_4 touch the boundary of B. [Figure 4 of Demaine et al. 2003b, by permission, Elsevier.]

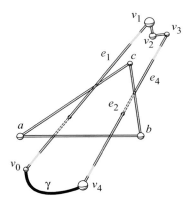

Figure 7.9. The link 4_1^2, formed when links e_1 and e_4 both pierce $\triangle abc$. [Figure 5 of Demaine et al. 2003b, by permission, Elsevier.]

Returning to Theorem 7.2.2, the configuration in Figure 7.4 represents the topological link 7_1^2, and defining a suitably large ball B with v_0 and v_4 connected outside, it can be argued similarly that separating the two chains requires a topological change in the link structure.

We have not been successful at extending this topological technique to handle two open chains. The difficulty is that the two ropes exterior to B interact and interfere with one another. For such problems, the only successful proof technique found is geometric and ad hoc. The most important example is this theorem, a strengthening of Theorem 7.2.2, establishing the 3f/4f cell of Table 7.1:

Theorem 7.3.2. *An open, flexible 3-chain can interlock with an open, flexible 4-chain.*

Figure 7.10 shows an interlocked configuration of a 3-chain (w, x, y, z) and a 4-chain (A, B, C, D, E). It is proved in Demaine et al. (2002b, Thm. 11) that the convex hull of the five interior joints B, C, D, x, and y can never change its combinatorial structure from

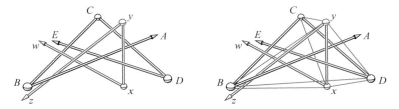

Figure 7.10. Interlocked open, flexible 3- and 4-chains. [Figure 6 of Demaine et al. 2002b, by permission, Elsevier.]

the bipyramid illustrated. The analysis tracks orientation determinants representing the relationship of one vertex with respect to the plane containing three others, and uses these to capture the configurations when one link pierces a particular triangle, for example, when xy pierces $\triangle BCD$. The initial piercings are then shown to be invariants, in that no event that could lead to a change in their status can happen first.

Similar reliance on orientation determinants is used to prove that a flexible 2-chain can interlock with an open, rigid 5-chain, a result mentioned earlier.

Many particular interlocking situations remain to be explored, and it seems fair to say that an overall theory has yet to clearly emerge.

8 Joint-Constrained Motion

We now turn to chains—and more generally linkages—that restrict joint motions in some way. Many physical linkages have some type of joint-angle constraint, for example, robot arms (Figure 1.2, p. 12). Among the variety of possible constraints, we concentrate in this section on just two: keeping all angles between incident links fixed, and keeping all angles convex. Both have applications, and both have led to a rich collection of results.

8.1 FIXED-ANGLE LINKAGES

Introduction and motivation. This section is concerned with *fixed-angle linkages*: linkages with a fixed angle assigned between each pair of incident edges. (We'll use the term *linkage* to include both general and fixed-angle linkages.) Fixed-angle chains are of particular interest, because, as mentioned in Section 1.2.4 (p. 14), they may serve as models for protein backbones: each vertex models an atom, and each link an atom-to-atom bond. The backbone of the protein (ignoring the "side chains") is an open chain. The bond angles are nearly fixed, leaving one degree of spinning motion.[1] As this is the primary motivation for fixed-angle chains, we will disallow self-crossings to match the physical constraints. The folding behavior of fixed-angle chains is of intense current interest, constituting the unsolved "protein folding problem." We will more directly address protein folding via fixed-angle chains in Section 9.1.

Fixed-angle trees arise in the same protein models, either when the side chains cannot be ignored, or for so-called "star" or "dentritic" polymers (Frank-Kamenetskii 1997; Soteros and Whittington 1988). Fixed-angle linkages with a more general graph structure have fewer direct applications, but they are a natural topic, and shed light on the behavior of chains and trees. We should note that we require the angle between each pair of links incident to a common vertex v to remain fixed, not only the angle between links consecutively adjacent around v. This implies that the edges incident to v are rigid with respect to one another under the action of a motion that maintains fixed angles. In particular, if the edges are coplanar, they remain coplanar, a fact we will find of use below.

Dihedral motions. We define the motion of any linkage that maintains all angles fixed a *dihedral motion*. The reason for this terminology (originating in biochemistry) is

---

[1] We will see later (p. 148) that spinning about all bonds is not equally free.

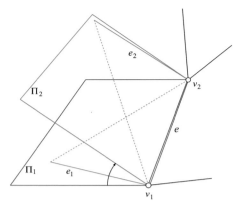

Figure 8.1. A local dihedral motion about edge e.

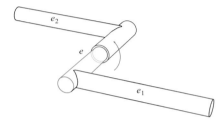

Figure 8.2. Dihedral motion about e viewed as constrained by nested cylinders.

that a dihedral motion can be "factored" into what we'll call *local dihedral motions* about individual edges of the linkage, each of which can be viewed as rotating two planes about their line of intersection. Let $e = (v_1, v_2)$ be an edge for which there is another edge e_i incident to each endpoint v_i. Let Π_i be the plane through e and e_i. A *(local) dihedral motion about e* changes the dihedral angle between the planes Π_1 and Π_2 while preserving the angles between each pair of edges incident to the same endpoint of e (see Figure 8.1). If we view e and $e_1 \in \Pi_1$ as fixed, then a dihedral motion spins e_2 about e; in Soss and Toussaint (2000) this motion is called an "edge spin." Thus, the focus of a dihedral motion is an edge, not the joint vertex. One can model each edge e as constructed from two concentric cylinders which can spin relative to one another, and to which are attached the incident edges (see Figure 8.2). Below we will selectively disallow local dihedral motions about certain *rigid* or *frozen* edges e, which can be achieved in this model by fusing the two cylinders that constitute e.

Our definition of "dihedral motion" includes rigid motions of the entire linkage, which could be considered unnatural because a rigid motion has no local dihedral motions. However, including rigid motions among dihedral motions does not change the results we will present. For a linkage of a single connected component, we can "modulo out" rigid motions; and for multiple connected components, we will always pin vertices to prevent relative rigid motions.

The investigation of fixed-angle chains from the point of view of computational geometry was initiated by Soss and Toussaint (2000, 2001), and we will start by presenting their hardness results on two problems they identified.

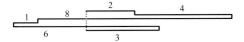

Figure 8.3. A chain whose minimum span is $5/6$; $2 + 4 + 6 = 1 + 8 + 3$ is the corresponding partition of S.

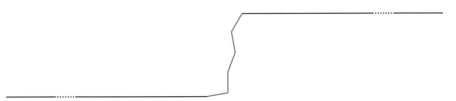

Figure 8.4. Internal turns of $10° + 80° − 20° + 30° − 40° − 60°$ result in parallel end links, corresponding to the $1 + 8 + 3 = 2 + 4 + 6$ partition. (Not to scale: the dots indicate omissions.)

8.1.1 Extreme Spans

The first problem is to find a configuration of a given fixed-angle chain that maximizes or minimizes the *span* of the chain, defined as the distance between the two endpoints, v_0 and v_n. Soss cites in his thesis 22 papers in polymer physics and other areas that study the mean squared-distance between endpoints, and argues that finding the full range for the span of a molecule is the first step toward computing the statistical distributions. Aside from this practical justification for the problem, we will see in Section 8.2 that the span plays an important role in Cauchy's "arm lemma." Say that a configuration of a chain is *flat* (or has reached a *flat state*) if all of its links lie in one plane without self-intersection, that is, it is a planar configuration. Define the maximum *flat span* of a chain as the largest span achieved in one of the chain's flat states, and define the minimum flat span analogously. Soss proves that computing either the minimum or the maximum flat span is NP-hard. The corresponding problems in 3D remain open. We will sketch both proofs.

Minimum flat span. Both are reductions from PARTITION (p. 25). Let $S = \{x_1, \ldots, x_n\}$ be the set of integers to be partitioned into two equal parts. For the minimum flat span problem, a chain is created to have "S-links" of length x_i, and between every pair of links, a short edge of length $1/n$ orthogonally connecting them. In any configuration, all the S-links are parallel, as are all the short edges. The sum of the lengths of the short edges is $(n − 1)/n < 1$. Then it is easily seen that the chain has a flat configuration in which the span is less than 1 if and only if S has a partition into two equal halves (see Figure 8.3.) Notice that this proof is basically the same as that for ruler rolding, Theorem 2.2.1 (p. 25).

Maximum flat span. The proof for the maximum flat span problem is more technically difficult, but conceptually equally simple. A chain of $n + 1$ unit-length links is created, with the n fixed turn angles at their joints given by $\alpha_i = x_i/\sigma$, where $\sigma = \sum_{i=1}^{n} x_i$ is the sum of the integers in S. Finally, two long links, each of length L, are appended to either end of the chain. The idea is that now the span is maximized when the two long links are parallel; any misalignment shortens the span. Alignment is achieved by arranging the

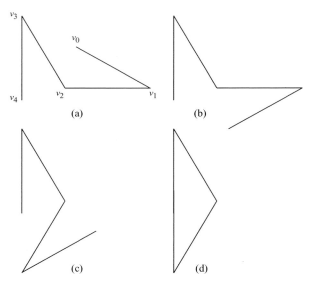

Figure 8.5. Four flat configurations of the same chain.

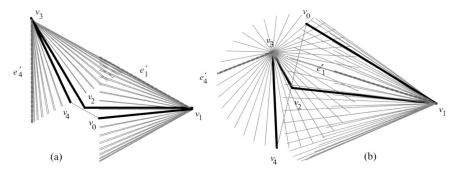

Figure 8.6. Chain achieving the maximum span in 3D is shown dark; dashed e'_1 and e'_4 edges achieve the maximum span in 2D. (a) View orthogonal to flat state, corresponding to Figure 8.5(a); (b) oblique view.

left turns between the unit-length links to exactly cancel the right turns, as illustrated in Figure 8.4. Soss computes that $L = 256n\sigma^2$ yields a bound $M = 2L - 8/n - 1$ such that the span is at least M if and only if set S has an exact partition.

Failure in 3D. It is common that a proof of NP-hardness in 2D immediately establishes the same result in 3D, usually because the solution in 3D is the same as the solution in 2D. But Soss showed that here this is not the case. Consider the 4-link chain shown in several flat states in Figure 8.5. The maximum planar span is achieved in the (congruent) configurations 8.5(a) and (c); in Figure 8.5(b) v_0 and v_4 are closer, and Figure 8.5(d) self-intersects. Now consider the same chain in 3D as illustrated in Figure 8.6. The distances $|v_0 - v_2|$ and $|v_4 - v_2|$ are fixed by the $(30°)$ fixed angles at v_1 and v_3 respectively, which constrain v_0 and v_4 to the circular rims of cones centered on $v_1 v_2$ and $v_2 v_3$ respectively. The maximum of $|v_0 - v_4|$ over all positions on these cones is achieved when $\{v_0, v_2, v_4\}$ are collinear (dotted in the figure), for it is then when the vectors $v_0 - v_2$ and $v_2 - v_4$ align to create their longest sum. This configuration is nonplanar, as a comparison of the dark edges to the dashed edges in Figure 8.6(b) shows. This leaves us with this intriguing open problem:

Open Problem 8.1: Extreme Span in 3D. [a] Is the problem of computing the maximum or the minimum span of a fixed-angle chain in 3-space NP-hard, as it is in 2D?

---

[a] Michael Soss, implicit in Soss (2001).

Recent work establishes that the problem can be solved in polynomial time for equiangular fixed-angle chains (Benbernou and O'Rourke 2006).

8.1.2 Flattening

The second problem identified by Soss and Toussaint is determining whether a given fixed-angle chain can be *flattened*: embedded in a plane without self-intersection. One can distinguish two versions of this question, depending on how the chain is "given": either abstractly, as a list of link lengths and fixed joint angles, or as a specific configuration in $\mathbb{R}^3$. We start with this latter version, which was the focus of Soss and Toussaint (2000).

Nonplanar chains. First, not all fixed-angle chains can be flattened. For example, the knitting needles Figure 6.2 (p. 88) cannot be, as it is locked in 3D even with all joints flexible. So restricting the joint angles leaves it locked and unable to be flattened. Recall that no chain of fewer than five links can be locked in 3D. However, we now show a fixed-angle 4-chain that cannot be flattened.

Coordinates for its five vertices, with $\varepsilon > 0$ a small number (e.g., $\varepsilon = \frac{1}{4}$) are given below:

v_0	$(0, -\frac{1}{2}, \varepsilon)$
v_1	$(0, 1, 0)$
v_2	$(0, 0, 0)$
v_3	$(1, 0, 0)$
v_4	$(0, 0, \frac{3}{2}\varepsilon)$

(see Figure 8.7). Think of $\{v_1, v_2, v_3\}$ as fixed and determining a coordinate system (for the angle at v_2 is fixed to $90°$). The two end vertices, v_0 and v_4, are free to rotate on circles. With ε small, the $v_0 v_1$ link is nearly parallel to the fixed $v_2 v_1$ link, and so all positions of v_0 leave the two links nearly parallel. The link $v_3 v_4$ is above $v_0 v_1$ in the initial configuration, and must become coplanar with $\{v_1, v_2, v_3\}$ in the xy-plane in the final, flattened configuration. But when $\varepsilon < \frac{1}{3}$, $\frac{3}{2}\varepsilon < \frac{1}{2}$, e_4 cannot "get around" e_1, and the two must cross in the xy-plane.

It is easy to see that any 3-link chain (e_1, e_2, e_3) can be flattened, by the following argument. Fix the plane Π to contain $\{e_1, e_2\}$. The third link e_3 moves on a cone centered

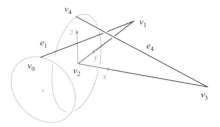

Figure 8.7. This fixed-angle 4-chain cannot be flattened, although if all joints are flexible it is not locked. Light colored circles constrain positions of v_0 and v_4.

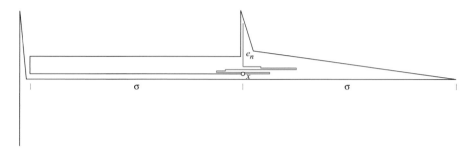

Figure 8.8. The "key" e_n fits in the "lock" if and only if S has a partition.

on e_2. Rotate e_3 onto Π to land on the side of e_2 opposite e_1. The symmetry of the cone guarantees this is always possible, and results in a nonintersecting configuration in Π.

Flattening NP-hard. Soss and Toussaint proved that deciding whether a given fixed-angle chain can be flattened is NP-hard, again by reduction from PARTITION. The full construction of the chain is shown in Figure 8.8. The central section is the same chain as used in Figure 8.3: links each the length of an integer in the set S, connected by short links at $90°$. It has the same property, permitting the start and end to be close if and only if S has a partition. In this case, this allows the longer, last vertical link e_n of the chain to fit, like a "key," inside a surrounding "lock." The lock chain, composed of nine links, is 2σ wide, where σ is the sum of the elements of S. As shown in Figure 8.9, it has only two flat configurations, both of which "seal off" the left end. The critical issue is where to place e_n in the lock structure. The width 2σ prevents the chain from reaching from x to outside of the lock. This only leaves the vertical spike as a "home" for e_n. Thus the chain can be flattened if and only if S can be partitioned.

8.1.3   Flat-State Connectivity

A *flat configuration/state* of a linkage is an embedding of the linkage into $\mathbb{R}^2$ without self-intersection. A linkage X is *flat-state connected* if, for each pair of its (distinct, i.e., incongruent) flat configurations x_1 and x_2, there is a dihedral motion from x_1 to x_2 that stays within the free space throughout. In general, this dihedral motion alters the linkage to nonflat configurations in $\mathbb{R}^3$ intermediate between the two flat states. If a linkage X is not flat-state connected, we say it is *flat-state disconnected.*

Flat-state disconnection could occur for two reasons. It could be that there are two flat states x_1 and x_2 which are in different components of free space but the same

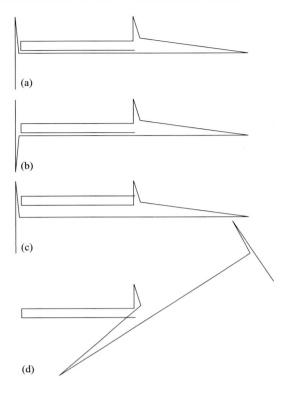

Figure 8.9. (a) and (b) Flat embeddings of the lock chain; (c) and (d) self-crossing configurations.

component of configuration space: they are connected by dihedral motions, but not by dihedral motions that avoid self-intersection. Or it could be that the two flat states are in different components of configuration space: there is no dihedral motion between them even permitting self-intersection. The former reason is the more interesting situation for this section; currently no nontrivial examples of the latter possibility are known.[2]

Here we report on work whose goal is to delimit the class of linkages that are flat-state connected. Research to date has identified several classes of linkages that are flat-state connected, and a few that are not, as summarized in Table 8.1.

The situation at this writing has left several gaps, and in particular, this basic problem is unresolved:

Open Problem 8.2: Flat-state Connectivity of Open Chains. [a] Is every open, fixed-angle chain flat-state connected?

---

[a] Aloupis et al. (2002a).

We will not detail all the special classes of linkages whose flat-state connectivity has been settled. Instead, we will select one thread through the available results, concentrating on *orthogonal linkages*, those in which any two edges incident to a common

---

[2] The only examples involve restricting motion so much that the two flat states are completely rigid.

Table 8.1: Summary of known results

Constraints on fixed-angle linkage		Flat-state connectivity	Reference
Connectivity	Chain Geometry		
Open chain	Has monotone state	Connected	Aloupis and Meijer (2006)
	α orthogonal or obtuse	Connected	Aloupis et al. (2002a)
	α equal acute	Connected	Aloupis et al. (2002b)
	unit; $\alpha \in (60°, 150°)$	Connected	Aloupis et al. (2002b); Aloupis and Meijer (2006)
Multiple pinned open chains	Orthogonal	Connected	Aloupis et al. (2002a)
Closed chain	Unit; orthogonal	Connected	Aloupis et al. (2002a)
Tree	Orthogonal; partially rigid	Disconnected	Aloupis et al. (2002a)
Graph	Orthogonal	Disconnected	Aloupis et al. (2002a)

"Unit" means with unit link lengths; a "monotone state" is a planar, strictly monotone configuration; α represents the joint angles.

vertex form an angle between them that is a multiple of $90°$. We will prove that open orthogonal chains are flat-state connected, but some orthogonal graphs might not be. It may help to start with the latter result, as it is not immediately clear how any linkage could be flat-state disconnected.

8.1.4   Flat-State Disconnected Linkages

All our examples revolve around the same idea, which can be achieved under several models. We start with partially rigid orthogonal trees, and then modify the example for other classes of linkages.

8.1.4.1 Partially Rigid Orthogonal Tree

An *orthogonal tree* is a tree linkage such that every pair of incident links meet at a multiple of $90°$. (Recall that the angle between two incident links is the smaller of the two angles between them in the plane they determine; thus the angle is always $\leq 180°$.) *Partial rigidity* specifies that only certain edges permit dihedral motions.

Figure 8.10(a–b) shows two incongruent flat states of the same orthogonal tree; we will call the flat configurations $X_{(a)}$ and $X_{(b)}$. All but four edges of the tree are frozen, the four incident to the central degree-4 root vertex x. It is through rigidifying much of the linkage that we can achieve flat-state disconnection. Call the 5-link branch of the tree containing a the *a-branch*, and similarly for the other 4-link branches. Label the vertices of the *a*-branch (a, a_1, a_2, a_3), and similarly for the others. Although perhaps not evident from the figure, $|aa_1| < |cc_1|$ and $|bb_1| < |dd_1|$.

We observe three properties of the example. First, as mentioned previously, fixed-angle linkages have the property that all links incident to particular vertex remain coplanar throughout all dihedral motions. In Figure 8.10, this means that $\{x, a, b, c, d\}$ remain coplanar; and we view this as the plane Π of the flat states under consideration.

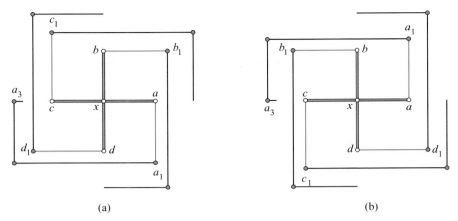

(a)   (b)

Figure 8.10. Two flat states of an orthogonal tree. The four edges incident to x are the only ones not rigid, permitting dihedral motions.

Note that, for example, a rotation of just a about bd would maintain the 90° angles between all edges adjacent consecutively around x, but would alter the 180° angle between xa and xc, and thus is not a fixed-angle motion.

Second, $X_{(a)}$ and $X_{(b)}$ do indeed represent incongruent flat states of the same linkage. The purpose of the extra pin attached to a_3 is to ensure this asymmetry. Without this pin, a flat state congruent to $X_{(b)}$ could be obtained by a rigid motion of the entire linkage: rotation of $X_{(a)}$ by 180° about a vertical line, followed by a 90° clockwise rotation, matches all details but the location of that pin.

Third, it is clear that state $X_{(b)}$ can be obtained from state $X_{(a)}$ by rotating the a-branch 180° about xa, and similarly for the other branches. Thus the two flat states are in the same component of configuration space, in that there are dihedral motions connecting them. We now show that no dihedral motions connect them without self-intersection, which establishes that they are in different components of the free space.

Theorem 8.1.1. *The two flat states in Figure 8.10 cannot be reached by dihedral motions that avoid crossing links.*

Proof: Each of the four branches of the tree must be rotated 180° to achieve state $X_{(b)}$. We first argue that two opposite branches cannot rotate to the same side of the Π-plane, either both above or both below. Without loss of generality, assume both the a- and the c-branches rotate above Π. Then, as illustrated in Figure 8.11, edge $a_1 a_2$ must properly cross the cc_1 edge, because $|aa_1| < |cc_1|$ (the figure illustrates the situation when these edges are the same length, which results in crossing but not proper crossing).

Now we argue that two adjacent branches cannot rotate to the same side of Π. Consider the a- and b-branches, as illustrated in Figure 8.12. As it is more difficult to identify an exact pair of points on the two branches that must collide, we instead employ a topological argument, similar to that used in Theorem 7.3.1. Connect a shallow rope R from a to a_3 underneath Π, and a rope S from b to b_3 that passes below R (see Figure 8.13). In $X_{(a)}$, the two closed loops $A = (R, a, a_1, a_2, a_3)$ and $B = (S, b, b_1, b_2, b_3)$ are unlinked, representing the link 0_1^0. But in $X_{(b)}$, A and B are topologically linked, representing the link 2_1^2. Therefore, it is not possible for the a- and b-branches to rotate above Π without passing through one another.

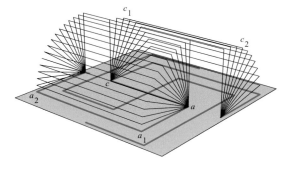

Figure 8.11. The a- and c-branches collide when rotated above. Here the linkage is drawn with $|aa_1| = |cc_1|$ and $|bb_1| = |dd_1|$.

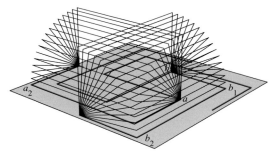

Figure 8.12. The a- and b-branches collide when rotated above.

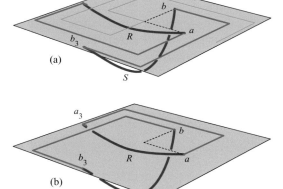

Figure 8.13. With the additions of the ropes R and S underneath, the a-chain is not linked with the b-chain in (a), but is linked in (b).

Finally, we apply the pigeon-hole principle. There are four branches, and two sides of Π. Thus one of the sides (say, above) must have (at least) two branches rotating through it. Whether these branches are opposite or adjacent, a collision is forced.   □

8.1.4.2 Orthogonal Graph

It is easy to convert the partially rigid tree to a completely flexible, orthogonal graph, simply by using extra "struts" to enforce rigidity. One method is shown in Figure 8.14. Note that a degree-3 vertex v on the interior of an edge does not permit a dihedral motion that "folds" the edge at v, for that would change the angle between the collinear links forming the edge from its initial 180°. The degree-4 vertices are rigid, with the

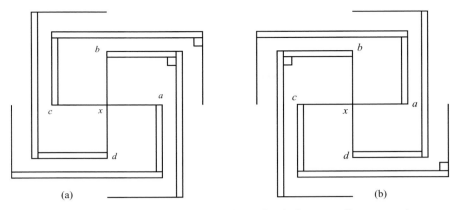

Figure 8.14. An orthogonal graph linkage that is flat-state disconnected.

result that only dihedral motions are possible around the four edges incident to x, just as in the tree example.

What is unclear is if disconnection is possible with orthogonal trees, no edges rigid:

Open Problem 8.3: Flat-State Connectivity of Orthogonal Trees.[a] Is there a fixed-angle, orthogonal tree that is flat-state disconnected?

---

[a] Aloupis et al. (2002).

We now turn to the positive result previously mentioned.

8.1.4.3 Open Orthogonal Chains

In this section we prove that any open, orthogonal chain can be reconfigured to a canonical form, establishing that this class is flat-state connected. Let $C = (v_0, v_1, \ldots, v_n) = (e_1, \ldots, e_n)$ be the chain in the xy-plane. We will call the z-direction "vertical." A *staircase* S is a planar, orthogonal chain, whose angle turns alternate between left and right; staircases are monotone with respect to either direction determined by its steps. The canonical form for C is a staircase S in a vertical plane containing its last edge e_n. At an intermediate stage of the algorithm, the first $i - 1$ edges of C have been picked up into S, which lies in a vertical "quarter" plane containing $e_i = (v_{i-1}, v_i)$. Vertex v_{i-1} at the foot of a staircase is its *base vertex* and the next edge of the chain e_i is the staircase's *base edge*. The first, z-edge of the staircase we call its *pivot edge*. Say that a staircase S is *aligned* with its base edge e_i if the base edge continues the monotone staircase in the sense that the first edge of S parallel to the xy-plane does not project onto e_i. Otherwise, the first two edges of the staircase form a nonmonotone "bracket" with its base edge.

The algorithm picks up one edge at a time, as follows:

1. Rotate S about its pivot edge until it is aligned with its base edge e_i.
2. "Pick-up" e_i by rotating it and the attached S by $90°$ about e_{i+1} until e_i is vertical.

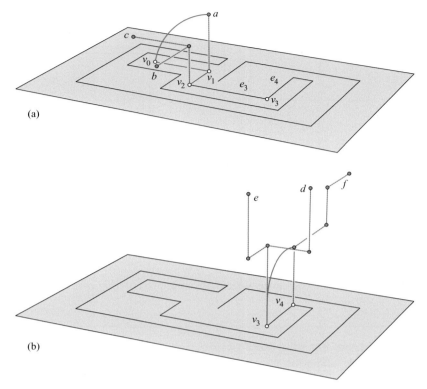

Figure 8.15. (a) Lifting edges $e_1 = (v_0, v_1)$ and $e_2 = (v_1, v_2)$: a, b, c; (b) lifting edges e_3 and e_4: d, e, f.

The first step is essential, for if S formed a bracket with e_i, the rotation in the second step could drive S through the xy-plane, risking collision with C. Alignment guarantees that S is lifted above the plane. The example shown in Figure 8.15 follows these detailed steps:

a. Rotate $e_1 = (v_0, v_1)$ about $e_2 = (v_1, v_2)$ until it is vertical.
b. Rotate e_2 about e_3 until it is vertical.
c. Rotate staircase about its pivot edge e_2 to align it with e_3.
d. Rotate e_3 about e_4 until it is vertical.
e. Rotate S about its pivot edge e_3 until it is aligned with e_4.
f. Rotate e_4 about e_5 until it is vertical.

Note that the half-space below xy is never entered.

This algorithm and result has been extended in several directions. One is that any collection of open, orthogonal chains, each with one edge pinned to the xy-plane, is flat-state connected. Under a "general position" assumption (no two pinned vertices from different chains have a common x- or y-coordinate), it is established in Aloupis et al. (2002a) that the chains may be reconfigured to a canonical form, with each chain a staircase in a plane parallel to the z-axis and containing its pinned edge.

That same paper extends the single orthogonal chain algorithm to all open chains with obtuse fixed angles:

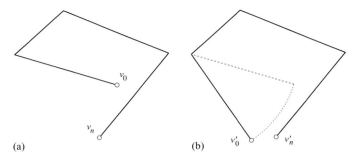

Figure 8.16. (a) A chain that fails to be convex (v_0 is not on the hull); (b) $|v_0' - v_n'| < |v_0 - v_n|$.

Theorem 8.1.2. *Every open chain all of whose fixed angles are nonacute is flat-state connected.*

Extending this theorem to acute angles (cf. Open Problem 8.2), or to closed chains, remains open.

8.2 CONVEX CHAINS

8.2.1 Cauchy's Arm Lemma

In this section we explore a simple but important joint constraint: maintaining all joint angles convex. We start in 2D, where the meaning of "convex" is clear, but will, in Section 8.2.3, generalize to $\mathbb{R}^3$. The source for this line of development is Cauchy's 1813 rigidity theorem (Cromwell 1997, p. 228), which we will study in Part III (Section 23.1). We will not pause here to explain his rigidity theorem, but instead concentrate on the key lemma, which has come to be known as *Cauchy's arm lemma*, and which we have already had occasion to employ (Lemma 5.3.1, p. 72).

Let $C = (v_0, v_1, \ldots, v_n)$ be a polygonal chain in the plane. We will view C as an open chain, with v_0 and v_n distinct vertices, although they may happen to be located at the same point. In particular, there is no link connecting v_0 to v_n. We require C to be nonintersecting (except for the possible overlap of v_0 and v_n). Define such a chain to be *convex* if all of its edges lie on the convex hull of C. For a convex chain, the internal angles $\alpha_1, \ldots, \alpha_{n-1}$ at the joints all lie in the range $(0, \pi]$. With this notation, here is one phrasing of Cauchy's lemma:

Lemma 8.2.1. *If a planar convex chain C is opened by increasing some or all of its internal angles, but not beyond π—angle α_i is replaced by α_i', with $\alpha_i \leq \alpha_i' \leq \pi$—then the distance between v_0 and v_n is increased: $|v_0' - v_n'| > |v_0 - v_n|$.*

We first note that the claim of the lemma might be false if all internal angles are convex but the chain as whole is not convex, as illustrated in Figure 8.16. In fact, the situation illustrated in this figure could arise at a substep in Cauchy's original induction proof, a flaw which was noticed and corrected by Steinitz a century later. Consequently, Lemma 8.2.1 is sometimes known as "Steinitz's lemma" (Cromwell 1997, p. 235). Now several different proofs simpler than Steinitz's are available, for example,

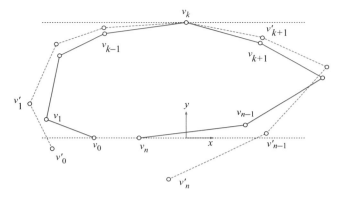

Figure 8.17. Opening C while keeping v_k fixed and highest.

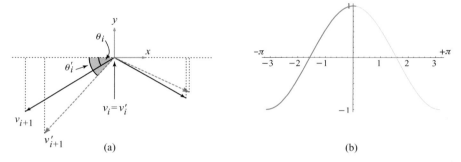

Figure 8.18. (a) Angles between rays and x-axis; (b) cosine function.

Singer (1997) and Aigner and Ziegler (1999, p. 64). Here we will present a beautiful proof of S. K. Zaremba (Schoenberg and Zaremba 1967).[3]

Proof: Establish a coordinate system in the plane of the chain as follows. The x-axis contains v_0 and v_n, with v_n right of v_0. The y-axis passes through the vertex v_k furthest from $v_0 v_n$. (If there are two vertices tied for furthest, let v_k be the right one.)

C is opened to C', and placed so that

1. v_k does not move: $v'_k = v_k$;
2. the new angle $\angle v'_{k-1} v_k v'_{k+1}$ contains the old angle $\angle v_{k-1} v_k v_{k+1}$; and
3. neither v'_{k-1} nor v'_{k+1} is placed above v_k.

Because $\alpha_k \leq \alpha'_k \leq \pi$, it is possible to satisfy these three conditions (see Figure 8.17).

For the right portion of the chain, from v_k to v_n, define θ_i to be the angle between the ray $v_i v_{i+1}$ and the positive x-axis. This angle lies in $[-\pi, 0)$, for all these rays point downward. The opening motion increases (or leaves unchanged) each θ_i, as illustrated in Figure 8.18(a), for each ray is turned counterclockwise by opening angles. The new angles never exceed 0, so we have $\theta_i \leq \theta'_i \leq 0$.

Let ℓ_{i+1} be the length of the edge $v_i v_{i+1}$. Then the x-coordinate x_n of v_n can be computed as the sum of the horizontal contributions of each edge:

$$x_n = \ell_{k+1} \cos \theta_k + \ell_{k+2} \cos \theta_{k+1} + \cdots + \ell_n \cos \theta_{n-1} \tag{8.1}$$

---

[3]   We thank Godfried Toussaint for introducing us to Zaremba's proof.

Each $\theta_i' \geq \theta_i$, and because the cosine is an increasing function of its argument in the range $[-\pi, 0]$ (see Figure 8.18(b)), it must be that $x_n' \geq x_n$, that is, v_n moves rightward.

By symmetry, a similar argument establishes that v_0 moves leftward: $x_0' \leq x_0$. By assumption, at least some angles open, which means that at least one of the two inequalities is strict. Therefore,

$$x_n' - x_0' > x_n - x_0 = |v_n - v_0|,$$

which, because $|v_n' - v_0'| \geq x_n' - x_0'$, implies that $|v_0' - v_n'| > |v_0 - v_n|$. $\qquad\square$

8.2.2 Uses and Generalizations of Cauchy's Lemma

Despite the simplicity of Cauchy's lemma—or perhaps, because of it—it has a variety of applications:

1. Cauchy's original motivation was to prove that convex polyhedra are rigid. As mentioned, his arm lemma is a key component in that proof (Part III, Section 23.1.1).
2. It can be used to prove that certain curves on polyhedra "develop" in the plane (by rolling the polyhedra) without self-intersection (Part III, Section 24.5).
3. It can also be used to prove that "edge-unfoldings" of restricted classes of polyhedra avoid overlap (Part III, Section 24.5.3).

Several of these applications rely on generalizations, and indeed the lemma can be generalized in several directions, to

1. convex *spherical polygons*, those drawn with great-circle arcs on the surface of a sphere;
2. smooth, convex curves in the plane, where angle opening is replaced by diminishing the curvature at each point;
3. convex chains in the plane, but permitting the angle to become reflex; and
4. chains in $\mathbb{R}^3$.

That Cauchy's lemma generalizes to smooth curves is not surprising, as a smooth curve can be viewed as a limit of polygonal chains. But the third and fourth generalizations (which turn out to be fundamentally the same) are perhaps of more interest, and more closely related to the research on fixed-angle chains described in Section 8.1. We turn to these now.

8.2.3 Straightening Convex Chains

To describe the third generalization in the list above, it will be convenient to alter notation to concentrate on the "turn angles" of the chain rather than on the "interior angles" as we did above. Let $A = (a_0, a_1, \ldots, a_n)$ be a planar chain. The *turn angle* α_i at joint a_i, $i = 1, \ldots, n-1$, is the angle in $[-\pi, \pi]$ that turns the vector $a_i - a_{i-1}$ to $a_{i+1} - a_i$, positive for left (counterclockwise) and negative for right (clockwise) turns. Let $B = (b_0, b_1, \ldots, b_n)$ be a reconfiguration of A, with turn angles $\beta = (\beta_1, \ldots, \beta_{n-1})$.

Cauchy's lemma (Lemma 8.2.1) may be rephrased in this notation as follows: If A is a convex chain and B a reconfiguration of it that satisfies $\beta_i \in [0, \alpha_i]$, then we must have $|b_n b_0| \geq |a_n a_0|$. The generalization is as follows:

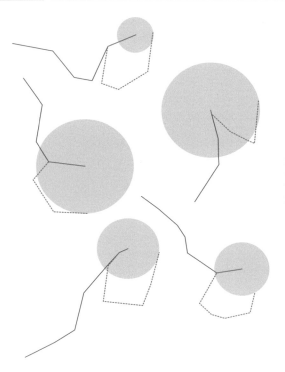

Figure 8.19. Reconfigurations of convex chain A (red) to B (blue) satisfying Equation (8.2). In each case, the forbidden disk is shown. [Based on Figure 8 of O'Rourke 2000c.]

Lemma 8.2.2. *If A is a convex chain (in the plane) and B a planar reconfiguration of it that satisfies*

$$\beta_i \in [-\alpha_i, \alpha_i] \tag{8.2}$$

then we must have $|b_n b_0| \geq |a_n a_0|$.

In other words, the Cauchy condition $0 \leq \beta_i \leq \alpha_i$ is replaced by $\beta_i \leq |\alpha_i|$. The conclusion is identical: the "span" (p. 133) of the chain increases. Another way to view the conclusion of the lemma is in terms of a "forbidden disk": the open disk of radius $|a_n a_0|$ (the length of the "missing" link) centered on the vertex a_0. This disk is forbidden in the sense that in any reconfiguration of A to B that satisfies Equation (8.2), the last vertex b_n cannot enter this disk. The precondition of the lemma can be viewed as demanding "straightening": all angles turn less sharply (or stay the same), that is, they become more nearly straight. These viewpoints perhaps render the claim of the lemma as less surprising. Figure 8.19 may provide more insight, and exploring the movement of chains under the straightening constraint with a Java applet[4] further strengthens intuition.

Three proofs of Lemma 8.2.2 are provided in O'Rourke (2000c), but none with quite the elegance of Zaremba's proof. Instead of proving the lemma, we describe the fourth generalization mentioned above, to chains in $\mathbb{R}^3$. The generalization is easy to describe: simply strike the word "planar" from the statement of Lemma 8.2.2. But first we need to define the meaning of "turn angle" in $\mathbb{R}^3$. It may be interpreted in the natural way as the angle between the relevant vectors within the plane they determine. Note that the distinction between left and right turns, $+$ and $-$ angles, disappears in $\mathbb{R}^3$: the

---

4   http://cs.smith.edu/~orourke/Cauchy/.

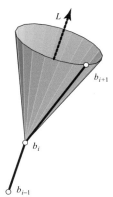

Figure 8.20. All b_{i+1} on the cone have the same turn angle at b_i.

turn angle β_i at b_i is the angle in $[0, \pi)$ between the line L containing (b_{i-1}, b_i) and the line containing (b_i, b_{i+1}). All positions of b_{i+1} on the cone centered on L with apex at b_i (Figure 8.20) have the same angle $\beta_i \geq 0$. The lemma is then:

Lemma 8.2.3 (Chern-Schur). *If A is a convex chain (in the plane) and B a reconfiguration of it into $\mathbb{R}^3$ that satisfies $\beta_i \leq \alpha_i$, then $|b_n b_0| \geq |a_n a_0|$.*

The most general form of the lemma, phrased in terms of smooth space curves and the absolute value of curvature rather than turn angle, is due to S. S. Chern (1967, p. 36), based on a much earlier proof by Axel Schur (1921). See O'Rourke (2001a) for the somewhat tangled history of this result. Chern captures its essence as follows: "if an arc is 'stretched,' the distance between its endpoints becomes longer."

It is interesting to interpret Lemma 8.2.3 in terms of fixed-angle chains (cf. p. 131) in $\mathbb{R}^3$, which satisfy the lemma because $\beta_i = \alpha_i$:

Corollary 8.2.4. *If a fixed-angle chain has a flat state A in which it forms a convex chain (in the sense defined on p. 143), then the span of any reconfiguration B into $\mathbb{R}^3$ exceeds the span of A.*

9 Protein Folding

A complete overview of protein folding is beyond our scope, being the topic of several other books (e.g., Creighton 1992, Pain 2000, Murphy 2001). Rather, we concentrate in the next three sections on three geometric approaches to protein folding.

9.1 PRODUCIBLE POLYGONAL PROTEIN CHAINS

As mentioned in Section 8.1 (p. 131), fixed-angle polygonal chains in 3D may serve as a model of protein backbones, with each vertex an atom and each link a covalent bond between atoms, with nearly fixed bond angles.[1] We detail this model a bit more before proceeding. Each amino acid residue in a protein backbone consists of three atoms: nitrogen followed by two carbons, conventionally named N–C_α–C'. As the name implies, the backbone is only the central core: H attaches to N, O attaches to C', and a "side chain" (which differs for each amino acid) attaches to C_α. We will only model the backbone without attachments. The residues are joined by peptide bonds, and the full structure is often called a polypeptide. All the bond lengths along the backbone (link lengths in the model) are approximately fixed, at values between 1.33 and 1.52 A, that is, roughly the same length. The bond angles are again approximately fixed angles:

$$\angle NC_\alpha C \approx 109.5°, \quad \angle C'_\alpha C' N \approx 116°, \quad \angle CNC_\alpha \approx 122°.$$

Again note that it is not too far wrong to treat all bond angles as roughly equal. In our polygonal chain model, we permit free spinning (dihedral motions, p. 131) about each link, but in fact the C'–N bond only permits two values: 0° and 180°, resulting in two planar configurations (known as *trans* and *cis*). Thus the chain is "partially rigid" in our earlier notation (p. 138). However, we will ignore this partial rigidity, modeling protein backbones as fixed-angle polygonal chains, and when convenient, with equal joint angles and equal link lengths.

In this section (based on Demaine et al. 2006c), we explore a crude model of the production of a fixed-angle chain and explore its consequences. Our inspiration derives from the ribosome, which is the "machine" that creates protein chains in biological cells. Figure 9.1 shows a cross section of a ribosome and its exit tunnel, based on a model developed by Nissen et al. (2000). We quickly deviate from the complex reality of the ribosome's structure, a topic of active investigation in laboratories around the world,

---

[1] Another approach is to model each residue by a link, with residues connected by a peptide bond vertex. In this model, the bond angle may take on a wide range of values.

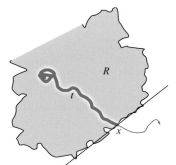

Figure 9.1. The ribosome R in cross section. The protein is created in tunnel t and emerges at x. [Figure 1 of Demaine et al. 2006c, by permission, Springer.]

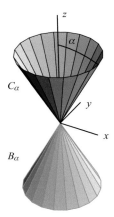

Figure 9.2. The chain is produced in C_α, and emerges at the origin into the complimentary cone B_α below the xy-plane. [Figure 2 of Demaine et al. 2006c, by permission, Springer.]

and replace the ribosome by a simple geometric constraint: the chains are produced inside a cone of half-angle $\alpha \le \pi/2$, emerging through its apex.

9.1.1 Notation

The fixed-angle polygonal chain P has $n+1$ vertices $v_0, \ldots, v_n$ and is specified by the fixed turn angle θ_i at each vertex v_i, and by the edge length ℓ_i between v_i and v_{i+1}. When all angles $\theta_i \le \alpha$ for some $0 < \alpha \le \pi/2$, P is called a $(\le\alpha)$-*chain*. Please note that in Section 8.1, we focused on the angle between adjacent edges, $\pi - \alpha$. Thus a "nonacute chain" corresponds to a $(\le\pi/2)$-chain here. The turn angle is more in consonance with cone production.

We distinguish the abstract chain from a *configuration* $Q = \langle q_0, \ldots, q_n \rangle$ of the chain P (see Figure 9.3): an embedding of P into $\mathbb{R}^3$, that is, a mapping of each vertex v_i to a point $q_i \in \mathbb{R}^3$, satisfying the constraints that the angle between vectors $q_{i-1}q_i$ and q_iq_{i+1} is θ_i, and the distance between q_i and q_{i+1} is ℓ_i, spanned by a straight segment e_i.

9.1.2 Chain Production

Our model of protein production in the ribosome is that the chain is produced inside an infinite open cone C_α with apex at the origin, axis on the z-axis, and half-angle (to the positive z-axis) $\alpha \le \pi/2$; (see Figure 9.2). Let $\overline{C}_\alpha$ be the corresponding closed cone. We similarly define the cone B_α, the mirror image of C_α with respect to the xy-plane.

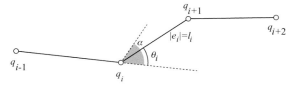

Figure 9.3. Notation for a configuration Q. [Figure 3 of Demaine et al. 2006c, by permission, Springer.]

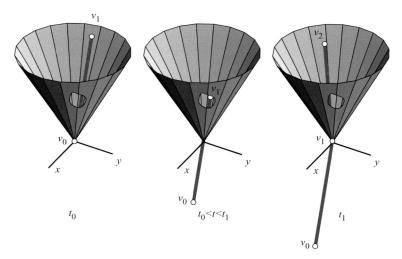

Figure 9.4. Production of e_0 and e_1 during $t \in [t_0, t_1]$. [Figure 4 of Demaine et al. 2006c, by permission, Springer.]

The vertices and edges are created inside $\overline{C}_\alpha$ and exit the machine at the apex of $\overline{C}_\alpha$. The portion of the chain already produced is allowed to move freely as long as it stays simple and never meets C_α. At time $t_0 = 0$, the machine creates v_0 at the apex of C_α, v_1 inside $\overline{C}_\alpha$, and the segment e_0 connecting them (see Figure 9.4). In general, at time t_i, vertex v_i reaches the apex of $\overline{C}_\alpha$, and v_{i+1} and e_i are created inside $\overline{C}_\alpha$. The vertex v_i stays in $\overline{C}_\alpha$ between times t_{i-1} and t_i and $0 = t_0 < t_1 < \cdots < t_n$.

This instantaneous creation of e_i at the moment that v_i reaches the apex is a crucial aspect of the model. Allowing time for e_{i-1} to rotate exterior to the cone prior to the creation of e_i leads to rather different results, as detailed in Demaine et al. 2006.

A configuration Q is said to be α-*producible* if there exists a continuous motion of the chain as it is created by the above model from within a cone C_α. It is nearly obvious from the production model that a chain produced in a cone $C_{\alpha/2}$ must be an $(\leq\alpha)$-chain, for a half-angle of $\alpha/2$ permits a total turn at the cone apex of at most α.

9.1.3   Main Results

There are two main results in Demaine et al. (2006c), which we first state and interpret, followed by a hint at the proofs.

The first result is roughly that the producible configurations are the same as the flattenable (cf. Section 8.1.2, p. 135) configurations:

Theorem 9.1.1. *A configuration of a $(\leq\alpha)$-chain is α-producible if and only if it is flattenable, provided $\alpha \leq \pi/2$.*

Many of the results are restricted to $\alpha \leq \pi/2$, which fortunately accords with the generally obtuse (about $110°$) protein bond angles between adjacent residues, which correspond to turn angles α of about $70°$. In fact, it is useful to concentrate for specificity on $\alpha = \pi/2$, orthogonal chains. Then the cone C_α is a half-space with a pinhole, which is not as an inaccurate model of the ribosome as one might think (cf. Figure 9.1). The theorem in this case says: The class of half-space-producible configurations of orthogonal chains is identical to the flattenable configurations. This connects two seemingly disparate classes of chains.

The second result, a consequence of the first, is that the α-producible chains are rare in a technical sense:

Theorem 9.1.2. *If a class of chains includes a lockable configuration, then the probability that a random configuration of a random chain is α-producible approaches zero geometrically as $n \to \infty$.*

The precondition here is important, for much remains unknown about which classes of chains can lock. However, because the locked "knitting needles" example (Figure 6.2, p. 88) can be built with chains satisfying $\alpha \leq \pi/2$ (by replacing the acute-angled joints with obtuse-angled chains of very short links), we know the precondition holds for $\alpha \leq \pi/2$. Thus, in this case, its conclusion follows: configurations of ($\leq \alpha$)-chains are rarely producible. The consequence is that the space of all possible configurations of a class of chains is vastly larger than the space of producible configurations, with the potential for reduction in the search space for protein foldings. We will return to this point below.

9.1.4 Producible $\equiv$ Flattenable

Theorem 9.1.1 is established in two main steps:

1. Every α-producible configuration can be moved to a canonical configuration and therefore to every other α-producible configuration.
2. All flat configurations of a ($\leq \alpha$)-chain have an α-production for $\alpha \leq \pi/2$.

The canonical configuration, called the *α-cone canonical configuration* or α-CCC, bears a resemblance to the helical form preferred by many proteins (see Figure 9.5 for an example). Its precise definition is a bit technical, and will only be sketched here. Let Q be a particular configuration of a ($\leq \alpha$)-chain, and normalize all edge vectors $q_i q_{i+1}$ to unit vectors $u_i = (q_{i+1} - q_i)/|q_{i+1} - q_i|$ which lie on the unit sphere. The α-CCC is constructed to have the property that all such vectors lie along a circle σ of radius $\alpha/2$ on that sphere. In other words, the vectors u_i lie on the boundary of a cone with half-angle $\alpha/2$, $\overline{C}_{\alpha/2}$.

Thus, the α-CCC is completely contained in $\overline{C}_{\alpha/2}$. The motivation for confining it inside a cone of half the size of C_α is that the proof that every α-producible configuration requires this. That proof runs the production movements backward and builds up the partial chain inside $\overline{C}_{\alpha/2} \subset \overline{C}_\alpha$ (see Figure 9.6). During this reverse production, the chain from v_{i+1} to v_n is inside as e_i enters the apex of C_α. The partial α-CCC swings around inside C_α to accommodate the motion. The invariant that is maintained is that σ is

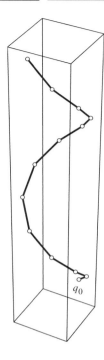

Figure 9.5. A chain in its α-CCC configuration. Here $\theta_i = \pi/4$ for all i. [Figure 7 of Demaine et al. 2006, by permission, Springer.]

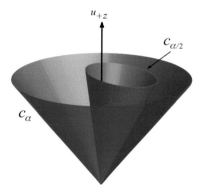

Figure 9.6. Cone $C_{\alpha/2}$ is nested inside C_α. [Figure 8 of Demaine et al. 2006, by permission, Springer.]

arranged to encircle the north-pole vector u_{+z} of C_α. This is possible because σ has radius $\alpha/2$ and diameter α, while C_α has half-angle α.

We now turn to the second point above that flat configurations of a $(\leq\alpha)$-chain have an α-production. The flat configuration will be drawn in the xy-plane. The basic idea of the proof is to create the edges e_i one after another, maintaining a moving cone C_α containing the edge, with its apex on the xy-plane. At the creation of e_i, $(e_0, \ldots, e_{i-1})$ already lie in the xy-plane. C_α is tangent to the xy-plane on the support line of e_{i-1} and with its apex at v_i. e_i is created in the vertical plane through e_{i-1} at angle θ_{i-1} with the xy-plane. It is then rotated down the plane, moving C_α accordingly. Throughout the motion, C_α never intersects the xy-plane.

Now, we know from Theorem 8.1.2 (p. 143) (Aloupis et al. 2002a) that any $(\leq\alpha)$-chain with $\alpha \leq \pi/2$ has a flat configuration. Thus, for $\alpha \leq \pi/2$, the class of α-producible configurations is exactly the class of flat configurations.

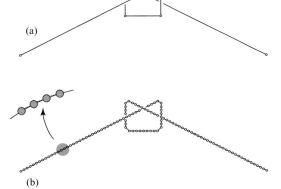

Figure 9.7. (a) The knitting needles; (b) a locked, fixed-angle, unit chain.

9.1.5   Random Chains

The result on random chains is somewhat technically contrived, but still interesting because of the potential search-space reduction. Recall we distinguish between an abstract chain and a configuration of a chain. A random chain of n links is defined by drawing the lengths and angles from any "regular" (e.g., uniform) distribution on any subset of the possible values. A random configuration of a chain embeds into $\mathbb{R}^3$ by, in addition, drawing the dihedral angles from any regular distribution. The result hinges on an "inheritance" property of locked configurations. Suppose the class has a locked configuration of k links. Then it can be established that there is a positive probability ρ that a random chain P_k of this length is locked: $\Pr[P_k$ is lockable$] \geq \rho > 0$. Consider a chain P_n of $n > k$ links. Break P_n into $\lfloor n/k \rfloor$ subchains of length k. P_n is lockable if any of the subchains are lockable, so the probability that P_n is not lockable is $< (1 - \rho)^{\lfloor n/k \rfloor}$, which approaches 0 geometrically as n grows. Thus, producible configurations of chains become rare as soon as one chain in the domain of the distribution is lockable.

Here we run into the limits of our knowledge on locked chains, for no nontrivial domain is known with no lockable chains. It was mentioned earlier that the class of chains with $\alpha \leq \pi/2$ definitely includes locked configurations, so the random result applies here. But because protein backbones are nearly equilateral (all edge-lengths equal), it is of particular interest to know whether there is a locked *unit* ($\leq \pi/2$)-chain, that is, one all of whose link lengths are 1 (obviously equilateral and unit chains are equivalent under a scale change). There is such a chain, as illustrated in Figure 9.7, but this result is achieved only by selecting many α's to be nearly 0, which differs from protein backbones which have α's roughly equal and closer to 90° than to 0. We end this section with four open problems motivated by this cutting-edge issue.

9.1.6   Open Problems on Unit Chains

> **Open Problem 9.1: Locked Unit Chains in 3D.** Can all unit chains in 3D, with universal joints, be straightened?

(See also Open Problem 6.1, p. 91.) A somewhat broader question, perhaps with more practical import, is the following:

Open Problem 9.2: Locked Length Ratio.[a] What is the smallest value of $L \geq 1$ for which there exists a locked open polygonal chain in 3D, again with universal joints, all of whose link lengths are in the interval $[1, L]$? (Call L the *length ratio* of the chain.) More precisely, to allow for the possibility that no specific value of L answers this question, define L to be the infimum of the ratios of the longest to the shortest link lengths over all locked chains. The knitting needles example (Figure 9.7(a)) shows that $L \leq 3$.

[a] Implicit in Cantarella and Johnston (1998).

This broad formulation concentrates on fixed-angle chains:

Open Problem 9.3: Locked Fixed-Angle Chains.[a] For a fixed-angle ($\leq \alpha$)-chain of n links and length ratio L, for which triples (α, n, L) does there exist a locked chain? Might it be that for any α, there is some n so that there is a locked chain of ratio $L = 1 + \varepsilon$, for arbitrarily small $\varepsilon > 0$?

[a] Stefan Langermann, January 2004.

We know, for example, that there are locked chains for $(\alpha, n, L) = (60°, 4, 1 + \varepsilon)$ and $(90°, 6, \sqrt{2} + \varepsilon)$. Open Problem 9.1 corresponds to asking whether $L = 1$.

Finally, this related question (Section 6.5) remains unresolved:

Open Problem 9.4: Locked Unit Trees in 3D?.[a] Can all 3D unit trees be flattened?

[a] Demaine (2001) and Poon (2006).

Poon conjectures that all unit trees can be flattened in both 2D and 3D, which would obviously imply that the answer to Open Problem 9.1 is YES.

9.2 PROBABILISTIC ROADMAPS

Introduction. As mentioned in Section 1.2.4, much of the work on the protein folding problem in the biochemistry and biophysics communities has focused on simulation of the energy minimization process. With careful choice of the configuration space ("torsional angle space" rather than Euclidean coordinate space) and careful numerical integration, researchers have succeeded in deriving some statistics relevant to conformational dynamics for chains of up to $n = 50$ links (He and Scheraga 1998). However, it seems unlikely these techniques can be used to compute detailed foldings for even these small values of n, let alone push on to the higher values (e.g., $n = 500$) most useful for drug design. In this section we will report on one approach to protein folding that has arisen from the computer science community.

PRM history. Recall from Section 2.2 (p. 19) that a roadmap is a network of curves that preserves the connectivity of the free space in configuration space, and is reachable from any configuration. Canny's silhouette roadmap is a powerful theoretical tool, but it is of primarily theoretical interest, leading to complexities exponential in k, the number of degrees of freedom (dof). For even moderately large values of k, say, $k = 6$, explicit computation of silhouette roadmaps is impractical. This led, in the early 1990s, to two groups[2] independently inventing a new structure called the *probabilistic roadmap (PRM)*, which was immediately successful in increasing the range of accessible dofs.[3] We now describe their technique, as laid out in their seminal joint paper on the topic (Kavraki et al. 1996).

Let $\mathcal{C}$ be the configuration space of a robot arm (or other movable object) with k dof, and let $\mathcal{F} \subset \mathcal{C}$ be the free space, the configurations that avoid collision with obstacles (including perhaps the arm itself). The PRM motion planning algorithm consists of two phases: roadmap construction and the query phase. The roadmap is constructed by generating random points in $\mathcal{C}$, checking to see if they are in $\mathcal{F}$, and if so, connecting nearby points into a graph structure G with a "local planner." The query phase takes start and goal configurations s and t, connects each to G, and uses G to construct a path from s to t. We now describe the roadmap construction steps in more detail.

Each parameter constituting a dof of the mechanism has a certain valid range, for example, a joint angle range. Random points in $\mathcal{C}$ can be generated by selecting each parameter uniformly within its valid range. A point $p \in \mathcal{C}$ might not also be in $\mathcal{F}$. This can be checked with a collision detection algorithm, determining whether the fixed configuration corresponding to p intersects any obstacles. Assuming $p \in \mathcal{F}$, the next step is to connect p into G (if possible). First, a set of "candidate neighbors" $N(p)$ of nodes of G are identified, usually as those closest to p under some metric. The distance function used is often some easily computed point-to-point distance in the workspace. Then an attempt is made to see if p can be connected to a node in $q \in N(p)$ using the "local planner." Typically, the local planner simply connects p and q via a segment $pq \subset \mathcal{C}$, and somehow checks the points along this segment for collisions. One method discretizes pq into points separated by $\varepsilon > 0$ and checks each point for collision when the robot is "grown" by ε. When it is determined that p can be connected to q, that arc is added to G; if no $q \in N(p)$ yields a connection, p is left as a separate component.

The PRM method is a general paradigm that embraces many options at each stage. Selecting amongst the options wisely often determines the effectiveness of the method. We mention here three choice points that complicate the above sketch.

1. *Distance Function.* The simple point-to-point distance function is inadequate for long and thin objects, for example, protein chains. One wants the distance

---

[2] At Stanford and Utrecht universities.
[3] More recent and accurate terminology for this approach is *sampling-based motion planning* (e.g., LaValle 2006).

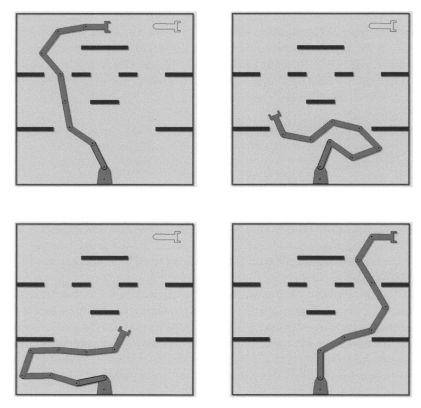

Figure 9.8. Four configurations of an 8-dof robot arm.

function to approximate the chance of a collision. A way to approach this is to compute the volume of the workspace region swept by the object as it moves along the path. As this computation can be quite difficult, it itself is often approximated.

2. *Connection in* $\mathcal{C}$. Connecting p and q with the straight segment $pq \subset \mathcal{C}$ is often inadequate. One method used for robot arms in Kavraki et al. (1996) is as follows. Let p_i be the position of joint i in the configuration p, and similarly for q_i. For each odd i, p_i and q_i are connected by a straight segment. But for each even j, the conforming rotation is computed via inverse kinematics, with p_{j-1}, p_j, p_{j+1} a 2-link chain (p. 63). The motion of all joints together then constitutes the connection between p and q in $\mathcal{C}$.

3. *Graph structure.* Although it makes sense to maintain G a forest at all times, for G only need connect the space, in terms of path quality it is useful to permit G to contain cycles. Otherwise the final solution path could be unnecessarily long and convoluted.

One of the challenges of the PRM technique is generating sufficiently many random points to accurately capture the connectivity of $\mathcal{F}$, which may have narrow, convoluted tunnels. Usually some heuristic (but theoretically guided) method is used to increase the density of sample points in these difficult regions (see, e.g., Boor et al. 1999).

The method was immediately successful at solving problems such as that illustrated in Figure 9.8, which illustrates initial and final configurations of an 8-dof planar robot

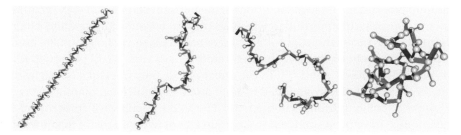

Figure 9.9. Folding snapshots of a polypeptide chain with 10 alanine amino acids (Song and Amato 2001, by permission).

arm in a cluttered environment. In fact, one of the inventors of the PRM technique, Overmars, incorporated it in a "Motion Planning Game"[4] which easily solved this problem instance.

Further development of PRM. The PRM technique has since developed into a method with both practical applicability (e.g., to industrial mechanical systems; Laumond and Simeon 2001) and theoretical power. There has been work on the key problem of increasing sampling density in narrow passages of $\mathcal{F}$ (Hsu et al. 1998), minimizing the number of collision checks made by the local planner (Bohlin and Kavraki 2000), and most theoretically important, a proof that PRM is "probabilistically complete" in the sense that if a path between s and t exists, the technique will find a path with high probability, if allocated enough processing time (Barraquand et al. 1997). More precisely, these researchers proved that, under a variety of assumptions, the number of collision detection checks needed grows only as the absolute value of the logarithm of the probability of an incorrect answer that could be tolerated.

Amato research. This leads us to the recent work of Amato and Song applying PRM to protein folding, reported in a series of papers (Amato and Song 2001; Song and Amato 2001, 2004). Under the assumption that the goal configuration is known, that is, the so-called "native state" of the minimum-energy folded protein is known, they have focused on the folding kinetics to reach this terminating configuration t from the primary structure s of the amino acid sequence. They have been able to push the technique to $n = 60$ and even $n \approx 100$, that is, 200 dof, with about 15 h of workstation computation time (see Figure 9.9 for an example). There are at least three major differences between the protein folding problem and the previously studied high-dof mechanisms to which PRM was applied. First, n is an order of magnitude (or several orders) larger. Second, collision detection is replaced by a preference for low potential energy configurations. Third, the PRM technique produces some path, but the "energetically favorable" folding paths are important in biology.

In response to these demands, Amato and Song alter the standard PRM paradigm in several ways. They bias their random configurations based on the so-called "Ramachandran plot" or "conformational map," which graphs pairs of adjacent backbone torsional angles (ϕ, ψ) of a protein structure. Such plots show clustering characteristic of secondary structures, for example, alpha helices. They further bias their random

---

[4] http://www.cs.uu.nl/people/markov/kids/motion.html. The solutions are precomputed off-line, not generated in real-time during the game.

sampling toward the known native fold of the protein. Potential energy is computed by known approximation formulas, which attempt to account for the main-chain bonds through terms inversely proportional to separation distance, $1/r_{ij}$, and the nonbonded van der Waals interaction forces between atomic pairs via a "12-6" potential, with terms proportional to $1/r_{ij}^{12}$ and $1/r_{ij}^6$. In the query phase, they use Dijkstra's algorithm with weight equal to a energy assigned to each arc during the local planning stage to find an approximately minimal energy path. The folding pathways constructed by their techniques[5] have been validated to a certain extent in comparison to known data: the intermediate, secondary structures (alpha helices and beta sheets) seem to form in the correct order, and energy potential landscapes and "RMSD plots" are reasonable. We can expect significant progress in this area even in the near future.

9.3 HP MODEL

Introduction. Because energy minimization is generally a difficult optimization problem, several researchers have explored simplified models of protein folding, with the hope that a deeper understanding of the simplified models reveals insight into reality. One of the most popular of these models is the *hydrophobic—hydrophilic* or *HP* model, introduced by biophysicist Ken Dill in 1990 (Dill 1990). This model intends to capture one prominent energy component that arises in protein folding: that some amino acids are repelled from water (hydrophobic) while others are attracted to water (hydrophilic or polar). The HP model differentiates amino acids only between these two types, denoted H and P respectively.

Model. A protein in the HP model is modeled by an abstract chain, usually open but possibly closed, where each link has unit length and each joint is marked either H or P. Thus, a protein can be thought of as a binary string of H's and P's in this model. In our figures, H nodes are red and P nodes are blue. Folding is not modeled as a continuous process, but rather we consider particular configurations (folded states) on the integer lattice in either 2D or 3D. Each joint must map to a unique point on the lattice, and each link must map to a single (unit-length) edge of the lattice. Such a non-self-intersecting configuration is called an *embedding* on the lattice.

It may seem of scant scientific value to consider the HP model in 2D when the real world is 3D. However, for short proteins (in particular those normally used in computations), the 2D model can be more realistic than the 3D model. The reason is that the perimeter-to-area ratio of a short 2D chain is a close approximation to the surface-to-volume ratio of a long 3D chain (Chan and Dill 1993; Hayes 1998).

The HP model of energy specifies that a chain desires to maximize the number of H–H *contacts*, that is, pairs of H nodes that are adjacent on the lattice but not adjacent along the chain. The hypothesis is that hydrophobic amino acids (H nodes) avoid the water surrounding the molecule by having attractive forces between one another. Indeed, experiments with maximizing contacts in the HP model seem to reproduce the real-world phenomena that hydrophobic amino acids (H) tend to cluster together to avoid the surrounding water, and hydrophilic amino acids (P) are frequently found on

---

[5] See http://parasol.tamu.edu/groups/amatogroup/foldingserver/ for folding movies.

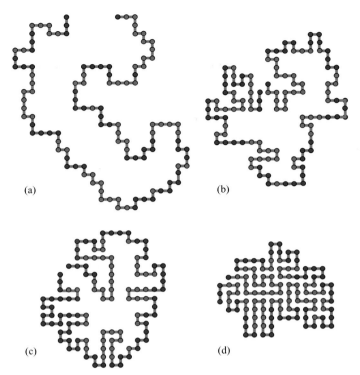

Figure 9.10. A four-frame animation of a protein folding continuously according to forces based on the HP model plus additional forces to avoid collisions. [Based on Figure 10B of Istrail et al. 1999.]

the outer surface of the folding in contact with the surrounding water molecules (Chan and Dill 1993). See Figure 9.10 for one such experiment.

9.3.1   NP-completeness

In the HP model, the protein folding problem—predicting the folded structure of a given sequence of amino acids—can be stated formally as a combinatorial optimization problem, corresponding roughly to energy minimization:

> **Optimal Folding Problem:** Given a sequence of H's and P's, find an embedding on the 2D lattice that maximizes the number of H—H contacts.

In an important paper, Crescenzi et al. (1998) proved that this problem is NP-complete. Their proof is highly intricate, and we offer only a crude sketch here. Their proof is by reduction from an NP-complete Hamiltonian cycle[6] problem: given a planar graph G, does G contain a Hamiltonian cycle? They prove that this problem remains NP-complete even for certain special graphs, which more easily match the HP model. They first map the nodes of G to the vertices of a hypercube, represented by strings in $\{0, 1\}^{8m}$. Trevisan (1997) devised a mapping, based on Hadamard codes, such that all nodes connected by an arc in G are close, in the Hamming distance[7] of this "code

---

[6] A *Hamiltonian cycle* in a graph is a cycle that passes through each node exactly once.

[7] The *Hamming distance* between two codes is the number of coordinate positions at which they differ.

space," and all nodes not connected by an arc in G are far apart in this code space. The result of this mapping is that Hamiltonian paths in G map to short Hamiltonian paths in the code space, whereas if there is no Hamiltonian path in G, any Hamiltonian path in the code space is long. They then construct strings with the property that they fold with a high number of contacts precisely when there is a Hamiltonian path between two particular nodes in G. Finally, they connect a set of strings with this property together into one long string that folds with a high number of contacts precisely when G has a Hamiltonian cycle.

Although their proof is technical[8] and divorced from the reality of proteins, it is interesting to note that their construction forces the string to fold to a large hydrophobic "sphere" (a square in the lattice) with "impurities" (P's) which must align precisely to maximize the number of contacts. (When two P nodes are adjacent, they necessarily reduce potential contacts between H nodes.) The high-level structure of these foldings, with a primarily hydrophobic (H) center of a sphere, matches the common physical phenomenon observed in real proteins (Chan and Dill 1993).

Crescenzi et al. (1998) also establish NP-completeness for optimal protein folding in the HP model in 3D. Their proof is simpler in 3D where planarity issues are moot. Berger and Leighton (1998) independently established the same result. Their proof is similar in that it folds the protein into a cube, but different in that it reduces from an exact bin packing problem.

One open problem in this area is whether the HP model remains NP-complete for other lattices. Two natural candidates are the triangular lattice in 2D (Agarwala et al. 1997) and the tetrahedral (cannonball) lattice in 3D. These lattices have been suggested as more realistic because they are not subject to parity issues, while the square and cubic lattices are vertex 2-colorable and so only H nodes with opposite parity can have a contact. On the other hand, the proofs of NP-hardness (Crescenzi et al. 1998; Berger and Leighton 1998) do not seem to rely directly on this parity issue.

Open Problem 9.5: Complexity of Protein Folding in Other Lattices. [a] What is the complexity of finding the optimal folding of a protein in the HP model in the 2D triangular lattice or the 3D tetrahedral lattice?

---

[a] The 3D version is posed in Berger and Leighton (1998).

9.3.2   Approximation Algorithms

Once NP-completeness is established for optimal protein folding, it is natural to ask how well we can fold proteins into approximately minimum-energy states.[9] In the HP model, how efficiently can we find a folding whose number of contacts is at least a factor of c of the maximum possible, for various values of $c < 1$? Such an algorithm is called a *c-approximation algorithm*. One major open problem in this area is the following:

---

[8] The proof in the direction (embedding with high number of contacts) $\rightarrow$ (Hamiltonian cycle) is 25 pages long!

[9] In fact, on a historical note, the approximation algorithms of both Hart and Istrail (1996) and Agarwala et al. (1997) precede the proofs of NP-hardness (Berger and Leighton 1998; Crescenzi et al. 1998).

> **Open Problem 9.6: PTAS for Protein Folding.** [a] Is there a polynomial-time
> c-approximation algorithm for protein folding in the HP model for all $c < 1$?
>
> ---
> [a] Implicit in Newman (2002).

Hart and Istrail (1996) gave the first constant-factor approximation algorithms for protein folding. Specifically, they achieved an approximation ratio of $\frac{1}{4}$ in 2D and $\frac{3}{8}$ in 3D. Mauri et al. (1999) experimented with three algorithms, achieving approximation ratio $\frac{1}{4}$ in 2D, and found some with an average observed ratio of $\frac{2}{3}$. Newman (2002) improved the approximation ratio in 2D to $\frac{1}{3}$ and proved that improving beyond $\frac{1}{3}$ requires a significantly different approach, namely a different a priori bound on contacts.

All of these approximation algorithms (Hart and Istrail 1996; Mauri et al. 1999; Newman 2002) use the following a priori bound on the maximum number of contacts: twice the number of odd-parity H nodes or twice the number of even-parity H nodes, whichever is smaller. The number of contacts is always at most this bound because each node except the two ends can have at most two contacts (two out of four of the neighbors are taken up by adjacent nodes along the chain) and because only two opposite-parity nodes can have a contact. Surprisingly, while the number of contacts is never larger than this bound, the number of contacts is also always at least one third of this bound (and sometimes that small) (Newman 2002). The algorithm achieving this third of the bound is therefore a $\frac{1}{3}$-approximation.

We sketch the ideas behind the $\frac{1}{3}$-approximation algorithm (Newman 2002). First, we convert the given open chain into a closed chain, which can only make our task harder, constraining our foldings and decrease the number of contacts (because of a potential uncounted contact between the two ends of the chain). Second, if there are more odd-parity H nodes than even-parity H nodes, or vice versa, we discard the excess; this change does not modify the desired bound, so it only makes the task only harder still. Third, we split the closed chain at an edge designated as the "top" such that, as we walk clockwise from the top, we always have seen at least as many odd-parity H's as we have even-parity H's, and the reverse: as we walk counterclockwise from the top, we always have seen at least as many even-parity H's as we have odd-parity H's. That such a top point exists is not obvious, but follows from an application of the intermediate-value theorem. Finally, the top edge is hung at the top of the folding, and the clockwise and counterclockwise halves are folded against each other with various local gadgets to align the odd-parity H nodes in the clockwise half with the even-parity H nodes in the counterclockwise half, usually two pairs at a time. As shown in Figure 9.11, there are four cases depending on whether the length of the chain connecting consecutive H nodes on a common half is 1 or at least 3 (it always being odd). In all cases it is argued that the number of H–H contacts is at least $\frac{1}{3}$ of the desired bound.

9.3.3 Unique Optimal Foldings

Protein design. A different approach to understanding protein folding, initiated by Aichholzer et al. (2003), is to consider *protein design*. Can we design a protein (in the HP model, a sequence of H's and P's) that folds into approximately any desired shape? Or what shapes are achievable? These questions have important applications in drug

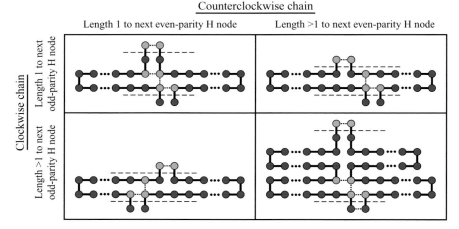

Figure 9.11. Gadgets used in the $\frac{1}{3}$-approximation algorithm of Newman 2002. [Based on Figures 5 and 6 of Newman 2002.]

design, where the goal is to design a synthetic protein with a specific function, so in particular with a specific shape.

Unique foldings. One natural desired property of a solution to the protein-design problem is that the synthetic protein folds "stably" into the desired shape. In the HP model, at least, the natural interpretation of this requirement is that the optimum folding be *unique*. Hayes (1998) asked whether such proteins even exist: Is there a protein in the 2D HP model of every sufficiently long length that has a unique folding with the maximum number of contacts? He also exhaustively verified the existence of such proteins for sizes up to 14, except for lengths 3 and 5 which have no uniquely foldable proteins. Aichholzer et al. (2003) extended these experiments to length 25, and more interestingly, proved their existence for all lengths divisible by 4. Thus, if any counterexamples other than 3 and 5 remain, they must be odd or congruent to 2 modulo 4; though the conjecture is that all lengths greater than 5 have uniquely foldable chains.

The construction of general uniquely foldable chains in Aichholzer et al. (2003) proceeds in two steps. First, they construct uniquely foldable closed chains of all possible lengths; these lengths are necessarily divisible by 2 in order to close the chain in the square grid. Second, they show that the open chain resulting from cutting the closed chain at a particular spot is still uniquely foldable for lengths divisible by 4, while for lengths divisible by 2 and not 4 the open chain has (at least) two optimal foldings. The second step is significantly harder to establish than the first step, so we concentrate here on just the case of closed chains.

Closed chains. Figure 9.12 shows the uniquely foldable closed chain of arbitrary length. The chain mainly alternates between H and P nodes, except at two roughly opposite ends where it has two P nodes in a row. The claimed optimal foldings, also shown in Figure 9.12, zigzag the folding diagonally so that all but the two extreme H nodes have two contacts. The total number of contacts is $h - 1$, where h is the number of H nodes.

First we argue that the zigzag folding is indeed optimal; that is, no more than $h - 1$ contacts are possible. In any folding, any H node on the bounding box of the folding can have only one contact to an adjacent H node, because one of the potential contacts

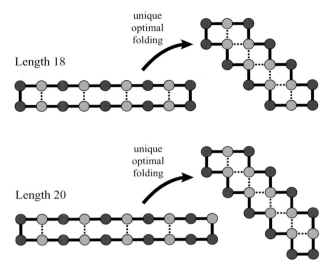

Figure 9.12. The uniquely foldable closed chains and their optimal foldings.

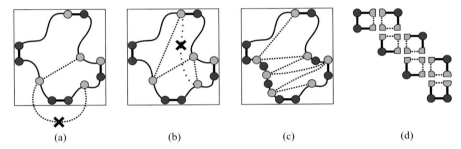

(a)   (b)   (c)   (d)

Figure 9.13. In an optimal folding of Figure 9.12, (a) there can be no contacts external to the closed chain, (b) the graph of contacts can have no cycles, (c) the graph of contacts must be a complete path, and (d) these contacts decompose the chain into uniquely embeddable squares.

is against the bounding box and therefore unused. Furthermore, there must be at least four nonconsecutive edges on the bounding box, one against each wall of the box. There are only two edges whose both endpoints are P nodes, so at least two remaining edges against the bounding box must have H nodes. The resulting loss of two potential contacts from individual H nodes translates to one lost contact because of double counting. Thus, the single lost contact in the zigzag folding ($h - 1$ instead of h) is indeed necessary.

Next we argue that the zigzag folding is the only optimal folding, that is, it is unique. We proceed by three claims. The first claim is that in any optimal folding, the closed chain does not wrap around and have a contact that is external to the loop (Figure 9.13(a)). This claim follows because contacts can occur only between nodes in opposite halves of the chain (by parity) and because the four edges against the bounding box prevent such connections exterior to the loop. The second claim is that the graph of all contacts between H nodes is acyclic (Figure 9.13(b)), which follows because the graph is bipartite and planar with the contacts drawn internally, so any cycle would necessarily lead to a crossing. The third claim is that this contact graph is a path through all the H nodes (Figure 9.13(c)), which is necessary to achieve enough contacts (edges) when the degrees are all at most 2. This path is uniquely specified by the

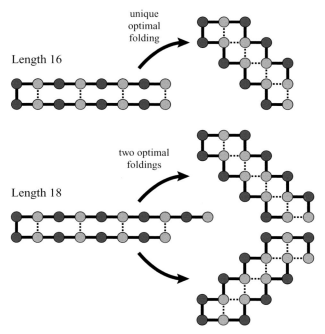

Figure 9.14. The uniquely and doubly foldable open chains and their optimal foldings.

chain, alternating sides of the chain from one end to the other. Fourth, these contacts decompose the graph of the chain and contacts into a sequence of unit squares, each of which has a unique embedding (Figure 9.13). Therefore, the entire embedding (optimal folding) is unique up to isometry.

Open chains. The construction of uniquely foldable open chains shown in Figure 9.14 is almost the same as the closed chains from Figure 9.12. We just remove the chain edge between one of the end pairs of P nodes and remove those two P nodes as well. The remaining ends are H nodes and the contact between them effectively forces the chain to remain tied closed. The formalization of this intuition is tedious (Aichholzer et al. 2003). Some insight into the difficulty is the surprise that for lengths divisible by 2 and not 4, there are two optimal foldings of the open chain (see the bottom of Figure 9.14). The difference between the two foldings can be seen by focusing on the graph of bonds, which is no longer a path in the second folding. The conjecture is that only these two foldings are optimal. For lengths divisible by 4, however, we have unique foldability (Aichholzer et al. 2003).

We reiterate the design-driven problem that opened this section and which remains unsolved:

Open Problem 9.7: Protein Design. Design a protein (in the HP model) with a desired shape as (roughly) its unique optimal folding.

Although this problem is abstracted far from the reality of proteins, it is directly inspired by an important practical problem, perfectly illustrating the interplay between theory and applications mentioned at the start of Part I (Section 1.2.5, p. 15).

Origami

10 Introduction

This second part concerns various forms of paper folding, often called *origami*. We start in this chapter with a historical background of paper and paper folding (Section 10.1), and of its study from mathematical and computational points of view (Section 10.2). This history can safely be skipped by the uninterested reader. Then in Section 10.3 we define several basic pieces of terminology for describing origami, before providing an overview of Part II in Section 10.4.

10.1 HISTORY OF ORIGAMI[1]

The word "origami" comes from Japanese; it is the combination of roots "oru," which means "fold," and "kami," which means "paper." While origami was originally popularized largely by Japanese culture, its origins are believed to be pre-Japanese, roughly coinciding with the invention of paper itself. Paper, in turn, is believed to have been invented by Ts'ai Lun, a Chinese court official, in 105 A.D. The invention of paper was motivated by the then-recent invention of the camel hair brush, from 250 B.C., which could be used for writing and calligraphy.

Paper, and presumably paper folding at the same time, spread throughout the world over a long period. Buddhist monks spread paper through Korea to Japan in the sixth century A.D. Arabs occupying Samarkand, Uzbekistan, from 751 A.D. brought paper to Egypt in the 900s, and from there continued west. The Moors brought paper (and at the same time, mathematics) to Spain during their invasion in the 700s. In the 1100s, paper making became established in Jativa, Spain. The Arab occupation of Sicily brought paper to Italy. Eventually, paper mills were built in Fabriano, Italy, in 1276; in Troyes, France, in 1348; and in Hertford, England, in the 1400s. By around 1350, paper was widespread for literary work in Europe. Finally, paper made its way to North America; the first paper mill was built in Roxboro, Pennsylvania, in 1690.

When and where during this history origami arose is a matter of some debate, but it is generally believed to have begun in China, Korea, and/or Japan. As the invention of paper traveled "by caravan" from country to country, one theory is that some origami models traveled too, and so the idea that paper could be folded followed naturally. Another theory is that folding of paper was independently discovered in several places, as it is a natural activity. In any case, origami sprung up in several countries after China,

---

[1] We rely here on historical notes by Smith (2005) and Lister (1998).

Korea, and Japan, most notably in Spain, Italy, England, and quite recently in North America.

The history within Japan is well-recorded. During the Heian period, 794–1185 A.D., ceremonial folds and some basic origami models developed. Origami globalized throughout Japan during the Kamakura period, 1185–1333 A.D. "Modern" origami is believed to have been developed during the Muromachi period, 1333–1573. By the Edo period, 1603–1867 A.D., origami had become a popular pastime. The earliest book to describe origami is from 1682. In 1797, the famous book *How to fold 1,000 cranes* was published (which much later inspired the Hiroshima memorial honoring a girl who died of radiation poisoning; Ishii 1997). By the Meiji period, 1868–1912 A.D., origami was in popular use in schools, as it is now in Japan.

The resurgence of interest in origami in the twentieth century, and its spread throughout the world, is often attributed to the influence of the origami artist Akira Yoshizawa (1911–2005), who introduced the origami notational system of dotted lines and arrows in his 1954 book (Yoshizawa 1954); the same system, slightly modified, remains in use today. Lillian Oppenheimer (1898–1992) played a key role in popularizing origami in the United States. She encouraged and organized interest in origami starting in the late 1940s, which led her to form *The Origami Center* in New York in 1958. This organization subsequently developed into *The Friends of The Origami Center of America* in 1980, which became *OrigamiUSA*[2] in 1994, an organization that remains active today. Parallel developments in other countries led to the eventual formation of origami societies in nearly every country throughout the world; one that has played an important role in the English-speaking world is the British Origami Society, founded in 1967 and active today.

In the past twenty or so years, origami has taken off to new heights and accomplished amazing technical and artistic feats. On the technical side, modern advanced origami has more intricate structure and limbs than was previously thought possible. On the artistic side, modern origami by expert folders resembles the art subject far more than a piece of paper. We believe that the incredible advance in origami design has been spurred largely by a growing mathematical and computational understanding of origami, the primary subject of Part II and the topic to which we now turn. See also Lang (2003) for the technical side of the development of origami design over the years, and how many of these techniques leverage mathematics.

10.2   HISTORY OF ORIGAMI MATHEMATICS

Origami has an intrinsic geometry that is a natural subject of study. The oldest known reference to origami in the context of geometry is an 1840 book by Rev. Dionysius Lardner (1840), which illustrates several geometric concepts using paper folding. A more influential and accessible work is an 1893 book by T. Sundara Row (1893), which illustrates a variety of geometric constructions, traditionally executed by straight edge and compass, via origami as a construction tool. Implicit in this work is that origami can simulate the constructions of a straight edge and compass, but it still only used origami as a tool for illustration, rather than analyzing origami itself. In 1936, origami was analyzed in terms of the power of its geometric constructions, according to a certain

---

[2] http://www.origami-usa.org/.

set of axioms, by Margherita Piazzolla Beloch (Piazzolla 1936). This work is probably the first contribution to "origami mathematics." It was followed later by the Huzita "axioms," which will be discussed in more detail in Chapter 19.

Several fundamental theorems on local crease patterns around a single flat-folded vertex were established by Jun Maekawa, Toshikazu Kawasaki, and Jacques Justin, and will be discussed in Section 12.2. Thomas Hull extended this work on flat foldings in a series of papers (Hull 1994, 2002, 2003a), and just recently published a book on the mathematics of origami (Hull 2006). Robert Lang, one of the premier origamists today, developed an algorithm around 1993 for designing origami, which will be the subject of Chapter 16. He also recently published his "magnum opus" *Origami Design Secrets* (Lang 2003) describing his practical and mathematical approach to origami design. Lang's work may be viewed as the start of the recent trend to explore *computational origami* (Demaine and Demaine 2002; Cipra 2001) (a term coined in Demaine et al. 2000c). This concentration was given a second impetus in the Bern–Hayes paper establishing that it is NP-hard to decide whether a given crease patterns has a flat folding (Bern and Hayes 1996), the focus of Section 13.1.

10.3 TERMINOLOGY

The intuitive notion of paper folding is straightforward. A *piece of paper* is a surface, primarily two-dimensional, and usually flat. In origami, the piece of paper is often assumed to be a square, while mathematically it is sometimes convenient to consider it an entire plane. Throughout this book, however, we allow any planar polygon P as our piece of paper. A *folding motion* of a piece of paper is a continuous motion of the paper from one configuration to another that does not cause the paper to stretch, tear, or self-penetrate. A snapshot of this motion at a particular time is called a *folded state*; in particular, we distinguish the *initial folded state* from which the folding begins and the *final folded state* at which the folding arrives.

It is important to capture the possibility of the paper touching but not penetrating itself. Of particular interest are *flat folded states*, where the layers of paper lie in a plane and therefore necessarily touch each other. We defer a precise definition of these touching possibilities, which are quite intricate, to Chapter 11.

We normally distinguish the final folded state from the folding motion that arrives there. Ideally, we would always have an explicit description of the folding motion, but this task is often difficult, so usually we settle for just an explicit construction of the final folded state. In Section 11.6 we will see that every origami can be attained by a continuous folding motion, so this concentration on folded states is justified.

A *crease* is a line segment (or, in some cases, a curve) on a piece of paper. Creases may be folded in one of two ways: as a *mountain fold*, forming a protruding ridge, or as a *valley fold*, forming an indented trough. The usual convention is to display mountain creases as a dash—dot pattern,[3] and valley folds as dashes only. Because these patterns are easily confused by the eye, we often use blue for valleys and red for mountains. A *crease pattern* is a collection of creases drawn on paper, meeting only at common endpoints, which may be viewed as a (usually planar straight-line) embedding of a graph. A *mountain–valley assignment* is a specification of which creases should

---

[3] Mnemonic: the dots are mountain peaks.

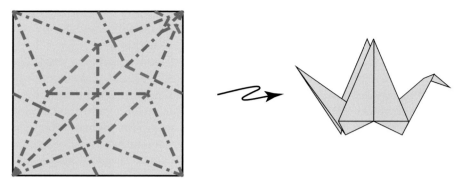

Figure 10.1. The classic crane mountain-valley pattern and corresponding flat origami.

be folded as mountains and which as valleys. Together, a *mountain–valley pattern* consists of both a crease pattern and a mountain–valley assignment. Figure 10.1 shows an example. Many books on origami include instructions (employing a stylized set of arrow notations) for describing a folding motion. As our goals are rather different, we usually only show the mountain–valley pattern and the final folded state.

Other definitions will be introduced as needed throughout the exposition.

10.4   OVERVIEW

As hinted at above, although the notions of paper and folding are both intuitive, capturing these intuitions formally has proved a challenge. Chapter 11 details one approach at some length, culminating in the equivalence between foldability by an actual motion and the existence of a final folded state (Section 11.6).

As mentioned in Chapter 0, we can partition mathematical origami research roughly around two foci: foldability and design.

Origami foldability generally asks which crease patterns can be folded into origami that use exactly the given creases. In reverse, our goal is to characterize which crease patterns derive from actual origami. We normally focus on flat origami, where this problem is particularly clean and well-studied: Is there a flat folding of a given crease pattern, possibly with a given mountain–valley assignment?

There are several variations on this flat-foldability problem. The simplest forms occur when all creases are parallel (Section 12.1), or when the crease pattern has a single vertex (Section 12.2). In these cases, we can completely characterize which crease patterns and mountain–valley assignments have flat folded states. Furthermore, all folded states of single-vertex crease patterns can be reached by corresponding folding processes without any additional creases (Section 12.3); this result is based on an equivalence to 2D linkage folding. A more general problem is to consider folding an arbitrary crease pattern, but requiring only that it works "locally" around each vertex (Section 13.1); here the main problem is to find a globally consistent mountain–valley assignment that is locally foldable at each vertex, and this problem can be solved efficiently. Testing true (global) foldability, however, is NP-hard (Section 13.2). Global foldability is, unsurprisingly, easy to test for 1D pieces of paper, equivalent to foldability by a sequence of "simple folds" (Section 14.1.1). The analogous simple-foldability problem in 2D is easy for "maps" (rectangles with horizontal and vertical creases) but NP-hard as soon as we introduce diagonal creases or more general shapes of paper (Chapter 14).

Turning now to design, *origami design* is, loosely, the problem of folding a given piece of paper into an object with certain desired properties, for example, a particular shape. Perhaps the most basic problem here is to design an origami model with a specified flat silhouette or 3D shape (Chapter 15), although the general algorithms developed so far do not lead to practical foldings. Most closely related to "traditional" origami design is Lang's tree method for designing origami "bases" (Chapter 16), which has brought modern origami design to a new level of achievement. Related to this work is the problem of folding a piece of paper to align a prescribed graph (Chapter 17), which can be used for a magic trick involving folding and one complete straight cut. A surprisingly related nonstandard form of origami is to start with a piece of paper that is not flat, but rather the surface of a polyhedron, and the goal is to flatten that surface (Chapter 18). We close Part II with a brief discussion of axiomatics and constructibility (Chapter 19), and touching on the largely unexplored area of curved and curved-fold origami (Chapter 20).

11 Foundations

We believe that research in mathematical origami has been somewhat hampered by lack of clear, formal foundation. This chapter provides one such foundation, following the work of Demaine et al. (2004, 2006a). Specifically, this chapter defines three key notions: what is a piece of paper, what constitutes an individual folded state (at an instant of time) of a piece of paper, and when a continuum of these folded states (animated through time) forms a valid folding motion of a piece of paper. Each of these notions is intuitively straightforward, but the details are quite complicated, particularly for folded states and (to a lesser extent) for folding motions. In the final section (Section 11.6, p. 189), this chapter also proves a relationship between these notions: every folded state can be achieved by a folding motion. At first glance, one would not normally even distinguish between these two notions, so it is no surprise that they are equivalent. The formal equivalence is nonetheless useful, however, because it allows most of the other theorems in this book to focus on constructing folded states, knowing that such constructions can be extended to folding motions as well.

While we feel the level of formalism developed in this chapter is important, it may not be of interest to every reader. Many will be content to skip this entire chapter and follow the rest of the book using the intuitive notion that mathematical paper is just like real paper except that the paper has zero thickness. Many others will be happy to read the first section or two of this chapter, which give a gentle introduction to some of the difficulties with the definitions, accept that there are some intricacies in the definitions, and take on faith that they can be worked out precisely. The remaining sections on definitions are for those interested in a detailed resolution of the intricacies (and perhaps interested in finding a simpler approach). The final section uses a surprisingly simple idea to transform folded states into folding motions, and the execution of that idea is not highly technical, so it may interest many mathematically inclined readers.

11.1 DEFINITIONS: GETTING STARTED

At a high level, a folded state is a mapping of a piece of paper into Euclidean space that satisfies two properties: "isometry" and "noncrossing." The isometry condition specifies that the distances between pairs of points, as measured by shortest paths on the surface of the paper, are preserved by the mapping; in other words, the mapping does not stretch or shrink the paper. One consequence of being isometric is that the folded state must be "continuous" in the sense that the mapping does not tear the paper. The noncrossing condition specifies that the paper does not cross through itself when mapped by the folded state. This latter condition is considerably more complicated, mainly because portions of paper are permitted to come into geometric contact as multiple overlapping layers, yet we must forbid these layers from actually penetrating each other (what is sometimes referred to as "proper crossing").

We start with the definitions of paper, folded states, and folding motions in two simpler contexts, which serve both as warmups and as necessary steps toward the final definition:

1. **One-dimensional paper.** To begin we specialize the piece of paper we are folding to a 1D line segment. (Equivalently, we can think of a thin strip of 2D paper always folded perpendicular to the direction of the strip.) In this context, essentially all of the same issues arise as with normal 2D paper, but they are considerably simpler to describe and understand. We also use many parts of the definitions for 1D paper in the definitions for 2D paper.

2. **Free folded states.** Initially we also specialize the folded state to a special class that is easier to define. Intuitively, a "free folded state" is a folded state in which no two points of paper come in contact, that is, the piece of paper does not touch itself in its folded form.[1] In this context, folded states are considerably simpler because the noncrossing condition becomes almost trivial, thus allowing us to focus on the isometry condition which will be identical to the isometry condition in the final definition.

At a high level, the definition of a folded state developed here generalizes Justin's definition of flat folded states (Justin 1994).

11.1.1 Free Folded States of 1D Paper

We start with the simplest case: restrictions to both the piece of paper and the folded state.

A *1D piece of paper* P is a line segment, including its endpoints and allowing the infinite cases of a ray or line. When present, we call the endpoints of P *boundary points*. We view the paper as lying along the x-axis, and distinguish the two *sides* of P: the *top side* visible from points with $y > 0$ and the *bottom side* visible from points with $y < 0$.

A *free folded state* of a 1D piece of paper P is a one-to-one isometric function $f : P \rightarrow \mathbb{R}^2$ mapping P into the Euclidean plane. The *one-to-one condition*—$f(p) = f(q)$ only when $p = q$—guarantees that no two points of P come in contact, and thus in particular the mapping causes no crossings. The *isometry condition* specifies that the intrinsic geodesic (shortest-path) distance between any two points of P is the same when measured either on P or on the surface of the folded state when P is mapped by f.

It remains to provide a formal definition of the isometry condition, where the main issue is how we define the distance between two points $p, q \in P$ measured on the piece of paper P and on the folded state $f(P)$. The distance $d_P(p, q)$ between the two points in P is simply the length $|p - q|$ of the interval $[p, q]$ they span along the x-axis.[2] The

---

[1] This terminology mimics the notion of a free configuration from motion planning (see, e.g., Sharir 2004). What we define to be a (normal) folded state is called *semifree* (p. 11) in this terminology.

[2] For notational convenience, $[p, q]$ denotes the interval between p and q independent of whether p or q is smaller.

distance $d_f(p, q)$ between the two points when mapped via f into the plane is the arc length of the curve defined by f on the interval $[p, q]$:

$$d_f(p, q) = \text{arclength}_{[p,q]} f = \left| \int_p^q \left\| \frac{df(x)}{dx} \right\| dx \right|,$$

where $\| \cdot \|$ denotes the Euclidean norm. Now a function $f : P \rightarrow \mathbb{R}^2$ is *isometric* if $d_P(p, q) = d_f(p, q)$ for all $p, q \in P$.[3]

This isometry condition also constrains f to be *continuous*. Suppose that f were not continuous, so there is at least some $\varepsilon > 0$ distance between $f(p)$ and $f(p \pm \delta)$ in $\mathbb{R}^2$ for arbitrarily small $\delta > 0$ and for one choice of sign for $\pm$. Then the isometry condition must be violated for the two points p and $p \pm \delta$ for any $0 < \delta < \varepsilon$, because $d_P(p, p \pm \delta) = |p - (p \pm \delta)| = \delta < \varepsilon$, yet $d_f(p, p \pm \delta) \geq \varepsilon$, so $d_P(p, p \pm \delta) \neq d_f(p, p \pm \delta)$.

11.1.2 Free Folding Motions of 1D Paper

A free folding motion is a continuum of free folded states. More precisely, a *free folding motion* is a continuous function mapping each $t \in [0, 1]$ to a free folded state f_t. The value $t \in [0, 1]$ represents the time at which f_t is the snapshot of the folding motion. The only additional constraint is that f_t varies continuously in t which, for free folding motions, is a purely geometric constraint.[4]

Time continuity of geometry. Continuity of a free folding motion f_t is defined somewhat differently from usual: for every bounded subset P' of P and for every $\varepsilon > 0$, there is a $\delta > 0$ such that, for any $t, t' \in [0, 1]$ with $|t - t'| < \delta$, we have $d(f_t|_{P'}, f_{t'}|_{P'}) < \varepsilon$, where $f|_{P'}$ denotes the function f restricted to the (bounded) subdomain P'. For this definition to make sense, however, we need a metric d on free folded states of bounded pieces of paper.

We use a metric called the *supremum metric d* (sometimes also called the "uniform metric"). For two free folded states $f, f' : P' \rightarrow \mathbb{R}^2$, their *supremum-metric distance* $d(f, f')$ is

$$d(f, f') = \sup_{p \in P'} \| f(p) - f'(p) \|.$$

Thus the supremum metric d measures distance as the maximum Euclidean displacement of a point in P' when comparing how that point is mapped by the two folded states f and f'. A standard result in metric spaces is that d is a metric (indeed, on the entire space of bounded functions from P' to $\mathbb{R}^2$) (Sutherland 1975, Sec. 2.2).

We require continuity only within every bounded subset P' of P to handle motions of an infinite piece of paper P. For example, a simple motion we would like to allow is rotating an infinite line about some point p. Any two snapshots of this motion have

---

[3] This definition is subtly different from the usual notion of an isometry, which is a (sometimes surjective) function $f : A \rightarrow B$ where the distance between every two points $p, q \in A$ is equal when measured both in A and mapped into B: $d_A(p, q) = d_B(f(p), f(q))$. The difference has no effect for free folded states, because we can use f^{-1} to define $d_B(p', q') = d_f(f^{-1}(p'), f^{-1}(q'))$, but later we will need to define distance in the target space B in terms of f and points in the source space A, because multiple points in the source space A will map to the same point in the target space B.

[4] Note that we refer to a free folding motion with the notation f_t to allow the use of more conventional function notation on the folded states: compare $f_t(p)$ for $p \in P$ and $t \in [0, 1]$ to, e.g., $[F(t)](p)$, where F is the function mapping $t \in [0, 1]$ to the folded state f_t.

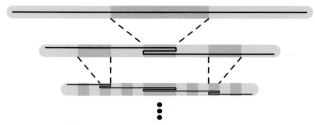

Figure 11.1. Lang's fractal flat folded state of a 1D piece of paper, based on the Cantor set. The length of the folded state is one third of the length of the piece of paper. [Personal communication from Robert Lang, August 2004.]

an infinite distance under the supremum metric: points farther from p are displaced more. Fortunately, the distance is finite between any two snapshots restricted to any bounded subset.

11.1.3 Smoothness and Creases of 1D Paper

Before we proceed to lifting the restriction that foldings be free, we define a basic smoothness restriction on functions mapping a 1D piece of paper into the Euclidean plane $\mathbb{R}^2$ (e.g., free folded states). In fact, we define a family of smoothness conditions parameterized by a positive integer k (typically 1 or 2). A function $f : P \to \mathbb{R}^2$ is *piecewise-C^k* if the first k derivatives f', f'', ..., $f^{(k)}$ exist and are continuous at all points of P except for finitely many exceptions c_0, c_1, ..., $c_m \in P$. These exceptional points c_i, at which f may lack a kth derivative or the kth derivative may be discontinuous, are the *crease points* of f. In particular, boundary points of P are crease points.[5]

The behavior of a function f is most complicated at crease points. The piecewise-C^k constraint forces crease points to be rare in the sense that every nonboundary crease point c_i is the center of a punctured neighborhood $(c_i - \varepsilon, c_i + \varepsilon) \setminus \{c_i\}$ of noncrease points for some $\varepsilon > 0$. One feature of this constraint (for any $k \geq 1$) is that it prevents *fractal* behavior involving infinitely many creases, such as the fractal flat folding shown in Figure 11.1. Although we expect that such fractal states could be handled by suitably augmented definitions, we do not consider them essential in capturing origami, and thus leave them as a direction for further study.

11.2 DEFINITIONS: FOLDED STATES OF 1D PAPER

To generalize beyond free folded states, we need to allow portions of paper to be geometrically collocated ("on top of each other"). In particular, we can no longer constrain the geometry f of the folded state to be one-to-one, so we need a different noncrossing constraint. The difficulty is that we cannot extract from the geometry f the relative stacking order of two portions of paper that are collocated, that is, which portion is "on top of" the other. This lack of information makes it impossible to determine whether a folded state crosses itself by a purely local test.

---

[5] We also allow finitely many points at which f has a continuous kth derivative to be marked as "crease points." This flexibility will be useful when actually constructing folded states, because we can omit certain information at crease points.

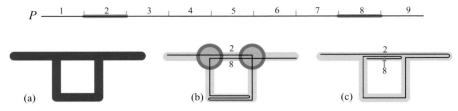

Figure 11.2. Two candidate folded states f with the same geometric image $f(P)$ (a). The f in (b) is invalid, and the f in (c) is valid.

Consider the examples in Figure 11.2. Here we imagine "thickening" the geometric image $f(P) = \{f(p) \mid p \in P\}$ so that we can see in what order f visits and revisits various points in $\mathbb{R}^2$. This viewpoint allows us to consider the relative stacking order of collocated layers of paper. For example, in Figure 11.2(c), Segment 2 is stacked on top of Segment 8, and this stacking is necessary for the folded state to avoid crossing itself. On the other hand, the function f in Figure 11.2(b) crosses itself no matter how we order the collocated layers of paper. This fact is nonlocal in the sense that it depends on the behavior of the folded state near two distant points in the plane, highlighted with circles. Each point can be locally satisfied by an appropriate stacking order of Segments 2 and 8, but these two satisfying orders conflict with each other, and therefore the folded state is globally invalid.

11.2.1 Order

In order to capture this nonlocal behavior, we need an additional mechanism beyond f for communicating this stacking-order information, so that a local consistency check guarantees a noncrossing folded state. This mechanism is another function λ, alongside f, which specifies the above/below relationship of every two regions of paper that are collocated according to f. More precisely, for every pair of distinct noncrease points $p, q \in P$ such that $f(p) = f(q)$, we assign a value of $+1$ or -1 to $\lambda(p, q)$, with the intent that λ disambiguates whether p is stacked "above" q ($\lambda(p, q) = +1$) or p is stacked "below" q ($\lambda(p, q) = -1$). We view λ as a *partial function* from $P \times P$ to $\{-1, +1\}$: if $p = q$, if $f(p) \neq f(q)$, or if either point p or q is a crease point, then $\lambda(p, q)$ is *undefined*.

It remains to define the meanings of "above" and "below" in a consistent way. Because f is piecewise-C^1 and p and q are not crease points, we can take the *unit normal vector* of the curve defined by f at either point p or q:

$$\mathbf{n}_f(r) = \left(\frac{f'(r)}{\| f'(r) \|} \right)^{\perp} \quad \text{where} \quad f'(r) = \left. \frac{df(x)}{dx} \right|_{x=r} \quad \text{and} \quad \langle x, y \rangle^{\perp} = \langle -y, x \rangle.$$

The normal vectors $\mathbf{n}_f(p)$ and $\mathbf{n}_f(q)$ at points p and q provide (two) well-defined notions of "above" and "below." In the context of $\lambda(p, q)$, we choose q's notion of "above" and "below": $\lambda(p, q) = +1$ when p is on the side of q pointed to by $\mathbf{n}_f(q)$, and $\lambda(p, q) = -1$ when p is on the opposite side of q, that is, the side pointed to by $-\mathbf{n}_f(q)$. In other words, $\mathbf{n}_f$ defines a consistent notion of the *top side* of the piece of paper P and follows this top side as the paper gets folded by f. Then $\lambda(p, q) = +1$ says that p touches q on q's top side, and $\lambda(p, q) = -1$ says that p touches q on q's bottom side. (Here we use the term "touch" loosely: p and q may in fact be intervened by other layers of paper,

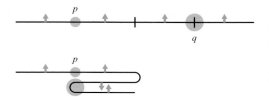

Figure 11.3. $\lambda(p, q)$ cannot always be defined in a continuous way when q is a crease point: points near q have opposite orientations in the folded state, so $\lambda(p + \varepsilon, q \pm \varepsilon) = \pm 1$ for $\varepsilon > 0$ small.

and thus only "touch" when considered in isolation; the key aspect is the side of q on which p lies.)

This intuitive notion of λ is not a definition, because there is no geometric sense (at p and q) in which p is above or below q: mapped by f, p and q are collocated. Nonetheless, the intuition specifies how we could draw a diagram of the folded state if we "zoomed" into the behavior of the folded state at a point in $\mathbb{R}^2$ to which several noncrease points of P map via f. In particular, it disambiguates the various scenarios of Figure 11.2, as desired, by specifying the order of the two layers of paper in the top middle.

A simple observation is that $\lambda(p, q)$ and $\lambda(q, p)$ measure slightly different information: $\lambda(p, q)$ is relative to $\mathbf{n}_f(q)$, while $\lambda(q, p)$ is relative to $\mathbf{n}_f(p)$. Thus we read $\lambda(p, q)$ as "the order of p relative to q." The antisymmetry condition below formalizes the relationship between $\lambda(p, q)$ and $\lambda(q, p)$.

We can also now look back at our as-yet unjustified requirement that p and q be noncrease points. The intuitive notion of $\lambda(p, q)$ does not necessarily require normal vectors $\mathbf{n}_f(p)$ and $\mathbf{n}_f(q)$ to be defined—sidedness can be defined more generally— but Figure 11.3 shows a situation where it is difficult to define a concept similar to $\lambda(p, q)$ unambiguously when q is a crease point. To avoid this ambiguity, we simply do not define $\lambda(p, q)$ when either p or q is a crease point. This lack of definition for λ will require special care to ensure that ordering information is not lost and crossings cannot be introduced, but this challenge is eased by the well-spacing of the crease points guaranteed by f being piecewise-C^1.

To solidify the intuitive notion of λ into a matching definition, we place several constraints on λ in relation to f. These conditions, together with f being piecewise-C^1 and isometric, define what makes a pair (f, λ) a (valid, semifree) *folded state* of a 1D piece of paper P. Thus a folded state consists of both a *geometry* f and an *ordering* λ. The first three conditions on (f, λ) essentially constrain λ to be a sort of partial-order relationship, which is relatively straightforward. The fourth condition constrains the folded state to be noncrossing and forces the geometric reality to match the intuition behind λ. This last condition is the most complicated.

11.2.2 Antisymmetry Condition

The antisymmetry condition relates $\lambda(p, q)$ and $\lambda(q, p)$ for any two points $p, q \in P$ at which $\lambda(p, q)$ is defined. Intuitively, the condition says that if p is above q, then q is below p, and vice versa, for a fixed notion of "above" and "below." The complication is that $\lambda(p, q)$ measures relative to q's notion of "above," $\mathbf{n}_f(q)$, while $\lambda(q, p)$ measures relative to p's notion of "above," $\mathbf{n}_f(p)$. To unify these notions, we introduce the products of λ's and $\mathbf{n}$'s: $\lambda(q, p)\,\mathbf{n}_f(p)$ is a vector pointing from p "toward" q, and $\lambda(p, q)\,\mathbf{n}_f(q)$ is a vector pointing from q "toward" p. The antisymmetry condition stipulates that these vectors point in opposite directions, that is, $\lambda(q, p)\,\mathbf{n}_f(p) = -\lambda(p, q)\,\mathbf{n}_f(q)$.

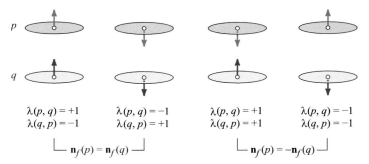

p

q

$\lambda(p,q)=+1$ $\lambda(p,q)=-1$ $\lambda(p,q)=+1$ $\lambda(p,q)=-1$
$\lambda(q,p)=-1$ $\lambda(q,p)=+1$ $\lambda(q,p)=+1$ $\lambda(q,p)=-1$

$\mathbf{n}_f(p)=\mathbf{n}_f(q)$ $\mathbf{n}_f(p)=-\mathbf{n}_f(q)$

Figure 11.4. Antisymmetry of λ.

One consequence of this condition is that, for any two points $p, q \in P$ at which $\lambda(p, q)$ is defined, the unit normal vectors are either equal or negations of each other: $\mathbf{n}_f(p) = \pm\mathbf{n}_f(q)$. This fact could also be derived from the noncrossing condition in Section 11.2.5, which in particular requires that neighborhoods of p and q, when mapped via f, do not properly cross (but may touch) and thus are tangent at $f(p) = f(q)$.

Figure 11.4 illustrates the antisymmetry condition in the various cases of the relative directions of the two normal vectors $\mathbf{n}_f(p)$ and $\mathbf{n}_f(q)$.

11.2.3 Transitivity Condition

The transitivity condition relates λ among three points $p, q, r \in P$ at which λ is defined pairwise, that is, for which p, q, r are distinct noncrease points with $f(p) = f(q) = f(r)$. Informally, the condition says that if p is above q and q is above r, then p is above r.

The formal version effectively depends on the relative directions of the normal vectors $\mathbf{n}_f(p), \mathbf{n}_f(q), \mathbf{n}_f(r)$, but can be stated without explicit use of these vectors. Specifically, the *transitivity condition* stipulates that if $\lambda(p, q) = -\lambda(r, q)$, then $\lambda(r, p) = \lambda(q, p)$. The precondition $\lambda(p, q) = -\lambda(r, q)$ specifies that p and r are on opposite sides of q, which effectively means that we are in the desired case that p, q, r are stacked sequentially. The consequence $\lambda(r, p) = \lambda(q, p)$ specifies that r and q are on the same side of p, which effectively duplicates the stacking relation between q and p onto the corresponding relation between r and p. Figure 11.5 shows the main cases after factoring out symmetry (assuming $\lambda(p, q) = +1$).

11.2.4 Consistency Condition

The consistency condition relates λ between "nearby" pairs of points at which λ is defined. Informally, the condition says that λ is continuous (has a single, constant value) over any pair of connected regions that map to the same location under f.

First we define the notion of "nearby" pairs of points, or equivalently, what form of connectivity we need of the two regions. Let dom λ denote the domain of λ, that is, the set of pairs (p, q) at which $\lambda(p, q)$ is defined. Call two pairs $(p, q), (p', q') \in$ dom λ *path-connected* if there is a path in dom λ from (p, q) to (p', q'), that is, there is a continuous function $C : [0, 1] \to$ dom λ with $C(0) = (p, q)$ and $C(1) = (p', q')$.

The *consistency condition* stipulates, for any two path-connected points $(p, q), (p', q') \in$ dom λ, that $\lambda(p, q) = \lambda(p', q')$. In other words, this condition specifies that λ is continuous (constant) in every path-connected component of its domain dom λ.

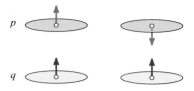

p

q

Figure 11.5. Transitivity of λ.

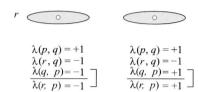

r

$$\begin{aligned} \lambda(p, q) &= +1 \\ \lambda(r, q) &= -1 \\ \underline{\lambda(q,\ p)} &= -1 \\ \hline \lambda(r,\ p) &= -1 \end{aligned} \Bigg] \qquad \begin{aligned} \lambda(p, q) &= +1 \\ \lambda(r, q) &= -1 \\ \underline{\lambda(q,\ p)} &= +1 \\ \hline \lambda(r,\ p) &= +1 \end{aligned} \Bigg]$$

The consistency condition prevents a particular kind of order crossing of λ's changing value in two touching regions. The noncrossing condition described next prevents all types of order crossings. In particular, it will turn out that the consistency condition is subsumed by part of the noncrossing condition (the first two subcases of Case 4). Nonetheless, we find it useful to state the consistency condition first, because of its intuitive importance and as a sort of warmup.

11.2.5 Noncrossing Condition

The final condition on (f, λ), noncrossing, is the most complicated. Intuitively, it forbids two types of crossings: *geometric crossings*, when two neighborhoods of paper properly cross each other, and *order crossings*, when layers of paper touch and their λ value specifies an invalid order relative to other λ values or relative to other local geometry. Geometric crossings are the easier type to handle; order crossings require a few additional cases to handle. The consistency condition already forbids a basic kind of order crossing: two layers of paper cannot change their order (as specified by λ) for the entire extent of their touching.

The noncrossing condition applies to every two distinct nonboundary points p and q for which $f(p) = f(q)$ (even if p and/or q are crease points), and to all sufficiently small $\varepsilon > 0$.[6] Intuitively, if the folded state (f, λ) has any crossings, there will be a pair of points p, q that are collocated according to f and that exhibit the crossing in a small, ε-radius neighborhood around that point $f(p) = f(q)$.

Next we characterize the local behavior around the point $f(p) = f(q)$. Let C denote the circle of radius ε centered at $f(p) = f(q)$. Let $p^+ \in P$ denote the smallest value larger than p such that $f(p^+)$ is on the circle C. (Such a point exists provided that ε is indeed small enough, because the isometry condition on f forces $f(p + \delta)$ to be distinct from $f(p)$ for some $\delta > 0$.) Similarly, let $p^- \in P$ denote the largest value smaller than p such that $f(p^-)$ is on C, and similarly define q^+ and q^- relative to q. Thus the open intervals (p^-, p^+) and (q^-, q^+) map via f to strictly inside the circle C, and their endpoints p^-, p^+, q^-, q^+ map onto the circle C.

To simplify the possible situations, we ensure that the punctured intervals $[p^-, p^+] \setminus \{p\}$ and $[q^-, q^+] \setminus \{q\}$ contain no crease points by choosing ε small enough. As ε decreases, p^+ and q^+ strictly decrease and p^- and q^- strictly increase, so the two

---

[6] More precisely, there must be a bound $E = E(p, q) > 0$ (possibly dependent on p and q) such that the noncrossing condition holds for all ε with $0 < \varepsilon < E$.

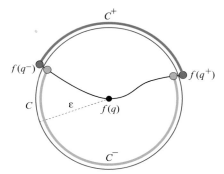

Figure 11.6. The interval (q^-, q^+) containing q and mapping via f to the interior of the ε-radius circle C splits C into two halves: C^+ and C^-.

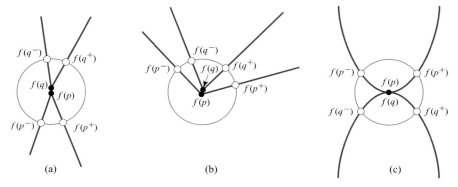

(a)   (b)   (c)

Figure 11.7. Valid local situations in which neither p^+ nor p^- map via f to the same location as q^+ or q^-. In (a) and (b), p and q are crease points. In (c), p and q are smooth.

punctured intervals shrink in the sense of strict subset containment. Furthermore, p^+ and p^- approach p, while q^+ and q^- approach q, so the punctured intervals approach length zero. Because f is piecewise-C^1, for sufficiently small ε, the two punctured intervals must avoid the finite number of creases other than p and q. As a consequence, if any two points from these punctured intervals are collocated according to f, we know that λ is defined on them.

We are now in the position to forbid geometric crossings (refer to Figure 11.6). The points q^+ and q^- divide the circle C into two arcs: C^+ is the arc counterclockwise from $f(q^+)$ to $f(q^-)$ and C^- is the arc counterclockwise from $f(q^-)$ to $f(q^+)$. We define both C^+ and C^- to include their endpoints q^+ and q^-. Intuitively, C^+ and C^- specify the points of C that are on the "+ side" of q and the "− side" of q, respectively, leaving the endpoints q^+ and q^- as ambiguously on both sides. If $f(q^+) = f(q^-)$, this definition of C^+ and C^- is ambiguous, and we use $\lambda(q^-, q^+)$ to disambiguate: if $\lambda(q^-, q^+) = +1$, then C^+ is the single point $\{f(q^+)\} = \{f(q^-)\}$ and C^- is the entire circle C; and if $\lambda(q^-, q^+) = -1$, then $C^+ = C$ and $C^- = \{f(q^+)\} = \{f(q^-)\}$. The noncrossing condition forbids geometric crossings by requiring either that both $f(p^+)$ and $f(p^-)$ are on C^+ or that both $f(p^+)$ and $f(p^-)$ are on C^-. This condition properly constrains the situations in Figure 11.7(a–b), for example.

Finally, the noncrossing condition relates λ to whether $f(p^+)$ and $f(p^-)$ are on C^+ or C^-, to ensure that the geometry corresponds to the ordering. This relation forbids the order crossings alluded to before. We have several cases, depending on which of $\lambda(p, q)$, $\lambda(p^+, q^+)$, $\lambda(p^+, q^-)$, $\lambda(p^-, q^+)$, and $\lambda(p^-, q^-)$ are defined.

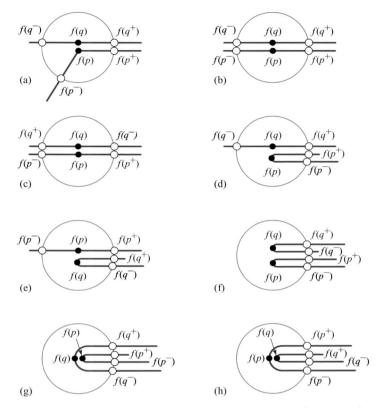

Figure 11.8. Valid local situations in which one or more pairs from $\{p^+, p^-\} \times \{q^+, q^-\}$ are collocated according to f. In (a), exactly one pair is collocated. In (b–e), exactly two pairs are collocated. In (f–h), all four points collocate.

Case 1: $\lambda(p, q)$ is defined. In this case, $\lambda(p, q)$ forces the direction: if $\lambda(p, q) = +1$, then the noncrossing condition requires that both $f(p^+)$ and $f(p^-)$ be on C^+; and similarly, if $\lambda(p, q) = -1$, then they must both be on C^-. This condition properly constrains the situation in Figure 11.7(c), for example.

In the remaining cases, we assume that $\lambda(p, q)$ is undefined (which happens whenever p or q is a crease point).

Case 2: None of $\lambda(p^+, q^+)$, $\lambda(p^+, q^-)$, $\lambda(p^-, q^+)$, and $\lambda(p^-, q^-)$ are defined. In other words, as in Figure 11.7, neither p^+ nor p^- map via f to the same location as q^+ or q^-. (Although not shown in the figure, we could still have $f(p^+) = f(p^-)$ and/or $f(q^+) = f(q^-)$.) In this case, there are no λ values of relevance, so the noncrossing condition imposes no additional conditions.

Case 3: Exactly one of $\lambda(p^+, q^+)$, $\lambda(p^+, q^-)$, $\lambda(p^-, q^+)$, and $\lambda(p^-, q^-)$ is defined, as in Figure 11.8(a). As a consequence, we must have $f(p^+) \neq f(p^-)$ and $f(q^+) \neq f(q^-)$. In this case, the single defined λ value forces the direction: if it is $+1$, then the noncrossing condition requires that both $f(p^+)$ and $f(p^-)$ be on C^+; and similarly, if the λ value is -1 (as in Figure 11.8(a)), then they must both be on C^-.

Case 4: Exactly two of $\lambda(p^+, q^+)$, $\lambda(p^+, q^-)$, $\lambda(p^-, q^+)$, and $\lambda(p^-, q^-)$ are defined, as in Figure 11.8(b–e). There are four subcases, the last of which is different from the first three. If $\lambda(p^+, q^+)$ and $\lambda(p^-, q^-)$ are defined, as in Figure 11.8(b), or if $\lambda(p^+, q^-)$

and $\lambda(p^-, q^+)$ are defined, as in Figure 11.8(c), or if $\lambda(p^\pm, q^+)$ are defined, as in Figure 11.8(d), then we require that these λ values are equal. However, if $\lambda(p^+, q^\pm)$ are defined, as in Figure 11.8(e), then we require that $\lambda(q^+, p^+) = \lambda(q^-, p^+)$, or equivalently that $\lambda(p^\pm, q^+) = -\lambda(p^\pm, q^-)$.

Case 5: All of $\lambda(p^+, q^+)$, $\lambda(p^+, q^-)$, $\lambda(p^-, q^+)$, and $\lambda(p^-, q^-)$ are defined, as in Figure 11.8(f–h). In this case, $f(p^+) = f(p^-) = f(q^+) = f(q^-)$, and there are essentially two behaviors we want to allow: side-by-side and nested "U's". In the first situation (Figure 11.8(f)), p^+ and p^- are on the same side of q^+, and of q^- (independent of the stacking order of the "U's"). In the second situation (Figure 11.8(g–h)), p^+ and p^- are on opposite sides of q^+, and of q^- (independent of the nesting order of the "U's"). To capture both situations, the noncrossing condition requires that $\lambda(p^+, q^+) = \lambda(p^-, q^+)$ if and only if $\lambda(p^+, q^-) = \lambda(p^-, q^-)$.

As mentioned before, the first two subcases of Case 4 (corresponding to Figure 11.8(b–c)) are in fact equivalent to the consistency condition of Section 11.2.4. Throughout Cases 4 and 5, there are no constraints on whether $f(p^+)$ and $f(p^-)$ should be on C^+ or C^-, because the geometry f can neither confirm nor deny the correctness of the order specified by λ. Indeed, in most subcases of these cases, both $f(p^+)$ and $f(p^-)$ are on (the endpoints of) both C^+ and C^-. The only exception is the fourth subcase of Case 4, as in Figure 11.8(e), where $f(q^+) = f(q^-)$, so the disambiguation of the definition of C^+ and C^- forces them to correspond to λ properly.

11.2.6   Extensions

This description completes the definition of a folded state (f, λ) for a 1D piece of paper P: f must be piecewise-C^1 and isometric, and (f, λ) must satisfy the antisymmetry, transitivity, consistency, and noncrossing conditions.

This definition easily extends in two directions, which we will need in our definition of folded states for two-dimensional pieces of paper:

1. **Disconnected pieces of paper.** P can consist of multiple intervals, which we can view as lying on separate horizontal axes.
2. **Circular pieces of paper.** The two endpoints of an interval of P can in fact be identified, forming a topological circle.

In both cases, the key property is that the piece of paper has an orientation, that is, every point p has a local notion of the positive direction ($p + \varepsilon$ and p^+) and the negative direction ($p - \varepsilon$ and p^-) for sufficiently small ε neighborhoods. In the case of circular pieces of paper, these notions wrap around at the endpoints of the interval. In particular, the shortest-path metric on a circular piece of paper takes the shorter of the two paths between any two points. For disconnected pieces of paper, shortest-path distances between different connected components are infinite and thus the isometry condition effectively ignores them.

11.3   DEFINITIONS: FOLDING MOTIONS OF 1D PAPER

Similar to free folding motions as described in Section 11.1.2, a *folding motion* of a piece of paper P is a continuum of folded states of P, that is, a continuous function mapping each time $t \in [0, 1]$ to a folded state (f_t, λ_t) of P. Other than each (f_t, λ_t) being a valid folded state for each $t \in [0, 1]$, the only constraint is that (f_t, λ_t) varies continuously in t. Now this constraint consists of two parts: continuity of f_t (geometry) and continuity of λ_t (order). Continuity of f_t is defined exactly as it was for free folded states in Section 11.1.2: f_t should vary continuously with respect to t under the supremum metric. Continuity of λ_t is a new, nontrivial constraint that builds on the technology developed in the noncrossing condition of Section 11.2.5.

Time continuity of order. Continuity of λ_t in t constrains any two points $p, q \in P$ and time $t \in [0, 1]$ for which $\lambda_t(p, q)$ is defined, that is, for which p and q are distinct noncrease points of f_t and $f_t(p) = f_t(q)$. We focus on the possible departure of p from q as time increases, which defines the *forward-continuity constraint*. By symmetry (by reversing time), we obtain the analogous *backward-continuity constraint* on the possible arrival of p at q as time decreases.

The forward-continuity constraint characterizes the local behavior of the motion of p relative to q, that is, $f_{t+\varepsilon}(p)$ relative to $f_{t+\varepsilon}(q)$ for all sufficiently small $\varepsilon > 0$. Let C denote the circle of radius $\| f_{t+\varepsilon}(p) - f_{t+\varepsilon}(q)\|$ centered at $f_{t+\varepsilon}(q)$, so that $f_{t+\varepsilon}(p)$ is on the circle C. Because f_t varies continuously in t and $f_t(p) = f_t(q)$, as ε approaches 0, the radius of C also approaches 0; so this definition mimics the definition of C in the noncrossing condition of Section 11.2.5. Let $q^+ \in P$ denote the smallest value larger than q such that $f_{t+\varepsilon}(q^+)$ is on the circle C, and let $q^- \in P$ denote the largest value smaller than q such that $f_{t+\varepsilon}(q^-)$ is on C. As in Section 11.2.5, by choosing ε small enough, both q^+ and q^- exist and we can assume that the punctured interval $[q^-, q^+] \setminus \{q\}$ contains no crease points of $f_{t+\varepsilon}$. The points $f_{t+\varepsilon}(q^+)$ and $f_{t+\varepsilon}(q^-)$ divide the circle C into two arcs overlapping at their endpoints: C^+ is the arc counterclockwise from $f_{t+\varepsilon}(q^+)$ to $f_{t+\varepsilon}(q^-)$ and C^- is the arc counterclockwise from $f_{t+\varepsilon}(q^-)$ to $f_{t+\varepsilon}(q^+)$. If $f_{t+\varepsilon}(q^+) = f_{t+\varepsilon}(q^-)$, we disambiguate this definition using $\lambda_{t+\varepsilon}(q^-, q^+)$ as in the noncrossing condition of Section 11.2.5.

With this setup, there are two cases of the forward-continuity constraint.

In the *contact case*, $f_{t+\varepsilon}$ collocates p with either q^+ or q^-, say $q^\pm$. Intuitively, this case means that p has slid along a neighborhood of q, always remaining in contact. In this case, we constrain that the λ value remains the same: $\lambda_{t+\varepsilon}(p, q^\pm) = \lambda_t(p, q)$. This constraint forbids λ from changing instantaneously from one moment in time to the next: p must leave contact from a neighborhood of q before it can reach the other side of q.

In the *departure case*, $f_{t+\varepsilon}$ collocates p with neither q^+ nor q^-. In this case, we constrain the geometry of the departure specified by $f_{t+\varepsilon}$ to match the order specified by λ_t. Namely, if $\lambda_t(p, q) = +1$, then $f_{t+\varepsilon}(p)$ must be on C^+ (the "+ side" of q); and if $\lambda_t(p, q) = -1$, then $f_{t+\varepsilon}(p)$ must be on C^- (the "− side" of q).

Together with the symmetrically defined backward-continuity constraint, these two constraints define continuity of λ_t in t.

11.4   DEFINITIONS: FOLDED STATES OF 2D PAPER

Our next task is to generalize the definitions of folded state and folding motion to the case of 2D pieces of paper. The extension mainly involves finding the correct higher-dimensional analogs of the concepts used in the 1D definition.

11.4.1   2D Paper

We define a *(2D) piece of paper* to be a "nice" closed subset P of $\mathbb{R}^3$ that, topologically, is an orientable 2-manifold. There are three essential constraints here: the manifold topology, the orientability of that manifold, and the "niceness" of the geometry.

The first constraint is the simplest. Topologically, P inherits the *subspace topology*: the open sets of P are the intersection of P with the open sets of $\mathbb{R}^3$. This topology must be a 2-*manifold* in the sense that every point in P either has a neighborhood homeomorphic to an open disk or, in the case of a *boundary point*, homeomorphic to a half-disk.

Next we define orientability. Given a subset B of the paper P that is homeomorphic to a closed disk—or a *topological disk B* for short—a *boundary orientation* of B is a homeomorphism from the (oriented) unit circle $\mathbb{S}^1$ to the boundary of B. There are two fundamentally different boundary orientations of a topological disk (up to isotopy[7]), which correspond to a clockwise and a counterclockwise traversal of the boundary. The paper P must be *orientable* in the sense that we can assign a consistent boundary orientation to each topological disk in P.[8]

Finally, we make a basic assumption about "niceness" of the geometry of P: every two points on P that are connected by a curve on P (i.e., *path-connected*) are connected by a piecewise-C^1 curve on P. Here a *curve* between two points $p, q \in P$ is a continuous function $C : [0, 1] \to P$ such that $C(0) = p$ and $C(1) = q$. A curve $C(t)$ is *piecewise-C^1* if it has a continuous derivative with respect to t except at finitely many exceptions. For such curves C, we can define its *arc length* by arclength $C = \int_0^1 \left\| \frac{dC(t)}{dt} \right\| dt$. Given the existence of such curves, the embedding of P in $\mathbb{R}^3$ defines a *shortest-path metric* on P: the distance $d_P(p, q)$ between two points $p, q \in P$ on P is the arc length of the shortest piecewise-C^1 curve from p to q on P:

$$d_P(p, q) = \inf\{\text{arclength } C \mid C : [0, 1] \to P \text{ is piecewise-}C^1, C(0) = p, C(1) = q\}.$$

(This definition makes P a "Riemannian manifold.")

This definition of a piece of paper is quite general. It includes simple polygons, polygons with holes, polyhedral surfaces, and curved surfaces such as spheres. The piece of paper can even be disconnected. On the other hand, this definition does not capture nonorientable surfaces, such as a Möbius strip, or nonmanifold surfaces, such as three triangles glued together along a common edge. We expect that our definitions can be extended to these cases as well, but only with additional complications that we leave to future work.

11.4.2   Free Folded States of 2D Paper

As with 1D paper, free folded states are relatively easy to define, but the isometry condition serves to define the geometry of a general folded state.

---

[7] An *isotopy* between two topological disks A and B in P is a continuum of homeomorphisms h_t, one for each $t \in [0, 1]$, from the (oriented) closed unit disk $\mathcal{B}^2$ to a subset of P, such that the image of h_0 is A and the image of h_1 is B. (This definition is the natural adaptation of the definition of isotopy between two topological spaces.)

[8] An assignment of boundary orientations to all topological disks in P is *consistent* if, whenever there is an isotopy between two topological disks A and B, there is an isotopy h_t between A and B with the property that h_0 restricted to the boundary of $\mathcal{B}^2$ is the boundary orientation of A and h_1 restricted to the boundary of $\mathcal{B}^2$ is the boundary orientation of B. In other words, isotopies on topological disks can also be made isotopies on their boundary orientations.

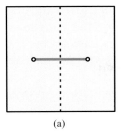
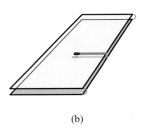

Figure 11.9. (a) The shortest path between two points measured on a square piece of paper; (b) a simple (semifree) folded state and the shortest path between the same points in that folding.

(a)   (b)

A *free folded state* of a 2D piece of paper P is a one-to-one isometric function $f : P \to \mathbb{R}^3$ mapping P into Euclidean 3-space. The one-to-one condition is as before: $f(p) = f(q)$ only when $p = q$. The isometry condition again specifies that the intrinsic geodesic (shortest-path) distance between any two points of P is the same when measured either on P or on the surface of the folded state when P is mapped by f. Now, however, instead of intervals between points, the definition must use shortest paths. Figure 11.9 shows a simple example of folding a square of paper in half, illustrating the subtlety in how we measure shortest paths: despite two points being collocated when mapped by f, their shortest path on the folded state defined by f remains of positive length. (Although this folded state is not free, our definition of the isometry condition must handle this case as well for future use.)

Precisely, $f : P \to \mathbb{R}^3$ is *isometric* if $d_P(p, q) = d_f(p, q)$ for all $p, q \in P$. As described in Section 11.4.1, the distance $d_P(p, q)$ between two points $p, q \in P$ on P is the arc length of the shortest piecewise-C^1 curve from p to q on P. The distance $d_f(p, q)$ between the points on the surface in $\mathbb{R}^3$ defined by f is similar, but we map the curve via f before measuring its arc length:

$$d_f(p, q) = \inf \{\text{arclength}(f \circ C) \mid C : [0, 1] \to P \text{ is piecewise-}C^1, C(0) = p, C(1) = q\}.$$

Using this definition of a free folded state of a 2D piece of paper P, a free folding motion of P can be defined the same as for 1D paper in Section 11.1.2: a continuous function mapping each $t \in [0, 1]$ to a free folded state f_t, where continuity is with respect to the supremum metric.

11.4.3 Smoothness and Creases of 2D Paper

The smoothness condition generalizes creases from points to 1D curves.

A function $f : P \to \mathbb{R}^3$ is *piecewise-C^k* if the 2-manifold P can be decomposed into finitely many pairwise-disjoint open 2-manifolds $P_1, P_2, \ldots, P_m \subseteq P$ such that

1. the closures $\overline{P_i}$ of the pieces P_i union to P: $\overline{P_1} \cup \overline{P_2} \cup \cdots \cup \overline{P_m} = P$;
2. f has continuous derivatives up to order k on each P_i; and
3. f is piecewise-C^k when restricted to the 1D curve $\overline{P_i} - P_i$ that bounds piece P_i, for $i = 1, 2, \ldots, m$. (This recursive application of the definition of course uses the topology of $\mathbb{R}^1$, instead of $\mathbb{R}^2$, to define open sets, C^k, etc.)

The union of the boundaries of all pieces, $P \setminus (P_1 \cup P_2 \cup \cdots \cup P_m)$, is the set of *crease points*. Note that this definition allows finitely many piecewise-C^k 1D curves to be added to the set of crease points, even if f has a continuous kth derivative; this fact is useful when defining a (semifree) folded state, because it allows us to avoid defining λ at such points.

The orientation of the paper P provides a well-defined orientation of the unit normal vector $\mathbf{n}_f(p)$ for a noncrease point p of f. The *tangent plane* $T_f(p)$ is the plane spanned by all rays emanating from $f(p)$ that are tangent to the surface defined by f at p, which exists because p lies in a C^1 piece of f. The normal $\mathbf{n}_f(p)$ lies on the line passing through $f(p)$ that is perpendicular to the tangent plane $T_f(p)$, but it remains to specify which of the two possible orientations the normal has. The *exponential map* provides a homeomorphism between a neighborhood N_T of $f(p)$ in $T_f(p)$ and a neighborhood N_P of p in P.[9] We can use this exponential map to transform the boundary orientation of N_P, given by the orientation of P, into a boundary orientation of the neighborhood N_T. This boundary orientation of N_T in turn determines an orientation of the plane $T_f(p)$: the boundary orientation of N_T should be a counterclockwise traversal around $f(p)$ when viewed from the positive side of $T_f(p)$. The *unit normal vector* $\mathbf{n}_f(p)$ is then the unit vector perpendicular to $T_f(p)$ on the positive side of $T_f(p)$.

11.4.4 Order

A *folded state* of a 2D piece of paper is a pair (f, λ) consisting of an isometric function $f : P \to \mathbb{R}^3$ (the *geometry*) and a partial function λ from $P \times P$ to $\{-1, +1\}$ defined on pairs (p, q) of distinct noncrease points $p, q \in P$ for which $f(p) = f(q)$ (the *order*). As with 1D paper, $\lambda(p, q) = +1$ specifies that p is on the "top side" of q, that is, in the direction pointed to by the normal vector $\mathbf{n}_f(q)$; and $\lambda(p, q) = -1$ specifies that p is "below" q, that is, in the direction $-\mathbf{n}_f(q)$. The order λ must satisfy four properties, all but the last of which are identical to the case of 1D paper described in Section 11.2.

11.4.4.1 Antisymmetry Condition

The antisymmetry condition stipulates, for any two points $p, q \in P$ at which $\lambda(p, q)$ is defined, that $\lambda(q, p) \, \mathbf{n}_f(p) = -\lambda(p, q) \, \mathbf{n}_f(q)$, that is, p and q must be on opposite sides of each other.

11.4.4.2 Transitivity Condition

The transitivity condition stipulates that, for any three points $p, q, r \in P$ at which λ is defined pairwise, if $\lambda(p, q) = -\lambda(r, q)$, then $\lambda(r, p) = \lambda(q, p)$; that is, if p and r are on opposite sides of q, then r and q are on the same side of p.

11.4.4.3 Consistency Condition

The consistency condition stipulates, for any two path-connected points (p, q), $(p', q') \in \mathrm{dom}\,\lambda$, that $\lambda(p, q) = \lambda(p', q')$. (The domain $\mathrm{dom}\,\lambda$ of λ and path-connectivity in that space are defined the same as for 1D paper in Section 11.2.4.) Intuitively, this condition says that λ is consistent (continuous) among nearby noncrease points.

11.4.4.4 Noncrossing Condition

The noncrossing condition is the only condition that differs from dimension to dimension. The main idea is to reduce the 2D condition to the 1D condition, using constraints similar to the 1D condition from Section 11.2.5.

---

[9] Roughly speaking, the exponential map curls a tangent vector onto a geodesic (locally shortest path) of the surface P emanating from p, preserving the length and initial tangent direction, and returns the endpoint of the resulting curve on $f(P)$ other than p.

The noncrossing condition applies to every two distinct nonboundary points p and q for which $f(p) = f(q)$ (even if p and/or q are crease points), and to all sufficiently small $\varepsilon > 0$. Let S denote the sphere in $\mathbb{R}^3$ of radius ε centered at $f(p) = f(q)$. Let I denote the set of points of $f(P)$ that are strictly inside S. Let N_p denote the path-connected component of $f^{-1}(I)$ that contains p, that is, the set of points path-connected to p in $f^{-1}(I)$. We claim this intuitively natural lemma without proof (Demaine et al. 2006a):

Lemma 11.4.1. *If f is piecewise-C^1, then for sufficiently small $\varepsilon > 0$, N_p is homeomorphic to an open 2D disk.*

Let $B_p = \overline{N_p} \setminus N_p$ be the boundary of N_p, which maps via f onto the sphere S. By Lemma 11.4.1, for sufficiently small ε, the orientation of P specifies a boundary orientation of $\overline{N_p}$, i.e., a homeomorphism from the unit circle $\mathbb{S}^1$ to B_p. Similarly, we define N_q to be the path-connected component of I that contains q, to which Lemma 11.4.1 applies equally as well, and we define and orient the boundary $B_q = \overline{N_q} \setminus N_q$. By choosing ε small enough, we can guarantee that N_p and N_q are disjoint and that B_p and B_q are disjoint. (As we decrease $\varepsilon \to 0$, N_p and N_q shrink in the sense of set containment toward $\{p\}$ and $\{q\}$, respectively, so for any $\delta > 0$, there is an upper bound on ε guaranteeing that N_p is contained in a δ-radius ball centered at p and similarly for N_q. Because $p \neq q$, we can choose $\delta < \frac{1}{2}d(p, q)$ and the corresponding upper bound on ε to force N_p and N_q to have disjoint closures, and therefore B_p and B_q are disjoint.)

We view the restriction $f|_{B_p}$ of f onto the subdomain B_p, together with the restriction $\lambda|_{B_p \times B_p}$ of the ordering λ, as a folded state $(f, \lambda)|_{B_p}$ of the subpiece of paper B_p which is homeomorphic to a circle. What is essential here is that B_p inherits the orientation of P, as described above; such an orientation is required for our recursive use of the definition of "folded state" to make sense. Similarly, define the lower-dimensional folded state $(f, \lambda)|_{B_q} = (f|_{B_q}, \lambda|_{B_q \times B_q})$, and define the combination folded state $(f, \lambda)|_{B_p \cup B_q}$ on the disconnected piece of paper $B_p \cup B_q$ consisting of two topological circles. Intuitively, the 2D noncrossing condition on f requires that the two 1D folded states $(f, \lambda)|_{B_p}$ and $(f, \lambda)|_{B_q}$ do not cross each other; it achieves this restriction by requiring that the combination folded state $(f, \lambda)|_{B_p \cup B_q}$ satisfies the 1D noncrossing condition.

11.5 DEFINITIONS: FOLDING MOTIONS OF 2D PAPER

As with 1D paper in Section 11.3, a *folding motion* of a 2D piece of paper P is a continuum of folded states of P, that is, a continuous function mapping each time $t \in [0, 1]$ to a folded state (f_t, λ_t) of P. As before, continuity of (f_t, λ_t) with respect to t consists of two parts. Continuity of f_t is defined exactly as it was for free folded states in Section 11.1.2: f_t should vary continuously with respect to t under the supremum metric. Continuity of λ_t is another nontrivial constraint that builds on the technology developed in the noncrossing condition of Section 11.4.4.4.

Time continuity of order. As with 1D paper in Section 11.3, continuity of λ_t constrains any two points $p, q \in P$ and time $t \in [0, 1]$ for which $\lambda_t(p, q)$ is defined, that is, for which p and q are distinct noncrease points of f_t and $f_t(p) = f_t(q)$. We focus on the possible departure of p from q as time increases, which defines the *forward-continuity constraint*. By symmetry (by reversing time), we obtain the analogous *backward-continuity constraint* on the possible arrival of p at q as time decreases.

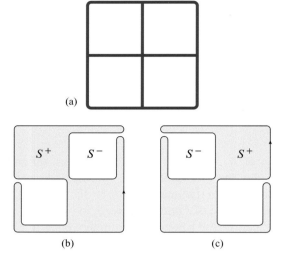

(a)

(b)   (c)

Figure 11.10. (a) The image of $f_{t+\varepsilon}$ applied to B_p; (b–c) two possible curves yielding the same image. Note that points interior to the four quadrants have opposite winding numbers in (b) and in (c).

The forward-continuity constraint characterizes the "local behavior" of p's motion relative to q, that is, $f_{t+\varepsilon}(p)$ relative to $f_{t+\varepsilon}(q)$ for all sufficiently small $\varepsilon > 0$. Let S denote the sphere in $\mathbb{R}^3$ of radius $\| f_{t+\varepsilon}(p) - f_{t+\varepsilon}(q)\|$ centered at $f_{t+\varepsilon}(q)$, so that $f_{t+\varepsilon}(p)$ is on the sphere S. Because f_t varies continuously in t and $f_t(p) = f_t(q)$, as ε approaches 0, the radius of S also approaches 0, so this definition mimics the definition of S in the noncrossing condition of Section 11.4.4.4. Let B_p be the boundary of the set of points that are path-connected to p and strictly inside the sphere S. By Lemma 11.4.1, for sufficiently small ε, B_p is homeomorphic to a circle, and we can obtain an orientation of B_p from the boundary orientation of $\overline{N_p}$ given by the orientation of P.

One new definition that we need is how the restriction $f_{t+\varepsilon}|_{B_p}$ of f onto the subdomain B_p divides the sphere S into two regions S^+ and S^- overlapping at their boundary. Figure 11.10 illustrates how this division relies on the orientation of B_p and the function $f_{t+\varepsilon}|_{B_p}$, not merely the image $f_{t+\varepsilon}(B_p)$ which loses information. Now S^+ is defined as the closure of the set of points x on S such that the winding number (cf. p. 200) of $f_{t+\varepsilon}|_{B_p}$ with respect to x is positive (in fact, $+1$); and S^- is the closure of the set of points x on S such that the winding number of $f_{t+\varepsilon}|_{B_p}$ with respect to x is negative (in fact, -1). The two regions S^+ and S^- intersect at precisely their shared boundary (i.e., the image $f_{t+\varepsilon}(B_p)$, the set of points relative to which the winding number of $f_{t+\varepsilon}|_{B_p}$ is undefined). Here we exploit that B_p inherits an orientation from P as described above, so that the sign of the winding number of $f_{t+\varepsilon}|_{B_p}$ is well-defined.

With this setup, there are two cases of the forward-continuity constraint.

In the *contact case*, $f_{t+\varepsilon}(p)$ is on the boundary of S^+ and S^-, that is, $f_{t+\varepsilon}$ collocates p with some point $x \in B_p$. Intuitively, this case means that p has slid along a neighborhood of q, always remaining in contact. In this case, we constrain that the λ value remains the same: $\lambda_{t+\varepsilon}(p, q) = \lambda_t(p, q)$. This constraint forbids λ from changing instantaneously from one moment in time to the next: p must leave contact from a neighborhood of q before it can reach the other side of q.

In the *departure case*, $f_{t+\varepsilon}(p)$ belongs to either S^+ or S^- but not both. In this case, we constrain the geometry of the departure specified by $f_{t+\varepsilon}$ to match the order specified by λ_t. Namely, if $\lambda_t(p, q) = +1$, then $f_{t+\varepsilon}(p)$ must be on S^+ (the "+ side" of q); and if $\lambda_t(p, q) = -1$, then $f_{t+\varepsilon}(p)$ must be on S^- (the "− side" of q).

This completes the forward-continuity constraint, which, together with the symmetrically defined backward-continuity constraint, define continuity of λ_t in t.

Therefore, we finally obtain formal definitions of the natural notions of 2D pieces of paper, their folded states, and their folding motions. Indeed all these notions extend to higher dimensions, as will be mentioned in Chapter 26.

11.6 FOLDING MOTIONS EXIST

With these definitions detailed, we may now prove the promised result: there is a folding motion (f_t, λ_t) between any two folded states (f_0, λ_0) and (f_1, λ_1) of a simple polygonal piece of paper P. The proof constructs a folding motion from the *unfolded state* of P, given by the trivial free folded state $u(p) = p$ for all $p \in P$, to any folded state (f, λ) of P. We then apply this construction to both (f_0, λ_0) and (f_1, λ_1) and piece together the two folding motions to obtain the desired result.

11.6.1 Rolling Between Flat Folded States

The first part of the construction uses a motion of "flat" folded states to bring P to an arbitrarily small form. A folded state (f, λ) is *flat* if the third (z) coordinate of $f(p)$ is zero for all $p \in P$. In particular, we view P as lying in the $z = 0$ plane, so the unfolded state u of P is flat. The *silhouette* of a flat folded state (f, λ) is the image $f(P)$ of the folded-state geometry. For example, the silhouette of the unfolded state u is P itself.

Lemma 11.6.1. *Let $T \subseteq P$ be a triangle that does not intersect any diagonal of some triangulation of a simple polygonal piece of paper P. Then there is a folding motion of P from the unfolded state of P into a flat folded state whose silhouette is T, such that the intermediate folded states are flat and their silhouettes nest by subset over time.*

Proof: The proof is by induction on the number of vertices of P.

We start with the general case, when P has $n > 3$ vertices. Because we have a triangulation of a simple polygon, there are at least two ears, that is, triangles of the triangulation each having two edges on the boundary of the polygon.

Let $\triangle abc$ be an ear of P that does not include T, with ear diagonal ab and ear tip c. Our goal is to "roll" $\triangle abc$ into $P' = P \setminus \triangle abc$, for once we achieve this goal, we can apply induction to the smaller polygon P'. Let the interior angle at a be $\alpha + A$, with α in $\triangle abc$, and let the interior angle at b be $\beta + B$, with β in $\triangle abc$. There are now two cases, depending on the angles α and β.

1. Both α and β are acute angles (see Figure 11.11(a)). Let $\alpha \geq \beta$. Continuously roll the edge ac, rotating around a, until it meets ab. This motion cuts α in half, so we remain in the both-acute case. Relabel if necessary so that again $\alpha \geq \beta$, and repeat. Continue this process until $\alpha < A$, $\beta < B$, and the height of the triangle from ab is less than ε, where ε is the minimum distance from ab to another visible vertex of P'. Then roll $\triangle abc$ into P', rolling a crease parallel to ab. This rolling can be accomplished so that the continuity definition in Section 11.5 is satisfied. Note that the silhouettes nest as time proceeds.
2. Either α or β, say α, is not acute and $|ac| \leq |ab|$ (see Figure 11.11(b)). Divide $\triangle abc$ into two right triangles by the perpendicular from a to bc. Roll c along

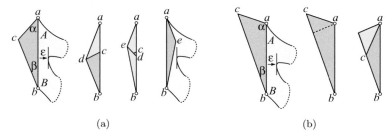

Figure 11.11. Illustration of the proof of Lemma 11.6.1. (a) α and β both acute; (b) α not acute.

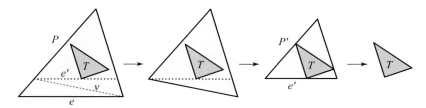

Figure 11.12. Rolling a triangle P into a subtriangle T.

    cb, forming a right triangle acute at both a and b. Continue as in the previous case.

3. If α is not acute but $|ac| > |ab|$, roll I along ac until c meets a. Thus ac is creased at its midpoint a'. Let c' be the other end of the crease, where the perpendicular bisector of ac meets bc. Now $\triangle a'c'b$ is an ear to which Case 1 applies, rolling that ear into P. The new ear is $\triangle aba'$, with the same α but with $|aa'|$ half the length of $|ac|$. Repeat until Case 2 applies.

This procedure eventually succeeds in rolling $\triangle abc$ into P' after a finite number of iterations, and so will achieve this goal with a finite number of creases.

    In the base case, P has three vertices. We reduce this special case to the general case above as follows (see Figure 11.12). Consider sliding an edge e of P perpendicularly inward until it reaches a vertex v of T, and let e' be the resulting chord of P. If $e' \neq e$, this chord bounds a trapezoid. Choose a diagonal and apply the induction step to roll the trapezoid into the remaining polygon P', at which point P' is bounded by e' flush against v. Apply the same procedure to the other edges of P until T is circumscribed by P. Apply the induction step to the three remaining triangles. Finally, all of the original P has been rolled into the silhouette of T.   □

    The condition that T avoid triangulation diagonals is artificial and assumed for convenience in the proof above. This additional assumption will be easy to arrange in the main argument below. However, a different proof establishes the lemma for any triangle $T \subseteq P$.

11.6.2   Unfurling onto the Target Folded State

We are now prepared for the main theorem:

Theorem 11.6.2. *If (f, λ) is a piecewise-C^2 folded state of a simple polygonal piece of paper P, then there is a folding motion of P from the unfolded state into (f, λ).*

Figure 11.13 provides a précis of the proof.

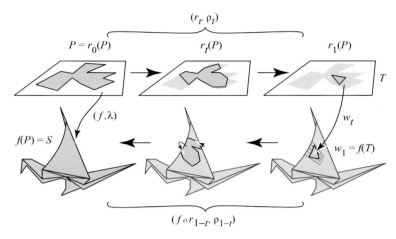

Figure 11.13. The construction of a folding motion of P into a folded state (f, λ) (not to scale). $S = f(P)$ is the image of the folded state. w_t is the free folding motion that wraps T onto its home $f(T)$ on S. r_t is the rolling motion that takes P to a flat folded state T within the plane. [Origami bird is based on a design by L. Zamiatina.]

Proof: Let (f, λ) be a folded state with image $S = f(P)$. Fix some triangulation of P; f maps its diagonals to curves on S. We now locate a triangle T in P (not necessarily a triangle of the triangulation), mapping to a patch $f(T)$ on S, that satisfies these properties:

1. The interior of T avoids all triangulation diagonals.
2. The interior of T avoids all crease points of f.
3. There is a direction in which the orthogonal projection of $f(T)$ is non-self-overlapping.

It is easy to achieve the first two properties by selecting a suitably small triangle in P. Any such patch is a "developable surface" (a C^2 surface whose Gaussian curvature is zero at every point) and "torsal ruled," which means that it may be swept out by lines generated by a well-behaved curve (Pottmann and Wallner 2001, p. 328). That the patch is C^2 suffices to ensure that a small enough piece will have a nonoverlapping projection. The ruling of the patch can be used to obtain a free folding motion w_t that "wraps" the flat triangle T onto this patch $f(T)$ of S. For example, one could bend the ruling lines of the ruled surface $f(T)$, interpolating from a straight segment to the generating curve of the ruling. (If the folded state (f, λ) is piecewise-linear, we can even arrange for w_t to be a rigid motion.)

Now we apply Lemma 11.6.1 to obtain a rolling motion of flat states (r_t, ρ_t) from P to a silhouette of T (i.e., with $r_1(P) = T$). After applying this rolling motion, we can apply a modified wrapping motion $(w_t \circ r_1, \lambda_1)$ to bring the multilayer flat folding of P from a silhouette of T to a silhouette of $f(T)$.

The last step of the construction is to "unravel" the folded state $(w_1 \circ r_1, \lambda_1)$, whose silhouette is $f(T)$, onto S. One can imagine S as a virtual scaffold, as depicted in Figure 11.13. The unraveling reverses the rolling motion (r_t, ρ_t) by considering (r_{1-t}, ρ_{1-t}) for $t \in [0, 1]$, but rather than progressing through the continuum of flat states, the motion unfurls on the surface S. Thus, at each time t, we compose the folded state (f, λ) with (r_{1-t}, ρ_{1-t}). The geometry f_t of this composition is simply $f \circ r_{1-t}$. The subset-nesting property from Lemma 11.6.1 ensures that f is applied only within its domain P. We define the creases of f_t to be the union of the creases of $f \circ r_{1-t}$ and the creases of r_{1-t}. (Recall that we can artificially add creases so long as they remain finite.) The ordering $\lambda_t(p, q)$ of this composition is defined to be $\rho_{1-t}(p, q)$ when that is defined and as $\lambda(f_{1-t}(p), f_{1-t}(q))$ when that is defined. At noncrease points p and q with $f(r_{1-t}(p)) = f(r_{1-t}(q))$, at least one of these alternatives is defined by definition, and at most one is defined because, if $\rho_{1-t}(p, q)$ is defined, then $r_{1-t}(p) = r_{1-t}(q)$, so $\lambda(r_{1-t}(p), r_{1-t}(q))$ is undefined. The various conditions required of this composed state (f_t, λ_t) follow from the corresponding conditions of both (f, λ) and (r_{1-t}, ρ_t). At the end we have continuously folded P into (f, λ).   □

An immediate consequence of this theorem is connectivity of the origami configuration space:

Corollary 11.6.3. *The configuration space of piecewise-C^2 folded states is connected via folding motions.*

Proof: Let (f_0, λ_0) and (f_1, λ_1) be any two folded states of the same polygonal piece of paper P. Let (f_t^i, λ_t^i) be the motion that folds P from the unfolded state into (f_i, λ_i), $i \in \{0, 1\}$, given by Theorem 11.6.2. Then the time-reversal of the motion (f_t^0, λ_t^0), followed by the motion (f_t^1, λ_t^1), is a motion that folds P from (f_0, λ_0) to (f_1, λ_1). In other words, the desired motion is given by

$$(f_t, \lambda_t) = \begin{cases} (f_{1-2t}^0, \lambda_{1-2t}^0) & \text{if } t \leq \frac{1}{2}, \\ (f_{2t-1}^1, \lambda_{2t-1}^1) & \text{if } t \geq \frac{1}{2}. \end{cases}$$ □

To reiterate the consequences of Theorem 11.6.2 at an intuitive level: if an origami folded state exists, it can be reached by a continuous folding motion. Although we proved the theorem only for polygonal paper and piecewise-C^2 folded states, we believe it can be extended to handle arbitrary paper (as defined in Section 11.4.1) and piecewise-C^1 folded states.

12 Simple Crease Patterns

In this chapter, we consider two of the simplest types of crease patterns, with the goal of characterizing when they arise as the crease patterns of origami, particularly flat origami (see Figure 12.1).

In the first type of crease pattern, all creases are parallel to each other. In this case, the shape of the paper is not important, so we can imagine a long thin strip of paper with all creases perpendicular to the strip. In fact, we can imagine the paper as one-dimensional, a line segment with points marking creases.

In the second type of crease pattern, all creases are incident to a single common vertex. Again, in this case, the shape of the paper is irrelevant, so we view it to be a unit disk centered at the sole vertex of the crease pattern. At a high level, the two types of crease patterns are related: the first type is a limiting case of the second type in which the radius of the disk is infinite.

We explore two natural problems about flat foldings of either type of crease pattern:

1. Characterize which of the crease patterns can be folded flat, that is, for which there is a flat folded state using precisely those creases.
2. For each such flat-foldable crease pattern, characterize which mountain–valley assignments correspond to flat foldings.

12.1 ONE-DIMENSIONAL FLAT FOLDINGS

The first question, characterizing flat-foldable crease patterns, has a simple answer for 1D pieces of paper: everything. Consider our piece of paper as a horizontal line segment (refer to Figure 12.2). Let $c_1, c_2, \ldots, c_n$ denote the *creases* on the segment (each of which is a point), oriented so that c_i is left of c_j for $i < j$. Let c_0 and c_{n+1} denote the left and right endpoints of the segment, respectively; they are not considered creases, but rather they are called *ends*. Now if we assign the odd creases $c_1, c_3, \ldots$ to be mountains, and we assign the even creases $c_2, c_4, \ldots$ to be valleys, then we obtain a flat folded state as in Figure 12.2(a). This "accordion" or "zigzag" folding avoids self-intersection by stacking the layers of paper in order.

The second question, characterizing flat-foldable mountain–valley patterns, turns out to be richer. Figure 12.2(b–c) shows two other examples of mountain–valley assignments for the same crease pattern, one flat foldable and one not. We describe a solution to this problem obtained by Arkin et al. (2004) in the context of (2D) map folding, a problem we will consider in more detail in Chapter 14.

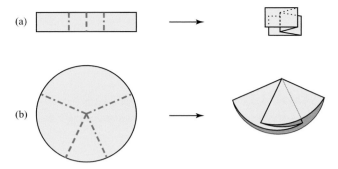

Figure 12.1. Two simple types of crease patterns. (a) Parallel creases and (b) creases incident to a single vertex.

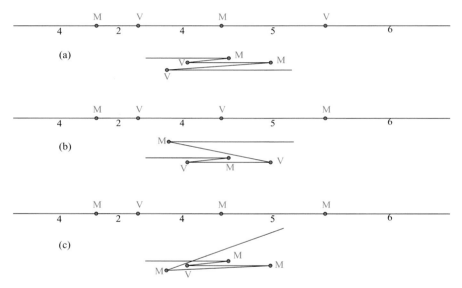

Figure 12.2. A 1D piece of paper with three different mountain–valley patterns. (a–b) are flat foldable; (c) is not.

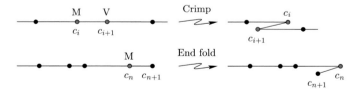

Figure 12.3. The two local operations for 1D folds.

As part of this characterization, we will show that any flat folded state can be reached by a sequence of only two local operations: *crimps* and *end folds*, illustrated in Figure 12.3.

Both operations are simple rotations about crease points: about one for end folds (c_n in the figure), and about two for crimps (c_i and c_{i+1}). If the segment is viewed as a polygonal chain with joints at the crease points, these motions can be viewed as a linkage reconfigurations. Origami usually permits bending the paper during a folding, and such bending is often necessary to reach the desired final folded state. But for this 1D folding problem, we will show that the paper may be kept rigid between creases.

Define a pair (c_i, c_{i+1}) of consecutive creases to be *crimpable* if c_i and c_{i+1} have opposite assigned directions (necessary to even permit the crimping operation), and in addition, the crimping only swaps the left—right order of c_i and c_{i+1}, that is, both adjacent subsegments are at least as long as the crimping subsegment:

$$|c_{i-1} - c_i| \geq |c_i - c_{i+1}| \leq |c_{i+1} - c_{i+2}|.$$

Similarly, define c_{n+1} to be a *foldable end* if the subsegment into which it is folded is at least as long:

$$|c_{n-1} - c_n| \geq |c_n - c_{n+1}|.$$

c_0 is defined to be a foldable end under symmetric conditions at the left end of the segment. The crimp and end-fold operations will be performed only under these subsegment length restrictions; both are illustrated in Figure 12.3.

Our immediate goal is to prove that any mountain–valley assignment that corresponds to a flat folded state permits one of the two operations to be executed. Once executed, we will consider the paper "sticky" in that the crimp or end fold will never again be undone. This permits recursive application of the rule until the final flat folded state is attained.

The key property is what is called "mingling." A 1D mountain–valley pattern is called *mingling* if, for every maximal sequence $c_i, c_{i+1}, \ldots, c_j$ of consecutive creases with the same direction, the subsegment on one end or the other is longer than its neighbor: either

1. $|c_{i-1} - c_i| \leq |c_i - c_{i+1}|$ or
2. $|c_{j-1} - c_j| \geq |c_j - c_{j+1}|.$

As Figure 12.4 illustrates, one of the end creases of the sequence is closer to a crease of the opposite direction than it is to a crease of its own direction. So there is a sense in which the mountain creases "mingle" with the valley creases and vice versa.

We now proceed to establish two implications: flat-foldable $\Rightarrow$ mingling, and mingling $\Rightarrow$ local operation.

Lemma 12.1.1. *Every flat-foldable 1D mountain–valley pattern is mingling.*

Proof: Consider a flat folding of a mountain–valley pattern, and let $c_i, \ldots, c_j$ be maximally consecutive creases with the same direction. First consider the case when the sequence contains just a single crease, that is, $j = i$. Then either the subsegment right of c_i is at least as long as the subsegment to the left, or vice versa. So the sequence satisfies the definition of mingling.

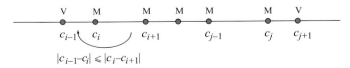

Figure 12.4. The mingling property. Here there is mingling at the c_i end and the c_j end of the sequence.

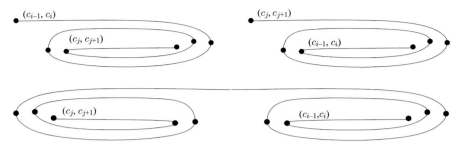

Figure 12.5. The innermost edge of a spiral cannot be longer than the adjacent edge.

For $j > i$, the folding must form one or two spirals, one of the three configurations illustrated in Figure 12.5:

1. One spiral with (c_{i-1}, c_i) the outermost edge and (c_j, c_{j+1}) the innermost; or
2. One spiral with (c_j, c_{j+1}) the outermost edge and (c_{i-1}, c_i) the innermost; or
3. Two spirals connected by a common outermost edge and with (c_{i-1}, c_i) and (c_j, c_{j+1}) being the two innermost edges.

Consider the first case, with (c_j, c_{j+1}) innermost. Then it must be that $|c_j - c_{j+1}| \leq |c_{j-1} - c_j|$ to fit in the spiral, so the sequence is mingling. Similarly for the other cases, the innermost subsegment must be no longer than its adjacent subsegment, which is precisely the mingling property. □

Next we show that having the mingling property suffices to imply the existence of a single crimpable pair or foldable end.

Lemma 12.1.2. *Any mingling 1D mountain–valley pattern has either a crimpable pair or a foldable end.*

Proof: Let $s_1, s_2, \ldots, s_m$ be the maximal sequences of creases with the same direction: s_1 is $c_1, \ldots, c_{i_1}$, s_2 is $c_{i_1+1}, \ldots, c_{i_2}$, $s_m = \ldots, c_n$. We know by the definition of the mingling property that each s_i is mingling at either the left or the right end (or at both ends). Let us represent a mingling end by a parenthesis, and a nonmingling end by a square bracket. Thus ( ] is mingling at the left but not the right.

If s_1 is mingling at the left, then c_0 is a foldable end, that is, $|c_0 - c_1| \leq |c_1 - c_2|$. Similarly, if s_m is mingling at the right, then c_{n+1} is a foldable end. In either case, we have established the claim of the lemma. So assume that neither case holds. Then the sequences appear something like

$$[ ) [ ) \cdots [ ) ( ] \cdots ( ]$$

Each subsequence must have at least one parenthesis representing mingling; thus there are at least m parentheses. Focus on the subsegments between s_i's, that is, the "cracks" between paired parentheses or brackets. There are just $m - 1$ such cracks. Now we must distribute the at least m parentheses into $m - 1$ cracks; note that s_0 and s_m must contribute their parenthesis to their adjacent crack. So some crack must receive two back-to-back parentheses:) (. This corresponds to two adjacent sequences both being

Figure 12.6. A mingling mountain–valley pattern that when crimped is no longer mingling.

mingling at their shared joining subsegment. This is exactly the condition for a crimp: a short subsegment c_i, c_{i+1} is surrounded on either side by subsegments at least as long. Thus, if neither end of the original pattern is end foldable, there must be a crimpable pair of creases.   □

It would be natural at this point to argue that either of the local operations guaranteed by this lemma preserves the mingling property. Unfortunately this claim is false, as shown by the example in Figure 12.6: after crimping and fusing the leftmost three segments, the new leftmost segment is longer than its neighbor to the right. Fortunately, this example is not flat foldable, so it is not a counterexample of relevance. Next, we show that the local operations preserve flat foldability, which, by Lemma 12.1.1, guarantees we always have the mingling property and continue to have local operations to perform, provided the original mountain–valley pattern is flat foldable.

Lemma 12.1.3. *The local operations (end fold and crimp) both preserve flat foldability.*

Proof:

1. End fold. Suppose c_n is a foldable end of a flat-foldable mountain–valley pattern P, and let F be a flat folding of P. Fold at c_n and absorb $c_n c_{n+1}$ into the penultimate subsegment $c_{n-1} c_n$ (cf. Figure 12.3.). This new mountain–valley pattern P' is just P with the end chopped off. Define F' to be identical to F from c_0 through c_n. Thus F' is a flat folding of P'. Finally, alter F' to F'' by adding back the short segment $c_n c_{n+1}$ folded under at c_n. Then F'' is a flat folding of P after the end fold.

2. Crimp. Let (c_i, c_{i+1}) be a crimpable pair of a flat-foldable mountain–valley pattern P, and assume by symmetry that c_i is a mountain and c_{i+1} is a valley. Consider a flat folding F of P, as depicted in Figure 12.7 (left). We orient our view to regard the segment (c_i, c_{i+1}) as flipping over during the folding, so that the remainder of the (unfolded) segment keeps the same orientation. Thus, (c_{i-1}, c_i) is above (c_i, c_{i+1}), which is above (c_{i+1}, c_{i+2}). It could be that F places some layers of paper in between (c_i, c_{i+1}) and (c_{i+1}, c_{i+2}). Our goal is to alter F by moving these layers of paper out of these spaces, leaving the crimp. These layers of paper can be moved to immediately above (c_{i-1}, c_i), because (c_{i-1}, c_i) is at least as long as (c_i, c_{i+1}), and hence there are no barriers closer than c_i (see Figure 12.7 right). In the notation of Chapter 11, we are just modifying the folded-state order given by λ. Similarly, we move material between (c_i, c_{i+1}) and (c_{i-1}, c_i) to immediately below (c_{i+1}, c_{i+2}). In the end, we have a flat folding F' of the object obtained from making the crimp (c_i, c_{i+1}).   □

Theorem 12.1.4. *Any flat-foldable 1D mountain–valley pattern can be folded by a sequence of crimps and end folds.*

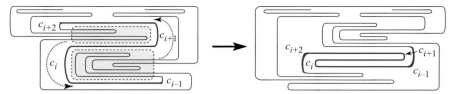

Figure 12.7. Moving layers of paper out of the zigzag formed by a crimp (c_i, c_{i+1}), highlighted in bold.

Proof: The proof follows nearly immediately from the previous lemmas. If the mountain–valley pattern P has no creases, the claim is trivially true. So suppose P has n creases. Because we assume P to be flat foldable, Lemma 12.1.1 shows that its mountain–valley pattern is mingling. Lemma 12.1.2 then assures that there is a local operation available. After performing that operation and fusing the end fold or crimp, the resulting mountain–valley pattern P' has $n-1$ or $n-2$ creases. Lemma 12.1.3 guarantees that this smaller pattern is flat foldable. The result then follows by induction. $\square$

Corollary 12.1.5. *Any flat-foldable 1D mountain–valley pattern can be folded as a reconfiguration of a non-self-intersecting linkage of $O(n)$ rotations.*

Theorem 12.1.6. *Determining whether a 1D mountain–valley pattern has a flat folding, and computing a sequence of folds that achieves it, can be solved in $O(n)$ worst-case time on a machine supporting arithmetic on the input lengths.*

Proof: First note that it is trivial to check in constant time whether an end is foldable or a pair of consecutive folds form a crimp. We begin by testing all such folds, and hence in linear time have a linked list of all possible folds at this time. We also maintain reverse pointers from each symbol in the string to the closest relevant possible fold. Now when we make a crimp or an end fold, only a constant number of previously possible folds can no longer be possible and a constant number of previously impossible folds can be newly possible. These folds can be discovered by examining a constant-size neighborhood of the performed fold. We remove the old folds from the list of possible folds, and add the new folds to the list. Then we perform the first fold on the list, and repeat the process. By Theorem 12.1.4, if the list ever becomes empty before reaching a complete folded state, it is because the mountain–valley pattern is not flat foldable. $\square$

12.2   SINGLE-VERTEX CREASE PATTERNS

Next we turn to the second simple type of crease pattern, consisting of several straight creases emanating from a single vertex, the center of a disk. Now a crease pattern is specified entirely by a cyclic sequence $\theta_1, \theta_2, \ldots, \theta_n$ of angles between creases around the vertex, which sum to $360°$ in a flat piece of paper (see Figure 12.8(a)).

We also consider a more general scenario in which the angle sum $\sigma = \theta_1 + \theta_2 + \cdots + \theta_n$ is not necessarily $360°$. When $\sigma < 360°$ (a vertex of "positive curvature" in the language of Part III, Section 21.2), the piece of paper can be viewed as a convex cone, which could be obtained by cutting and removing an angle of $360° - \sigma$ from a flat piece of paper and gluing together the two cut edges (Figure 12.8(b)). When $\sigma > 360°$ (a vertex of "negative

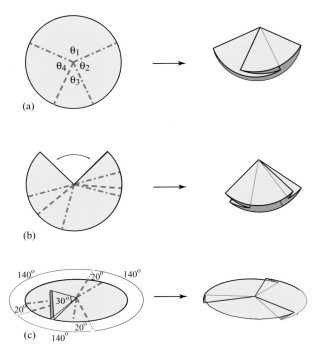

Figure 12.8. Flat foldings of single-vertex crease patterns. (a) Flat paper: $\sigma = 360°$; (b) cone: $\sigma = 270°$ (the two straight edges are glued together); (c) negative curvature: $\sigma = 420°$.

curvature"), there is more material than a flat piece of paper, so the paper naturally "buckles" or "ruffles" around the vertex (Figure 12.8(c)). Nonetheless, we can consider folding such pieces of paper, with one vertex surrounded by σ angle of material, and all creases emanating from that vertex. (We assume every other point on the paper is locally flat, surrounded by 360° of material.) Another way to view such a piece of paper is a little snippet of paper around a vertex of a polyhedron. This level of generality will in fact be useful for analyzing normal flat pieces of paper, as well as being useful directly when considering foldings of polyhedra as in Chapter 18.

Both of the characterization problems mentioned above—flat-foldable crease patterns and flat-foldable mountain–valley patterns—have been solved in the case of flat paper by results of Hull (2002, 1994, 2003a), Bern and Hayes (1996), Justin (1989a, 1994), Kawasaki (1989b), and Maekawa (Kasahara and Takahama 1987, p. 29). Here we extend these results to nonflat paper, inspired in particular by Hull's description and proofs for the flat case (Hull 1994, 2002).

12.2.1 Flat-Foldable Single-Vertex Crease Patterns

The solution to the first characterization problem is a fairly simple condition on the angles. For flat paper, the result is often called Kawasaki's Theorem(Kasahara and Takahama 1987, p. 29; Kawasaki 2005, p. 143), although it was discovered independently by Justin (1989a).

Theorem 12.2.1 (Kawasaki 1989b, Justin 1989a, Hull 1994). *A single-vertex crease pattern defined by angles $\theta_1 + \theta_2 + \cdots + \theta_n = 360°$ is flat foldable if and only if n is even* *and the sum of the odd angles θ_{2i+1} is equal to the sum of the even angles θ_{2i}, or equivalently, either sum is equal to* 180°:

$$\theta_1 + \theta_3 + \cdots + \theta_{n-1} = \theta_2 + \theta_4 + \cdots + \theta_n = 180°.$$

Instead of proving this theorem, we first generalize it to nonflat pieces of paper:

Theorem 12.2.2. *A single-vertex crease pattern defined by angles $\theta_1, \theta_2, \ldots, \theta_n$ is flat foldable if and only if n is even and the alternating sum of the angles θ_i is equal to 0, 360°, or* −360°:

$$\theta_1 - \theta_2 + \theta_3 - \theta_4 + \cdots + \theta_{n-1} - \theta_n = \sum_{i=1}^{n}(-1)^i\,\theta_i \in \{0, 360°, -360°\}.$$

In particular, Theorem 12.2.2 implies Theorem 12.2.1 for flat pieces of paper, because each $\theta_i > 0$ and the θ_i's sum to 360°, so the alternating sum has absolute value less than 360°, leaving 0 as the only possibility.

   Proof: Refer to Figure 12.8. If we imagine walking along the boundary of the paper in a flat folded state, we alternate between traveling clockwise and counterclockwise about the central vertex, so the number of (angles between) creases must be even. Furthermore, the total angular motion is

$$\pm(\theta_1 - \theta_2 + \theta_3 - \theta_4 + \cdots + \theta_{n-1} - \theta_n),$$

which must equal a multiple of 360° because any closed walk on a circle must bring us back to our starting point. The multiple of 360° is called the *winding number* of the closed walk with respect to the central vertex. There are two possibilities: the closed walk encloses the central vertex (and therefore encloses the entire disk of paper, as in Figure 12.8(c)), or the closed walk does not enclose the central vertex (and therefore it encloses zero area as in Figure 12.8(a–b)). In the former case, the winding number is +1 or −1, depending on orientation; the winding number cannot be larger in absolute value because the closed walk does not self-intersect. In the latter case, the winding number is 0. Therefore the alternating sum of angles is 0, 360°, or −360° in any flat folded state, that is, the condition in the theorem is necessary for flat foldability.
   To prove that the condition is also sufficient for flat foldability, we need to construct a valid mountain–valley assignment and overlap order. Refer to Figure 12.9. First we observe that if we cut our cycle of faces of paper along one of the creases to give the regions a linear order, then the problem is easy: fold the creases alternately mountain and valley, accumulating the faces into an "accordion." The linear order on the faces defines their stacking total order, guaranteeing no self-intersection. This folded state is valid even without the condition of the theorem. The condition of the theorem guarantees the additional property that the two copies of the cut edge align geometrically on the circle. However, the two copies are at opposite extremes of the stacking total order, and in some cases it is not possible to suture them back together because of intervening layers of paper (see Figure 12.9(a)).
   To more easily visualize the situation, we conceptually unroll the angles to horizontal line segments of the same length, as shown in Figure 12.9(right). This view lacks the wrap-around feature of the circle—in reality, the line should be treated modulo

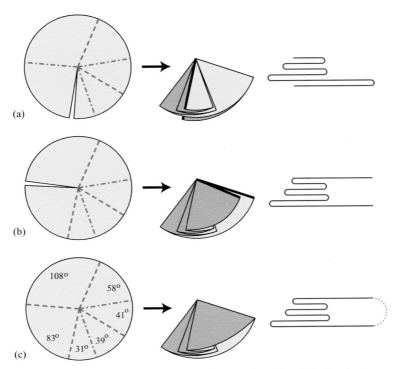

Figure 12.9. Constructing a mountain–valley assignment for a flat-foldable single-vertex crease pattern with angles $31° + 39° + 41° + 58° + 108° + 83° = 360°$, which have alternating sum $31° - 39° + 41° - 58° + 108° - 83° = 0°$. (a) Cutting an arbitrary edge and folding alternately mountain and valley; (b) cutting at an edge that folds to an extreme; (c) regluing the cut.

$360°$—but this simplification actually helps us find a solution. The accordion folding of the linear sequence of angles now corresponds to a zigzag chain of line segments. By the condition of the theorem, the two endpoints of the chain are either horizontally aligned or horizontally separated by $360°$.

We change the original cut crease to be the one corresponding to a vertex in the horizontal representation with maximum x coordinate (Figure 12.9(b)). This change in the cut affects the horizontal representation in a straightforward way, cutting the linear order of horizontal segments into two parts and exchanging their vertical stacking order, possibly shifting one part horizontally to make the ends meet, but keeping fixed the relative horizontal placement of segments in each part. Thus at least one of the two copies of the vertex with extremal x coordinate remains extremal. If the two endpoints of the chain are horizontally aligned, then they are both extremal, and therefore we can join them without crossing other horizontal segments (Figure 12.9(c)). Otherwise, we can attach a "temporary" horizontal segment of length $360°$ connecting the two endpoints, again without crossing other horizontal segments. In Figure 12.10, this segment is red. This temporary segment actually just extends the horizontal segment incident to the nonextremal endpoint by $360°$, bringing the two endpoints into alignment so that we can join them. Note that the temporary segment is at the top or the bottom of the linear stacking order.

Now that we have a flat folding in the linear view, we continuously "roll" the flat folding onto the circle. More precisely, we place the left end of the linear flat folding at

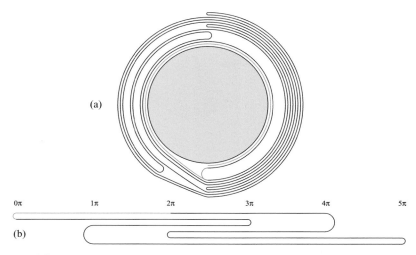

Figure 12.10. (a) The wrapping of the linear flat folding (b) [not to scale]. The ends are separated by $360° = 2\pi$, joined after wrapping once by the red arc. This single vertex has 16π of paper incident to it. The resulting flat folding has $M = V = 3$.

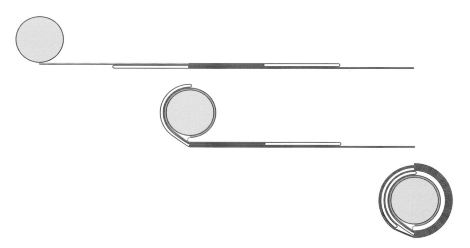

Figure 12.11. The rolling process that produces the wrapping shown in Figure 12.10 [to scale].

one point on the circle, and wrap the linear flat folding counterclockwise around the circle, spiraling around itself if the total length is more than $360°$. Figures 12.10 and 12.11 illustrate the rolling/wrapping process for a complex example.

In this way, we obtain a flat folding of the augmented segments onto the circle. Because any temporary horizontal segment of length $360°$ was on the top or bottom extreme in the linear view, the spiral rolling will place the temporary segment at the innermost or outermost layer of the spiral, making the two endpoints adjacent, so we can remove the temporary segment and join the two endpoints directly. In this way, we obtain a flat folding of the boundary of the paper on the circle, which can be translated directly into a flat folding of the piece of paper on the disk. □

12.2.2 Flat-Foldable Single-Vertex Mountain–Valley Patterns

Implicit in the proof of Theorem 12.2.1 is a construction of a mountain–valley assignment. On the other hand, crease patterns often have more than one possible mountain–valley assignments, different from those that (largely) alternate between mountain and valley. We now consider the second problem of characterizing all valid mountain–valley assignments for each flat-foldable crease pattern.

There are two main necessary properties for mountain–valley assignments. The first such property concerns just the numbers of mountain and valley creases. This property is often known as Maekawa's Theorem, first mentioned as such in a book called *Origami for the Connoisseur* (Kasahara and Takahama 1987, p. 29); it was also discovered independently by Justin (1994).

Theorem 12.2.3 (Maekawa (Kasahara and Takahama 1987), Justin 1994). *In a flat-foldable single-vertex mountain–valley pattern defined by angles $\theta_1 + \theta_2 + \cdots + \theta_n = 360°$, the number of mountains and the number of valleys differ by ± 2.*

Again we generalize the theorem to nonflat pieces of paper, and then prove this more general form:

Theorem 12.2.4. *In a flat-foldable single-vertex mountain–valley pattern defined by angles $\theta_1, \theta_2, \cdots, \theta_n$, the number of mountains and the number of valleys differ by ± 2 if the alternating angle sum $\sum_{i=1}^{n}(-1)^i \theta_i$ is 0, and otherwise the numbers of mountains and valleys are equal.*

Proof: As in the proof of Theorem 12.2.2, imagine walking around the boundary of the paper in a flat folded state, as shown in Figure 12.9. Now consider the total turn angle τ during this walk, measuring counterclockwise turn as positive and clockwise turn as negative. There are two parts to this total turn angle, which we analyze separately. The first part is the integral of the turning from clockwise or counterclockwise travel around the circle, ignoring the creases in between. By Theorem 12.2.2, this integral is the alternating angle sum $A = \sum_{i=1}^{n}(-1)^i \theta_i \in \{0, 360°, -360°\}$. The second part is the sum of the turn angles at the creases, where we reverse the direction of travel around the circle. Each mountain crease introduces a turn angle of $180°$, while each valley crease introduces a turn angle of $-180°$, or vice versa, depending on the orientation of the paper. Thus, if M denotes the number of mountain creases and V denotes the number of valley creases, then the total turn angle τ is

$$\tau = A \pm (180° \, M - 180° \, V) = A \pm 180°(M - V).$$

On the other hand, the total turn angle τ in a planar non-self-intersecting closed walk is $\pm 360°$, depending on whether the walk is oriented clockwise or counterclockwise. If $A = 0$, then dividing both sides by $180°$, we obtain $M - V = \pm 2$, as desired (Figure 12.8(a–b)). If $A = \pm 360°$, then $\tau = A$, that is, the signs on the $360°$'s match, because in both cases the sign is determined by whether the walk is oriented clockwise or counterclockwise. Thus we obtain $\tau = A = A \pm 180°(M - V)$, so $M - V = 0$, as claimed (Figure 12.8(c)). $\square$

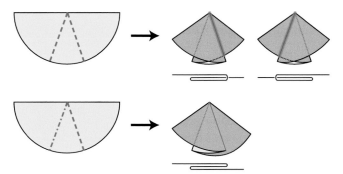

Figure 12.12. If an angle is a strict local minimum, then the two incident creases must have opposite orientation or else the folding is forced to self-intersect, no matter what the choice of overlap order.

The second main necessary property of mountain–valley assignments is purely local. Specifically, it considers the case when an angle θ_i is a *strict local minimum*, that is, strictly smaller than its two neighboring angles θ_{i-1} and θ_{i+1}.

Lemma 12.2.5 (Kawasaki 1989b, Justin 1994). *If an angle θ_i is a strict local minimum (i.e., $\theta_{i-1} > \theta_i < \theta_{i+1}$), then the two creases bounding angle θ_i have an opposite mountain–valley assignment in any flat-foldable mountain–valley pattern.*

Proof: Refer to Figure 12.12. Suppose the two incident creases were assigned the same direction, for example, valley. The neighboring angle θ_{i-1} of paper extends beyond the crease between angles θ_i and θ_{i+1}, and symmetrically θ_{i+1} extends beyond the crease between angles θ_i and θ_{i-1}. Because both creases are valley, both angles θ_{i-1} and θ_{i+1} of paper must be above the angle θ_i of paper. For the angle θ_{i+1} of paper to avoid penetrating the valley crease between θ_i and θ_{i-1}, we must have θ_{i+1} above θ_{i-1}. Symmetrically, we must have θ_{i-1} above θ_{i+1}. Thus, neither θ_{i-1} nor θ_{i+1} can be above the other, a contradiction. □

These two necessary conditions can be applied recursively as follows. Suppose we find a strict local minimum angle θ_i. We know from Lemma 12.2.5 that the incident creases must be assigned opposite directions in any flat folding. Apply a crimp fold (p. 194) and fuse together the angles of paper that come together, thereby removing the original local minimum angle θ_i and its incident creases, and combining the two adjacent angles θ_{i-1} and θ_{i+1} into a single angle $\theta_{i-1} - \theta_i + \theta_{i+1}$. This operation works even if the original angle θ_i is a *nonstrict* local minimum (i.e., $\theta_{i-1} \geq \theta_i \leq \theta_{i+1}$), so that subtracting θ_i from either one of θ_{i-1} and θ_{i+1} leaves a nonnegative deficit.

The crimp fold changes the piece of paper by reducing the amount of material at the central vertex, for example, turning an originally flat piece of paper into a convex cone. We claim that this reduced paper with the remaining creases and mountain–valley assignment has the same ability to fold flat as the original, so that the necessary conditions must also apply to it.[1] This claim is similar to a lemma already proved in the context of flat foldings of 1D paper, and our proof uses that result.

---

[1] Despite the necessity of this claim, to our knowledge it has not been stated or proved in the literature in this context. However, a closely related form of the lemma was stated and proved by Arkin et al. (2004), in the context of 1D folding instead of 2D disk folding. We use that proof.

Lemma 12.2.6. *A single-vertex mountain–valley pattern is flat foldable if and only if the result of a crimp fold around any nonstrict local minimum angle is flat foldable.*

Proof: One implication is easy: if the reduced cone is flat foldable, then so is the original cone, because we can crimp the original cone into the reduced cone and then compose the two foldings. The other implication is nearly identical to the analogous result about crimps in 1D flat folding, Lemma 12.1.3 (p. 197), which shows that any intervening layers can be removed from between the three crimped layers by a suitable change to the stacking order, and therefore the crimp could be performed "first." The only difference is that, now, the piece of paper has circular topology instead of linear topology, and we fold the paper onto a circle instead of a line. The first difference has no effect; the same argument as that in Lemma 12.1.3 would work if the two ends of a 1D segment paper were attached to one another. To handle the second difference, we can use the same trick as in the proof of Theorem 12.2.2: unroll the flat folding from the circle to the line, change the stacking order according to Lemma 12.1.3, and then roll it back onto the circle. (Note that the unrolling and rolling may themselves cause changes to the stacking order, forcing a spiral wrapping.)   □

Combining what we know about flat-foldable single-vertex mountain–valley patterns, we obtain a recursive characterization for the generic case of such patterns when two neighboring angles are never equal. More precisely, call a single-vertex crease pattern *generic* if, for any two sequences $\theta_i, \theta_{i+1}, \ldots, \theta_{j-1}$ and $\theta_j, \theta_{j+1}, \ldots, \theta_{k-1}$, each consisting of an odd number of consecutive angles, the alternating sums differ:

$$\theta_i - \theta_{i+1} + \theta_{i+2} - \theta_{i+3} + \cdots + \theta_{j-1} \neq \theta_j - \theta_{j+1} + \theta_{j+2} - \theta_{j+3} + \cdots + \theta_{k-1}.$$

A crimp fold preserves genericity because it replaces three angles with their alternating sum. Thus the following theorem characterizes flat-foldable mountain–valley assignments for generic single-vertex crease patterns:

Lemma 12.2.7. *In a generic single-vertex flat-foldable crease pattern, a mountain–valley assignment is flat foldable if and only if, for every strict local minimum angle, the incident creases have opposite mountain–valley assignments and the result of crimping these creases is flat foldable.*

Proof: Because the crease pattern is generic, no two neighboring angles are equal. Because the angles are connected in a cycle, the angles cannot always increase around the cycle. Therefore, some angle must be a strict local minimum; for example, the globally smallest angle is. By Lemma 12.2.5, every strict local minimum angle must have incident creases with opposite assignments. By Lemma 12.2.6, the result of any such crimp is flat foldable. Conversely, by Lemma 12.2.6, if any such crimp result is flat foldable, then so is the original mountain–valley pattern.   □

The remaining situation we need to handle is when two neighboring angles are equal. The following result strengthens Lemma 12.2.5 from a single strict local minimum angle to multiple equal strict local minimum angles:

Lemma 12.2.8 (Hull 2002, 2003a). *Consider a flat-foldable single-vertex mountain–valley pattern, and suppose that there are k equal angles surrounded by strictly larger angles:*

$$\theta_{i-1} > \theta_i = \theta_{i+1} = \theta_{i+2} = \cdots = \theta_{i+k-1} < \theta_{i+k}.$$

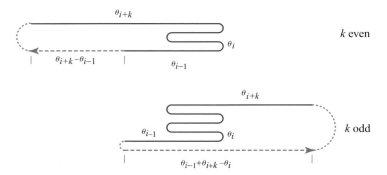

Figure 12.13. Closing a chain of equal angles surrounded by strictly larger angles.

*Then, among the creases between neighboring pairs of these angles, the number of moun-
tains and number of valleys differ by 0 if k is odd and ±1 if k is even.*

Proof: We apply Theorem 12.2.4 to a closed version of the open chain of angles
$\theta_{i-1}, \theta_i, \dots, \theta_{i+k}$. If k is even, we extend the smaller extreme angle $\min\{\theta_{i-1}, \theta_{i+k}\}$ to
match the larger extreme angle $\max\{\theta_{i-1}, \theta_{i+k}\}$ and attach θ_{i-1} to θ_{i+k} as neighbors. If
k is odd, we add a final angle of $\theta' = \theta_{i-1} - \theta_i + \theta_{i+k}$ neighboring both θ_{i-1} and θ_{i+k}
(see Figure 12.13). The resulting piece of paper consisting of this closed chain of angles
can have an unusual amount of material at its central vertex; for example, an angle set
of $90°$, $45°$, $45°$, $180°$ (which sum to $360°$) results in an angle set of $180°$, $45°$, $45°$, $180°$
(which sum to $450°$). It is here that we use the level of generality developed earlier, even
if our goal is to characterize flat foldings of flat pieces of paper.

We claim that this closed chain can be folded flat with the same mountain–valley
assignment as the original single-vertex crease pattern on the shared creases between
θ_{i-1} and θ_i, between θ_i and θ_{i+1}, and so on, between θ_{i+k-1} and θ_{i+k}. Because the original mountain–valley pattern is flat foldable, there is a flat folding of just the open
chain of angles $\theta_{i-1}, \theta_i, \dots, \theta_{i+k}$. Because the interior angles $\theta_i, \theta_{i+1}, \dots, \theta_{i+k-1}$ are all
equal, and the surrounding angles θ_{i-1} and θ_{i+k} are larger, the latter angles must ex-
tend beyond the bundle of interior angles, so their previously unattached edges do
not overlap any other angles of paper. Therefore we can extend and join θ_{i-1} and θ_{i+k}
(when k is even) or attach the extra angle θ' (when k is odd) without self-intersection.
In the end, we obtain a flat folding of the newly constructed closed chain whose
only difference in mountain–valley assignment is for the newly added one or two
creases.

By construction, the alternating sum of angles in the new closed chain is 0, so by
Theorem 12.2.4, the numbers of mountains and valleys differ by ±2. If k is odd, we
added two new creases both with the same orientation. Without these creases, the total
turn angle (as analyzed in the proof of Theorem 12.2.4) is 0, and therefore the numbers
of mountains and valleys are equal. If k is even, we added a single crease. Without
this crease, the total turn angle is ±180°, and therefore the numbers of mountains and
valleys differ by ±1.   □

Finally we obtain a recursive characterization of all flat-foldable single-vertex
mountain–valley patterns:

Theorem 12.2.9. *Given a flat-foldable single-vertex crease pattern with n creases, a mountain–valley assignment is flat foldable if and only if*

1. *$n = 2$, the angles are equal, and the two creases have the same mountain–valley assignment; or*

2. *$n = 2$, the angles differ by $\pm360°$, and the two creases have opposite mountain–valley assignments; or*

3. *$n \geq 4$ and some nonstrict local minimum angle has incident creases with opposite assignments (one mountain and one valley) and the result of this crimp is flat foldable.*

Proof: The condition for the second base case ($n = 2$) is necessary by Theorem 12.2.4 and sufficient by inspection. (The first base case arises when the alternating sum of angles is 0; the second base case arises when the alternating sum is $\pm360°$.) Outside the base cases, the crimp condition is certainly sufficient to conclude flat foldability; this fact was the easy half of Lemma 12.2.6. The remaining question is whether every flat-foldable mountain–valley assignment has such a crimp.

First suppose that all angles are equal, so that all angles are nonstrict local minima. By Theorem 12.2.4, there are two consecutive creases with opposite assignments (beyond the base case), defining a crimp. By Lemma 12.2.6, the result of this crimp is flat foldable precisely when the original mountain–valley pattern is flat foldable.

Now suppose that not all angles are equal. Consider a maximal interval of equal angles $\theta_i = \theta_{i+1} = \cdots = \theta_{i+k-1}$ that equal the globally smallest angle. Because not all angles are equal, two (possibly identical) angles θ_{i-1} and θ_{i+k} surround these k equal angles, and by global minimality these two angles are strictly larger than the k equal angles. By Lemma 12.2.8 and because $k \geq 1$, there are two consecutive creases between angles $\theta_{i-1}, \theta_i, \ldots, \theta_{i+k}$ with opposite mountain–valley assignments, defining a crimp. By Lemma 12.2.6, the result of this crimp is flat foldable precisely when the original mountain–valley pattern is flat foldable. $\square$

This recursive characterization has several consequences, both algorithmic and combinatorial. We start with the algorithmic consequence, which was established for flat paper by Bern and Hayes (1996):

Corollary 12.2.10. *Flat foldability of a single-vertex mountain–valley pattern can be decided in linear time.*

Proof: First we check the condition from Theorem 12.2.2 by computing the alternating sum of the angles. Then we maintain a list of all nonstrict local minimum angles whose two incident creases have opposite mountain–valley assignments. This list can be initialized in linear time by scanning the angles and creases in order. Then we repeatedly crimp an angle extracted from the list. Whenever we perform a crimp, we check whether the resulting angle and/or its neighbors have become nonstrict local minima with adjacent creases having opposite mountain–valley assignments, in which case we add such angles to the list. At each step we spend constant time and remove two angles, so the total work is linear. If the list empties, by Theorem 12.2.9 we simply check that zero or two creases remain, and in the latter case, that these creases have the same assignment.	□

Before we proceed to the combinatorial consequence, we give a simpler characterization of flat-foldable mountain–valley patterns:

Corollary 12.2.11. *If the angles of a flat-foldable single-vertex crease pattern are equal, then a mountain–valley assignment is flat foldable precisely if it satisfies Theorem 12.2.4. Otherwise, a mountain–valley assignment is flat foldable precisely if any maximal sequence of k equal angles surrounded by strictly larger angles satisfies Lemma 12.2.8 and we obtain a flat-foldable mountain–valley pattern after folding all of the creases between these angles, except the last such crease if there are an odd number of creases (i.e., k is even).*

Proof: In the base cases with $n = 2$ angles, the claim is equivalent to Theorem 12.2.9. Outside this base case, if all angles are equal, Theorem 12.2.9 establishes that the condition from Theorem 12.2.3 is necessary and sufficient because such crease patterns always have crimps and because crimps preserve the difference between the numbers of mountains and valleys. Finally suppose that not all angles are equal and that $\theta_i, \theta_{i+1}, \ldots, \theta_{i+k-1}$ is a maximal sequence of equal angles surrounded by strictly larger angles θ_{i-1} and θ_{i+k}. Theorem 12.2.9 establishes that the condition from Lemma 12.2.8 is necessary and sufficient for local folding of the creases between $\theta_{i-1}, \theta_i, \ldots, \theta_{i+k}$ because such crease patterns always have crimps and because crimps preserve the difference between the numbers of mountains and valleys.	□

The following combinatorial result was established for flat paper by Hull (2003a).

Corollary 12.2.12 (Hull 2003a). *The number $C(\theta_1, \theta_2, \ldots, \theta_n)$ of flat-foldable mountain–valley assignments for a flat-foldable single-vertex crease pattern defined by angles $\theta_1, \theta_2, \ldots, \theta_n$ is given recursively as follows. If all angles are equal, then*

$$C(\underbrace{\theta, \theta, \ldots, \theta}_{n}) = 2\binom{n}{n/2 - 1}.$$

Otherwise, consider k equal angles surrounded by strictly larger angles:

$$\theta_{i-1} > \theta_i = \theta_{i+1} = \theta_{i+2} = \cdots = \theta_{i+k-1} < \theta_{i+k}.$$

Then

$$C(\theta_1, \theta_2, \ldots, \theta_n) = \begin{cases} \binom{k+1}{\frac{k+1}{2}} C(\theta_1, \ldots, \theta_{i-2}, (\theta_{i-1} - \theta_i + \theta_{i+k}), \theta_{i+k+1}, \ldots, \theta_n) & \text{if } k \text{ is odd,} \\[2mm] \binom{k+1}{\frac{k}{2}} C(\theta_1, \ldots, \theta_{i-2}, \theta_{i-1}, \theta_{i+k}, \theta_{i+k+1}, \ldots, \theta_n) & \text{if } k \text{ is even.} \end{cases}$$

In particular, $C(\theta_1, \theta_2, \ldots, \theta_n)$ can be computed in linear time. If the alternating sum of angles $\sum_{i=1}^{n}(-1)^i \theta_i$ is 0, then

$$2^{n/2} \leq C(\theta_1, \theta_2, \ldots, \theta_n) \leq 2\binom{n}{n/2 - 1};$$

and if the alternating sum is $\pm 360°$, then

$$2^{n/2} \leq C(\theta_1, \theta_2, \ldots, \theta_n) \leq 2\binom{n}{n/2};$$

and all of these bounds are tight.

Proof: By Theorem 12.2.9, the number of flat-foldable mountain–valley assignments of n equal angles is the number of ways to choose $n/2 - 1$ mountains and $n/2 + 1$ valleys or vice versa, which is $2\binom{n}{n/2-1}$. (Because n is even, the alternating sum of angles must be 0.) If not all angles are equal, Corollary 12.2.11 establishes that the condition on $M - V$ from Lemma 12.2.8 is necessary and sufficient for local folding of the creases between $\theta_{i-1}, \theta_i, \ldots, \theta_{i+k}$ because such crease patterns always have crimps and because crimps preserve the $M - V$ condition. If k is odd, the condition is that $M - V = 0$, and there are $\binom{k+1}{(k+1)/2}$ ways to choose which half of the $k + 1$ creases are mountains. If k is even, the condition is that $M - V = \pm 1$, and there are $\binom{k+1}{k/2}$ ways to partition the $k + 1$ creases into a group of $k/2$ creases and a group of $k/2 + 1$ creases. In the latter case, the recursively obtained assignment for the new crease between θ_{i-1} and θ_{i+k} determines which group contains mountain creases and which group contains valley. After all creases between $\theta_{i-1}, \theta_i, \ldots, \theta_{i+k}$ are folded, we have either a remaining angle of $\theta_{i-1} - \theta_i + \theta_{i+k}$ (if k is odd) or two remaining angles of θ_{i-1} and θ_{i+k}. Thus we obtain the desired recursions.

The recursions can be evaluated in linear time by coalescing equal angles into single items in a list, where each item specifies its multiplicity, as well as storing a list of local minima (items corresponding to angles surrounded by larger angles). If the first list contains just one item, then all n angles must be equal, so we apply the first equation. Otherwise, we take the first item from the list of local minima, apply the recursive rule of the appropriate parity, check whether the modified angles and/or their neighbors can be coalesced into a single item, and update the list of local minima accordingly. Thus, at each step, we spend constant time and remove at least one item, so the total work is linear. In fact, if we are given an implicit form of the input that specifies multiplicities, the work is linear in the size of this implicit encoding.

The $2^{n/2}$ lower bound follows by taking a folding according to $n/2$ crimps, and for each crimp, picking each of the two ways to assign its mountain and its valley. This bound is tight in the generic case, where there is only one folding according to $n/2$ crimps. The $2\binom{n}{n/2-1}$ and $2\binom{n}{n/2}$ upper bounds follow because Theorem 12.2.4 is necessary. The first upper bound is tight in the case of all equal angles. The second upper bound is tight in the case of an angle of $360° + \theta$ followed by $n - 1$ angles of θ. □

As simple examples, the $n = 6$ crease patterns shown in Figures 12.8(c) and 12.9(c) can both be folded in $2^3 = 8$ ways, the lower bound.

Corollary 12.2.13. *A uniformly random flat-foldable mountain–valley assignment for a specified flat-foldable single-vertex crease pattern can be chosen in linear time.*

Proof: As is often the case, uniform random generation follows from the ability to count. As in Corollary 12.2.12, we maintain a list in which each item represents a maximal coalescing of consecutive equal angles, and we maintain a list of local minima. If the first list has just one item, we flip a fair coin to determine whether there are more mountains than valleys or vice versa, and then randomly select $n/2 - 1$ creases without replacement to assign the lesser direction. Otherwise, we take the first item from the list of local minima and check the parity of its multiplicity k. If k is odd, we randomly select $(k + 1)/2$ creases without replacement among the $k + 1$ creases and assign them mountain. If k is even, we randomly select $k/2$ without replacement among the $k + 1$ creases and assign them the lesser direction determined recursively. The recursion proceeds as in Corollary 12.2.12.   □

There is a striking similarity between the characterizations of flat-foldable mountain–valley patterns for single-vertex crease patterns (as above) and for 1D paper (from Section 12.1). Both state that essentially the only necessary operation is a crimp. However, the details differ slightly because of the wrap-around effect of the circular piece of paper considered here, instead of the line segment with two ends. The proof techniques differ substantially; in particular, the notion of mingling does not seem useful in the circular case. It would be interesting to find a unified proof technique that handles both cases with equal ease.

12.2.3   3D Foldings of Single-Vertex Crease Patterns

If we generalize from single-vertex flat foldings to single-vertex 3D foldings, we obtain a generalization of the problem of characterizing foldable mountain–valley assignments: what assignments of fold angles to creases have corresponding 3D folded states? Here we desire foldings in which all creases fold to some given angles, and the faces between creases remain flat in their folded state, as in Figure 12.14. When all fold angles are $\pm 180°$, the folding is forced to be flat, so the question becomes the same as the one answered by Theorem 12.2.9. This section describes what is known for general fold angles, following belcastro and Hull (2002a,b). For simplicity, we focus on the case of flat paper, which is all that is considered in their work.

As before, we can assume that the piece of paper is a disk and that the creases radiate from the center of the disk. The input now consists of two parallel circular sequences of angles: the face angles θ_i in the crease pattern and the fold angles ϕ_i by which each crease folds.

The main result is a generalized necessary condition similar to Theorem 12.2.1 on the alternating sum of angles. As before, the necessary condition stems from traversing the boundary of the paper in the folded state, and because this traversal is circular, we should return to the start. This idea leads to a necessary condition on the face angles and fold angles because we can view the traversal as repeated rotation in various planes: each face angle we visit corresponds to rotating the point by that angle in the current

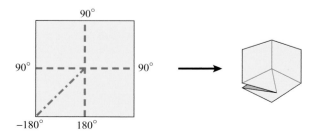

Figure 12.14. The crease pattern and fold angles ϕ_i of a corner fold.

face plane, and each fold angle we visit corresponds to rotating the current face plane by that angle around the corresponding crease.

To determine the exact form of the necessary condition, we view the same actions from a different point of view. Instead of considering the folded state as fixed and a point traversing the boundary of the paper in this folding, we consider the point as fixed and consider the folded state as rolling around so that the point again traverses the boundary of the paper. The motion of the point with respect to the folded state is identical, but simpler to compute. We maintain the invariants that the single vertex is at the origin, that the current face plane is the xy-plane, and that the current crease lies along the positive x-axis.

Now we consider the face angles and fold angles in order. When we encounter a face angle θ_i, we rotate within the xy-plane by θ_i to place the next crease on the x-axis. When we encounter a fold angle ϕ_i, we rotate about the x-axis by ϕ_i to bring the next face to lie in the xy-plane. Thus the total motion can be written as a composition of rotation matrices:

$$R_{yz}(\phi_n) \cdot R_{xy}(\theta_n) \cdots R_{yz}(\phi_2) \cdot R_{xy}(\theta_2) \cdot R_{yz}(\phi_1) \cdot R_{xy}(\theta_1)$$

where

$$R_{xy}(\theta_i) = \begin{pmatrix} \cos\theta_i & -\sin\theta_i & 0 \\ \sin\theta_i & \cos\theta_i & 0 \\ 0 & 0 & 1 \end{pmatrix}$$

and

$$R_{yz}(\phi_i) = \begin{pmatrix} 1 & 0 & 0 \\ 0 & \cos\phi_i & -\sin\phi_i \\ 0 & \sin\phi_i & \cos\phi_i \end{pmatrix}.$$

This composition of rotations defines a linear motion not only of the original point on the paper boundary but of the entire space. Because the linear motion simply rotates the unit sphere, and the original point must be a fixed point of the total motion, the entire motion must in fact be the identity transformation. Therefore we obtain the following necessary condition:

Theorem 12.2.14 (belcastro and Hull 2002a,b). *In a foldable crease pattern with face angles $\theta_1, \theta_2, \ldots, \theta_n$ and fold angles $\phi_1, \phi_2, \ldots, \phi_n$, where ϕ_i corresponds to the crease between faces corresponding to θ_i and θ_{i+1}, the following condition must hold:*

$$R_{yz}(\phi_n) \cdot R_{xy}(\theta_n) \cdots R_{yz}(\phi_2) \cdot R_{xy}(\theta_2) \cdot R_{yz}(\phi_1) \cdot R_{xy}(\theta_1) = I = \begin{pmatrix} 1 & 0 & 0 \\ 0 & 1 & 0 \\ 0 & 0 & 1 \end{pmatrix}.$$

In the special case of flat foldings ($\phi_i = \pm 180°$), there are no $R_{yz}(\phi_i)$ rotations because the face plane is always the xy-plane. Thus we obtain a simple composition of planar rotations about the origin, and this composition equaling the identity is equivalent to the condition in Theorem 12.2.1.

Unfortunately, this condition is not sufficient for single-vertex 3D foldings, unlike Theorem 12.2.1 for single-vertex flat foldings. A more complex form of self-intersection can arise as shown in Figure 12.15. Such intersections can be detected in polynomial

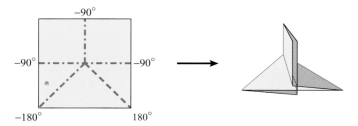

Figure 12.15. An unfoldable crease pattern that satisfies the necessary condition of Theorem 12.2.14. Dotted lines denote self-intersections in the attempted folded state.

time by constructing the potential folded state and testing for intersection between any two faces. Finding a cleaner characterization of single-vertex 3D foldings remains open.

Open Problem 12.1: 3D Single-Vertex Fold. [a] Provide necessary and sufficient conditions on (θ_i, ϕ_i) sequences to be realizable as a 3D single-vertex fold, both for flat paper ($\Sigma\theta_i = 360°$), and for arbitrary incident paper angle.

---

[a] belcastro and Hull (2002a,b).

12.3   CONTINUOUS SINGLE-VERTEX FOLDABILITY

While the previous section does much to characterize single-vertex foldability, it is concerned only with the existence of folded states, not folding motions. We already know from Section 11.6 that these two notions are equivalent, but that construction of folding motions is not particularly clean. In the context of single-vertex foldings, there is an explicit description of motions from unfolded to folded state if the total angle of paper at the central vertex is at most 360°. Even more interesting is that this motion does not fold the paper except at creases: the regions between creases move rigidly.

More precisely, Streinu and Whiteley (2001) prove the following:

Theorem 12.3.1 (Streinu and Whiteley 2001). *Consider a single-vertex crease pattern defined by angles $\theta_1 + \theta_2 + \cdots + \theta_n \leq 360°$. The space of 3D free folded states of the crease pattern, including the original unfolded state, is connected by folding motions whose restrictions to regions between creases are rigid motions.*

The proof of this theorem uses a connection between folding motions of single-vertex crease patterns and continuous reconfiguration of polygon linkages on the sphere. Specifically, if we consider just the boundary of the piece of paper, we obtain a spherical polygon linkage whose edge lengths equal the angles θ_i in the crease pattern, and whose vertex angles represent the dihedral angles between faces in the folded crease pattern. Any folding motion of the crease pattern corresponds to a continuous

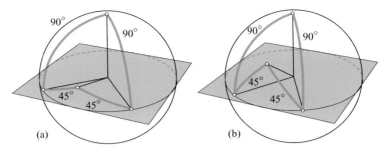

Figure 12.16. Flexing of a spherical quadrilateral composed of $(90°, 90°, 45°, 45°)$ arcs.

reconfiguration of the spherical polygon that preserves the edge lengths but can vary the vertex angles (see Figure 12.16).

Thus we obtain a spherical analog to the carpenter's rule problem considered in Part I, Section 6.6 (p. 96).

Streinu and Whiteley (2001) prove that every spherical polygon whose total edge length is at most 360° can be convexified. Their proof uses a spherical analog of the pseudotriangulation technique used in the plane (Section 6.7.2, p. 107). One reason that this result cannot be extended to spherical polygons with total edge length more than 360° is that, in this case, there is no convex configuration of the spherical polygon.

Once we reach a convex configuration, we need to establish that all convex configurations are interreachable by continuous motions. In the special case of a flat piece of paper (total edge length of exactly 360°), there is only one such convex configuration, the unfolded state, so there is nothing to prove. In the general case, we generalize results on planar reconfiguration of convex polygons from Part I, Section 5.3.1 (p. 70). This generalization is straightforward because the convex-polygon theorem relies mainly on the Cauchy–Steinitz lemma (Lemma 5.3.1, p. 72), which was already generalized from the plane to the sphere for proving Cauchy's celebrated theorem about rigidity of convex polyhedra (Section 23.1, p. 341).

This approach is limited to non-self-touching configurations of the polygon, as are all approaches to convexifying chains in the plane. This limitation forces us to require that the 3D folded states of the piece of paper are free, that is, non-self-touching, which unfortunately does not include flat folded states (other than the unfolded state). This requirement is likely not necessary, but removing it would probably require solving the self-touching analog of the carpenter's rule problem, Open Problem 6.5 (Section 6.8, p. 119).

13 General Crease Patterns

13.1 LOCAL FLAT FOLDABILITY IS EASY

Our knowledge about single-vertex flat foldability from Section 12.2 (p. 198) allows us to solve a more general problem involving local foldability of vertices in a general crease pattern (on a flat piece of paper). A vertex in a crease pattern or mountain–valley pattern is *locally flat foldable* if we obtain flat foldability when we cut out any region containing the vertex, portions of its incident edges, and no other vertices or edges. We can make the canonical choice for the cut-out region of a small disk centered at the vertex. Thus local flat foldability of a vertex is equivalent to single-vertex foldability in a crease pattern or mountain–valley pattern obtained by restricting the larger pattern to the creases incident to the vertex.

We can easily determine whether every vertex in a crease pattern or a mountain–valley pattern is locally flat foldable: just test each vertex according to the algorithms in Section 12.2. The catch is that while every vertex in a crease pattern may be locally flat foldable, each vertex might require its own local mountain–valley assignment. The more interesting question, the *local flat foldability problem*, is to determine whether a general crease pattern has one, global mountain–valley assignment for which every vertex is locally flat foldable. Bern and Hayes (1996) showed how to solve this problem by finding one such mountain–valley assignment, or determining that none exist, in linear time. We will see in Section 13.2 that such an assignment does not imply that there exists a flat folded state with that crease pattern; it is only a necessary condition for the existence of such a folded state.

13.1.1 Generic Case

As with single-vertex flat foldability, the algorithm is simpler for the generic case in which no two neighboring angles become equal by a sequence of crimps (except at the very end when only two angles remain). In this case, the mountain–valley assignment is essentially unique up to certain symmetries. Specifically, for each vertex, Lemma 12.2.7 (p. 205) tells us that the mountain–valley assignment of creases around that vertex is determined by a sequence of crimps of strict local minimum angles. We call two creases *oppositely paired* if they are both folded during a common crimp, because in this case their directions must be opposite. We call two creases *equally paired* if they are the two final creases after all crimps, because in this case their directions must be equal. (Here we are interested only in the case of flat paper.)

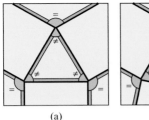 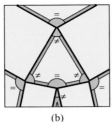

Figure 13.1. Examples of determining local foldability. (a) No locally flat-foldable mountain–valley assignment: the central odd cycle cannot alternate; (b) no locally flat-foldable mountain–valley assignment: the central even cycle must alternate. The symbol "$\neq$" denotes two oppositely paired creases, and "$=$" denotes an equal pairing.

(a) (b)

Lemma 13.1.1 *In a generic single-vertex crease pattern, the pairing of creases into opposite and equal pairs is independent of the sequence of the crimps of strict local minimum angles.*

Proof: Consider the globally smallest angle θ_i, which is necessarily a strict local minimum. Its two surrounding angles θ_{i-1} and θ_{i+1} are strictly larger, so neither θ_{i-1} nor θ_{i+1} is itself a local minimum. If we crimp one of the surrounding angles, θ_{i-2} or θ_{i+2}, we may replace θ_{i-1} or θ_{i+1} with a different angle, but that angle can only be larger than the original value. Thus, θ_{i-1} and θ_{i+1} remain not local minima (and uncrimpable) until we crimp the two creases neighboring θ_i. Therefore the pairing between these two creases is forced, and the crimp results in a single-vertex crease pattern with fewer creases. By induction, the entire pairing is forced, that is, unique. □

If we consider the pairings formed by all vertices in a general crease pattern, each crease belongs to at most two pairs, one for each end that is a vertex. Thus, if we consider only the connections between paired creases, we decompose the crease pattern into paths and/or cycles. Each path has exactly two valid mountain–valley assignments: once we set one crease's direction, the rest follow by propagating the same direction to equally paired creases and propagating the opposite direction to oppositely paired creases. Similarly, each cycle has at most two valid mountain–valley assignments. The only difference is that a cycle might not have a valid mountain–valley assignment because the parity of the cycle length does not match the parity of the number of equal pairings in the cycle. For example, if all pairings in the cycle are opposite and the cycle has odd length, then no mountain–valley assignment can alternate along the cycle as necessary. These parity mismatches are exactly what can prevent locally flat-foldable mountain–valley assignments in the generic case. A simple linear-time algorithm checks all parities and, if they all match, finds a valid mountain–valley assignment (see Figure 13.1).

13.1.2 General Case

The nongeneric case is more difficult because the natural generalization of Lemma 13.1.1 does not hold: the pairing depends on the choice of the crimp sequence (see Figure 13.2). Indeed, one crimp sequence can lead to a valid pairing, whose constraints are satisfied by a mountain–valley assignment, while another crimp sequence can lead to an invalid pairing, whose constraints cannot be satisfied by any mountain–valley assignment, as in Figure 13.1(b). One way to view the difference between the generic and nongeneric cases is that, in the nongeneric case, there is more than one way

 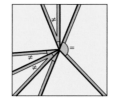 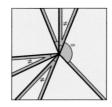 

Figure 13.2. All possible pairings of creases at a nongeneric vertex with several equal angles.

to "break the tie" between two equal angles and decide which is first considered a local minimum.

To solve the nongeneric case, we define a *merging operation* that modifies the arbitrary choices made during the recursive construction of the pairing at each vertex. Consider the recursive construction defined by the second characterization of flat-foldable mountain–valley assignments, Corollary 12.2.11 (p. 208). Suppose that, at some time during this construction, we crimped an angle θ_i equal to the globally smallest angle and there was an adjacent equal angle θ_{i+1}. Thus we could have crimped the adjacent angle θ_{i+1} instead of θ_i and obtained the same recursive crease pattern. This change leads to a different pairing among creases: the crease neighboring θ_i but not θ_{i+1}, and the crease neighboring θ_{i+1} but not θ_i, swap their pairing partners. This swap is called a *merge* if it combines two components (paths or cycles) of the connections between paired creases in the entire crease pattern.

We perform as many merge operations as possible. Each merge operation only increases the possibilities of a component (path or cycle) for merging again, so the particular sequence of merge operations is irrelevant. Each merge also can only fix parity problems. If we merge a path and a cycle, the resulting path has no parity problem. If we merge two cycles, the resulting cycle has a parity problem if and only if exactly one of the original cycles has a parity problem; thus, merging resolves a double parity problem and does not create a problem if none existed before. Furthermore, the configurations reachable by swaps encompass all possible choices for crimping, so by Corollary 12.2.11, all possible mountain–valley patterns. Therefore, if all parity problems can be resolved, repeated merging will resolve them.

Theorem 13.1.1 (Bern and Hayes 1996). *In linear time, we can determine whether a crease pattern is locally flat foldable, and if so, compute a global mountain–valley assignment that is locally flat foldable.*

Proof: The algorithm for exhaustively merging components uses some sophisticated data structures to attain a linear time bound. First we run the recursive pairing construction of Corollary 12.2.11, always crimping an angle equal to the globally smallest angle, breaking ties among such angles arbitrarily. Then we construct the graph of pairing connections and compute the connected components (paths and cycles). Next, for each possible swap (of which there are two per crease), we check whether the two incident components are indeed distinct, and therefore whether the swap corresponds to a merge, and if so, perform the merge. To determine whether the components are equal and then merge two components, we need a union-find structure. Because the unions occur only between edges incident to a common vertex, the structure of Gabow and Tarjan (1985) uses linear total time.   □

13.2 GLOBAL FLAT FOLDABILITY IS HARD

In this section we address the main question concerning flat foldability: given a crease pattern, does it have a flat folded state? Bern and Hayes (1996) proved that this problem is strongly NP-hard, implying that there is likely no efficient algorithm for solving the problem. They also proved a superficially easier problem NP-hard: given a flat-foldable mountain–valley pattern (i.e., creases plus mountain–valley labels), construct a flat folded state. Both NP-hardness results hold even when the piece of paper is a square. In the rest of this section we sketch the first NP-hardness reduction and discuss the significance of the second result.

13.2.1 All-Positive Not-All-Equal 3-Satisfiability

To prove that flat foldability is strongly NP-hard, we show that it is at least as hard as another strongly NP-hard problem. Specifically, in *All-Positive Not-All-Equal 3-Satisfiability* (Schaefer 1978), we are given a collection of *clauses* of the form Not-All-Equal(x_i, x_j, x_k) where $1 \leq i, j, k \leq n$. A truth assignment to the variables $x_1, x_2, \ldots, x_n$ is called *satisfying* if the three variables in each Not-All-Equal clause are assigned three values that are not equal: two are false and one is true, or vice versa. The goal in All-Positive Not-All-Equal 3-Satisfiability is to decide whether a given collection of clauses has a satisfying assignment.[1]

13.2.2 Reduction Overview

To prove that flat foldability is at least as hard as All-Positive Not-All-Equal 3-Satisfiability, we show that any instance of All-Positive Not-All-Equal 3-Satisfiability can be efficiently converted into an equivalent instance of flat foldability, and therefore All-Positive Not-All-Equal 3-Satisfiability is a special case of flat foldability. Such a conversion from a formula to a crease pattern is called a *reduction*. The reduction is based on a collection of local patches of crease patterns called *gadgets* that combine to represent an arbitrary formula. Gadgets are interconnected by *wires* which communicate truth values.

We begin with the problem in which we are given only a crease pattern; later we will consider the case in which we are also given a mountain–valley assignment.

13.2.3 Wire

A wire is represented by two parallel creases as shown in Figure 13.3, separated by a distance smaller than the distance to any other creases on either side of the strip. Thus, in order to avoid local self-intersection between the regions of paper on either side of the strip, the two parallel creases must be assigned opposite directions, one mountain and one valley. These two different assignments communicate a truth value from one

---

[1] The "all-positive" modifier represents the property that the Not-All-Equal clauses involve just variables x_i; in the more general Not-All-Equal 3-Satisfiability problem, clauses can also involve negated variables $\neg x_i$. This assumption is not made in Bern and Hayes (1996), but it simplifies their NP-hardness proof. Schaefer (1978) proved that All-Positive Not-All-Equal 3-Satisfiability is NP-hard, though he called the problem "Not-All-Equal Satisfiability."

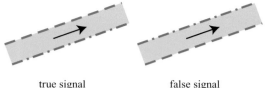

Figure 13.3. A wire carries a Boolean signal based on whether the crease on the left is a valley, where the left side is determined by the orientation denoted by an arrow.

true signal   false signal

end to the other. To determine this truth value, wires also have an associated orientation specifying a "forward" and "backward" direction, which determines which crease is "on the left" and which crease is "on the right." We fix the convention that TRUE is represented by a left valley crease and a right mountain crease, and the reverse represents FALSE.

The truth assignment of a variable can be represented by a wire coming in from the edge of the paper, so that the wire's truth assignment is unconstrained on that side. This truth assignment can then be fed into other gadgets to force the truth assignment of all variables to satisfy not-all-equal constraints.

13.2.4   Not-All-Equal Clause

The most important gadget in the reduction represents a Not-All-Equal clause by forcing three incoming wires to have truth assignments that are not all equal. The gadget achieves the forcing by being not flat foldable precisely when all three truth assignments are equal. The gadget is a slight variation on what is called the *triangular twist* in the origami community (see Figure 13.4). The crease pattern is locally flat foldable according to many mountain–valley assignments, one for each possible truth assignment of incoming wires. Exactly two of these mountain–valley assignments lead to self-intersection, namely, when all truth assignments are equal. In a standard triangular twist, the 35° angle is 30°, and this self-intersection does not arise; instead, the relevant faces of paper come together at a point but do not properly intersect. With the angle at 35°, however, self-intersection of three mountain creases occurs in the center of the triangle, an intersection that cannot be removed by stacking order because each mountain crease must be on top of the other cyclically.

13.2.5   Splitting and Routing

To complete the reduction, we need to show how wires representing truth assignments of variables can be (1) split so that the same value can be sent to all Not-All-Equal clauses involving that variable and (2) routed to connect one gadget to another. One gadget called the *reflector*, shown in Figure 13.5, achieves these goals, with some help from *crossover* gadgets shown in Figure 13.6. The reflector produces two copies of the input signal, one equal and one negated, because the local equal/opposite pairing forces the two outer obtuse angles to have neighboring creases with equal mountain–valley assignment. By chaining two reflectors together at the negated signal, we obtain two equal copies and one negated copy of the original signal. Thus, if we are able to effectively "discard" the extra negated copy, the reflector allows us to split a signal. Because the reflector gadget works for any nonacute angle $90° \leq \theta < 180°$ in the central isosceles triangle, we can use the gadget to effectively turn a wire as well.

Along the way to their destination, wires may cross. We arrange for wires to cross at only two possible angles: 90° and 135°. In these cases we can use the crossover gadgets

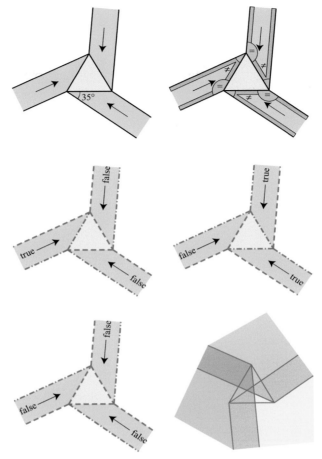

Figure 13.4. Not-All-Equal clause gadget: the crease pattern, the local equal/opposite pairing, all valid mountain–valley assignments up to rotational symmetry, an invalid mountain–valley assignment, and a shadow of the intersection that arises.

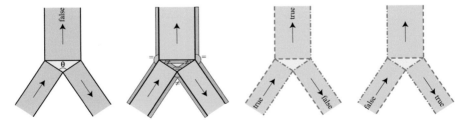

Figure 13.5. Reflector gadget: the crease pattern with parameter $90° \leq \theta < 180°$, the local equal/opposite pairing, and all valid mountain–valley assignments up to inversion of the central isosceles triangle.

shown in Figure 13.6 to cross the wires. These gadgets satisfy the desired property that the signal along a wire is preserved through the gadget. Although not drawn as such, both of these crossover gadgets work equally well when the two crossing wires do not have the same width.

In the 90°-crossover gadget, the crease pattern of the central square is uniquely determined up to inversion by local foldability conditions (Theorem 12.2.3, p. 203). If

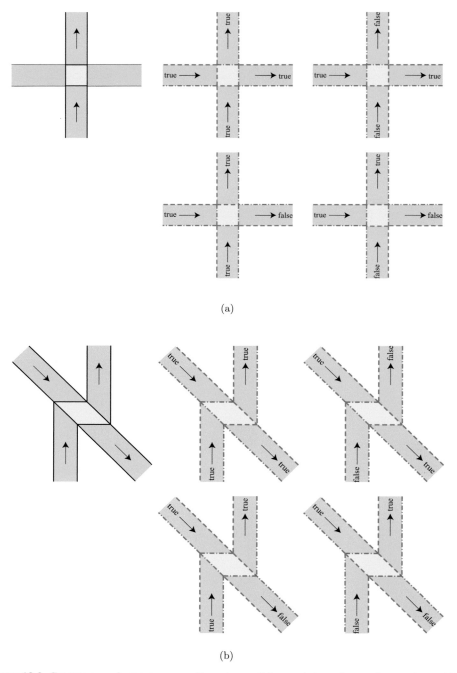

Figure 13.6. Crossover gadgets: crease pattern, two valid mountain–valley assignments, and two invalid mountain–valley assignments. (a) 90° crossover; (b) 135° crossover.

a wire signal is inverted, as in the bottom of Figure 13.6(a), then two opposite sides of the square are assigned the same direction, forcing one of the two incident strips to intersect the other crease, preventing global flat foldability. On the other hand, if both wire signals pass through correctly, then the crease pattern corresponds to folding one crimp after the other crimp. (Inverting the order of crimps corresponds to inverting the directions of creases in the central square.)

The 135°-crossover gadget behaves similarly because the underlying graph of the crease pattern is identical; the only difference is that the central square is replaced by a parallelogram. As before, if a wire signal is inverted so opposite sides of the parallelogram are assigned the same direction, as in the bottom of Figure 13.6(b), then one of the incident strips intersects the other crease. When both wire signals are preserved, one of the correct flat foldings (as in the top of Figure 13.6(b)) again corresponds to folding one crimp after the other crimp; inverting the parallelogram also leads to a flat folding, but it is not such a composition of crimps.

13.2.6 Putting It Together

Now that we have all the necessary gadgets, we describe how they fit together to represent any instance of All-Positive Not-All-Equal 3-Satisfiability (refer to Figure 13.7). We represent a variable by a zigzagged wire at ±45° to horizontal, where each 90° turn between a zig and a zag is made by a 90° reflector gadget. Thus, the zigs carry the opposite signal from the zags; we define the upward-sloping zigs to carry the value of the variable.

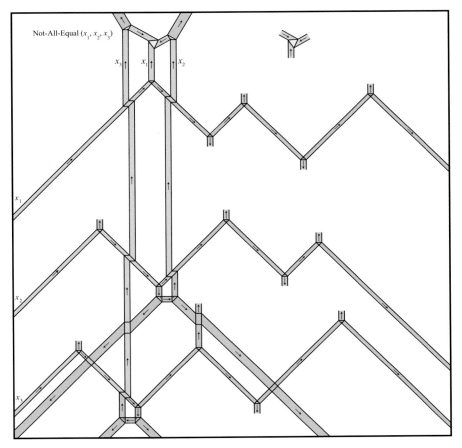

Figure 13.7. Overview of the reduction from All-Positive Not-All-Equal 3-Satisfiability to flat foldability of a crease pattern.

Not-All-Equal clause gadgets are at the top of the square of paper, oriented so that one input wire comes from above, and scaled to have input wires of width $\sqrt{2}$ times larger than the variable wires. For each clause Not-All-Equal(x_i, x_j, x_k), we make a vertically aligned zig-to-zag bend in x_i wires to feed a widened copy along the vertical input of the clause gadget. The x_j and x_k wires each make a zag-to-zig bend, are negated, and turn upwards via two $90°$ reflectors, which are then unnegated and fed into an input of the clause gadget using a $\pm120°$ reflector. Each of these three reflector gadgets per input produces an excess copy of the signal at a different size; we propagate these wires to the boundary of the paper using crossover gadgets and by arranging so that the wires do not intersect any gadgets.

This crease pattern simulates the All-Positive Not-All-Equal 3-Satisfiability problem in the sense that the crease pattern is flat foldable if and only if the given formula has a satisfiable truth assignment. Thus we can solve the satisfiability problem by converting the formula into a crease pattern and applying a hypothetical solution to flat foldability, implying that flat foldability is at least as hard as satisfiability and therefore flat foldability is NP-hard:

Theorem 13.2.1 (Bern and Hayes 1996). *Deciding whether a crease pattern is flat foldable is strongly NP-hard.*

13.2.7 Overlap Order from Valid Mountain–Valley Assignment

The second hardness result of Bern and Hayes (1996) considers when we have even more information: in addition to a crease pattern, we have a mountain–valley assignment and a promise that there is a flat folded state with the given mountain–valley pattern. The problem is to find an actual flat folded state. In the terminology of Chapter 11, the folded-state geometry f—the locations in the plane of each face of the crease pattern after folding—is already determined by the crease pattern. The challenge is to construct the folded-state partial order λ, called the *stacking order* in the context of flat foldings: for each pair of faces that overlap in the plane, which one is on top? Bern and Hayes show that this ordering is the heart of the hardness of deciding global foldability:

Theorem 13.2.2 (Bern and Hayes 1996). *Constructing a flat folded state matching a given flat-foldable mountain–valley pattern is strongly NP-hard.*

The NP-hardness reduction for this theorem follows the same overall structure as the one just presented, but the gadgets are substantially more complicated. The basic idea is to use structures that always fold with the same mountain–valley assignment, but with multiple, different stacking orders. We do not describe the details here, but rather turn to the impact of this result on practical origami.

The exact scenario of this theorem is common in complex origami instructions, particularly in the context of origami tessellations (touched upon in Section 20.2): the origami designer gives a mountain–valley pattern for the model, and simply states "fold as indicated." In contrast, most origami up to the high-intermediate level specifies a sequence of steps, each of which involves primarily only a constant number of creases at a time. But origami designers find it difficult to find such a decomposition into steps, and sometimes it may not even be possible, in which case the current best general descriptive notation for origami diagrams is a mountain–valley pattern. While a reasonable origamist can fold a reasonable mountain–valley pattern, it poses an intimidating challenge to a novice folder or when the pattern is extremely intricate. Bern and Hayes's NP-hardness result shows that this practical difficulty is also a computational difficulty. Effectively, a mountain–valley pattern does not give enough information to specify exactly how to fold. From a theoretical point of view, such origami diagrams should also specify the entire folded state. Unfortunately, we lack a good notation for such a specification; this is an interesting challenge for bringing together the theory and practice of origami.

14 Map Folding

Map folding is a problem everyone encounters during a road trip: how should you refold your road map into a compact packet, folding each crease just as it was originally folded? More formally, two-dimensional (2D) "map folding" problems seek to determine when a given mountain–valley pattern on a piece of paper can be realized by a flat folding of the paper. The name comes from the most basic version:

> **2D Map Folding Problem.** Given a rectangle (the *map*) partitioned into an $n_1 \times n_2$ regular grid of squares, with each nonboundary grid edge assigned to be either a mountain or valley crease, can the map be folded flat into one square, respecting the creases?

Figure 14.1 shows an example. (The additional information in the figure, provided by the labels within each square, is not part of the input to the problem.) Jack Edmonds asked for the computational complexity of deciding an instance of the map folding problem, and this problem remains unsolved today:

Open Problem 14.1: Map Folding. [a] What is the computational complexity of deciding whether an $n_1 \times n_2$ grid can be folded flat (has a flat folded state) with a specified mountain–valley assignment?

---

[a] J. Edmonds, personal communication, August 1997.

The main issue is whether this problem is NP-hard (intractable) or polynomial-time (tractable). We have already seen how to solve $1 \times n$ maps in $O(n)$ time, because in this case all creases are parallel, so we can use the solution to the 1D foldability from Section 12.1 (p. 193). Remarkably, the problem remains open even for $2 \times n$ maps.

Arkin et al. (2004) consider a variation on the map folding problem in which the folding is restricted to "simple folds." In this setting, the complexity depends on the type of simple folds and on the shape of the paper and the creases. Before describing these results, we turn to the definition of simple folds.

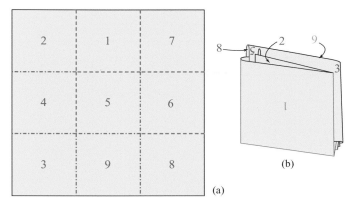

Figure 14.1. (a) A map-folding puzzle. (b) A crude depiction of a solution, with several squares labeled (lightly shaded labels are facing away from viewer).

SIMPLE FOLDS

The folding depicted in Figure 14.1(b) requires "tucking" squares into pockets and considerable bending of the paper before the final flat state is achieved. This is an instance where "complex origami folds" are necessary. For our purposes, we define a *complex fold* to be an unrestricted folding motion: one that of course does not tear the paper or cause the paper to self-intersect, but is otherwise unrestricted. In contrast, a *simple folding* consists of a sequence of simple folds, each of which is restricted as follows:

1. Simple folds apply only to flat folded states, and map one flat folded state to another.
2. Each simple fold is along a segment on the top of the folded state (where "top" refers to one of the two sides of the paper, which is permitted to flip over). This segment may or may not extend all the way across the silhouette of the paper.
3. The fold is a rigid rotation of some layers of paper under the segment, avoiding self-intersection throughout 180° of rotation.
4. The fold must respect the given crease pattern, folding only at creases, and according to the specified mountain–valley assignment.

In origami, a simple fold is sometimes called a "valley fold" or "mountain fold," or occasionally a "book fold."

Three classes of simple folds may be distinguished, depending on how the layers are folded:

1. *One-layer simple fold*: Just folds the top layer of paper.
2. *All-layers simple fold*: Simultaneously folds all layers of paper under the crease segment.
3. *Some-layers simple fold*: Folds some layers beneath the creasing segment, perhaps a different depth of layers along different portions of the crease.

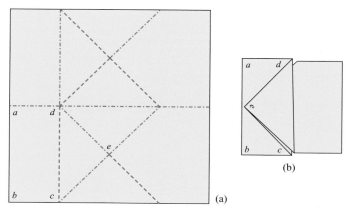

Figure 14.2. (a) A crease pattern foldable by some-layers simple folds, but not by one- or all-layers simple folds. (b) Top view of a flat folding.

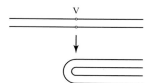

Figure 14.3. An all-layers fold of 1D paper.

It is established in Arkin et al. (2004) that these models are all different, in that there are crease patterns that can be folded flat using one model but not another. For example, the pattern in Figure 14.2 can be flat folded under the some-layers model but not under either the one- or all-layers model. The only general relationship is that some-layers simple folds are strictly more powerful than one-layer and all-layers simple folds.

14.1.1 Simple Folds in 1D

In Section 12.1, we showed that all flat-foldable 1D crease patterns ("maps") can be folded flat by crimps and end folds. Both these types of folds are one-layer simple folds. Thus, for 1D paper, flat foldability by complex folds is equivalent to flat foldability by one-layer simple folds. Another consequence is that one-layer simple folds and some-layers simple folds have the same power in 1D. As mentioned above, neither of these properties extend to 2D paper.

The all-layers simple folds are more restrictive than the other models with 1D paper. Figure 14.3 shows an example of such a fold. In this context, folds can never be "un-made," so any crease that we fold via an all-layers simple fold must have the property that any valley creases in one direction have matching mountain creases at the same distance in the other direction, and vice versa. Finding these creases can be viewed as a string matching problem. Arkin et al. (2004) show how to determine, in $O(n \log n)$ worst-case time or $O(n)$ expected time, whether a given mountain–valley pattern can be folded flat by a sequence of all-layers folds, and find a sequence if so.

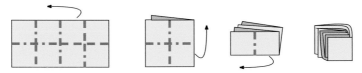

Figure 14.4. Folding a 2×4 map via a sequence of three all-layers simple folds.

14.2 RECTANGULAR MAPS: REDUCTION TO 1D

We now explore the 2D map folding problem for a rectangular map under the three varieties of simple folds, and show that in each case the problem reduces to 1D. We need to distinguish between the types of simple folds only at the end of this section.

If a mountain–valley pattern can be folded flat by simple folds, there must be at least one *crease line*, a crease that extends all the way across the rectangle that is labeled entirely valley or entirely mountain. Otherwise, the pattern would not permit a first simple fold. Furthermore, all such crease lines must be parallel; otherwise, the vertex of intersection between two crossing crease lines would violate Maekawa's Theorem 12.2.3 and so not be locally flat foldable: there would be either two mountains and two valleys, four mountains and no valleys, or no mountains and four valleys. Without loss of generality, assume that these crease lines are horizontal, and let $\mathcal{H}$ denote the set of them.

Lemma 14.2.1. *All crease lines in $\mathcal{H}$ must be folded before any other crease.*

Proof: We first show that no vertical crease can precede folding the lines in $\mathcal{H}$. Let $h \in \mathcal{H}$ and let $x \in h$ be a grid-line junction. Then the vertical creases incident to x must be labeled mountain and valley, to satisfy Maekawa's Theorem. Suppose, in contradiction to the claim, that h is not folded before the vertical crease through x. Then folding the vertical mountain crease misfolds the vertical valley crease, and vice versa.

It remains to show that no horizontal crease not in $\mathcal{H}$ can precede all those in $\mathcal{H}$. Any such horizontal crease cannot be labeled uniformly across its entire length. Suppose that it includes a crease labeled mountain incident to x from the left, and valley to the right. We have just shown that no vertical crease could have been folded yet. Thus we are in the same situation: folding mountain misfolds valley and vice versa. □

This lemma reduces the problem to 1D, with a bit of additional checking for consistency of the creases. First we fold all the creases in $\mathcal{H}$ according to the 1D problem. All the non-$\mathcal{H}$ folds must match up appropriately after making all the folds in $\mathcal{H}$; otherwise, the map is not foldable via simple folds. After folding all of $\mathcal{H}$, the map size has reduced, and we can continue the folding with the vertical crease lines $\mathcal{V}$. The folding continues to alternate direction until we completely fold the map, or we reach a contradiction in matching crease labels. See Figure 14.4 for a simple example.

Finally we examine the three separate models of simple folds:

1. *One-layer simple folds.* Here we use the crimp and end-fold algorithm of Theorem 12.1.6 to solve the 1D problem for $\mathcal{H}$. After folding the creases in $\mathcal{H}$, no further folding is possible, because the next step, folding some crease in the resulting $\mathcal{V}$, would require folding several layers at once. So the only 2D maps

foldable by one-layer simple folds are in fact 1D maps, with all creases parallel. (An example is the traditional folding of a letter in thirds.)

2. *All-layers simple folds.* Here we use the 1D all-layers folding algorithm that was mentioned (but not detailed) in Section 14.1.1; (p. 226).

3. *Some-layers simple folds.* As with one-layer simple folds, here we use the 1D crimp and end-fold algorithm. The first step, folding $\mathcal{H}$, is accomplished via one-layer folds, and subsequent steps use some-layers folds.

Some-layers simple folds are the most natural in this context, the equivalent of the folds we make to sheets of paper in the workplace nearly every day. The corresponding algorithm can be designed to make the additional consistency checks, and maintain the successive $\mathcal{H}$ and $\mathcal{V}$ sets, in time linear in $n_1 n_2$, which is the size of the input crease pattern. Together with the linear-time 1D map folding algorithm (Theorem 12.1.6), the entire algorithm runs in linear time.

14.3   HARDNESS OF FOLDING ORTHOGONAL POLYGONS

In dramatic contrast to the previous result, which might be characterized as saying that rectangular map folding via simple folds is easy, next we prove that folding a map with a more complicated shape—an orthogonal polygon—is NP-hard. The creases are still restricted to horizontal and vertical grid lines (parallel to the polygon's edges), and the folds are still restricted to be simple folds.

The proof is based on a reduction from the NP-hard PARTITION problem (cf. p. 25): given a set X of n integers $a_1, a_2, \ldots, a_n$ whose sum is A, does there exist a set $S \subset X$ such that $\sum_{a \in S} a = A/2$? For convenience, let $\bar{S} = X \setminus S$. Also, without loss of generality, assume that $a_1 \in S$.

We transform an instance of the PARTITION problem into an orthogonal 2D crease pattern on an orthogonal polygon, as shown in Figure 14.5. All creases are valleys. There is a staircase of width ε, where $0 < \varepsilon < 2/3$, with one step of length a_i corresponding to each element a_i in X. In addition, there are two final steps of length L and $2L$, where L is chosen greater than $A/2$. The total width W_1 of the staircase is chosen to be less than the width W_2 of the frame attached to the staircase.

The main mechanism in the reduction is formed by the vertical creases v_0 and v_1. Basically, the first time we fold one of these two creases, the staircase must fit within the frame, or else the second of these two creases is blocked. Fitting in the frame requires solving the PARTITION instance. Then when we fold the other of these two creases, the staircase exits the frame, enabling us to fold the remaining creases in the staircase.

Lemma 14.3.1. *If the* PARTITION *instance has a solution, then the crease pattern in Figure 14.5 can be folded flat by simple folds.*

Proof: For $2 \leq i \leq n$, valley fold v_i if exactly one of a_{i-1} and a_i is in S. After these folds, as we travel along the steps corresponding to $a_1, \ldots, a_n$, we travel in the $-y$ direction for elements that belong to S and in the $+y$ direction for elements that belong to $\bar{S}$. Because the sums of elements of both S and $\bar{S}$ are $A/2$, the point P_5 has the same y-coordinate as the point P_4 after these folds. Because $L > A/2$, the steps corresponding

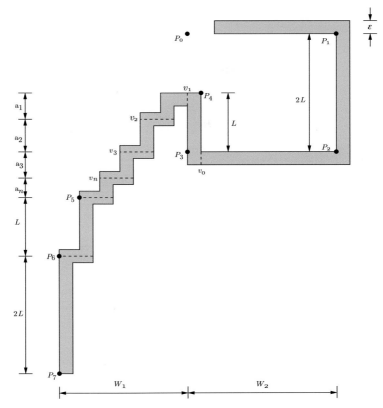

Figure 14.5. Hardness reduction from the PARTITION problem.

to a_i's are confined to remain in between the y-coordinates of points P_1 and P_2. Because P_5 has the same y-coordinate as that of P_4 and because the vertical distance between P_5 and P_6 is L, point P_6 will have the same y-coordinate as that of either P_1 or P_2.

Now valley fold v_{n+2}. Because the vertical distance between P_6 and P_7 is $2L$, the y-coordinate of P_7 will be same as that of P_1 or P_2 and the step between P_6 and P_7 will lie exactly between the y-coordinates of P_1 and P_2. This situation is illustrated in Figure 14.6.

Now valley fold v_1. Because $W_2 > W_1$, the partly folded staircase, which currently lies between the y-coordinates of P_1 and P_2, fits within the rectangle $P_0 P_1 P_2 P_3$. Now valley fold v_0. We now have the semi-folded stairs on the right and the rectangular frame $P_0 P_1 P_2 P_3$ on the left. Finally, valley fold all the remaining unfolded creases in the staircase. This can be done because the rectangular frame is now on the left of P_4 and all steps of the staircase are on the right of P_4.   □

Lemma 14.3.2. *If the crease pattern in Figure 14.5 can be folded flat by simple folds, then there is a solution to the PARTITION instance.*

Proof: If either v_0 or v_1 is folded without having the staircase confined between the y-coordinates of P_1 and P_2, the rectangular frame $P_0 P_1 P_2 P_3$ would intersect with the staircase and would make the other of v_0 and v_1 impossible to fold. Hence the staircase

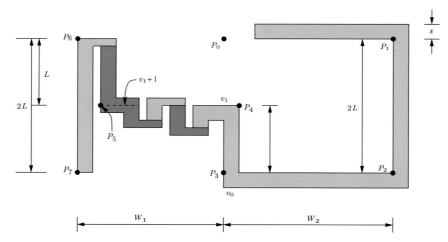

Figure 14.6. Semi-folded staircase confined between *y*-coordinates of P_1 and P_2.

must be brought between the *y*-coordinates of P_1 and P_2 before folding either v_0 or v_1. Because the last and the second-last steps of the staircase are of size $2L$ and L, respectively, point P_5 must have the same coordinate as that of the point P_4 when the staircase is confined between the *y*-coordinates of P_1 and P_2.

As we travel from P_4 to P_5 along the staircase, we travel equally in $+y$ and $-y$ directions along the steps corresponding to the elements of X. Hence the sum of elements along whose steps we travel in $-y$ direction is same as the sum of elements along whose steps we travel in the $+y$ direction. Thus there is a solution to the PARTITION instance if the crease pattern in Figure 14.5 is foldable.   □

These lemmas rely only on the folds being simple, and not on the number of layers included in the folding: every simple fold considered involves a single layer, and there are no other layers underneath. So they imply this result:

Theorem 14.3.3. *It is (weakly) NP-complete to decide simple foldability of an orthogonal piece of paper with an orthogonal mountain–valley pattern, for all-layers, some-layers, and one-layer simple folds.*

14.4   OPEN PROBLEMS

Arkin et al. (2004) show that deciding simple foldability of a rectangular map is (weakly) NP-hard if the creases are permitted along 45° diagonal lines as well as along the horizontal and vertical, in the some- and all-layers models. This problem variant is open, however, for the one-layer model.

The complexity of simple foldability of maps with orthogonal creases is settled only for creases aligned with the polygon's edges. Generalizing to nonaligned creases and/or to specialized map shapes remains open:

> **Open Problem 14.2: Orthogonal Creases.**[a] (a) What is the complexity of simple foldability of a rectangular map with orthogonal creases, where the creases are not aligned with the rectangle sides? (b) What is the complexity for *orthogonally convex polygon* maps, with orthogonal creases aligned with the polygon edges? Such maps meet every vertical or horizontal line in a single segment or not at all.
>
> ----
>
> [a] Arkin et al. (2004).

The map-folding hardness results are all based on reduction from PARTITION, which is only weakly NP-hard, leaving open the possibility for an efficient algorithm when the n map vertices and crease endpoints are given as points of a small $N \times N$ grid:

> **Open Problem 14.3: Pseudopolynomial-Time Map Folding.**[a] Can simple map folding be solved in pseudopolynomial-time (polynomial in n and N), or is it strongly NP-hard?
>
> ----
>
> [a] Arkin et al. (2004).

But surely the most attractive unsolved problem in this area is the one that opened this chapter (Open Problem 14.1): the complexity of deciding whether a rectangular map, with horizontal and vertical creases, can be folded flat employing arbitrary (complex) folds.

15 Silhouettes and Gift Wrapping

Perhaps the simplest geometric formulation of origami design is to fold a desired 2D or 3D shape from a specified shape of paper (typically square). For example, we may be given the polygonal region shown in Figure 15.1(a) representing the silhouette of a horse, and our goal is to fold a square piece of paper into a flat origami with this silhouette. The origami literature provides countless examples of 2D and 3D shapes foldable from a square piece of paper. Folding a 3D shape can also be thought of as wrapping a general-shape gift.

The desired shape should be made of flat sides—polygonal (in 2D) or polyhedral (in 3D)—to be achievable by finitely many folds of a polygonal piece of paper, but other than this basic constraint, it is conceivable that any shape is foldable from a sufficiently large piece of paper. Although this problem is implicit throughout the origami literature, the problem was not formally posed until 1999 by Bern and Hayes (1996) and then only in the 2D case. The 3D version—a kind of "gift-wrapping" for complex polyhedral gifts—was implicitly studied as early as 1960, for example, by Gardner (1990, 1995a), but the general problem appears not to have been formally posed in the literature.

A further variation on these problems, introduced by Demaine et al. (2000c), is to suppose that the original piece of paper is *bicolored*: a different color on each side. For example, origami paper is often white on one side and colored on the other; wrapping paper is often white on one side and patterned on the other. When we fold a 2D or 3D shape out of such paper, different parts of the surface may have different sides of the paper exposed. A natural question is which two-color patterns can be achieved by folding. For example, we might specify the two-color pattern in Figure 15.1(b), which has the same silhouette as in Figure 15.1(a), but imposes additional constraints to form a zebra. This particular two-color pattern can be achieved by an efficient folding by John Montroll, as shown in Figure 15.1(c). Can all two-color patterns be achieved from sufficiently large pieces of bicolor paper?

Answering these questions, Demaine et al. (2000c) proved the following universality result:

Theorem 15.0.1 (Demaine et al. 2000c). *Every 2D polygon and every 3D polyhedron can be folded from a sufficiently large square of paper (or sufficiently large scaling of any shape of paper). Furthermore, the surface can be colored according to any two-color pattern with bicolor paper.*

The proof of this theorem provides an algorithm for constructing such solutions. The only catch is that the foldings are usually highly inefficient; they can use unnecessarily

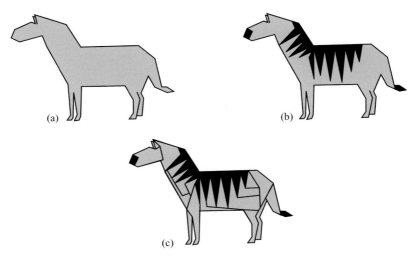

Figure 15.1. (a) Polygonal silhouette of a horse; (b) polygonal two-color pattern of a zebra; (c) origami zebra folded from a bicolor square of paper, designed by Montroll (1991, pp. 94–103).

large paper to fold relatively simple shapes. Nonetheless, the result is general, and opens the door for more efficient solutions. We return to this concern of efficiency in Section 15.4, after describing the general method.

15.1 STRIP FOLDING

The basic idea is to start by folding the piece of paper into a long, narrow *strip* (rectangle). This strip can then be folded according to various small *gadgets*, in particular the turn gadgets in Figure 15.2, to "navigate" the strip around the surface of the desired shape. At various times we can "flip over" the strip using the gadget in Figure 15.3 to exchange the visible color and achieve the desired two-color pattern. While there are certainly details in how these gadgets are put together, the figures should provide enough intuition to make the claim plausible.

15.2 HAMILTONIAN TRIANGULATION

One natural way to navigate the desired surface with a strip is to triangulate the surface, and cover each triangle in turn. Figure 15.4 shows how to cover each triangle: zigzag the strip parallel to the edge opposite where the strip begins, choosing the initial direction of the zigzag so that the final vertex reached is opposite the edge shared by the next two triangles. If the strip width does not evenly divide the height of the triangle, then the last two parallel strips partially overlap.

Next we use a turn gadget with overhang from Figure 15.2(d), as shown in Figure 15.5, to change the direction of the strip to be parallel to the opposite edge in the next triangle. The strip may also have the wrong initial direction for the triangle, in which case we turn it around using two 90°-turn gadgets from Figure 15.2. We reverse the color of the strip if the next triangle has the opposite color.

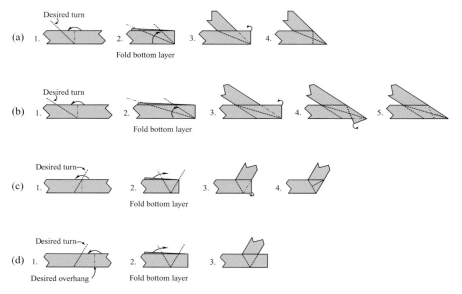

Figure 15.2. Turn gadgets. (a) Sharp turn ($> 90°$); (b) very sharp turn ($> 135°$); (c) blunt turn ($< 180°$); (d) (blunt) turn with overhang.

Figure 15.3. Color-reversal gadget.

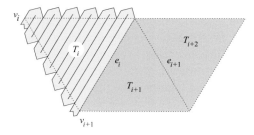

Figure 15.4. Covering a triangle T_i by zigzagging parallel to the edge e_i incident to the next triangle T_{i+1}, choosing the initial direction so that we end at the vertex v_{i+1} opposite the next edge e_{i+1}.

The only problem remaining is that our folding may lie slightly outside the desired shape because of the zigzag overhang in Figure 15.4 and/or the turn-at-vertex overhang in Figure 15.5. This overhang can be removed after completing each triangle using the simple "hide" gadget shown in Figure 15.6. The basic idea is to repeatedly fold along the edges of a triangle (a special case of a convex polygon to which the gadget applies) to fold excess paper underneath the triangle. We may fold along the line extending the edge e that connects to the rest of the folding from previous triangles, but we do not fold along that connection, and the folds are all underneath the existing folding, so we do not disturb the existing folding. Also, the strip that is about to be folded into the next

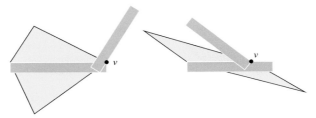

Figure 15.5. Silhouette of turn gadget needed to effect a turn at vertex v between two triangle zigzags.

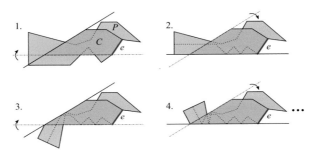

Figure 15.6. First three folds in "hiding" a polygon P of paper underneath a convex polygon C, without disturbing a specified edge e.

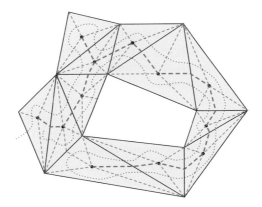

Figure 15.7. Hamiltonian refinement of a triangulation.

triangle remains on top of the folding, so it also is not disturbed by repeatedly folding other paper underneath.

We can visit the triangles in a triangulation of the surface in any order that eventually visits every triangle. These visits can be made particularly efficient by refining the triangulation so that it becomes *Hamiltonian*, that is, so that there is a path through the triangles that visits each triangle exactly once. Then the zigzag path can visit triangles in this order. Hamiltonian refinements of triangulations always exist, as shown by Arkin et al. (1996). One simple refinement method is shown in Figure 15.7: take a spanning tree of the triangles (red dashed lines), cut each triangle from its corners to

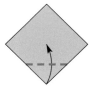 

Figure 15.8. A simple example of a reflex region between seams in a flat folding.

the spanning-tree node (green dashed lines), and walk around the spanning tree (blue dotted line). With this choice of zigzag path, we obtain the following theorem:

Theorem 15.2.1 (Demaine et al. 2000c). *Every 2D polygon or 3D polyhedron, with or without two-color pattern, can be folded from an $\varepsilon \times x$ strip of (bicolor) paper for sufficiently large x. As $\varepsilon \to 0$, the minimum $x = x(\varepsilon)$ that suffices for the method has the property that the strip area εx approaches the optimal bound of the polygon's area or polyhedron's surface area.*

15.3   SEAM PLACEMENT

A further strengthening of the problem (Demaine et al. 2000c) is to consider the folding's placement of *seams*—visible creases or paper boundary on the surface of the folding. Seams are often carefully placed in origami models to emphasize features of the subject. Any folding certainly needs a seam at each edge of a target polyhedron and along each edge of color reversal, but in the folding described above, each turn gadget introduces many additional, perhaps superfluous, seams. Can we somehow minimize the number or length of seams, or arrange seams into interesting patterns?

We have seen one method for folding shapes without seams: the hide gadget of Figure 15.6 can be used to fold any convex polygon from any containing polygon of paper without any seams. In one sense, this is the best we can hope for: convex polygons of paper can be folded seamlessly only into convex polygons. This limitation arises because any locally seamless reflex vertex of a flat folding of a convex piece of paper must come from an interior point of the paper, and hence must be folded flat according to the rules in Section 12.2 (p. 198), but no angle between two creases can be reflex by Theorem 12.2.1. However, this limitation does not require the region decomposition formed by seams to be made entirely of convex polygons. Figure 15.8 shows a simple example of a reflex region between seams.

An interesting open question is to characterize precisely which seam patterns can be folded:

Open Problem 15.1: Seam Patterns. Characterize which polygonal decompositions of a polygon or polyhedron (consistent with the decompositions of the polyhedron into faces and each face into colors) can be formed by seams in a folding of a sufficiently large scaling of a convex shape of paper.

This characterization must be independent of the particular convex shape used for the piece of paper, because the hide gadget of Figure 15.6 can convert between any two convex polygons without seams.

What is known about this problem is that any seam pattern forming a convex decomposition of the shape can be folded:

Theorem 15.3.1 (Demaine et al. 2000c). *Every 2D polygon or 3D polyhedron, with or without two-color pattern, can be folded from a sufficiently large scaling of any (bicolor) shape of paper, with the additional property that the seams decompose each single-color face into a specified convex decomposition.*

The proof of this result uses the same general approach of routing strips around the shape, but uses another gadget to dynamically vary the width of the strip to match the width of each convex piece. We omit the details.

15.4 EFFICIENT FOLDINGS

The solutions described above, while general, are rarely practical because of their inefficient use of, for example, a square piece of paper. An apparently difficult problem is to improve on this inefficiency:

Open Problem 15.2: Efficient Silhouettes and Wrapping. What is the complexity of finding the smallest scaling of a square that can be folded into the shape of a given 2D polygon or 3D polyhedron? If the shape is additionally marked by a two-color pattern, the pattern should be matched using bicolor paper. What about an arbitrary polygonal shape of paper?

Several special cases of this problem have been studied. Most notably, Lang's algorithms for designing "origami bases," described in the next chapter, constitute a solution to a special case of this problem where the 3D polyhedron forms a kind of stick figure. The rest of this section describes a few specific instances of the problem that have been considered, some of which have been solved (only recently) and some of which remain unsolved.

15.4.1 Cube Wrapping

Perhaps the simplest gift-wrapping problem is the following: how large a cube can be folded from a unit (1×1) square of paper? This problem was recently solved by Catalano-Johnson and Loeb (2001). The optimal solution of a $\sqrt{2}/4 \times \sqrt{2}/4 \times \sqrt{2}/4$ cube is shown in Figure 15.9. This folding was probably known much earlier—for example, it appears in Gardner (1990)—but its optimality was first established in 2001 (Catalano-Johnson and Loeb 2001). The proof is surprisingly simple: for any point on an $x \times x \times x$ cube, there is another "antipodal" point whose shortest-path distance measured along the surface of the cube is at least $2x$. Thus, the center of the square must have a point (namely, any of the corners) of distance at least $2x$, so the distance between opposite corners of the square must be at least $4x$. Therefore, the unit square has side length at least $2\sqrt{2}\,x$, so $x \le 1/(2\sqrt{2}) = \sqrt{2}/4$.

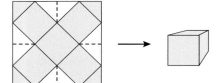

Figure 15.9. Folding of a unit square to largest possible cube (side length $\sqrt{2}/4$) from Catalano-Johnson and Loeb (2001). The white faces are tucked underneath the surface, leaving only the shaded faces visible. Dashed segments are valley folds; solid segments are mountain folds.

Table 15.1: Ratios of checkerboard side length to original paper side length for best-known foldings and the limit predicted by color-reversal perimeter

Checkerboard	2×2	3×3	4×4	8×8
Predicted limit	1	5/6	1/2	1/4
Folding	$2/3^a$	$3/5^a$	$1/2^{a,b}$	$1/4^a$
Seamless fold	$1/2^c$	$1/3^c$	Open	$1/5^d$

[a]   Dureisseix 2000.
[b]   Guy and Venables 1979.
[c]   Personal communication from Eric McNeill.
[d]   Casey 1989; Kirschenbaum 1998.

A simple but unsolved generalization of this problem is to characterize the optimal folding of an $x \times y$ rectangle into the largest possible cube, for all possible ratios y/x. At one extreme, when $x = y$, the optimal folding is Figure 15.9. At the other extreme, as the rectangle gets longer and longer, we can use the strip techniques described above to fold a cube whose surface area is arbitrarily close to the area of the polygon (Akiyama et al. 1997). (Indeed, these foldings of the cube were the inspiration for the strip techniques.) However, in between these two extremes, it seems difficult to characterize optimal foldings of rectangles into cubes.

15.4.2   Checkerboard Folding

Perhaps the simplest two-color folding problem is the following: how large a $k \times k$ checkerboard can be folded from a unit square of bicolor paper? In the simplest form of this problem, visible seams are permitted interior to the checker squares, and the reverse side of the checkerboard is arbitrarily colored. Ideally, of course, no seams would be visible, and the reverse side would be a solid color. The latter constraint seems to have no effect, while the former constraint does. In the origami literature, the best foldings known of a 1×1 square of paper into an 8×8 checkerboard produce a $1/4 \times 1/4$ board with seams (Dureisseix 2000)[1] or a $1/5 \times 1/5$ board without seams (Casey 1989; Kirschenbaum 1998). See Table 15.1 for a summary of other known foldings.

All efficient checkerboard foldings known so far have the property that every "color reversal" (edge along which two opposite visible colors meet) is achieved with a portion of the boundary of the piece of paper. This property may not be necessary.[2] For example,

---

[1]   Also achieved by an unpublished design by John Montroll (personal communication, June 1997).
[2]   Personal communication with Robert Lang, June 1997.

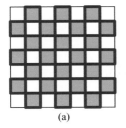
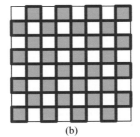

Figure 15.10. Shortest possible Eulerian tours that visit all color-reversal edges: (a) k odd: $2k(k-1)+2(k-1)$ edges; (b) k even: $2k(k-1)+2k$ edges.

(a)

(b)

Figure 15.11. The color-reversal pattern of a 2×2 checkerboard can be achieved without decreasing the size of the square, but two cells of the checkerboard are half absent.

Lang (2003, Sec. 9.11) shows that a square of paper can be folded into a flat shape with arbitrarily large perimeter. (Whether it is possible to increase the perimeter of a square at all by flat folding was known as the Margulis Napkin Problem, which circulated among mathematicians in the mid-1990s; Lang found the surprising answer that it could be increased arbitrarily.) Then this large-perimeter shape might conceivably be used for extensive color reversal. However, the perimeter-increasing foldings known so far are extremely wasteful in terms of area, so it seems impossible to obtain an efficient checkerboard by these methods.

Barring such foldings that create extra perimeter for color reversal, we obtain an efficiency limit on checkerboard foldings by considering the shortest possible Eulerian tour that visits all color-reversal edges (Dureisseix 2000)[3] (see Figure 15.10). The boundary of the paper is a cycle, so its length must be at least the length of this tour if all color reversals are achieved by the paper boundary. The number of edges along the tour is

$$2k(k-1)+2k-[2 \text{ if } k \text{ is odd}] \;=\; 2k^2-[2 \text{ if } k \text{ is odd}],$$

so for a $k \times k$ checkerboard of side length c, the tour has length

$$2ck-[2c/k \text{ if } k \text{ is odd}].$$

Under the assumption that color reversal is achieved by paper boundary, this tour length can be at most 4 for a unit square of paper. Hence, the maximum $k \times k$ checkerboard from a unit square has side length

$$c \le \frac{1}{k/2-[2/k \text{ if } k \text{ is odd}]}.$$

If correct, this limit would prove optimality of existing foldings for $k=4$ (Dureisseix 2000; Guy and Venables 1979, p. 32) and $k=8$ (Dureisseix 2000), but would not be tight for $k \le 2$. For $k=2$, the limit predicts that the optimal 2×2 checkerboard has the same size as the original square. Figure 15.11 shows that indeed the desired color reversals are possible without any reduction in square size, but obviously not while attaining a full checkerboard.

---

[3] Also in personal communication with Robert Lang and John Montroll, June 1997.

16 The Tree Method

The *tree method of origami design* is a general approach for "true" origami design (in contrast to the other topics that we discuss, which involve less usual forms of origami). In short, the tree method enables design of efficient and practical origami within a particular class of 3D shapes. Some components of this method, such as special cases of the constituent molecules and the idea of disk packing, as well as other methods for origami design, have been explored in the Japanese technical origami community, in particular by Jun Maekawa, Fumiaki Kawahata, and Toshiyuki Meguro. This work has led to several successful designs, but a full survey is beyond our scope (see Lang 1998, 2003). It suffices to say that the explosion in origami design over the last 20 years, during which the majority of origami models have been designed, may largely be due to an understanding of these general techniques.

Here we concentrate on Robert Lang's work (Lang 1994a,b, 1996, 1998, 2003), which is the most extensive. Over the past decade, starting around 1993, Lang developed the tree method to the point where an algorithm and computer program have been explicitly defined and implemented: TreeMaker is freely available and runs on most platforms.[1] Lang himself has used it to create impressively intricate origami designs that would be out of reach without his algorithm. Figure 16.1 shows one such example.[2]

16.1 ORIGAMI BASES

The first idea, known to most origamists, is to decompose the folding process into two phases: (1) fold a "base" that puts roughly the right amount of paper in roughly the right places, and (2) "shape" the base into the actual origami model. There is no precise definition of what constitutes a base; rather it is a useful intermediate folding. Most simple-to-intermediate origami designs, and most origami designs before 20 years ago, use only a handful of standard origami bases, shown in Figure 16.2. It is surprising the extent to which these simple bases have been stretched to accommodate relatively complicated models, although these models are fundamentally limited to having only four long "limbs" unless the bases are modified.

This limitation inspired development of the tree method, which enables design of arbitrarily complex bases with many limbs of varying lengths to form any desired

---

[1] http://www.langorigami.com/science/treemaker/.
[2] See http://www.langorigami.com/ for more of Lang's creations.

Figure 16.1. Lang's *Scorpion varileg*, opus 379.

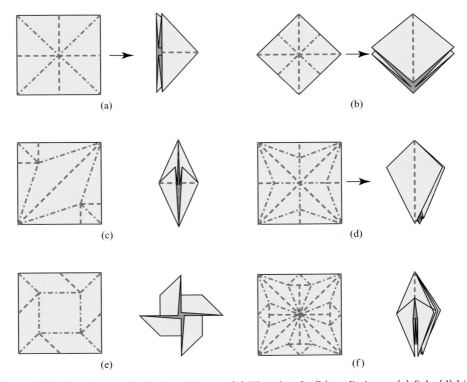

Figure 16.2. The six standard origami bases. (a) Waterbomb; (b) preliminary; (c) fish; (d) bird; (e) windmill; (f) frog.

"stick figure" (metric tree). Given such a powerful base, the second phase of the folding— shaping—becomes relatively "easy," at least to an experienced origamist. Figure 16.3 shows a simple example. It is the shaping that introduces most of the artistic quality of an origami model, so naturally this phase is difficult to capture algorithmically. Thus, the tree method focuses on the design of the origami base, which is what humans find the most difficult, and which turns out to have a rich geometric and algorithmic structure.

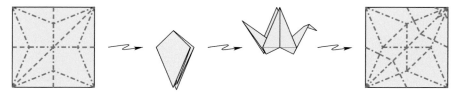

Figure 16.3. Folding a bird base, and "shaping" it into a crane.

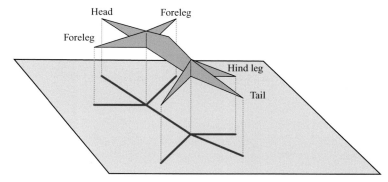

Figure 16.4. Uniaxial base for a lizard with four legs, head, body, and tail. The bottom of the base is placed against the xy-plane, but here drawn above to illustrate the projection. [Based on Figure 5.1 of Lang 1998.]

16.2   UNIAXIAL BASES

The tree method allows us to design a class of bases called "uniaxial bases." See Figure 16.4 for an example. Roughly speaking, a uniaxial base is a folding that lies above a plane, except where it touches the plane, which is exactly its projection into the plane, and that projection is a tree—the "shadow tree." The region of paper projecting to an edge of the shadow tree are "flaps" that can be rotated around "hinges"—vertical creases projecting to an internal vertex of the tree. While this description matches the intuition behind uniaxial bases, the precise definition is more complicated in order to allow more complicated flap structure: distinct flaps that project to the same line segment, flaps that touch but do not share a hinge, and flaps that wrap around other flaps.

Precisely, a folded state is a *uniaxial base* (Lang 1996) if the faces of the crease pattern can be partitioned into connected groups called *flaps* such that the following properties hold:

1. The entire base lies in the $z \geq 0$ half-space.
2. The projection of the base into the xy-plane is precisely the intersection of the base with the xy-plane.
3. Each flap of the base projects to a line segment in the xy-plane, called a *shadow edge*. In particular, every face of the crease pattern is perpendicular to the xy-plane in its folded position.
4. Any crease shared by faces from two different flaps, called a *hinge crease*, projects to a single point in the xy-plane, and that point is an endpoint of both shadow edges of the two flaps. In particular, the hinge creases (forming the boundary of all flaps, other than the paper boundary) are all vertical in their folded position.

5.   The following graph is a tree, called the *shadow tree*: the edges of the graph are the shadow edges (i.e., one per flap), and two shadow edges are incident in the graph at a pair of endpoints if the corresponding flaps share a hinge crease that projects to the same point as the two endpoints. The vertices of the shadow tree are called *shadow vertices*. Each internal (nonleaf) shadow vertex corresponds to a connected collection of hinge creases that project to a common point, which together form a *hinge*.

6.   Every leaf of the shadow tree is the folded position of only one point of paper.

Intuitively, a flap is a region of paper that can be "manipulated relatively independently" (Lang 2003, p. 52), and the hinges are the mechanism for this manipulation. These bases are called *uniaxial* because if we fold along the hinges so that the base lies in a single vertical plane, collapsing the shadow tree down to a single line, then the flaps all have edges along a common "axis," the shadow line.

For example, all but two of the bases in Figure 16.2 are uniaxial in their "natural" partition into flaps. The waterbomb base (a) is one of the simplest uniaxial bases; its shadow tree is an equilateral star with four points. The fish base (c), the bird base (d), and the frog base (f) are uniaxial after folding the paper in half along the vertical line of symmetry. On the other hand, the preliminary base (b) has four flaps, each with two faces, so the hinges are the vertical creases, forcing the xy-plane to slice along a horizontal line, violating either Condition 1 or Condition 2. The windmill base has four flaps that naturally form "two axes."

16.3   EVERYTHING IS POSSIBLE

Given the general results from Chapter 15 (p. 232), we already know that every *metric tree* (unrooted, unordered tree with a specified length along each edge) is an achievable shadow tree of a uniaxial base. However, uniaxial bases constructed with these methods are extremely inefficient and impractical. In contrast, Lang solves the problem via a completely different method with the following, much stronger (albeit qualified) result:

Theorem 16.3.1 (Lang 1996). *There is an algorithm that, given any convex piece of paper P and any metric tree T, constructs a crease pattern that folds P into a uniaxial base whose shadow tree is the largest possible scaled copy of T, ignoring possible self-intersection of the paper.*

Thus, the crease pattern has a "folded state" that corresponds to an optimal uniaxial base, but it remains open whether this folded state avoids self-intersection. The conjecture is that indeed the construction avoids self-intersection, and this conjecture has been verified experimentally on many examples of real-world origami design (see Lang 1998, 2003):

Conjecture 16.3.1 (Lang 1996, 2003). *Self-intersection never arises in Theorem 16.3.1.*

The running time of the algorithm is also large. As we will see, the algorithm uses a general form of nonlinear optimization, forcing exponential running time for optimal solutions. Fortunately, there are many practical heuristics that obtain reasonable approximate solutions, making the theorem practical (in a suboptimal form).

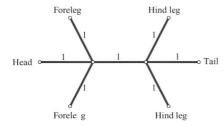

Figure 16.5. The metric tree that should be specified as the desired shadow tree to obtain the base in Figure 16.4. [Based on Figure 5.2 of Lang 1998.]

Modulo these issues, Lang's result says that you can design an efficient origami base that matches any desired "stick figure," with arbitrarily many limbs of appropriate length in appropriate arrangement. For example, the tree in Figure 16.5 can be used for a lizard (or other animals with these features and proportions), a star of eight long edges and two short edges can be used for a spider, a more complex tree can be used for a deer with antlers, etc. See again Figure 0.2 (p. 3) for such a complex example.

In the remainder of this chapter, we detail Lang's method.

16.4   ACTIVE PATHS

Consider a hypothetical uniaxial base folded from a convex piece of paper P and having the desired shadow tree. Each point on the shadow tree, viewed as an intersection point between the base and the xy-plane, can be mapped back to one or more points on P that fold to v (by Condition 2). Suppose that we draw a line segment s on P between two points that fold to two different points of the shadow tree. When we fold P into the base, the segment s maps to some continuous polygonal curve on the surface of the base. If we project this polygonal curve to the xy-plane, and possibly remove some sections of the projected curve, we obtain the *shadow path* (i.e., the path in the shadow tree) connecting the two points. Because projection and removal only shorten a curve, the shadow path is no longer than the original polygonal curve, which was the same length as the line segment s on the unfolded paper P. Thus we obtain the following key lemma in tree theory:

Lemma 16.4.1 (Lang 1996). *In any uniaxial base folded from a convex piece of paper P, the distance between any two points on the shadow tree is at most the distance between corresponding points on P.*

This lemma specifies a collection of necessary constraints between all pairs of points of the shadow tree. These constraints form an *invariant* that must be maintained throughout various reductions made during the algorithm.

We will concentrate on the invariant for all pairs of leaves of the shadow tree, in particular because these inequalities imply that the invariant holds for all pairs of points. Leaves have the additional property that only one point of paper maps to a leaf of the shadow tree (Condition 6). A path between two leaves of the shadow tree is called *active* if the inequality in Lemma 16.4.1 holds with equality, that is, the length of the shadow path is precisely the distance between the corresponding points on the unfolded paper. In this case, an *active path* can refer to either the path in the shadow tree (as usual) or the corresponding line segment on the unfolded piece of paper.

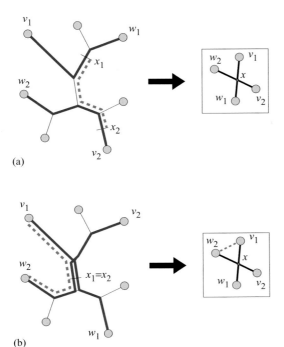

Figure 16.6. Illustration of the proof of Lemma 16.4.2. (a) The cross point x of two active paths on the paper must derive from a single point $x_1 = x_2$ on the paths, or else the path (x_1, x_2) violates Lemma 16.4.1; (b) crossing active paths violate Lemma 16.4.1 on the cross-path (v_1, w_2).

One basic fact about active paths is that they form a planar subdivision of the piece of paper:

Lemma 16.4.2. *If two active paths intersect on the piece of paper, then either they are collinear or they intersect only at common endpoints.*

Proof: Suppose for contradiction that (v_1, w_1) and (v_2, w_2) are active paths whose corresponding line segments on the piece of paper are noncollinear and intersect at a point x different from both v_1 and w_1. (Also, x must be distinct from either v_2 or w_2.) Let x_i denote the point along the path (v_i, w_i) in the shadow tree corresponding to the intersection point x on the piece of paper (see Figure 16.6(a)). Because x_1 and x_2 map to the same point x on the piece of paper, Lemma 16.4.1 implies that they must be the same point in the shadow tree. Thus the two active paths in the shadow tree intersect at $x_1 = x_2 = x$, as in Figure 16.6(b).

We claim that one of the cross-paths (v_1, w_2), (v_2, w_1), (v_1, v_2), (w_1, w_2) violates Lemma 16.4.1. Because the tree path from v_i to w_i passes through x, v_i and w_i must be on "opposite sides" of x in the sense that they would be in different components if we split the tree at x. By possible relabeling, we can make v_i and w_j on opposite sides of x, for all i, j. Because x is distinct from three of v_1, w_1, v_2, and w_2, even after relabeling we can find a v_i and w_j distinct from x, say v_1 and w_2 (refer to Figure 16.6(b)). The cross-path (v_1, w_2) of the shadow tree passes through x, so the cross-path length is the sum of the tree distance from v_1 to x plus the tree distance from x to w_2. Because (v_1, w_1) and (v_2, w_2) are both active paths, this sum equals the sum of the unfolded distance from v_1 to x plus the unfolded distance from x to w_2. By the triangle inequality, and because (v_1, x) and (x, w_2) are not collinear, this sum is strictly greater than the unfolded distance from v_1 to w_2. Combining this chain of inequalities, the tree length of the cross-path (v_1, w_2) is strictly greater than the unfolded distance from v_1 to w_2, contradicting Lemma 16.4.1.   □

16.5   SCALE OPTIMIZATION

The first step in tree theory, called *scale optimization*, is to minimize the extent to which the shadow tree must be scaled to "fit" on the piece of paper. The invariant of Lemma 16.4.1 plays a key role here. Suppose we arbitrarily chose distinct points on the unfolded piece of paper, and assigned each point a shadow leaf to which it will fold. This assignment may violate Lemma 16.4.1: some distances on the piece of paper might be smaller than the distances in the shadow tree. Such violations can be fixed by uniformly scaling the shadow tree smaller by a factor of

$$\max_{i \neq j} \frac{\text{distance in shadow tree between leaf } i \text{ and leaf } j}{\text{Euclidean distance between point } i \text{ and point } j}. \tag{16.1}$$

Based purely on the information from Lemma 16.4.1, our goal for scale optimization is thus to assign points of paper to leaves of the shadow tree to minimize this objective. This optimization problem can be phrased with a system of constraints as follows:

> maximize λ
> subject to   Euclidean distance between point i and point j
> $\geq \lambda \cdot$ distance in shadow tree between leaf i and leaf j,   (16.2)
> for all $i \neq j$,
> and   every point i lies within the convex piece of paper.

Here the variables are λ and the coordinates of each point i. The shadow tree, the distances it defines, and the convex piece of paper are all given and fixed.

Unfortunately, the first family of constraints is nonconvex: each constraint requires a point j to be outside a ball centered at i, which leaves a nonconvex "anti-ball." In general, such problems can be computationally intractable, although the general exponential-time polynomial-space algorithms from Part I, Section 2.1.1 (p. 17) can solve them. Despite the lack of efficient algorithmic results, reasonable solutions can be found by a variety of known techniques for nonlinear optimization. For example, TreeMaker 4 (Lang 1998) uses the CFSQP code developed at the University of Maryland, while other versions use an implementation of the Augmented Lagrangian Multiplier algorithm. Applied to the shadow tree of Figure 16.5 (p. 244), we obtain the mapping shown in Figure 16.7 with a scale factor of about 0.267.

The scale-optimization problem includes as a special case the well-studied problem of *equal-radius disk packing*: given n disks of equal radii, how small must they be scaled to fit in a unit square (say)? This problem arises when the shadow tree is a star graph with n equilateral edges. The optimal equal-radius disk packings are known for n up to 20, but not beyond (Friedman 2002, Nurmela and Östergøard 1997, Peikert 1994). Scale optimization is also more general than general disk packing (not just equal-radius), making it particularly challenging to solve to optimality.

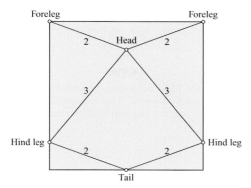

Figure 16.7. A likely optimal mapping of leaves from the shadow tree in Figure 16.5 to points on a square of paper. Line segments denote active paths or paper boundary. [Based on Figure 5.6 of Lang 1998.]

16.6   CONVEX DECOMPOSITION

What good is a solution to the particular formulation of scale optimization above if these mappings of leaves to points cannot be realized by a base? Tree theory establishes a surprisingly powerful strengthening of Theorem 16.3.1. Essentially, it says that the invariant on shadow-leaf placement is all that is needed, modulo one extra constraint that the resulting active paths decompose the unfolded piece of paper into only *convex* polygons:

Theorem 16.6.1. *Given any convex piece of paper P, any metric tree T, and any assignment of leaves to points on P satisfying the invariant of Lemma 16.4.1 and such that the active paths decompose the paper into convex polygons, there is a uniaxial base folded from P whose shadow tree is T and in which the specified points fold to shadow leaves, ignoring possible self-intersection of the paper.*

Given this theorem (whose proof we discuss in Section 16.8), all that remains is to somehow arrange for the active paths to decompose the piece of paper into only convex polygons. There are multiple ways to achieve this property from a given layout. One approach that is useful in practice is to permit modification of edge lengths in the shadow tree. It can be shown that such length modifications always suffice to "convexify" the active-path decomposition. By incorporating such relaxation into scale optimization and minimizing the resulting "strain," the amount of change in the shadow tree is often fairly small in practice. In addition, the error introduced by the change in edge lengths can be later corrected in the folded uniaxial base by "crimping" the corresponding flaps to their correct size. However, this crimping effectively reduces the scale factor below optimal.

Another approach to achieving a convex decomposition, while preserving optimality of the scale factor, is to add extra leaves to the shadow tree with particular assignments to points on the unfolded piece of paper. By careful placement, these extra shadow leaves "take up the slack" and make any decomposition convex without disturbing the mapping of existing shadow leaves to points of paper. The extra flaps connecting these leaves to the rest of the shadow tree can later be ignored (and effectively removed) by folding them against the rest of the uniaxial base.

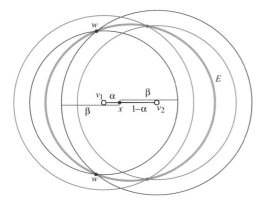

Figure 16.8. Satisfying placements of leaf w lie along ellipse E. The two points labeled w correspond to the subdivision point x.

Lemma 16.6.2. *Given any convex piece of paper P, any metric tree T, and any assignment of leaves to points on P satisfying the invariant of Lemma 16.4.1, there is another metric tree $T' \supseteq T$ and an assignment of the new leaves to points on P satisfying the invariant of Lemma 16.4.1, such that the resulting active paths decompose P into convex polygons.*

Proof: We prove the stronger result that the active paths can be arranged to decompose the piece of paper into triangles, as argued in Lang (2003, Sec. 11.7) and in Lang (1998). Consider any polygonal region R formed by more than three active paths and paper boundaries. Suppose that (v_1, v_2) is one of the active paths, and conceptually scale so that the length of this shadow path is 1. Imagine subdividing an edge at point x along this shadow path, with x at distance α from v_1 (and $1 - \alpha$ from v_2), and adding a new leaf w at distance β from x. For each choice of α and β, with $0 \le \alpha \le 1$ and $\beta \ge 0$, there are at most two points in the plane at which w could be placed that make w active with both v_1 and v_2, corresponding to the intersection of two circles centered at v_1 and v_2 of radii $\alpha + \beta$ and $1 - \alpha + \beta$, respectively. For example, with $\alpha = \beta = 0$, the satisfying placement of w is at v_1. If we increase α but keep $\beta = 0$, the satisfying placement of w moves along the active line segment from v_1 to v_2. More generally, the satisfying placements of w are precisely the points along the ellipse of major axis $1 + 2\beta$ and foci v_1 and v_2 (see Figure 16.8). Therefore we can vary α and β to make any point in the plane (or R) a satisfying placement for w, and then make an active triangle v_1, v_2, w.

We now consider when w first becomes active with other vertices while moving along some path inside R. For w sufficiently close to the line segment (v_1, v_2), w is not active with any other vertices. By Lemma 16.4.2, w cannot touch another active path before becoming active with another vertex. Thus, if R has another active path in addition to (v_1, v_2), then w can be made active with another vertex v_3 while remaining inside R. If w cannot be made "triply active" in this way, we place w at any vertex of R other than v_1 and v_2. This placement subdivides R into the triangle v_1, v_2, w and two other pieces, each with at least one original edge of R and with one new (active) edge. Hence, each piece has fewer sides than the original polygon R because the new edge is compensated by the two removed edges from the other pieces.

Now suppose w can be made triply active with vertices v_1, v_2, and v_3. Then we move w along the circle that makes w active with v_3. If w reaches the boundary of R without becoming active with a fourth vertex, then this placement of w subdivides R into four pieces each with at least one original edge of R (not counting the edge that w lies along, which is shared by two pieces) and with at most two new (active) edges. Hence, each piece has fewer sides than the original R because the two new edges are compensated by the three removed edges from the other pieces. Otherwise, at some point along the motion of w, w becomes active with a fourth vertex v_4 while remaining inside R. For this placement of w, we have at least four active paths, decomposing the polygon R into at least four pieces each with at least one original edge of R and with at most two new (active) edges, so again we decompose R into pieces each with fewer edges than R.

By induction on the maximum number of sides of any polygon in the active-path decomposition, multiplied by the number of polygons with that many sides, we eventually decompose the polygon into triangles. $\quad\square$

Together, Theorem 16.6.1 and Lemma 16.6.2 establish Theorem 16.3.1. The remainder of this section is therefore devoted to the proof of Theorem 16.6.1.

16.7   OVERVIEW OF FOLDING

Suppose we have already obtained a mapping of leaves of the shadow tree to points on the piece of paper. We can then find (some of) the points on the piece of paper that correspond to internal nodes of the shadow tree, using the correspondence between active paths in the piece of paper and paths between leaves in the shadow tree. Namely, we can mark points along the active paths in the piece of paper corresponding to internal nodes in the shadow tree, by measuring corresponding lengths in the shadow tree (see Figure 16.9).

The key property of active paths as a decomposition tool is that they localize incident creases as particular points on the shadow tree, causing incident creases from either side to "line up." For example, each of the marked points along an active path corresponding to internal nodes of the shadow tree must have an incident crease, perpendicular to the incident active path, corresponding to a hinge in the uniaxial base. In our recursive construction, such creases will already be present on either side of the active path, and the active path forces these creases to line up. More generally, if we can fold each region between active paths independently, then we will see how to attach these foldings along the active boundaries to form the desired uniaxial base.

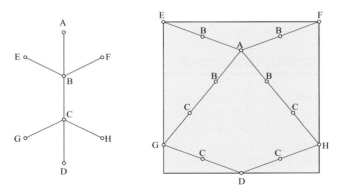

Figure 16.9. Location of all vertices of the shadow tree on a square of paper, using the leaf mapping from Figure 16.7. Line segments denote active paths or paper boundary. [Based on Figure 5.7 of Lang 1998.]

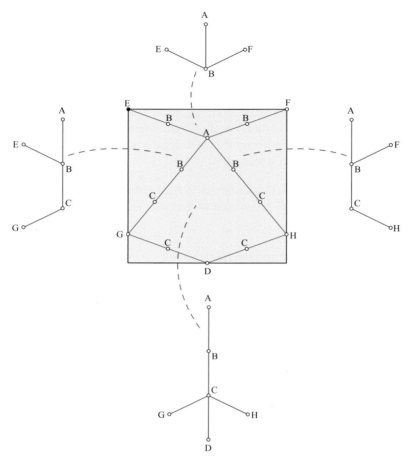

Figure 16.10. Subtrees induced by the convex decomposition of Figure 16.9, each solved by a universal molecule. [Based on Figure 5.8 of Lang 1998.]

Therefore, our problem decomposes into subproblems of building uniaxial bases for each of the convex polygons outlined by active paths and/or the boundary of the piece of paper (see Figure 16.10). This restriction to convex polygons enables us to use the "universal molecule" construction described in the next section.

16.8 UNIVERSAL MOLECULE

Given any convex polygon of paper whose vertices correspond to leaves in a specified shadow tree, the *universal molecule* folds the desired uniaxial base from that convex piece of paper. In fact, the universal molecule has the additional property that the intersection of the base with the *xy*-plane (what forms the shadow tree) consists of precisely the entire boundary of the piece of paper.

The basic idea is to construct the uniaxial base as a continuum of vertical slices or cross sections, starting at $z = 0$ where the cross section is precisely the paper boundary, and continuously increasing z (see ahead to Figure 16.11(b)). We define the uniaxial base by specifying that sliding this horizontal plane corresponds to *shrinking* or

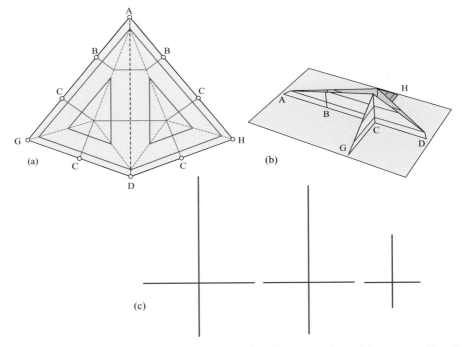

Figure 16.11. (a) Convex polygon shrinking: original and two snapshots; (b) corresponding slices in 3D; (c) shadow trees: original plus the two corresponding trimmed trees.

offsetting the convex polygon so as to keep each edge parallel to the original edge and to keep the perpendicular distance between offset edges and corresponding original edges equal across all edges. Each of these reduced polygons corresponds to higher z cross sections of the desired uniaxial base. Indeed, because all facets are vertical in a uniaxial base, the z-coordinate of the cross section is exactly the perpendicular distance between offset edges and original edges. At the same time as this shrinking, we imagine *trimming* the corresponding shadow tree near the leaves to represent the lengths of the projection already achieved. More precisely, we shorten the lengths of shadow-tree edges incident to each leaf by the same amount that a polygon edge decreases in length locally around the corresponding polygon vertex. (Thus, different shadow-tree edges shrink by different amounts, depending on the angle of the corresponding polygon vertex.) At each level of cross section, we define the folding to map the current shrunken polygon to the current trimmed shadow tree according to a simple doubling (Euler tour) around the tree.

All nonhorizontal creases in the uniaxial base must be single points in cross section, that is, must be vertices of the slices. Conversely, if we track the trajectories of the original polygon vertices, these form the main *ridge* creases in the uniaxial base. All ridge creases are mountains because they locally connect facets that touch the xy floor and hence these facets must be below the crease.

Two types of discrete *events* can arise during this cross-section-sliding/polygon-shrinking/shadow-tree-trimming process. (Several events can happen simultaneously, but we can consider them one at a time.)

The first type of event is that an edge of the polygon may shrink to zero length, causing two vertices to coincide. This event coincides precisely with when we completely trim off

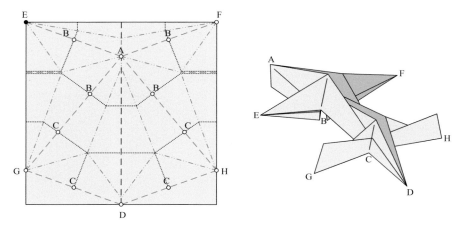

Figure 16.12. The crease pattern and resulting uniaxial base with the tree specified in Figure 16.4 by filling universal molecules into the active polygons in Figure 16.9. [Based on Figure 5.18 of Lang 1998.]

two leaves and the incident edges of the shadow tree, exposing a new leaf (previously the common parent of the two trimmed leaves). In this case, the shadow tree and horizontal cross section become combinatorially simpler, with one fewer vertex, but otherwise the shrinking process continues as before. In the shrinking polygon, the new coinciding vertex will move along a new trajectory, an angular bisector between the two edges that just became incident, creating a new crease.

The second type of event is that the line segment between two nonadjacent vertices of the polygon may become an active path in the trimmed shadow tree, so that any further shrinking would violate the constraints of Lemma 16.4.1. (For adjacent vertices, active paths will remain active and hence no constraints will be violated.) In this case, we introduce the active path as a crease, split the cross section, polygon, and shadow tree into two pieces, and continue the shrinking process with the two pieces in parallel. (In fact, there is still only one cross section, but it becomes disconnected, and we can treat each component separately.)

Figure 16.11 illustrates this shrinking process on the central quadrilateral of Figure 16.9 to form the crease pattern in Figure 16.12. Two slice snapshots are shown. The first slice corresponds to the second type of event above, where the vertical line segment becomes active, cutting the quadrangle into two triangles. The second slice is halfway through the remaining slices. At the final topmost slice, not shown because then everything disappears, we effectively have three events of the first type.

The final component of the construction of a universal molecule is to introduce *perpendicular creases*. For each vertex formed by the first type of event, which includes the final case when the polygon shrinks to a single vertex, we introduce three perpendicular creases between each consecutive pair of ridge creases from the vertex perpendicular to the three polygon edges that now coincide, including one edge of zero length and two previously separated edges of positive length. (In fact, if more than one event of the first type happen in the same area, more than three perpendicular creases are introduced, but the construction is otherwise the same.) In addition, from each marked point along an active boundary of the convex polygon (corresponding to an internal node of the shadow tree), we add a perpendicular crease to the first intersection with a ridge crease. If that hinge crease bisects the angle between two active boundaries of the convex polygon, then two perpendicular creases corresponding to the same internal shadow vertex come together at that point. On the other hand, if one of the bisected boundaries is nonactive, we have to explicitly add a crease perpendicular to the nonactive boundary incident to each perpendicular crease touching the ridge crease. In this way, perpendicular creases always come in pairs on either side of a ridge crease.

This completes the construction of the universal molecule. Correctness follows from the property that we simultaneously track cross sections of the folded state, the piece of paper, and the shadow tree. However, this property requires a careful analysis of the details, which we do not attempt here. Also, when we join two universal molecules along an active path, we may need to introduce additional perpendicular creases in one molecule to match the endpoints of perpendicular creases in the other molecule. Together, this completes our description of the proof of Theorem 16.6.1 and thus Theorem 16.3.1.

For example, if we apply the universal molecule construction to the convex polygons in Figure 16.10, we obtain the crease pattern in Figure 16.12. The triangles are creased by three angular bisectors and three perpendiculars; no new paths can become active in a triangle. In the quadrangle, as shown above, the vertical line segment becomes active, resulting in two triangles, which are then creased like the others.

Ongoing work with Robert Lang and Martin Demaine pursues a proof that there is no self-intersection (Conjecture 16.3.1) as well as the following technical advance to avoid modifying the shadow tree as in Section 16.6:

Conjecture 16.8.1. *The universal molecule can be generalized to nonconvex polygons to prove Theorem 16.6.1 without the proviso of convex decomposition, thereby proving Theorem 16.3.1 (and Conjecture 16.3.1) without Lemma 16.6.2.*

17 | One Complete Straight Cut

Suppose you take a sheet of paper, fold it flat however you like, and then make one complete straight cut wherever you like. The result is two or more pieces of paper which, when unfolded, form some polygonal shapes. Figure 17.1 shows how to make the simple example of a 5-pointed star. More surprisingly, it is possible to obtain a polygonal silhouette of a swan, angelfish, or butterfly, or to arrange five triangular holes to outline a star (see Figure 17.2). The *fold-and-cut problem* asks what polygonal shapes are possible, and how they can be arranged.

History. This problem was first posed in 1960 by Martin Gardner in his famous Mathematical Games series in *Scientific American* (Gardner 1995a). Being attuned to the magic community, Gardner was aware of two magicians who had experimented with fold-and-cut magic tricks: Harry Houdini, whose 1922 book *Paper Magic* (Houdini 1922) includes one page on how to fold and cut a regular 5-pointed star; and Gerald Loe, whose 1955 book *Paper Capers* (Loe 1955) is entirely about the variety of (largely symmetric) shapes that Loe could fold and cut. In fact, the fold-and-cut idea goes farther back: a 1721 Japanese puzzle book by Kan Chu Sen (1721) poses and later solves a simple fold-and-cut puzzle; and an 1873 article about the American flag (National standards and emblems 1873) tells the story of Betsy Ross convincing George Washington to use a regular 5-pointed star on the American flag because it was easy to produce by fold and cut.

Result. The surprising outcome to the fold-and-cut problem is a universality result: every plane graph of desired cuts can be made by folding and one cut. Here "plane graph" refers to any embedded planar graph, or in other words, any collection of (straight) line segments in the plane (possibly intersecting). Thus, with a single cut, we can produce a desired polygon, or several polygons arranged in a desired pattern, or polygons with polygonal holes, or an arrangement of slits, and so on—produced exactly, cutting all the edges and nothing more.

The fold-and-cut problem is equivalent to the following problem: given a plane graph drawn on a piece of paper, find a flat folding of the paper in which all the vertices and edges of the graph map to a common line, and nothing else maps to that line. Thus, cutting along the line would produce precisely the desired graph of cuts. It is this form of the problem that is most closely aligned with the tree method described in Chapter 16.

In addition to alignment of the plane graph, we can also specify which faces of the graph go above the line and which faces go below. When the graph is face two-colorable, which is the case for one or more polygons with holes for example, then it is most

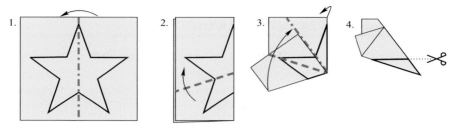

Figure 17.1. How to fold a square of paper so that one cut makes a regular 5-pointed star.

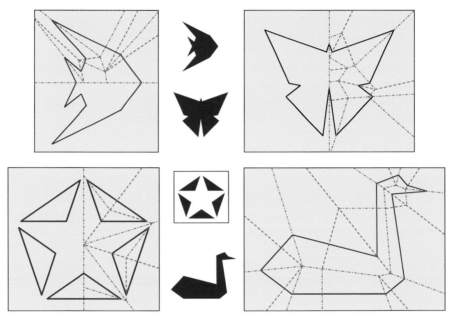

Figure 17.2. Examples of cut patterns achievable by folding and one complete straight cut (via the straight-skeleton method). All but the swan starts by folding in half along the line of symmetry. [Based partly on Figure 3–4 of Demaine et al. 2000a.]

natural for the above/below specification to match such a coloring, because this makes scissors have material on both sides of the cut. But when the graph has vertices of odd degree (or isolated points), it is unavoidable to have to cut right along a crease; in this case, no above/below specification is perfect, and just setting all faces to "above" is conceptually simplest.

There are two methods for folding plane graphs into alignment. The first method (Demaine et al. 1999a, 2000) is based on a geometric structure called the straight skeleton, which captures the local symmetries of the plane graph, and is related to the shrinking process used in the universal molecule of Lang's tree method (Section 16.8, cf. Figure 16.11, p. 251). The second method (Bern et al. 1998, 2002) is based on a different geometric structure known as disk packing, which effectively decomposes the faces of the graph into triangles and "nice" quadrangles, and is related to the scale-optimization aspect of the tree method (Section 16.5). The first method produces easier foldings in practice, but does not apply universally; the second method is better theoretically, applying universally. We describe each method in turn.

17.1   STRAIGHT-SKELETON METHOD

The straight-skeleton method was developed by Demaine et al. (1999a, 2000). Our description begins with the crease pattern of the folding. Then we sketch the proof of foldability for a special case, and mention some of the difficulties in the general case.

17.1.1   Straight Skeleton

The intuition behind the straight-skeleton method can be seen from an almost trivial example: the graph of desired cuts is just two boundary-to-boundary line segments, and for simplicity not meeting on the piece of paper (see Figure 17.3). In this case, one fold suffices—along the *bisector* of the two lines, that is, the line bisecting the angle where the two lines meet. Intuitively, this fold works because folding by 180° along a line corresponds to reflecting through that line.

    The *straight skeleton* generalizes this simple idea to arbitrary plane graphs. In general, the straight skeleton consists of several line segments (edges), each of which is a subsegment of a bisector of two graph edges. (We use the term *graph edges* to refer to the edges of the graph of desired cuts.) Intuitively, each of these straight-skeleton edges locally aligns a pair of graph edges, and together the entire straight skeleton aligns all the graph edges.

    The straight skeleton was introduced by Aichholzer et al. (1995) for the special case of simple polygons, and was later generalized by Aichholzer and Aurenhammer (1996) to arbitrary plane graphs. Its definition is as follows (refer to Figure 17.4). Suppose we simultaneously shrink each face of the graph in such a way that the edges retain their orientation, and the perpendicular distance from every shrunken edge to the corresponding original edge is the same for all shrunken edges. When an edge e shrinks to a

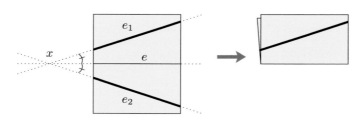

Figure 17.3. One fold aligns two graph edges.

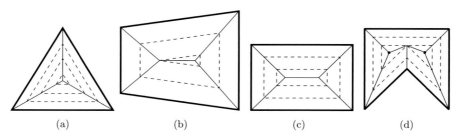

(a)   (b)   (c)   (d)

Figure 17.4. Shrinking a face of the graph to form the straight skeleton. (a) Triangle; (b) convex quadrangle; (c) rectangle; (d) nonconvex pentagon. Graph edges are thick; shrunken copies are dashed; skeleton edges are thin and solid.

Figure 17.5. The straight skeleton of a turtle. Graph edges are thick; skeleton edges are thin.

 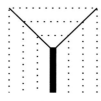

Figure 17.6. The behavior of the shrinking process (dotted) and the straight skeleton (thin) locally around a degree-0 or degree-1 cut vertex (thick).

point, we continue shrinking elsewhere but henceforth ignore e. When a face collapses to a point, we stop shrinking that face. When a face becomes nonsimple, because a vertex touches an edge as in Figure 17.4(d), we split the face into two components and continue shrinking each piece. The *straight skeleton* is the union of the trajectories of the vertices during this shrinking process.

Figure 17.5 shows a more complicated example: the straight skeleton of a plane graph with two faces (the interior and exterior of a turtle). In this example, we can see what happens when we shrink the outside face of a plane graph: the polygon itself "grows" (in area), but the exterior shrinks (in area if we clip to a bounding rectangle, say). A face of the skeleton—a maximal region bounded by skeleton edges—spans multiple faces of the graph of cuts, because the straight skeleton does not include the graph edges.

To build intuition about the straight skeleton, we now present some structural results and describe how the skeleton relates to bisecting graph edges. The proofs are all straightforward, and somewhat beside our goal, so we omit them.

Lemma 17.1.1 (Aichholzer and Aurenhammer 1996). *The straight skeleton of a plane graph with n vertices has $O(n)$ vertices, edges, and faces.*

Lemma 17.1.2 (Demaine et al. 2000a, Lem. 2). *There is a one-to-one correspondence between graph edges and the skeleton faces that contain them.*

For this last lemma to hold, there are some technical details that arise when graph vertices have degree 0 or 1. Essentially, we view such a vertex locally as a small square or the end of a thin rectangle, respectively, so that when we shrink the face bounded by this vertex, short graph edges are introduced (see Figure 17.6).

Lemma 17.1.3 (Demaine et al. 2000a, Lem. 3). *Every skeleton edge is a subsegment of the bisector of the two graph edges contained in the two skeleton faces sharing the skeleton edge.*

17.1.2   Perpendiculars

While the straight skeleton is effective for aligning the graph edges, for most graphs, the straight skeleton by itself is not a flat-foldable crease pattern; there is no reason to expect each vertex to satisfy the flat-foldability lemmas from Section 12.2 (p. 198). For this reason, we introduce another type of crease, called a *perpendicular*, at least one of which emanates from every skeleton vertex. That these perpendiculars do lead to flat foldability will not be clear until Theorems 17.1.6 and 17.1.7. Naturally, to avoid interfering with the alignment already achieved by the straight skeleton, perpendicular edges always meet graph edges at a right angle. In fact, each perpendicular edge is contained in one skeleton face, and is perpendicular to the graph edge contained in that skeleton face.

We construct a connected component of perpendiculars from every skeleton vertex, by attempting to draw a perpendicular edge into every incident skeleton face (with the angle of the perpendicular edge determined by the unique graph edge contained in that skeleton face). In some cases, the perpendicular edge immediately exits the skeleton face (before strictly entering), or hits a skeleton vertex dead on, in which case we stop following the edge. Otherwise, when the perpendicular hits a skeleton edge, we reflect the perpendicular edge through the skeleton edge to obtain the beginning of a new perpendicular edge. This reflection process repeats until the perpendicular edge hits a skeleton vertex or goes to infinity.

Figure 17.7 shows an example of this construction. Not all of these perpendiculars are needed to fold the turtle. If the reader cares to experiment, we recommend following the crease pattern shown in Figure 17.8 instead, which eliminates many perpendiculars and deviates slightly from the exterior straight skeleton to achieve additional crease savings.

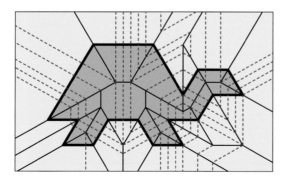

Figure 17.7. The straight skeleton (thin solid) and all perpendiculars (dashed) for the turtle from Figure 17.5.

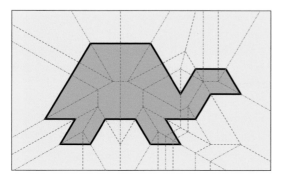

Figure 17.8. Mountain–valley pattern for the turtle crease pattern from Figure 17.7, slightly simplified.

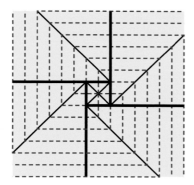

Figure 17.9. A simple example of spiraling. [Based on Figure 8 of Demaine et al. 2000a.]

17.1.3 Strange Behavior

Two interesting phenomena can happen just at the level of the crease pattern described thus far.

17.1.3.1 Spiraling

In some cases, the perpendicular construction algorithm never finishes, when given the freedom to act in the entire plane. Figure 17.9 shows a simple example. This behavior is not degenerate; perturbations of the graph in Figure 17.9 have the same property. There are also polygons with the same behavior. However, this behavior can be simply characterized: in most cases, after a finite number of edges, a perpendicular path repeatedly cycles around the same sequence of skeleton edges, spiraling outward toward infinity. In such cases, there are only finitely many creases in any finite-area piece of paper; however, the number of creases depends on the size of the paper.

17.1.3.2 Density

Unfortunately, there is a rare scenario in which a perpendicular path actually spirals indefinitely within a bounded region of the plane, trapped by a carefully placed circular corridor. This scenario is the only case in which the creases are *dense* in a region of the plane. See Figure 17.10 for the simplest known example, where perpendicular edges never meet each other head on because the original "corridor" widths are irrational multiples of each other.[1] We say that this situation is rare in the following sense:

Conjecture 17.1.1. *Given any graph, a random embedding of the graph is dense in a region of the plane with probability 0.*

If this conjecture were true, it would imply that almost any tiny perturbation of a given plane graph would avoid this problem. However, this issue dooms the straight-skeleton method to working only in limited cases (albeit only slightly limited).

17.1.4 Corridors

The straight skeleton, perpendiculars, and any graph edges between faces of the same above/below assignment together define the crease pattern for the straight-skeleton

---

[1] This example was found by E. Demaine, M. Demaine, and A. Lubiw in 2001 when trying to prove that it could not exist. Unfortunately, this example contradicts a claim in Demaine et al. (2000a, Lem. 7). For this reason, Demaine et al. (2000a, Thm. 1) claims that the straight-skeleton method can align an arbitrary graph, whereas the reality is the somewhat weaker result described here.

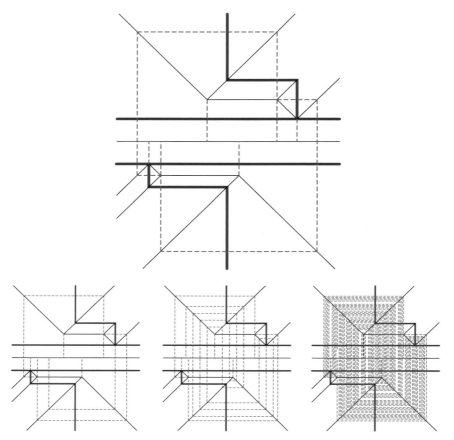

Figure 17.10. Example of dense behavior. Top shows plane graph (thick), straight skeleton (thin), and beginning of all perpendiculars (dashed). Bottom shows the progression of one perpendicular path, never ending.

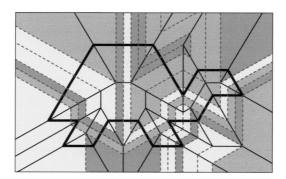

Figure 17.11. A three-coloring of the corridors resulting from the turtle.

solution to the fold-and-cut problem (aside from a few extra perpendicular creases arising in certain cases to be discussed later). It only remains to show that this crease pattern is flat foldable, or more precisely, has a flat folded state. A key structure in this proof is called a "corridor." A *corridor* is a region of the plane as decomposed solely by perpendiculars (see Figure 17.11 for an example).

Corridors have a relatively simple structure. While we will not go into the details here, an informal description is that a corridor is bounded by one or two *walls*, and if there

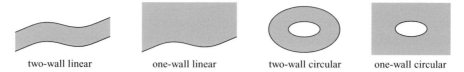

Figure 17.12. The four possible shapes of corridors. [Based on Figure 9 of Demaine et al. 2000.]

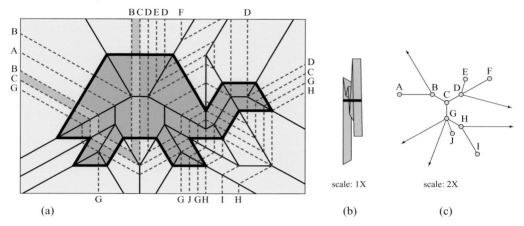

Figure 17.13. (a) Crease pattern for a turtle with perpendiculars labeled; (b) the shaded corridor folded into an accordion; (c) the shadow tree.

are two walls they track each other in parallel and maintain a constant perpendicular distance. Topologically, the interior of the corridor may be homeomorphic to a strip, a halfplane, an annulus, or a punctured plane (see Figure 17.12). The half plane and punctured plane correspond to the one-wall case, while the strip and annulus correspond to the two-wall case. If the walls of the corridor extend to infinity, we call the corridor *linear*; and if each wall cycles back to meet itself, we call the corridor *circular*.

17.1.5   Folded State for Linear Corridors

We describe a folded state of the entire piece of paper by first describing a flat folding of each corridor and then piecing the foldings together. We focus here on the special case in which all corridors are linear, as in the turtle. This situation seems to be common, in the following sense:

Conjecture 17.1.2. *Given any graph of maximum degree 2, a random embedding of the graph induces only linear corridors with probability 1.*

Each corridor folds by itself into a simple structure called an *accordion*. Figure 17.13(b) shows an example. In an accordion, each wall folds onto a line, because the straight-skeleton creases bisect the two relevant graph edges and hence also the two incident perpendicular edges. Thus, the entire accordion lies within an infinite strip, which we orient vertical. Any graph edges met by the accordion are perpendicular and thus horizontal, and furthermore by Lemma 17.1.3 they all align.

What remains is to describe how the accordions (i.e., the folded corridors) fit together in a consistent folded state. In the linear-corridor case, the connections between accordions is fairly straightforward. Because each accordion lies in a vertical strip, we can focus on the vertical projection of the accordions, resulting in a 1D structure (see Figure 17.13(c)). We call this structure the *shadow tree*, by analogy with the notion of shadow tree described in Chapter 16. Each two-wall corridor maps to a line segment in this tree; each one-wall corridor maps to a ray in this tree; and each connected component of perpendiculars maps to a vertex in this tree. The shadow tree is indeed a tree, because the removal of any line-segment edge corresponds to the removal of a two-wall corridor in the crease pattern, which causes disconnection.

Our claims so far about the connection between the shadow tree and the original piece of paper can be formalized as follows:

Lemma 17.1.4. *In the linear-corridor case, any flat folding of the shadow tree corresponds to a flat folding of the crease pattern that uses all creases except possibly for some perpendicular edges.*

Indeed, from any flat folding of the shadow tree, we can read off the mountain–valley assignment for perpendicular edges. Angles of 180° in the tree correspond to an unfolded portion of perpendiculars. Angles of 0° (valley) and 360° (mountains) specify the parity of a path of perpendicular edges, which alternate between mountain and valley.

The main difficulty in proving Lemma 17.1.4 is the analysis of a single point of connection between several accordions, that is, at a vertex in the shadow tree. It can be shown that the various faces of an accordion join in suitable ways to other faces along the wall edges. In particular, the joined edges are aligned vertically, the faces have matching orientations, and the collection of all joins is noncrossing. The noncrossing property relies on an additional detail that we have glossed over above: the shadow tree is not just a metric graph, but the edges around each vertex also have a natural orientation. A flat folding of such an oriented tree must obey these orientation constraints.

The final ingredient to constructing the folded state is the following lemma:

Lemma 17.1.5. *Every oriented metric tree has a flat folding.*

Proof: Root the tree at an arbitrary vertex, and hang the tree so that every edge points downward from the root (see Figure 17.14). □

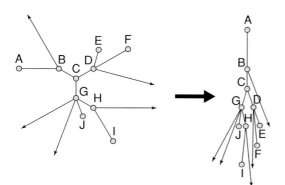

Figure 17.14. Flattening the tree from Figure 17.13.

This concludes the proof of the following theorem:

Theorem 17.1.6. *For any plane graph inducing only linear corridors, there is a flat folding of the crease pattern that uses all creases except possibly for some perpendicular edges, aligns the graph, and satisfies any specified above/below assignment.*

17.1.6 Circular Corridors

Circular corridors are much more complex to handle. The shadow tree no longer captures enough foldings to be appropriate in this situation, essentially because the accordions corresponding to circular corridors must be stuffed in the middle of another accordion (i.e., between two adjacent faces of an accordion). Even worse, the accordion corresponding to a single circular corridor may not be flat-foldable by itself, because the "wraparound" edge can cause a crossing, in some cases irrespective of which edge is chosen as the wraparound edge.

To handle this case, we allow only "normal" circular corridors. A circular corridor is *normal* if it satisfies two properties:

1. the graph edges contained in the skeleton faces crossed by the circular corridor all share a common endpoint, and
2. each wall of the corridor contains just one skeleton vertex.

Such circular corridors seem common in the following sense:

Conjecture 17.1.3. *Given any graph, a random embedding of the graph induces only normal corridors with probability 1.*

This conjecture implies Conjecture 17.1.1 (p. 259), implying that density is not an issue in this context.

The solution in this case involves introducing a more complex shadow model and more involved arguments that this model can be folded flat and that this folding corresponds to a valid flat folding of the piece of paper. This time, we need to introduce some extra perpendicular creases. In the end, we obtain the following theorem:

Theorem 17.1.7. *For any plane graph inducing only normal corridors, there is a flat folding of the crease pattern minus possibly some perpendicular edges and plus additional perpendicular edges that aligns the graph.*

17.2 DISK-PACKING METHOD

The disk-packing method was developed by Bern et al. (1998, 2002) shortly after the straight-skeleton method.[2] We first describe the method to cut out a graph that constitutes a single polygon P that does not touch the paper boundary, and generalize

---

[2] The descriptions of this method in Bern et al. (1998, 2002) have a common flaw, allowing molecules to "point" in opposite directions, which can cause the constructed folded state to self-intersect. In collaboration with the authors (September 2006), we have corrected the flaw, by forcing all molecules to point in the same direction and by adding a sixth and seventh step to the algorithm. We present this corrected algorithm.

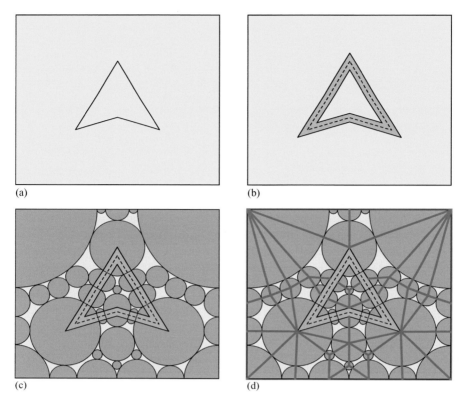

(a)   (b)

(c)   (d)

Figure 17.15. (a) Graph, a nonconvex quadrangle; (b) offsetting by $\pm\varepsilon$; (c) disk packing; (d) decomposition into triangles and quadrangles.

later (Section 17.2.7) to arbitrary graphs. The procedure follows five steps, the first three of which Figure 17.15 illustrates by example:

1. Parallel-offset the graph by a small ε, thickening the edges into a "ribbon" of width 2ε. Consider the *offset graph* formed by these offset edges and by the paper boundary.
2. Find a disk packing such that, in the offset graph,
   (a) every vertex is the center of a disk,
   (b) every edge is the union of radii of disks, and
   (c) every "gap" between disks has three or four sides.
3. Use the dual of this disk packing to decompose the faces of the offset graph into triangles and quadrangles with special properties. Label these triangles and quadrangles as *interior* or *exterior* according to their relation to the polygon P.
4. Fold each triangle and quadrangle by itself into a "molecule" with the property that the boundary of the triangle or quadrangle folds to a common line (see Figure 17.17).
5. Glue these individual folded states together into a global flat folding in which all exterior molecules' boundary edges lie along a common horizontal line, all interior molecules' boundary edges lie along a horizontal line 2ε above, and all molecules sit above their respective line (see Figure 17.16(a)).
6. Sink-fold each exterior molecule repeatedly in half along horizontal lines until its height is smaller than ε (see Figure 17.16(b)).

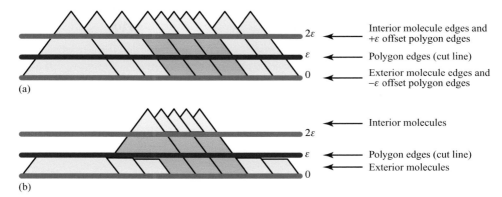

Figure 17.16. Schematic diagram of the flat folding of a polygon P, with exterior (blue) and interior (sand). The ribbon offset $\pm\varepsilon$ around P is shown pink. (a) Before and (b) after sink folding.

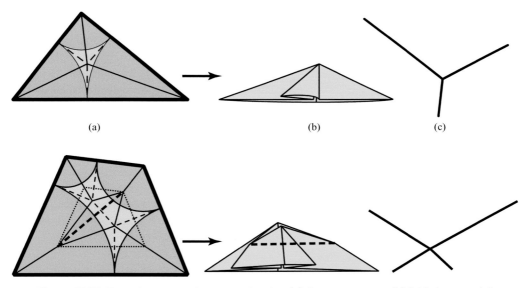

Figure 17.17. Triangle and quadrangle molecules. (a) Crease patterns; (b) folded states; (c) overhead views.

Because of the initial parallel offset of the graph in Step 1, the original graph edges align along a horizontal line ε above the base of the exterior molecules (see Figure 17.16(b)). The interior molecules are all above the horizontal line 2ε above, but in the folded state constructed in Step 5, the exterior molecules may rise above and therefore intersect the ε line. The purpose of Step 6 is to fold such molecules out of the way, so that only the original graph edges lie along the ε line. As a result, cutting along the line cuts precisely the graph edges, and none of the molecule boundary edges nor the molecule interiors.

Now that the desired outcome is clear, we revisit each of the six steps outlined above and explain how they are accomplished. As mentioned above, we focus first on the case of cutting out a single simple polygon, and then turn to the generalization to arbitrary graphs in Section 17.2.7.

17.2.1   Parallel Offset

The parallel offset is precisely the shrinking process during which the straight skeleton is formed. We just need to specify the allowable offset amount ε. In choosing this amount, we need to enforce the restriction that the ribbon folds simply, meaning that no creases terminate strictly interior to the ribbon. As a result, the ribbon folding will come "for free" as extensions of the molecules.

To achieve this restriction, we place an upper bound on the allowable offset ε, namely, it must be strictly less than two terms:

1. the smallest offset amount that causes a pair of vertices and/or edges to meet (when the offset first degenerates) and
2. the smallest offset amount that causes a vertex to touch the paper boundary (so that the offset does not penetrate the paper boundary).

17.2.2   Disk Packing

Disk packing has been developed as a powerful technique in computational geometry, largely in the context of mesh generation (Bern and Eppstein 2000). In general, a *disk packing* is collection of disks (of arbitrary radii) whose interiors are pairwise disjoint. A *gap* in the disk packing is a maximal connected planar region exterior to all disks. We distinguish gaps by how many *sides* they have, that is, how many different disk boundaries bound them. A disk packing is *compatible* with a plane graph if there is a disk centered at every vertex of the graph, and every edge of the graph is the union of radii of disks. The following theorem about compatible disk packing is based on technology developed in the disk-packing literature (Bern et al. 1995):

Theorem 17.2.1 (Bern et al. 1998, 2002). *Any plane graph has a compatible disk packing in which each gap has three or four sides. Furthermore, such a packing can be found in $O(N \log N)$ time, where N is the number of disks in the packing.*

We sketch one way to construct such a disk packing. First we place a disk centered at each vertex of the graph, with radius of half the minimum distance from the vertex to a nonincident edge. By this choice of radius, no disks overlap. Then we place disks along the remaining portion of each graph edge, by first attempting to cover the entire remaining portion of the edge with a single disk diameter, but if overlap occurs, subdividing the remaining portion of the edge at its midpoint and repeating. Now the disk packing is compatible with the graph. Finally, for each gap with more than four sides, we partition it into gaps with fewer sides by placing the largest possible disk inside the gap; algorithmically, we center the disk at the "deepest" vertex of the medial axis[3] of the gap.

We need a slight variation on this type of disk packing in which disks "pass through" the thick edges produced in the previous step. More precisely, disks centered at points along the graph edges get split in half, with each half centered on one offset of the graph edge. In addition, disks centered at graph vertices get split into two or more nonequal halves, each centered on one offset vertex, and with slightly different radii so that they

---

[3] The *medial axis* of a shape is the locus of centers of disks that touch the boundary of the shape in more than one point (see, e.g., O'Rourke 1998, p. 179).

meet as in Figure 17.15(c). This variation does not modify the properties of the disk packing; it only requires that the graph vertices and edges be packed by "half disks" in compatible ways on both sides of the offset.

17.2.3 Decomposition into Triangles and Quadrangles

Next we construct the *dual* of the disk packing, which has a vertex placed at the center of each disk, and an edge between two vertices corresponding to touching disks. Because each gap in the disk packing has three or four sides, each face of the dual is a triangle or quadrangle. (Here we ignore the outside face and the ribbon.) Because the disk packing has a disk centered at each offset vertex, the dual vertices are a superset of the offset vertices. Because the disk packing covers each offset edge by a union of radii, the dual only subdivides the offset edges. Overall, the dual is a *refinement* of the offset graph (see Figure 17.15(d)).

17.2.4 Molecules

With the disk-packing machinery in place, we now return to paper folding. Locally, each triangle and quadrangle of the decomposition folds to a *molecule* that aligns its boundary. Figure 17.17 illustrates both molecules. The triangle molecule is the classic "rabbit ear" fold in origami (named so because of its common use in forming ears in origami models). The quadrangle molecule is a special case of Lang's universal molecule described in Section 16.8. Each of these molecules fold into a *starfish* with either three or four "arms" that can independently rotate around a common vertical axis.

The molecule designs have the property that the emanating perpendicular creases hit the boundary at precisely the intersections of the disks. As a result, perpendicular creases from adjacent molecules align, preventing the type of further propagation that plagued the straight-skeleton method. It is here that we use the powerful properties of the disk packing, as opposed to any other decomposition into triangles and quadrangles.

17.2.5 Gluing Molecules

Next we show that the molecules can be glued together at their boundaries without self-intresection. We consider this process step by step in a simple example, a single square polygon interior to a square piece of paper. Figure 17.18 shows the disk packing of this example after expanding the ribbon. Despite the simplicity of this example, it serves to illustrate many of the key concepts. The packing in Figure 17.18(b) leads to nine quadrangle molecules connected in a grid as shown in Figure 17.19(a). In the remainder of Figure 17.19, we abstract away the original graph from the fold-and-cut problem and focus just on the constituent molecules and their connections. We will return to the fold-and-cut problem after the preliminaries are clear.

17.2.5.1 Molecule Tree

Define the *molecule graph* to be the dual of the decomposition into triangles and quadrangles, in which nodes correspond to molecules (triangles and quadrangles) and arcs correspond to two molecules sharing an edge. The *molecule tree* is a spanning tree of this molecule graph, shown in blue in Figure 17.19(b). We pick one *root molecule* that has at least one edge on the boundary of the original piece of paper. In the example in

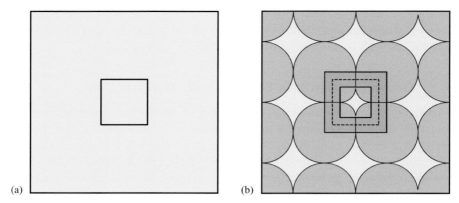

Figure 17.18. (a) Square polygon; (b) disk packing and ribbon expanding the polygon (now dashed) by $\pm\varepsilon$.

Figure 17.19, the upper-left molecule is the chosen root, and its left edge is the chosen boundary edge. Now each molecule except the root has a well-defined *parent molecule*. We choose the molecule tree to cross the ribbon only once, so that one entire rooted subtree of the molecule tree consists of precisely the molecule interior to the ribbon.

We initially imagine this molecule tree as constituting all connections between molecules. In other words, we imagine all other connections between molecules as being cut by scissors, as shown in Figure 17.19(b). The result is a connected piece of paper, but with more boundary than before. This piece of paper is easier to fold than the original; later we will show that these cut edges can be sewn back up without crossing. Each tree in the forest of these cuts is rooted where it meets the overall paper boundary.

17.2.5.2 Mountain–Valley Assignment

At this point we can describe the mountain–valley assignment of the crease pattern, which is determined by the rooted molecule tree. In Figure 17.19(c), solid edges denote mountains (or boundary edges) and dashed edges denote valleys. All angular bisector creases (thin solid) are mountains, and any "active path" creases from the quadrangle molecules are valleys (though such creases do not occur in this example). The remaining perpendicular creases are primarily valleys (thin dashed), but for each molecule, we assign mountain to the one perpendicular crease that points toward the parent molecule (red). In addition, for the root molecule, we assign mountain to the one perpendicular crease that is incident to the chosen boundary edge (also red). The edges between molecules, and the remaining edge on the outer boundary of the ribbon, are all valley creases. When we later sew up a tree of cut edges between molecules and the cut edges on the outer boundary of the ribbon, as in Figure 17.19(e), each edge is a valley crease in the portion incident to the parent and a mountain crease in the portion incident to the child.

17.2.5.3 Three Proof Parts

The gluing proof consists of three main parts:

1.   Glue together the *inner molecules*, that is, the molecules interior to the ribbon. This part has two subparts:
     (a) Glue according to the inner subtree of the molecule tree.
     (b) Glue the remaining edges between inner molecules.

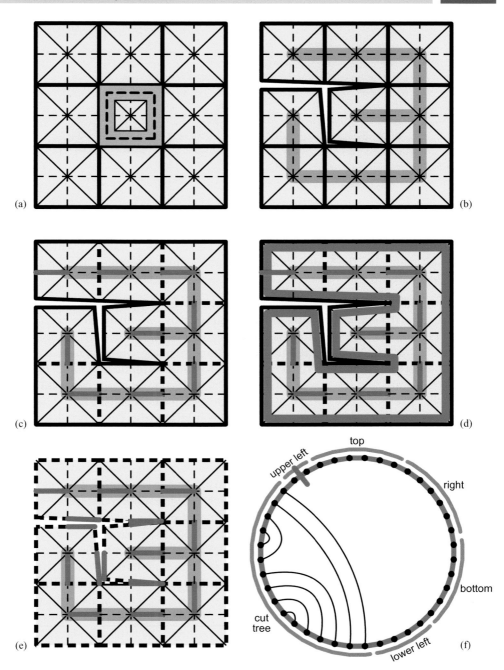

Figure 17.19. Cutting open a spanning forest to produce a molecule tree (blue), assigning mountain creases (red), and the ordering (green) of the molecule edges (thick dashed edges). (a) The polygon (a square), its ribbon, and suitable molecules; (b) the molecule tree after cutting a spanning forest; (c) mountain–valley assignment with cuts; (d) Euler tour of the molecule tree (boundary edges); (e) mountain–valley assignment after regluing cuts; (f) order and gluings of molecule boundary edges.

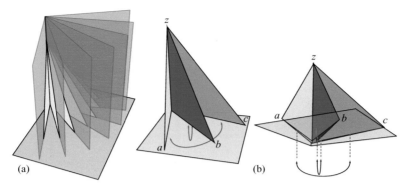

Figure 17.20. (a) A 7-arm starfish with quarter planes; (b) two views of nested underneath gluings of starfish arms, from above (left) and below (right) the base plane.

2.   Attach the ribbon to the folded state.
3.   Glue together the folded state so far and the outer molecules. This part has two subparts, just like Part 1.

At the end of each part and subpart (and inductively within these), we maintain the invariant that the folded state constructed so far is a *starfish*, with one arm per boundary edge of the region considered so far. More precisely, an n-armed starfish consists of n "arms," each of which lies in a different vertical quarter plane, where all n quarter planes are bounded by a common vertical boundary line and a common horizontal plane (see Figure 17.20(a)). The two lower edges of an "armpit" between two consecutive such flaps will correspond to a boundary edge of the region considered so far.

In the sequel, we refer to the three parts of the proof above as Part 1, Part 2, and Part 3.

17.2.5.4 Part 1a: Folding the Inner Molecule Subtree

Each inner molecule folds into a starfish with three or four arms, as in Figure 17.17. Our first goal is to show how to nest these starfish inside each other according to the molecule tree, or more precisely, the subtree restricted to inner molecules.

We build the starfish folded state incrementally from the root molecule down in a preorder traversal of the molecule tree. Initially, we have just a single starfish with three or four arms; at a general step, we have an arbitrary starfish. To add the next molecule, we simply slide the new starfish with three or four arms into the appropriate armpit of the parent, and glue along the two creases that form the shared edge of the two molecules. Thus two arms of the added starfish merge with two arms of the previous starfish, and the remaining one or two arms in the added starfish become new arms of the overall starfish, thus creating two or three new armpits corresponding to the two or three new boundary edges in the constructed folded state. We can locally check that this gluing of one starfish into another is valid, introducing no crossings and gluing together two faces with proper orientations, and that the resulting creases are indeed valleys. Also, the gluing aligns the bottom planes of the starfish, and therefore aligns all the inner molecule edges.

The square example of Figure 17.19 has only a single inner molecule, but as we will see below, the foldings of the inner and outer molecules follow the same procedure. We illustrate this process for the full molecule tree of the square example in Figures 17.21 and 17.22. The former figure illustrates the folding "halfway" to the final folded state. The long path of the molecule tree folds to a linear sequence of molecules A, B, ...,

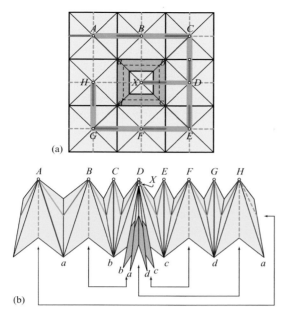

(a)

(b)

Figure 17.21. (a) Molecule tree from Figure 17.19, labeled: $abcd$ is the central square; (b) depiction of side-view of a partial folding; arrows underneath indicate later gluings to resuture cut edges.

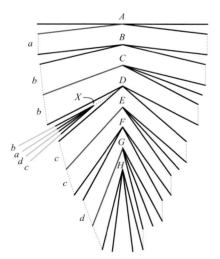

Figure 17.22. Creasing the partial folding in Figure 17.21 along the central red/blue molecule path turns each molecule into a starfish, here shown in an overhead view, separated for clarity (dashed lines connect repeated segments). Note the left/right asymmetry at the corner molecules C, E, and G.

H, with the branch of X nestled between the b and c arms of molecule D. Figure 17.22 shows the complete nesting pattern of the folded starfish.

17.2.5.5 Part 1b: Sewing the Cut Edges Between Inner Molecules

Next we consider the edges between the inner molecules that were cut to reduce down to the subtree of the molecule tree—edges ab, ad, and cd in Figure 17.21(a). These cut edges must be reglued "underneath" the rest of the model. As shown in Figure 17.21(b), ab is glued to molecule B, ad to H, and cd nested within that gluing to molecule F.

We consider regluing these edges in a postorder traversal of each tree of the cuts. At a general step of the induction, we have glued some of the edges and obtained a starfish folded state with one armpit for each edge of the current boundary. Next we glue one of the leaf cuts, that is, we glue together two incident boundary edges, which corresponds to two nested gluings between two adjacent starfish armpits: one gluing between the two sides of the shared arm, and the other gluing between the nearer sides of the two unshared arms (see Figure 17.20(b)). This gluing seals up the two armpits, reducing the number of arms and armpits by 2, which matches the two-edge reduction in boundary. The gluings occur underneath the model, do not interesect each other, and cannot interesect any future gluings because the two armpits are no longer accessible. Therefore, the gluings do not cause self-intersection.

The folding of the molecule tree defines a total order on the faces of the crease pattern, according to how they are stacked. (In particular, this total order determines the partial order function λ of Chapter 11.) This ordering can be read off from the nesting procedure described above. In fact, the ordering is precisely the Eulerian tour surrounding the molecule tree, except that the cyclic order is broken into a total order at the perpendicular crease incident to the chosen boundary edge. If we consider an example in which all nine molecules of Figure 17.19 are inner molecules, and the ribbon and outer molecules are exterior and not shown, then the Euler tour is the green cycle in Figure 17.19(d). We can stretch out the tour topologically into a circle, and represent the edges of the tour by dots and represent gluings by connections between the dots, as shown in Figure 17.19(f); the breakpoint is drawn in red. This view illustrates two important consequences of the inductive argument above: the gluings of cut edges (black arcs) do not interesect each other, and do not nest around any uncut edges (dots not incident to gluing arcs).

17.2.5.6 Part 2: Attaching the Ribbon

We now have a starfish folded state of the inner molecules, with one armpit for each boundary edge, that is, each inner edge of the ribbon. We can extend the folded state to include the ribbon by extending each such boundary edge vertically downward by the parallel offset 2ε. (During this offset, the edge may grow or shrink on either end, according to the shrinking process that formed the ribbon.) The key property we are using here is that the boundary edges are exposed on the bottom, free to have paper attached below. This property follows from the armpit correspondence; all gluings made underneath the folding in the previous section are contained within arms.

The resulting folded state is again a starfish with the same number of arms. Now the two bottom edges of each armpit correspond to an outer edge of the ribbon. Also, the original inner starfish has been raised by 2ε, putting the inner molecule edges 2ε above the outer ribbon boundary (see again Figure 17.21(b)).

17.2.5.7 Part 3: Outer Molecules

Finally, we consider adding the outer molecules. The argument mimics the inner molecule construction: first we just consider the connections made by the molecule tree, and then we consider regluing the other connections. The only difference is that one leaf "molecule" in the induction is the starfish representing all inner molecules and the ribbon. The induction goes through as before, because all it requires at each step is a starfish with one armpit per boundary edge, which we have constructed. Therefore we obtain a starfish folded state with all outer molecule edges aligned on one horizontal plane, and all inner molecule edges aligned on a horizontal plane 2ε above.

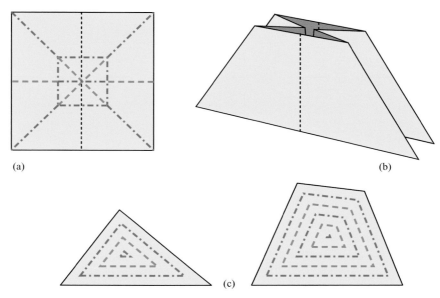

Figure 17.23. Sink-folding the starfish of a square molecule. (a) Crease pattern; (b) folded state; (c) nested M/V creases for generic molecules (not all creases shown).

17.2.6 Sink-Folding Molecules

The last step is to shrink the height of the outer molecules to less than ε, so that the plane midway between the outer molecule edges and the inner molecule edges contains only the original graph, as in Figure 17.16(b). This shrinkage is obtained by repeatedly halving the height of each molecule with a *sink fold*.

The high-level idea is to fold along a horizontal line halfway up the molecule. However, we cannot make a simple fold along that line, because this would rigidify the starfish, preventing the arms from rotating independently. The sink fold makes the same creases, but makes them all mountains and reverses all the creases in the portion of paper above the line, causing this upper portion of paper to sink "inside" the lower portion, as illustrated in Figure 17.23. The resulting folded state is a starfish with the same flexibility as before.

This process can be repeated, each time reducing the height in half. The first sink fold adds mountain creases along a shrunken copy of the polygon, meeting the perpendicular creases at their midpoints. The second sink fold reverses these creases to valleys, and adds mountain creases along two shrunken copies of the polygon, meeting the perpendicular creases at one-quarter marks. After k sink folds, we have $2^k - 1$ shrunken copies of the polygon, alternating mountain and valley. (For a square, this is precisely the crease pattern of a pleated hyperbolic paraboloid; see Section 20.2, p. 293.) For quadrangles, some of the inner shrunken copies may be two disconnected triangles. In all cases, the resulting folding is a fully flexible starfish with height $1/2^k$ times the length of a perpendicular crease.

Because these starfish act identically to the original molecule starfish, we can substitute them into the folded-state construction. Thus, setting k large enough, we obtain the desired folded state where a single plane (or, if flattened, a single line) at height ε contains precisely the original graph.

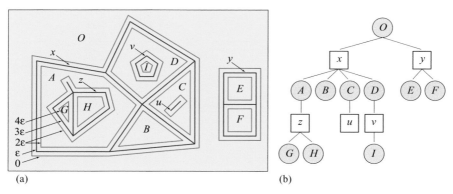

Figure 17.24. (a) A collection of polygons including all three generalizations; (b) the corresponding region–boundary tree.

17.2.7   Arbitrary Graphs

Most of the proof generalizes directly to arbitrary graphs. The parallel-offset construction follows the general offset algorithm from the straight-skeleton construction; the only difference is that the result consists of multiple connected components of ribbon. The disk packing and molecule folding assumed nothing of the graph. The primary changes in the proof are restricted to the gluing of molecules into a starfish folded state.

We describe the extensions to this part of the proof in three stages. First we generalize to collections of disjoint, nonnested polygons. Second, we consider collections of polygons whose boundaries do not interesect, this time permitting nesting—in other words, 2-regular graphs (every vertex has degree 2). Third, we generalize to arbitrary graphs, including touching polygons.

Each of these successive extensions fills in a piece of one general gluing argument with one unifying structure. Figure 17.24 illustrates this overarching structure: a rooted tree representing the disjointness, nesting, and touching relationships between the polygons. This *region–boundary tree* has two types of nodes: each *boundary node* (drawn as a square) represents a connected component of the graph of desired cuts and the corresponding connected component of the ribbon; each *region node* (drawn as a circle) represents a face of the graph and the corresponding connected family of molecules between ribbons. The root node is the outermost region, incident to the paper boundary.[4] A general region node corresponds to a connected region in the plane, which has one or more connected components of boundary: an outer component, which corresponds to the parent boundary node in the tree (and which exists for all but the root node), and zero or more inner components, which correspond to the children boundary nodes in the tree. A general boundary node corresponds to a connected boundary (graph) in the plane, which divides the paper into one or more

---

[4]   We assume in our description that none of the graph edges touch the paper boundary, and hence there is a well-defined such outermost region. In the case where a graph edge touches the paper boundary, the paper boundary must be "cut out" as well, so we can add the paper boundary to the set of graph edges at the beginning of the algorithm. The only difference in the proof is that the root node of the region–boundary tree becomes a boundary node, corresponding to the connected component of graph edges containing the outer paper boundary.

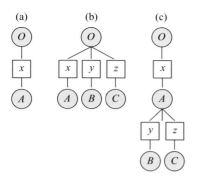

Figure 17.25. Region–boundary trees. (a) One polygon; (b) several disjoint, nonnested polygons; (c) one polygon with two disjoint polygons nested inside (corresponding to Figure 17.26(a) with $x = P$, $y = P_1$, $z = P_2$).

faces: the outer face, which corresponds to the parent region node in the tree, and zero or more inner faces, which correspond to the children region nodes in the tree.

So far, we have shown how to handle the one type of region–boundary tree corresponding to a single polygon, shown in Figure 17.25(a): a root region node corresponding to the polygon exterior, a single child boundary node corresponding to the polygon boundary, and a single grandchild region node corresponding to the polygon interior. This three-node tree mimics the three-part proof of molecule gluing described in Section 17.2.5. Part 1 constructs the starfish folded state of the polygon interior, corresponding to the inner region node; Part 2 attaches the ribbon, corresponding to the boundary node; and Part 3 constructs a starfish folded state containing both the starfish built so far and (the molecules of) the polygon exterior, corresponding to the root outer region node.

In the next few sections, we successively generalize the family of region–boundary trees that our proof can handle, until we reach the general case.

17.2.7.1 Disjoint Nonnested Polygons

To handle multiple disjoint polygons that neither nest nor touch, we must handle region–boundary trees as shown in Figure 17.25(b): a root region node corresponding to the exterior, $k \geq 1$ child boundary nodes corresponding to the polygon boundaries, and one grandchild region node for each child boundary node. The gluing proof for this case is nearly identical to the single-polygon case; we simply piece together the same components of the proof in a different combination.

First, we constrain the molecule tree to cross each ribbon exactly once, so that each leaf region node corresponds to a subtree of the molecule tree. Using Part 1 of the proof, we build a separate starfish folded state for each of the k leaf region nodes. Then, using Part 2 of the proof, we attach to each such starfish the surrounding ribbon, corresponding to the parent boundary node. Each of the k resulting starfish has the outer ribbon boundary on the bottom horizontal plane, the graph edges ε above, and the inner ribbon boundary 2ε above. Finally, we use a modified Part 3 to combine these k starfish with the outer-molecule starfish: simply apply the argument of Part 1 on the subtree of outer molecules augmented with k additional leaves corresponding to the k already constructed starfish. Just as Part 1 was general enough to handle one recursively constructed starfish, it can also handle k such starfish.

Once the gluing is complete, we have a folded state like Figure 17.16(a). We sink-fold the molecules in the outer region as before, and obtain the desired folded state in the form of Figure 17.16(b).

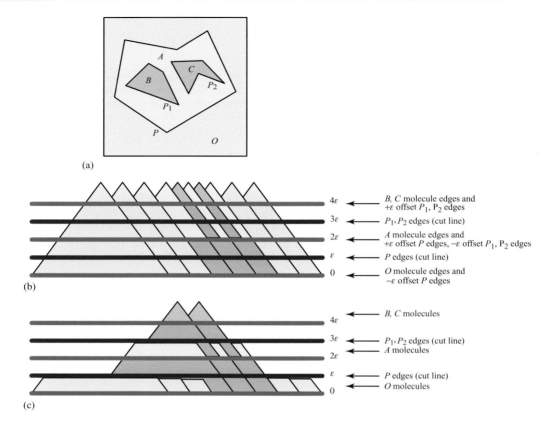

Figure 17.26. Schematic diagram of the flat folding for (a) a polygon P containing two polygons P_1 and P_2, (b) before and (c) after sink folding. The ribbon offsets $\pm\varepsilon$ around P, P_1, and P_2 are shown pink (cf. Figure 17.16).

17.2.7.2 2-Regular Graphs: Nontouching Polygons

To handle multiple polygons whose boundaries do not intersect, we must handle region–boundary trees such as the one shown in Figure 17.25(c). Specifically, in this case, each region node can have zero or more children, but each boundary node has exactly one child. This constraint corresponds to each boundary component (which is a single cycle) defining a single interior region (possibly with holes).

The gluing proof for this case is essentially identical to the previous case. The only difference is that the recursion can be arbitrarily deep. Specifically, the recursion works as follows. For each leaf region node, we construct a starfish folded state by Part 1. For each boundary node whose child has been constructed, we add the corresponding ribbon to the folded state by Part 2. For each nonleaf region node whose children have been constructed, we merge the molecules in that region together with the children nodes by the same modified Part 3. We apply these rules whenever possible until we obtain a folded state at the root node, which encompasses all molecules and ribbons.

The only catch is that this folded state does not align all the graph edges. Consider the example in Figure 17.26(a) (whose region–boundary tree is that of Figure 17.25(c)). The folded state constructed so far looks like Figure 17.26(b). When we attach the ribbons of polygons P_1 and P_2, their graph edges lie at height ε above the bottom plane. When we attach the ribbon of polygon P, we raise the graph edges of P_1 and P_2 another 2ε to height 3ε, while the graph edges of P lie at height ε. Thus we obtain two parallel lines that contain all of the cut edges.

Now we sink-fold the molecules in all but the leaf regions down to height less than ε, resulting in a folded state like Figure 17.26(c). The result is that the two lines containing the graph edges, at heights ε and 3ε, contain only graph edges. Finally, as a new seventh step in the algorithm, we make a simple fold along the line at height 2ε, bringing the graph edges into alignment with each other and nothing more, as desired.

More generally, if the deepest region node has k boundary nodes as ancestors, the graph edges will map to k different lines, with heights at odd multiples of ε: $\varepsilon, 3\varepsilon, \ldots, (2k-1)\varepsilon$. The seventh step of the algorithm makes $k-1$ simple folds, along lines with heights at even multiples of ε: $2\varepsilon, 4\varepsilon, \ldots, 2(k-1)\varepsilon$.

17.2.7.3 General Graphs

To handle general graphs as in Figure 17.24, it remains to show how to handle boundary nodes with more than one child, which arise when multiple regions touch on their exterior boundaries. First, we change the constraint on the molecule tree, because it is no longer possible to cross each ribbon exactly once. Rather, for each boundary node, we pick an arbitrary spanning tree of the parent region and the child regions of the boundary node, and constrain the molecule tree to cross the boundary at precisely the same edges as this spanning tree. The key property is that the molecules in each region node correspond to a connected subtree of the molecule tree. Now all that remains is replacing the rule for boundary node, which currently just adds a ribbon to the child region node. We need to change how we add the ribbon, because it now outlines several child regions, and we need to describe how to join these regions together.

The new rule applies to each boundary node whose children have been constructed. First, we add the inner half of the ribbon to each of the children starfish. This addition is like Part 2 of the usual proof, but only extending the folded state down by ε instead of the full 2ε. The resulting regions surrounded by half ribbons now touch at their boundaries to form a connected group. We can treat each of these as a single "molecule," each with an already constructed starfish folded state, and combine these foldings together using Part 1. The molecule tree again defines a "subtree" on these "molecules," allowing them to be combined into a single starfish, followed by underneath gluings. Finally, we attach to the resulting starfish folded state the outer half of the ribbon, extending down by another ε. The result is a starfish folded state for the boundary node and all its descendants, with graph edges at odd multiples of ε as usual. As before, we sink-fold the molecules in all but the leaf regions down to height less than ε, and then simple-fold at even multiples of ε to align all graph edges.

In this construction, some of the graph edges become valley creases, namely, those edges where two regions touch on their exterior boundaries. As a result, the final cut will have to be along a crease. However, as mentioned above, this feature is unavoidable in the general case. It may be possible to avoid it when the graph is face two-colorable, that is, when all vertex degrees are even, but we do not attempt this here; we have already handled the most interesting case of 2-regular graphs.

17.2.8   Summary

The proof of flat foldability of the disk-packing solution to fold-and-cut problem concludes the proof of the claim at the opening of this chapter:

Theorem 17.2.2. *Every plane graph drawn with straight segments on a piece of paper may be folded to a flat origami in such a way that one straight cut completely through the flat origami cuts exactly the vertices and edges of the graph, and nothing more. In particular, any collection of polygons may be so cut out.*

Both algorithms are rather profligate in their use of creases. It would be interesting to find a way to reduce the number of creases and obtain a reasonable upper bound.

Another way to view a consequence of this theorem is this: a polygonal piece of paper P may be folded to a (common) plane, with all the edges on P being folded to a common line. This view suggests that it would be natural to, in addition, map all the vertices of P to a common point. This additional condition is feasible for simple examples, but seems difficult to achieve in general. Whether the theorem can be strengthened in this way remains an open problem. See Section 26.2 (p. 438) for further open problems in this direction.

18 Flattening Polyhedra

A cereal box may be flattened in the familiar manner illustrated in Figure 18.1: by pushing in the two sides of the box (with dashed lines), the front and back of the box pop out and the whole box squashes flat.

This process leads to a natural mathematical problem: which polyhedra can be *flattened*, that is, folded to lie in a plane? This problem is a different kind of paper folding problem than we have encountered before, because now our piece of paper is a polyhedron, not a flat sheet. Our goal is merely to find *some* flat folding of the piece of paper, whereas normally our piece of paper is flat to begin with!

18.1 CONNECTION TO PART III: MODELS OF FOLDING

In Part III we will address the rigidity or flexibility of polyhedra from first principles (Sections 23.1 and 23.2, p. 341ff). In particular, Cauchy's rigidity theorem establishes that all convex polyhedra—so in particular a box, or a box with additional creases—cannot be flexed at all. So how is it that we are able to flatten the box? Even for nonconvex polyhedra, any flattening of a polyhedron necessarily decreases its volume to zero; yet the Bellows theorem (Section 23.2.4, p. 348) says that the volume of a polyhedron is constant throughout any flexing.

This seeming contradiction highlights an important aspect of our model of flattening: while Cauchy's rigidity theorem and the Bellows theorem require the faces to remain rigid plates, here we allow faces to curve and flex. In fact, we only concern ourselves with finding the flat folded state of the polyhedron (e.g., the right half of Figure 18.1). We believe that the existence of such a folded state implies the existence of a continuous folding motion:

Open Problem 18.1: Continuous Flattening. [a] Can every folded state of a polyhedral piece of paper be reached by a continuous folding process?

---

[a] Demaine et al. (2001a).

Note that this question is not answered by the results of Section 11.6 (p. 189), which only apply to flat pieces of paper, not polyhedral paper. Henceforth we concentrate on the existence of flat folded states.

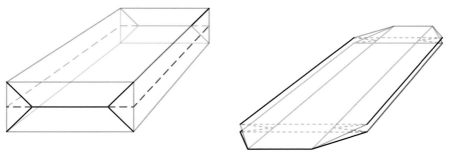

Figure 18.1. Flattening a cereal box.

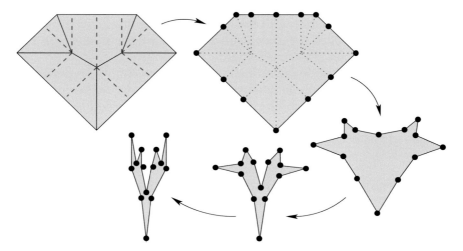

Figure 18.2. Solving the 2D polygon-flattening problem via the 2D fold-and-cut problem.

18.2   CONNECTION TO FOLD-AND-CUT PROBLEM

The flattening problem has two major connections to the fold-and-cut problem from the previous section. First, the disk-packing solution to fold-and-cut (Section 17.2, p. 263) can be applied almost directly to polyhedra homeomorphic to a sphere, solving a major case of the flattening problem. This approach is described in the next section. Second, the flattening problem can be viewed as the boundary of a 3D version of the fold-and-cut problem, as described now.

Let us start with the standard 2D fold-and-cut problem. If we apply a fold-and-cut algorithm to the interior of a simple polygon, we obtain a flat folding of the polygonal piece of paper that brings the boundary to a common line. If we ignore the folding interior to the polygon, we obtain a way to fold the 1D boundary down to a line (see Figure 18.2).

The analogous correspondence in 3D is as follows. Suppose that we could fold a 3D polyhedral solid, by creasing along 2D facets, to bring the surface of the polyhedron to a common plane. (During the folding, the solid would extend into 4D, but at the end the solid should return to 3D, to constitute a "flat" folding.) Then, if we ignore the folding of the polyhedron's interior, we obtain a flattening of the polyhedron (remembering that we have permitted it to fold through 4D). In this sense, flattening is a special case of 3D fold-and-cut. While we do not know general solutions to 3D fold-and-cut, we will discuss some approaches to flattening based on this connection in Section 18.4.

18.3   SOLUTION VIA DISK PACKING

The disk-packing solution to the 2D fold-and-cut problem applies equally well to polyhedral surfaces that are homeomorphic to a disk or a sphere.[1] Now the disk packing must satisfy the additional constraint that there is a disk centered at every vertex of the polyhedron, because a molecule could not span a point of curvature. The curvature causes no other difficulties, because the molecules glue together arbitrarily, lining up all their edges. Edges of the polyhedron do not have to be packed by the radii of disks, because they play no intrinsic role on the polyhedral surface, and folded states are entirely intrinsic structures. If the surface is homeomorphic to a sphere, the only "difference" is that every boundary edge of the molecule tree is glued to another boundary edge, and the gluing diagram in Figure 17.19(f) is a complete matching; but this extremeness poses no additional difficulty, because we could already handle arbitrary amounts of gluing.

As a result, we can find a flat folded state of a polyhedral surface that is homeomorphic to a disk or a sphere. In particular, this flat folding solves the flattening problem for these polyhedra. In addition, the flat folding has the property that it aligns the edges of the molecules. By choosing the molecule decomposition to be compatible with a given planar graph drawn on the polyhedral surface and applying the edge-thickening idea, we can align precisely the edges of the given graph. The only catch is that the graph cannot pass through vertices of the polyhedron; otherwise, the edge thickening is not well defined. Alternatively, we could align the edges of the polyhedron together with other molecule edges, while simultaneously aligning the faces of the polyhedron (in a common plane). This method solves a tiny piece of the multidimensional fold-and-cut problem (posed in Part III, Section 26.2, p. 437): the piece of paper is a polyhedron in $\mathbb{R}^3$, we care only about 1D faces (aligning edges) and 2D faces (aligning facets), and we allow extra segments to align with the 1D edges.

However, these methods leave the flattening problem unsolved for polyhedra not homeomorphic to a disk or sphere:

> **Open Problem 18.2: Flattening Higher Genus.**[a] Which polyhedra of genus $g > 0$ can be flattened (i.e., have flat folded states)?
>
> ---
>
> [a] Posed in joint work with Barry Hayes (2001).

Higher-dimensional generalizations are mentioned in Part III, Chapter 26.2.

18.4   PARTIAL SOLUTION VIA STRAIGHT SKELETON

Although we do not know how to solve the 3D fold-and-cut problem, its connection to flattening described in Section 18.2 makes it tempting to apply what we know about 2D fold-and-cut and attempt to extract just enough to solve the flattening problem.

---

[1] The sketch in this section was developed in collaboration with Barry Hayes.

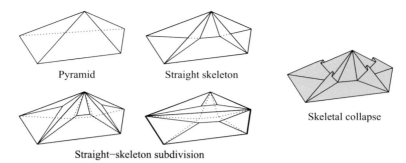

Pyramid   Straight skeleton

Skeletal collapse

Straight−skeleton subdivision

Figure 18.3. The 3D straight skeleton of a pyramid and the subdivision of its surface resulting from dropping perpendiculars.

Demaine et al. (2001a) have followed these lines, but so far have succeeded only for a small class of polyhedra, for example, pyramids. We sketch the main ideas in their approach.

To start out, we need a definition of the straight skeleton in 3D. This definition is almost identical to 2D: we offset all facets in parallel so that the perpendicular offset distance is equal among all inset facets. Now we trace the trajectories of the *edges* of the polyhedron, forming 2D facets of the straight skeleton which bisect pairs of polyhedron facets (see Figure 18.3, top).

One key structure in any flattening of a polygon's boundary or a polyhedron's surface is the *gluing*, that is, which points are glued to each other by connections interior to the polygon/polyhedron. In other words, suppose we color each facet red on the exterior and blue on the interior of the polygon/polyhedron. If we drill perpendicular to a flat folding at a particular point, we first see a point from its red side, then we see a point from its blue side, and so on. In all, we see an even number of points, if we count points along creases as seeing both their red side and their blue side (or neither). We can pair up each red point with its following blue point, and this pairing defines the *gluing* of the boundary/surface of the polygon/polyhedron.

In 2D, the gluing that results from the fold-and-cut method can be extracted directly from the structure of perpendiculars—not just the perpendiculars that would normally be creased, but applying the same perpendicular construction from every point on the straight skeleton (see Figure 18.4). For convex polygons such as the one shown, all but finitely many perpendiculars pass through exactly two polygon points, and these two points are paired. For nonconvex polygons, the situation is somewhat more complicated: a perpendicular may pass through multiple polygon points, but it will pass through an even number, and the gluing simply pairs off adjacent points in the sequence.

In 3D, the idea is roughly the same. Facets of the straight skeleton are formed by bisecting pairs of polyhedron facets, which allows us to drop perpendiculars from points on the straight skeleton toward polyhedron facets. For convex polyhedra, these perpendiculars reach polyhedron facets directly. Starting from a point interior to a facet of the straight skeleton, there are precisely two perpendiculars and the two corresponding points on the polyhedron's surface are paired in the *straight-skeleton gluing*. For nonconvex polyhedra, this definition can be generalized to capture reflecting perpendiculars and involve evenly many points on the polyhedron's surface, as in the 2D case.

The straight-skeleton gluing actually pairs up regions of the polyhedron's surface. These regions subdivide the facets of the polyhedron in what we call the

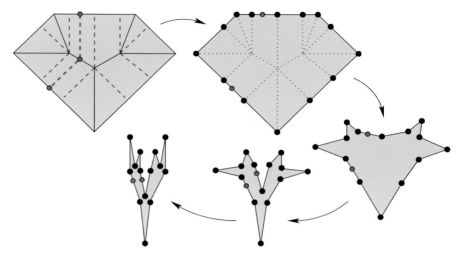

Figure 18.4. Extracting the gluing in 2D polygon-flattening from perpendiculars in the 2D fold-and-cut problem.

straight-skeleton subdivision, as shown in the bottom of Figure 18.3. Each such subdivision point corresponds to the orthogonal projection of some straight-skeleton edge onto that polyhedron facet.

We can imagine deflating the polyhedron so that the glued faces come together, reducing the polyhedron's volume to zero and resulting in a "wrinkled" structure called the *skeletal collapse,* as shown to the right in Figure 18.3. For convex polyhedra, this collapsed structure is topologically equivalent to the straight skeleton itself, but geometrically it has more material than the straight skeleton to be embedded in the same way with flat faces. Flattening a polyhedron according to the straight-skeleton gluing is equivalent to flattening this zero-volume skeletal collapse, which intuitively seems simpler because the skeletal collapse has a simpler topology than the original polyhedron. However, in general, this step remains an open problem:

Open Problem 18.3: Flattening via Straight Skeleton. [a] Which polyhedra can be flattened according to the straight-skeleton gluing?

---

[a] Demaine et al. (2001a).

Demaine et al. (2001a) solve this problem positively for two special classes of convex polyhedra: pyramids (consisting of one convex polygon whose vertices are all connected to a single apex) and thin convex prismatoids (roughly, what results from slicing a convex polyhedron into many thin parallel wafers). They conjecture that the answer to the open problem includes all convex polyhedra. It may well extend to nonconvex polyhedra as well.

We give a hint of the argument for pyramids. The main idea is to exploit the tree structure of a skeletal collapse formed by the straight-skeleton subdivision on the bottom face and follow that tree structure recursively. At each step of the recursion, we

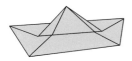

Figure 18.5. Flattening the skeletal collapse of a pyramid.

Last step of the
flattening algorithm

Flattened
pyramid

show that the subtree beyond one tree edge can be folded into the plane containing the bottom face and a perpendicular plane containing the tree edge. At the end, we obtain a folding of the entire skeletal collapse into the plane containing the base and a perpendicular plane, and we simply fold over the perpendicular plane, as shown in Figure 18.5.

19 Geometric Constructibility

19.1 TRISECTION

"Trisecting an angle" is one of the problems inherited from Greek antiquity. It asks for a series of constructions by straight edge and compass that trisects a given angle. This problem remained unsolved for 2,000 years and was finally shown to be algebraically impossible in the nineteenth century by Wantzel (1836). We saw earlier, in Figure 3.6 (p. 33), that it has been known for over a century that a linkage can trisect an angle. More recently it was established that it is possible to trisect an angle via origami folds; Figure 19.1 illustrates the elegant construction of Abe.[1]

The reader may sense that it is not clear that this construction actually works, nor what are the exact rules of the game. We will attempt to elucidate both these issues.

19.2 HUZITA'S AXIOMS AND HATORI'S ADDITION

What is "constructible" by origami folds was greatly clarified by Humiaki Huzita in 1985, who presented a set of six *axioms* of origami construction (Hull 1996; Huzita and Scimemi 1989; Murakami 1987).[2]

These axioms intend to capture what can be constructed from origami "points" and "lines" via a single fold. A *line* is a crease in a (finite) piece of paper or the boundary of the paper. A *point* is an intersection of two lines. Initially, the (traditionally square) paper has lines determined by the boundary edges. Crease the paper in half and you construct two points where the medial crease hits the paper boundary (see Figure 19.2). Huzita's axioms are intended to capture which single new creases can be constructed from existing points and lines. Figure 19.3 illustrates his six rules.

A1. Given two points, one can fold a crease line through them.

A2. Given two points, one can fold a crease along their perpendicular bisector, folding one point on top of the other.

---

[1] Abe discovered his trisection method in the 1970s (Abe 1980; Husimi 1980). The method has been described many times since, e.g., Hull (1996). Alperin (2004) observes that the method is similar to a classic trisection method using "Maclaurin's trisectrix" (a curve that can be aligned with features of a diagram to trisect an arbitrary angle). Other origami trisection methods have been discovered independently by Justin (Brill 1984) and Huzita (1994).

[2] Similar axioms were independently formulated by Justin (1989b) and Geretschläger (1995).

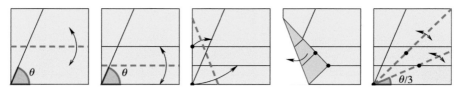

Figure 19.1. Abe's method for trisecting an acute angle using origami.

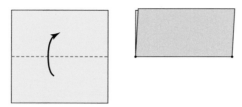

Figure 19.2. Folding a square to create two new points.

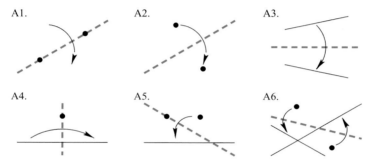

Figure 19.3. Huzita's six axioms. Solid lines are existing lines; dashed lines are the new creases.

A3. Given two lines, one can fold their bisector crease, folding one line on top of the other.

A4. Given a point and line, one can crease through the point perpendicular to the line, folding the line onto itself.

A5. Given two points and a line, one can fold a crease through one point that maps the other point onto the line.

A6. Given two points and two lines, one can fold a crease that simultaneously maps one point to one line and the other point to the other line.

Let us walk through Abe's construction explaining the axioms used at each step. It starts from a given line at angle θ.

1. The first fold is by A3, bisecting the square sides.
2. The second fold is again by A3.
3. The third fold is by A6, bring two points to two lines.
4. The last two folds are by A1, creasing through the corner and the two constructed points.

For *why* Abe's construction trisects the angle, see Box 19.1.

The ability to trisect an angle shows that these origami axioms are stronger than straight edge and compass constructions. It turns out that all the additional power resides in A6, the most complex axiom, for it is known that straight edge and compass

Box 19.1: Abe's trisection

Let a, b, and c be three points at the left ends of the bottom three parallel lines: the bottom side of the square and the first two fold lines (see Figure 19.4). Point b is the midpoint of ac by construction. Now reflection through the A6 crease line L maps the three points to a', b', and c', with again b' the midpoint of $a'c'$. Because a' lies on the horizontal line through b and because b is the midpoint of the vertical line segment ac, length $a'c$ equals length $a'a$, so $\triangle aa'c$ is isosceles. Therefore its reflection through crease line L, $\triangle aa'c'$, is also isosceles. Thus, $\angle a'ac'$ is cleverly split in half: $\angle a'ab' = \angle b'ac' = \alpha$. Because $\triangle aa'c'$ is isosceles, its base angle $\angle aa'c' = 90° - \alpha$. Because a' is a reflection of the corner of the square, it is a right angle, leaving $\angle xa'a = \alpha$, and therefore by reflection $\angle xaa' = \alpha$. Thus the angle θ at a is trisected into $\alpha + \alpha + \alpha$.

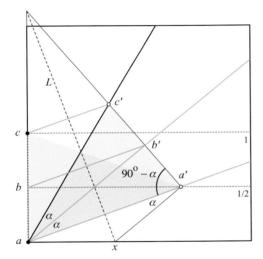

Figure 19.4. Abe's trisection analyzed.

constructions can construct rational numbers, the roots of quadratic equations with constructible coefficients, and nothing more. The key word here is "quadratic," not surprising because circles are quadratics. Axioms A1–A5 have been shown to be equivalent to straight edge and compass constructions (Alperin 2000; Geretschläger 1995; Justin 1989b). However, A6 cannot be accomplished (in general) using straight edge and compass; it adds the power to solve certain cubic equations. To see this, let us revisit A5, in Figure 19.5(a). A5 maps a point to a line over a crease. Because the point reflects across the crease, its distance to the crease is the same as the distance from the crease to the reflected point. This shows that the crease is tangent to the parabola whose focus is the point and whose directrix is the line, for a parabola is the locus of points equidistant from the focus and the directrix.

Now it should be clear that the crease in A6 is tangent to two parabolas, as shown in Figure 19.5(b). Finding a common tangent between two parabolas can be phrased as solving a cubic equation (Emert et al. 1994).[3]

---

[3] Alperin (2004) describes another geometric view of this axiom and its connections to "pedals of conics."

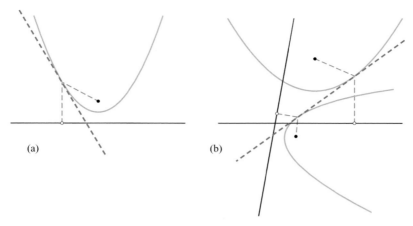

(a)   (b)

Figure 19.5. Huzita's axioms A5 and A6.

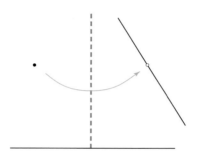

Figure 19.6. Hatori's axiom A7.

19.2.1   Hatori's Seventh Axiom

Recently Hatori suggested a seventh axiom (Hatori 2004; Hull 2003b):

> **A7.** Given one point and two lines, one can fold a crease perpendicular to one line so that the point maps to the other line.

See Figure 19.6.

This axiom is again fundamentally quadratic, but not equivalent to any one of the others. Lang proved that these seven axioms together are complete in the sense that "these are all of the operations that define a single fold by alignment of combinations of points and finite line segments" (Lang 2004, p. 234).

19.3   CONSTRUCTIBLE NUMBERS

We mentioned that origami, as specified by the Huzita–Hatori axioms, can go beyond the quadratic limitation of straight edge and compass constructions, in having at least some cubic capabilities, leading to the ability to trisect an angle. It is then natural to ask if origami can do more than trisect. There is a characterization of what is achievable with a straight edge, compass, and a tool that only trisects a given angle: one can then

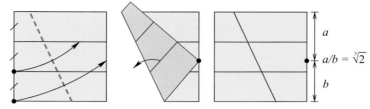

Figure 19.7. Messer's method for doubling a cube using origami.

solve a real cubic equation with constructible coefficients if and only if all its roots are real (Gleason 1988). In particular, because $z^3 - 2 = 0$ has only one real root, one cannot "double a cube," that is, construct $\sqrt[3]{2}$, using such tools.[4]

However, Messer (1986) showed in 1985 that you can double a cube using origami (see Figure 19.7).

Justin (1989b), Geretschläger (1995), and Alperin (2000) independently showed that all cubic equations with constructible coefficients can be solved with origami. Unknown to these authors, a proof of this result was published already in 1936 by Margherita Piazzolla Beloch (1936) (see also Huzita 1994). Because quartic equations can be solved with cubics, the consequence of this result is that all quartic equations can be solved with origami.

It remained an open problem for several years (Hull 1996; Murakami 1987) to determine whether origami can go beyond solving quartics. The answer is NO. This was first established by Huzita and Scimemi (1989), and independently by Alperin (2000). Here we rely on a connection to other work first noted by Demaine and Demaine (2000).

The basic origami fold that the axioms use is a "book fold," which amounts to a reflection through a line. A related geometric-construction tool with this property is a semireflective mirror allowing you to simultaneously see a reflection and see through the mirror. Originally, such a mirror was made by coating plate glass with a thin layer of evaporated metal. Today, one can use a hard thin piece of dark-colored transparent plastic with a smooth finish. Such a device is commonly referred to by the trade name Mira.[5]

Many papers have considered the geometric power of the Mira, starting in 1963 when Hochstein (1963) showed how to trisect an angle with a semireflective mirror (before the term "Mira" was coined). In 1994, Emert et al. (1994) were the first to give a full axiomatic characterization of the Mira. And it turns out that the operations are in fact equivalent to the origami axioms. Emert et al. provided a complete characterization of Mira-constructible (and hence origami-constructible) numbers; they proved that every constructible number is the root of a quadratic or cubic equation with constructible coefficients. Thus, origami can solve all quartic equations but no more.

19.4 FOLDING REGULAR POLYGONS

A classic motivation for determining the power of a particular geometric device is to characterize the constructible regular polygons. For example, it is well known that

---

[4] This is one of the three classical Euclidean construction questions: given the side of a cube, construct the side of a cube that has twice the volume. If the given side has length 1, the goal is to construct $\sqrt[3]{2}$.

[5] Registered trademark of the Mira-Math Company in Willowdale, Ontario, Canada.

straight edge and compass can construct a regular n-gon exactly when n is of the form $2^r p_1 \cdots p_k$, where the p_i's are distinct Fermat primes (i.e., prime numbers of the form $2^u + 1$). Origami has a similar characterization, essentially by adding "3" in addition to "2," because origami can solve cubic equations in addition to quadratic equations. More precisely, origami can construct a regular n-gon exactly when n is of the form $2^r 3^s p_1 \cdots p_k$, where the p_i's are distinct primes of the form $2^u 3^v + 1$ (Emert et al. 1994). Curiously, these are precisely the same regular polygons that are constructible with straight edge, compass, and a tool that just trisects (Gleason 1988). In other words, the ability to solve general cubic equations, instead of just those with all real roots, does not help to construct any new regular polygons.

In comparison to straight edge and compass, origami can construct many more regular polygons. Below we list the first several unconstructible polygons for each. In particular, $n = 11$ is the smallest unconstructible regular polygon by paper folding.

Straight edge and compass: 7, 9, 11, 13, 14, 18, 19, 21, 22, 23, 25, 26, 27, 28, 29, 31, 33, 35, 36, 37, 38, 39, 41, 42, 43, 44, 45, 46, 47, 49, 50, 52, 53, 54, 55, 56, 57, 58, 59, . . .

Origami: 11, 22, 23, 25, 29, 31, 33, 41, 43, 44, 46, 47, 49, 50, 53, 55, 58, 59, 61, 62, 66, 67, 69, 71, 75, 77, 79, 82, 83, 86, 87, 88, 89, 92, 93, 94, 98, 99, 100, 101, 103, 106, 107, . . .

19.5   GENERALIZING THE AXIOMS TO SOLVE ALL POLYNOMIALS?

The results on the power of paper folding rely on the restricted set of Huzita–Hatori axioms, which only permit simple folds. It is conceivable that more powerful axioms would permit solving polynomial equations of arbitrary degree via origami. Here we give the germ of an idea from Demaine and Demaine (2000) that might lead to this result. The plan is to exploit the universality theorems from linkages (Part I, Section 3.2, p. 31) that say that any polynomial curve can be traced out by a linkage. We sketch the method first and then discuss what additional axioms might be needed.

The first step is to use the methods described in Chapter 15 to fold the paper into a shape with several *rulers*, as shown in Figure 19.8(a), making n long rulers available for manipulation, plus the shaded *construction area*. The edges of this silhouette are horizontal and vertical and have integral length, and so are easily constructible.

The second step is to "draw" (crease) in the construction area the bars (line segments) of a linkage, conceptually connected by hinges at their endpoints. The example in Figure 19.8(b) shows a 4-bar linkage, but in fact the linkage will have to follow the complete Kempe or Kapovich–Millson universality proofs. We suppose that the linkage has n bars, the same as the number of rulers we construct.

The third step is to fold each ruler into position, atop a corresponding bar in the linkage, as in Figure 19.8(c)–(d). In general, a ruler can be folded into position by first folding it vertically upward, then horizontally left (for bars with negative slope), and then at an appropriate angle to match the bar. By first folding a ruler upward, we never fold a ruler on top of a lower ruler. Thus, we can assign each ruler a "layer" that it occupies, and these layers of paper do not interfere, so the rulers can be folded flexibly and independently.

The fourth step is to reconfigure the rulers so as to simulate the linkage. Each ruler can be marked (by creasing) with the length of the corresponding bar. The resulting

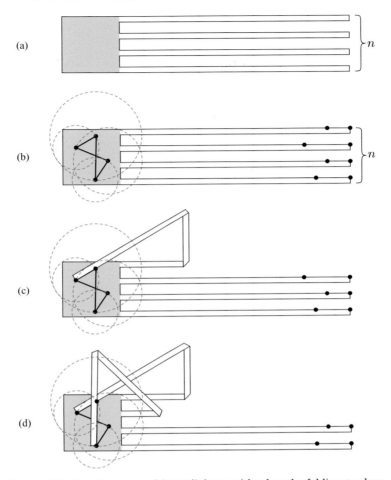

Figure 19.8. Simulating an arbitrary linkage with n bars by folding n rulers.

lengths are denoted by dots in Figure 19.8. In this way, we can simulate a linkage in which one vertex follows an arbitrary polynomial curve.

The fifth and last step is to reconfigure the linkage until this vertex crosses a particular line, which permits us to find the zeros of a polynomial of arbitrary degree.

The challenge is to formulate clean new axioms that permit continuous "rolling" (in a manner similar to that described in Section 11.6) of a set of creases to bring the folded state to one in which a given set of pairs of points are aligned. We leave this as an unresolved problem.

20 Rigid Origami and Curved Creases

In this chapter we gather a few miscellaneous results and questions pertaining to "curved origami," in either the folded shape or the creases themselves, and to its opposite, "rigid origami," where the regions between the creases are forbidden from flexing. In general, little is known and we will merely list a few loosely related topics.

20.1 FOLDING PAPER BAGS

We have seen that essentially any origami can be folded if one allows continuous bending and folding of the paper, effectively permitting an infinite number of creases (Theorem 11.6.2). Recall this result was achieved by permitting a continuous "rolling" of the paper (Section 11.6.1, p. 189). In contrast one can explore what has been called *rigid origami*, which permits only a finite number of creases, between which the paper must stay rigid and flat, like a plate (Balkcom 2004; Balkcom and Mason 2004; Hull 2006, p. 222). One example of the difference between rigid and traditional origami is the inversion of a (finite) cone. Connelly (1993) shows how this can be done by continuous rolling of creases, but he has proved that such inversion is impossible with any finite set of creases.[1]

One surprising result in this area is that the standard grocery shopping bag, which is designed to fold flat, cannot do so without bending the faces (Balkcom et al. 2004). Consider the shopping bag shown in Figure 20.1. Crease ab is the valley fold that permits collapse of the bag, after which ab is tucked behind the bc mountain fold. We assume that $h > w/2$, so that the two 45° valley creases from the bag corners meet (at c). We call such an object a *tall shopping bag*. The result is this:

Theorem 20.1.1. *If the faces of a tall shopping bag are rigid, then the configuration space of the bag consists of two isolated points: one with the bag fully opened (as illustrated) and one fully collapsed. In particular, the bag is rigid in either configuration.*

We only sketch a proof. Consider vertex a, at which four creases meet at right angles. The dihedral angles at the two vertical creases must be the same, as must the dihedral angles at the two horizontal creases. Suppose the angle of the vertical creases is neither 0° nor 180°. Then the angle of the horizontal creases must be either 0° or 180°. Suppose it is 180°. Then this forces the front and side faces incident to a to be flat. This

---

[1] Personal communication, June 2000.

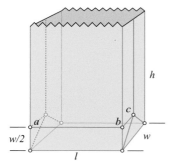

Figure 20.1. Model of a "tall" grocery shopping bag. The standard flat state uses valley folds along the blue creases and mountain folds along the red creases.

constraint propagates to vertex b, and from there to c, forcing the right side faces to be flat. Continuing the propagation forces the entire bag to be open with the angles as shown. Similar reasoning applied to the case when the horizontal crease at a has angle 0 leads to the fully collapsed configuration. Finally, if the angle of the vertical crease at a is either 0 or $180°$, we are forced into the same two configurations. Thus these are the only configurations for the bag, and they are isolated: the bag cannot move between its two configurations.

Several related results are described in Balkcom et al. (2004): if $h < w/2$ and $h \leq l/2$, the bag (which now looks more like a gift box) can collapse; and tall grocery bags are rendered collapsible by the addition of a finite number of additional creases (but not collapsible to the standard grocery bag flat folded state). One can pose many questions here, as there is as yet little understanding of the configuration space of rigid origami, even ignoring self-intersection of the faces.

20.2 CURVED SURFACE APPROXIMATION

One can always form a polyhedral approximation to a smoothly curved surface, and this is possible via origami. A spectacular example is the 25-ft. long Ukrainian Easter egg built by Ron Resch in 1975, a tourist attraction on the Trans-Canada Yellowhead Highway.[2] Although this sculpture is constructed out of aluminum, it relies on a technique he patented for folding paper into flexible surfaces.[3]

Tessellations. A related topic is that of "origami tessellations," which are, roughly, flat foldings of paper based on a *tessellation* or *tiling* of the plane (Grünbaum and Shephard 1987). This is a growing and as-yet largely unformalized area that we will not attempt to survey (see, e.g., Bateman 2002). We only mention that, although origami tesslations are generally flat, there are pleated tesselations, such as Hull's "nested hexagonal collapse" model,[4] which also approximate curved surfaces. The "Miura Map Fold" (Hull 2006, p. 215) is a particularly interesting combination of a tesselation crease pattern, but employing rigid origimi, which has been used to fold solar panel arrays for satellites in space. We turn now to a particular version of the general idea of using pleating to approximate a curved surface.

---

[2] http://www.ronresch.com.
[3] US Patent 4,059,932, November 1997; see http://patft.uspto.gov/netacgi/nph-Parser?TERM1= 4059932&u=/netahtml/srchnum.htm&Sect1=PTO1&Sect2=HITOFF&p=1&r=0&l=50&f=S&d= PALL.
[4] http://www.merrimack.edu/~thull/gallery/geomgallery.html.

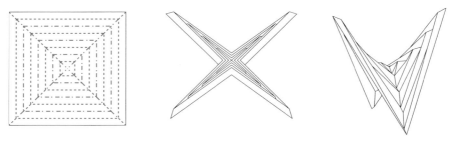

Figure 20.2. Folding a hyperbolic paraboloid.

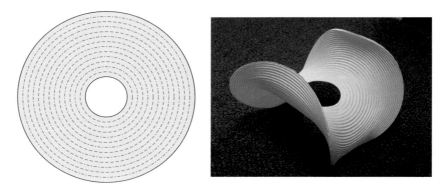

Figure 20.3. An origami design studied at the Bauhaus in the 1920s.

Figure 20.4. Two views of a curved origami construction designed with Martin Demaine.

Pleated hyperbolic paraboloids. An approximation to a hyperbolic paraboloid (e.g., $z = xy$) can be folded from the simple pattern of nested squares, alternating mountain and valley folds, as shown in Figure 20.2. This model was invented by John Emmet in 1989 (see Demaine et al. 1999b; Hull 2006, p. 216).

One variation on this "pleating" technique for approximating a curved surface is to take a disk of paper and crease concentric circles. Figure 20.3 shows the resulting form, studied at the Bauhaus in the 1920s. (To make the model easier to fold, we have cut out a central hole, but this modification is not necessary.) The mathematical surface approximated by this model remains unknown. Figure 20.4 shows a variation involving

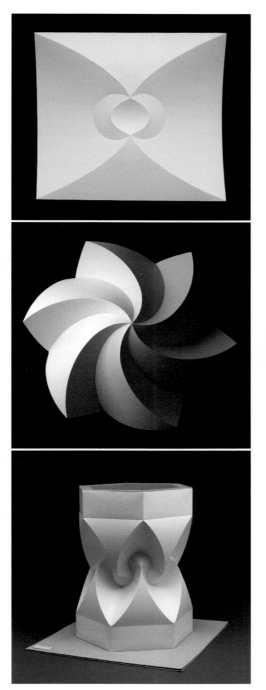

Figure 20.5. Three David Huffman constructions. [Credits: E. Huffman and T. Grant; Permission, E. Huffman, 2005.]

a spiral of paper formed by gluing two such annuli together in a loop. Both of these models are folded from "Elephant Hide" paper prescored by a laser cutter. It remains an open problem to develop general methods for designing pleated origami models that approximate desired surfaces in 3D.

20.3   DAVID HUFFMAN'S CURVED-FOLDS ORIGAMI

David Huffman (1925–1999), famous for the invention of "Huffman codes" in 1952, was a pioneer in mathematical origami. He studied the geometric conditions that must be obtained in the neighborhood of an origami vertex created by a nonboundary point of the paper (a "zero-curvature vertex" in his notation) (Huffman 1976), independently discovering an instance of Kawasaki's theorem (Theorem 12.2.1). He was also apparently among the first, along with Ron Resch, to explore the properties of curved creases. Resch discovered the following beautiful result: every smooth space curve C can be realized as the folded state of a curved crease γ in the interior of a piece of paper.[5] Huffman went on to establish two fundamental and attractive results:

1. Let γ be a curved crease. Then at any point x on γ, the osculating plane[6] of γ at x bisects the two planes tangent to the surface at x from either side.
2. Let γ be a curved crease that lies in a plane, which is necessarily the osculating plane of all the points of γ. The surfaces S_L and S_R to the left and right of γ are developable surfaces, which means they have *generating lines*—lines which when swept out generate the surface. He proved that, at any point x on γ, the generating line of S_L through x makes the same angle with the osculating plane as does the generating line of S_R through x: "that is, the generating lines are reflected from the plane as rays of light would be reflected from a mirror."

Impressive as these results are, even more impressive are the amazing models he built, which resemble elegant sculptures more than folded paper. A small sample is shown in Figure 20.5. See Wertheim (2004) for a fascinating history. As far as we know, his work has not been since equalled let alone surpassed.

Beyond. As we mentioned, curved origami remains a largely unexplored territory. We will knock against our ignorance again in Part III when we explore "D-forms" (Section 23.3.3, p. 352) and when we touch on the "teabag problem" (p. 419).

---

[5]   Personal communication, June 2004.
[6]   The *osculating plane* at x is the plane that best fits γ at x. It is the plane spanned by the three points (x_1, x, x_2) on γ, as $x_1 \to x$ and $x_2 \to x$.

Polyhedra

21 Introduction and Overview

21.1 OVERVIEW

In 1525 the painter and printmaker Albrecht Dürer[1] published a book, *Underweysung der Messung* (later translated as "The Painter's Manual"), in which he explained the methods of perspective, which he had just learned himself on a trip to Italy[2] (see Figure 21.1).

Dürer's book includes a description of many polyhedra, which he presented as surface unfoldings, what are now called "nets." For example, Figure 21.2 shows his net for a cuboctahedron. Even when he drew a more complex net, such as that for a truncated icosahedron (the shape of a soccer ball; Figure 21.3), he always chose an unfolding that avoided self-overlap. Although there is no evidence that Dürer distilled this property into a precise question, it is at least implicit in the practice of subsequent generations that every convex polyhedron may be so unfolded.[3] More precisely, define an *edge unfolding* as a development of the surface of a polyhedron to a plane, such that the surface becomes a flat polygon bounded by segments that derive from edges of the polyhedron. One may view such an unfolding as obtained by slicing the surface along a collection of edges. We would like an unfolding to possess three characteristics, enjoyed by all Dürer's drawings:

1. The unfolding is a single, simply connected piece.
2. The boundary of the unfolding is composed of (whole) edges of the polyhedron, that is, the unfolding is a union of polyhedron faces.
3. The unfolding does not self-overlap, that is, it is a "simple polygon."

We call a simple polygon that satisfies these conditions a *net* for the polyhedron.[4] This leads to the following open problem:

---

[1] 1471–1528, Nürnberg, Germany.
[2] Here we rely on George Hart's history at http://www.georgehart.com/virtual-polyhedra/durer.html.
[3] This point was first made to us by Branko Grünbaum (personal communication, 1987).
[4] Here we follow Shephard's (1975) and Grünbaum's usage (2001) in insisting that a net is derived from an edge unfolding. We will also explore polygons that fold to a polyhedron but are not a union of polyhedron faces. (Alexandrov (2005, p. 61) uses what is translated to English as "net" for another concept. What is translated to German as "das Netz" in Alexandrov (1958) is translated to English as "development" in Alexandrov (2005).) Dolbilin calls a net a "regular edgewise development" (Dolbilin 1998).

Vnderweysung der messung/mit dem zirckel vñ richt
scheyt/in Linien ebnen vnnd gantzen corporen/
durch Albrecht Dürer zů samen getzogē/
vnd zů nutz allē kunstlieb habenden
mit zů gehörigen figuren/in
truck gebracht/im jar.
M. D. X X v.

Mit begnadung Kayserlicher im end eyngeleibter Frey-
heyt damit sich ein yglicher vor scha
den zů hüten wyß zc.

Figure 21.1. The title page of Dürer's book: "Underweysung der messung . . . im jar MDXXV."

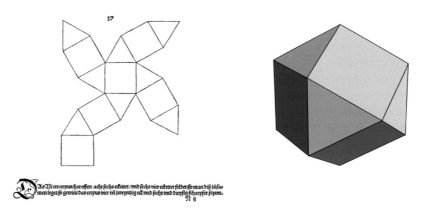

Als Vi ett corpus/hat offen acht sechs ecketter/vnd sechs vier ecketter felder/so man diß zůsa-
men legt/so gewint das corpus vier vñ zwentzig eck/vnd sechs vnd dreyssig scharpfer seyten.

Figure 21.2. Dürer's net (left) for a cuboctahedron (right).

Open Problem 21.1: Edge-Unfolding Convex Polyhedra. [a] Does every convex polyhedron have an edge unfolding to a simple, nonoverlapping polygon,
i.e., does every convex polyhedron have a net?

---

[a] Dürer (1525).

This has remained unsolved over the 475 years since Dürer's drawings. (The first explicit statement of the problem of which we are aware was formulated by Shephard in
1975 (Shephard 1975): "Does every 3-polytope possess a net?") This problem can be
viewed as a central motivation for the third part of this book.

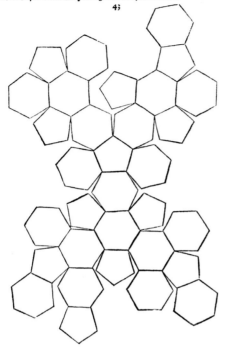

In anders das mach auß zweinßig sechseckter flachen feldern/gleichseitig vnd wincklich/ so man darzu thut zwölf fünfeckter flacher felder/ so die gleichseitig gegen den sechseckten feldern sind/ vnnd in jnen selbs auch gleich wincklich vnd ordenlich an cynander gesetzt wer; den / wie ich das offen im plano hernach hab aufgeriffen / So man dann das alles zusamen schleust / so würt ein corpus darauß/ das gewinnet zwey vnd sechßig ecke/ vnnd neünßig scharpffr seiten/ dis Corpus rüret in einer holen kugeln mit allen seinen ecken an.

4;

Figure 21.3. Dürer's net for a truncated icosahedron.

21.2   CURVATURE

The concept of "curvature" will play an important role in this part of the book. Although we will employ it primarily for polyhedra, its origins lie in the differential geometry of smooth curves and surfaces. After describing curvature in these contexts, the definition for polyhedral surfaces will be motivated.

21.2.1   Smooth Curves

Let $c(t)$ be a smooth[5] curve in the plane, having a unit-speed parametrization by t. The curvature of $c(t)$ at a point p measures how sharply the curve turns at p by the magnitude of the acceleration $|\ddot{c}|$ felt at p by a particle (or a car) traveling at unit speed along $c(t)$. This acceleration is the same[6] as the reciprocal of the radius of the *osculating circle* at p: the unique circle that matches at p both the velocity vector (the tangent) and the acceleration vector (which is normal to the curve), in both magnitude and direction (see Figure 21.4). If $c(t)$ is straight at p, the circle radius is effectively infinite and the curvature is zero.

---

[5] For our purposes, *smooth* means C^∞: possessing a continuous kth derivative for all $k = 1, 2, 3, \ldots$, although usually the assumption of C^2 suffices.

[6] The same under the assumption that $|\ddot{c}| = 1$.

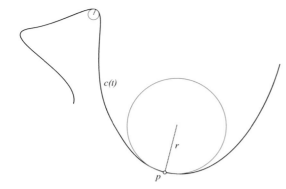

Figure 21.4. The curvature at a point is $1/r$, where r is the radius of the osculating circle.

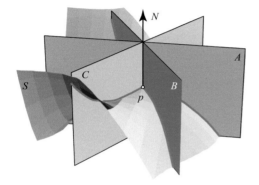

Figure 21.5. Spinning a plane about N through $p \in S$ yields the principal curvatures at p. The curvature of the curves for plane A and B are positive at p, but negative for plane C.

21.2.2 Smooth Surfaces

The curvature of a surface $S \subset \mathbb{R}^3$ at a point p cannot be captured by the osculating sphere, for this sphere does not distinguish between a spherical surface and a cylindrical surface, which intuitively have different curvatures (e.g., they have different flattening properties). Two numbers are needed at each point rather than one.

Principal curvatures. These two numbers are known as the *principal curvatures* and may be defined as follows (Lee 1997, p. 5). Let N be the unit normal vector to an oriented surface S at p, and P a plane through p containing N. Then $P \cap S$ is a one-dimensional curve whose curvature at p is defined as described above. It can be given a sign according to whether the osculating circle is locally inside S (+) or outside (−), with "outside" defined by the direction in which N points. Now imagine spinning P about N. The principal curvatures κ_1 and κ_2 at p are the minimum and maximum signed curvatures over all intersection curves obtained (see Figure 21.5). Notice that the principal curvatures do distinguish between a sphere and a cylinder: $\kappa_1 = \kappa_2 = 1/r$ for a sphere of radius r, but $\kappa_1 = 0$ and $\kappa_2 = 1/r$ for a cylinder of the same radius. There remains a flaw of sorts, however: the principal curvatures are not "intrinsic" properties of the surface, invariant under isometric embeddings. A flat piece of paper and the same paper rolled to a cylinder have the same intrinsic properties (to, say, flatlanders living in the surface), but have different principal curvatures.

Gaussian curvature. Gauss discovered in 1827 that the product $K = \kappa_1 \kappa_2$ is intrinsic. He called this result the "*Theorema Egregium*"—roughly, the distinguished, extraordinary theorem—and said that it "ought to be counted among the most elegant in the theory of curved surfaces" (Spivak 1979, II: p. 114). K, known as the *Gaussian Curvature*, will be used frequently in the sequel. Note that both the flat paper and the cylinder have the same Gaussian curvature K: for the cylinder, $\kappa_1 \kappa_2 = 0 \cdot 1 = 0$.

Curvature and normals. A useful extrinsic view of the Gaussian curvature is that "K measures the total spread of normal directions per unit surface area": it is "the ratio between the solid angle subtended by the normals on a small patch divided by the area" (Koenderink 1989, p. 215). Positive K spreads the normals, and negative K reverses the orientation of the spherical image of the normals.

Curvature and perimeter and area. Another equivalent—but this time intrinsic—definition of the Gaussian curvature at $p \in S$ is in terms of the perimeter or area of a small disk around p. Define a *geodesic disk* centered on p of radius r to be the locus of points of S at most a distance r of p, where distance to a point x is measured by the shortest path on S from p to x. (We will discuss shortest paths and geodesics in Chapter 24.) Such a disk is sometimes said to have "intrinsic radius" r. Let the circumference of the disk be $C(r)$. Bertrand and Puiseux proved in 1848 (Berger 2003, p. 110; Spivak 1979, II: p. 130) that

$$K = \lim_{r \to 0} \frac{3(2\pi r - C(r))}{\pi r^3}. \tag{21.1}$$

From this equation, we can see that points of positive curvature have a perimeter deficit with respect to the flat $2\pi r$ and those of negative curvature have a perimeter excess. Points of positive curvature are called *elliptic* and those of negative curvature *hyperbolic*.

Integrating around the perimeter of a geodesic disk leads to a similar equation for K in terms of the area deficit, discovered by Diquet:

$$K = \lim_{r \to 0} \frac{12(\pi r^2 - A(r))}{\pi r^4}, \tag{21.2}$$

where $A(p)$ is the area of the disk. Again we have the same classification: area deficit corresponds to positive curvature, and area excess to negative curvature.

21.2.3 Polyhedral Surfaces: Angle Deficit

The area-deficit viewpoint of Gaussian curvature makes the definition of Gaussian curvature on a polyhedral surface S natural, or at least plausible: the *Gaussian curvature* (or simply "the curvature") of a point p on a polyhedral surface S is the *angle deficit*[7] at p: 2π minus the total face angle incident to p. Thus, a point in the interior of a face of S has curvature zero. Less obvious is that a point on the relative interior of a polyhedron edge e also has curvature zero, for it has π face angle incident from either side of e. All the curvature of a (closed) polyhedron is concentrated at the vertices: A vertex v has curvature 2π minus the sum of the face angles (the *total angle*) incident to v.

For example, each vertex of a cube has curvature $\pi/2 = 90° = 360° - 3(90°)$; and each vertex of a dodecahedron has curvature $\pi/5 = 36° = 360° - 3(108°)$.

---

[7] Sometimes called the "angle defect" or "jump angle." This notion of angle deficit was introduced by Descartes in his "Treatise on Polyhedra," which he wrote in the mid-1600s, a century prior to Gauss.

If the surface is cut open and flattened in the neighborhood of v, there will be an angle gap equal to the curvature at v. It is this notion of curvature that will be used in the sequel.

21.3   GAUSS–BONNET THEOREM

If a person walks around the boundary of a simple polygon, by the time they revisit their starting point, they have turned or twisted in total (with negative turnings canceling positive turnings) exactly 2π. This is an easy consequence of the fact that the sum of the internal angles of a polygon of n vertices is $(n-2)\pi$. The sum of the complete angles at the n vertices is $n(2\pi)$, and so the sum of the exterior angles is $(n+2)\pi$. The turn at each vertex is the excess of the exterior angle there over π (e.g., no turn at all corresponds to an exterior angle of π). So the total turn is $(n+2)\pi - n\pi = 2\pi$. The same holds true for any Jordan curve in the plane, with the discrete sum of angles replaced by an integral of the angle turn at each point of the curve. Call this total angle turn around a curve or polygon τ. This turn τ is sometimes called the *geodesic curvature* of the curve, as it measures the deviation from a straight geodesic.

On an arbitrary (nonflat) surface, one can define a notion of the turn of a curve on the surface, which intuitively is the turn experienced by a person walking along the curve while vertically aligned with the surface normal. Under this notion of surface angle turn, a closed curve does not in general have an angle turn of 2π. For example, consider a spherical triangle on a sphere, with a corner on the equator, a corner at the north pole, and, after a $90°$ turn there, a third corner on the equator. The total turn of this triangle is $\tau = 3\pi/2$. The reason τ is less than 2π on the sphere is that each curve encloses some positive curvature. There is a tradeoff between turn and enclosed curvature, which is captured precisely in the *Gauss–Bonnet formula*, one of the jewels of differential geometry[8]:

$$\tau + \gamma = 2\pi, \tag{21.3}$$

where τ is the total turn around the boundary of the curve and γ is the integral of the Gaussian curvature enclosed in the "polygon" P:

$$\int_P K\,dA. \tag{21.4}$$

On the plane, $\gamma = 0$, so $\tau = 2\pi$. On positively curved surfaces, polygons turn less than 2π, and on negatively curved surfaces, they turn more than 2π.

Because the equator of (or any great circle on) a sphere is a geodesic, it has $\tau = 0$. A west–east walk around the equator implies via Equation (21.3) that the northern hemisphere must contain 2π of curvature, as must the southern hemisphere via an east–west walk. Thus the total curvature of a sphere is 4π. This result generalizes to any surface homeomorphic to a sphere (i.e., of genus zero), by partitioning it into regions and applying Equation (21.3) to each region:

Theorem 21.3.1. *The total sum of the Gaussian curvature over all points of a surface homeomorphic to a sphere is 4π.*

---

[8]   See, e.g., Lee (1997, Ch. 9) and Morgan (1991, Ch. 8).

Several examples:

1. *Sphere.* The curvature at each point of a sphere of radius $r = 2$ is $1/r^2 = 1/4$, and the integral over the entire surface area of $4\pi r^2 = 16\pi$ is 4π.
2. *Cube and dodecahedron.* The curvature sum for a cube is $8(\pi/2) = 4\pi$, and for a dodecahedron is $20(\pi/5) = 4\pi$.
3. *Cylinder.* Curvature is concentrated at the points on the top and bottom rim of a cylinder. If it has radius $r = 1$, each rim point has curvature 1, which, integrated over the $2\pi r = 2\pi$ length of the top and bottom rims, is 4π.

The reach of this theorem is breathtaking, as it holds for any surface, convex or non-convex, smooth or polyhedral,[9] and connects the intrinsic, local curvature to a global constant. Intuitively it says that a dent in a surface at one location must be compensated by an equal bump at another. The theorem generalizes to surface of genus g with a sum of $2\pi(2 - 2g)$; the number $\chi = 2 - 2g = V - E + F$ is known as the *Euler characteristic* of the manifold.

One immediate consequence that we will exploit in Section 24.4 is the generalization of our observation about the great circles on the sphere: a closed geodesic (which does not turn, and so has $\tau = 0$) must enclose $\gamma = 2\pi$ of curvature.

---

[9] This result for convex polyhedra was proved by Descartes, and is sometimes known as Descartes' theorem (Phillips 1999).

22 Edge Unfolding of Polyhedra

INTRODUCTION

We return to Open Problem 21.1 (p. 300): can every convex polyhedron be cut along its edges and unfolded flat into the plane to a single nonoverlapping simple polygon, a *net* for the polyhedron? This chapter explores the relatively meager evidence for and against a positive answer to this question, as well as several more developed, tangentially related topics.

22.1.1 Applications in Manufacturing

Although this problem is pursued primarily for its mathematical intrigue, it is not solely of academic interest: manufacturing parts from sheet metal (cf. Section 1.2.2, p. 13) leads directly to unfolding issues. A 3D part is approximated as a polyhedron, its surface is mapped to a collection of 2D flat patterns, each is cut from a sheet of metal and folded by a bending machine (Kim et al. 1998), and the resulting pieces assembled to form the final part. Clearly it is essential that the unfolding be nonoverlapping and great efficiency is gained if it is a single piece. The author of a Ph.D. thesis in this area laments that "Unfortunately, there is no theorem or efficient algorithm that can tell if a given 3D shape is unfoldable [without overlap] or not" (Wang 1997, p. 81). In general, those in manufacturing are most keenly interested in unfolding nonconvex polyhedra and, given the paucity of theoretical results to guide them, have relied on heuristic methods. One of the more impressive commercial products is TouchCAD by Lundström Design,[1] which has been used, for example, to design a one-piece vinyl cover for mobile phones (see Figure 22.1).

22.1.2 Problem Features

We now explore briefly five features of the open problem as stated: *edge* unfolding, *one* piece, *non*overlap, *convex* polyhedra, and *simple* polygon.

General unfoldings. Restricting the cuts to edges of the polyhedron produces what is called an *edge unfolding*. This is a natural restriction, but not essential in all

---

[1] http://www.touchcad.com/.

Figure 22.1. Lundström Design vinyl phone. [Lundström Design, by permission.]

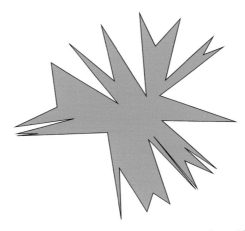

Figure 22.2. The star unfolding of a polyhedron of $n = 18$ vertices.

circumstances. General, unrestricted cuts on the surface of the polyhedron lead to (for lack of a better term) *general unfoldings*, for which the open problem dissolves. There are two known techniques for unfolding any convex polyhedron to a single polygon: the "source unfolding" and the "star unfolding."[2] Both are based on the same shortest paths construction and can be viewed as inverses of one another. The star unfolding is constructed as follows. Start with a point x on the surface of the polyhedron P and cut from x to every vertex of P. This always unfolds to a simple polygon, a nonobvious claim. Figure 22.2 shows an example. Both the source and star unfolding will be explored in more detail in Chapter 24.

Number of pieces. We have insisted that the unfolding be a single, connected piece. Suppose that this constraint is loosened:

---

[2] It is an interesting problem to identify a different, third technique for general unfolding.

> **Open Problem 22.1: Fewest Nets.**[a] Given a convex polyhedron of n vertices and F faces, what is the fewest number of pieces, each of which unfolds to a simple polygon, into which it may be cut by slices along edges? Provide an upper bound as a function of n and/or F.
>
> _____
>
> [a] Demaine and O'Rourke (2004).

Cutting out each face leads to the trivial upper bound F. We can easily improve this bound to a fraction of F in two special cases by employing graph-matching results:

1. The first cuts the bound in half for triangulated (*simplicial*) polyhedra by employing Petersen's theorem, a classic result in graph matching. This theorem leads, in this case, to a perfect matching in the dual graph, which matches pairs of triangles to quadrilaterals, each of which unfolds without overlap. This leads to the claimed $F/2$ pieces.
2. The second improvement holds for *simple polyhedra*, those for which every vertex has degree 3. In this case, the dual graph of the polyhedron is formed of triangles. Now we can apply the result that every planar triangulated graph on F nodes has a matching of size at least $(F + 4)/3$ (Biedl et al. 2001b), which leads (after some calculation) to a bound of $(2/3)(F - 2)$ pieces.

These arguments cover only the extreme cases of simplicial and simple polyhedra, but at least hint at the type of arguments that might succeed in the general case.

The first bound, in the general case, of cF for $c < 1$ was obtained by Michael Spriggs, who established $c = 2/3$.[3] His argument depends on the result that every 3-connected planar graph (and polyhedral duals are such graphs, by Steinitz's Theorem 23.0.3 (p. 339)) admits a spanning tree of maximum degree 3 (Czumaj and Strothmann 1997). One can then match faces by plucking them off from the leaves of this binary tree. This alternately matches a pair of faces, or leaves one face by itself, leading (for sufficiently large F) to an upper bound of $(2/3)F$.

The smallest value of c obtained so far is $1/2$, via an as-yet unpublished argument.[4] It is a reflection of our ignorance that although there is reason to suspect the answer to Open Problem 22.1 is "1," we cannot prove any upper bound sublinear in F. We remark that this same problem remains open for nonconvex polyhedra, although there are interesting heuristic algorithm implementations, for example, Mitani and Suzuki (2004).

Overlap penetration. In a similar vein, we may define a notion of the degree of *overlap penetration* $\omega(U)$ of an overlapping unfolding U. Although one could define the degree of overlap as the area of doubly covered points, this does not seem to capture the more relevant notion of how much one piece of the unfolding "penetrates" another. We therefore define the degree of overlap by the ratio of the largest overlap penetration in U to the diameter of the polyhedron P. The latter is the maximum (shortest path) distance between any two points on the surface of P and is used merely to normalize

_____

[3] Personal communication, August 2003.
[4] Personal communication from Vida Dujmović, Pat Morin, and David Wood, February 2004.

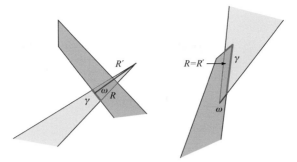

Figure 22.3. R is a region on the blue section that unfolds on top of the pink section. The cut curve γ is shown in red; the xy segment that determines the penetration ω is shown in blue.

the measure. The "overlap penetration" takes more explanation. Let R be a connected region on the polyhedron surface whose interior contains no cut edges and unfolds to be doubly covered (or covered to greater depth); that is, it is a region of overlap of U. The overlap penetration of U is the largest penetration of any R.

Focus now on a specific R. Let γ be a boundary-to-boundary curve that cuts off a portion R' of U that includes R (perhaps $R' = R$). That is, if we cut along γ, as well as the cut tree, and discard the disconnected R', the overlap depth in the unfolding has one fewer layer of paper at the planar position of R. Then the *penetration ω* of R with respect to γ is

$$\min_{x \in \gamma} \max_{y \in R'} |y - x| \,,$$

that is, the largest distance of a point in R' from the cut γ. Thus ω measures the depth beyond the cut that alleviates overlap (see Figure 22.3). The penetration of R is the smallest penetration over all γ that cut off R. Failing a resolution of Open Problem 21.1, it is natural to seek unfoldings that minimize overlap penetration. Little is known here:

Open Problem 22.2: Overlap Penetration. Find an algorithm that results in small overlap, for example, one that guarantees unfoldings with overlap penetration no worse that a constant fraction of ω_{max}.

A candidate example that might achieve ω_{max} will be shown in Figure 22.8.
We now turn to nonconvex polyhedra.

Nonconvex polyhedra. Nonconvex polyhedra is an important topic, for its applications to manufacturing mentioned above if for no other reason. There has been little theoretical progress, aside from two papers. The first (Biedl et al. 1998b) concentrated on orthogonal polyhedra, all of whose faces meet at right angles. We will discuss the more substantive results elsewhere (Section 22.5.4), but here we mention that this paper established that the open problem dissolves for nonconvex polyhedra: There are nonconvex orthogonal polyhedra with no edge unfoldings. One example, Figure 22.4(a), is nearly trivial: the box on top must unfold to fit inside the hole of the top face of the lower box, but there is clearly insufficient area to do so. But this example somehow feels

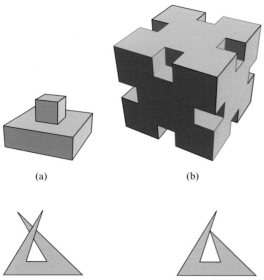

(a)   (b)

Figure 22.4. Orthogonal polyhedra with no edge unfoldings. [Based on Figure 9 of Biedl et al. 1998b.]

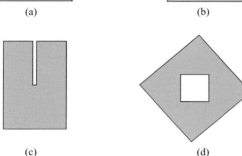

(a)   (b)

Figure 22.5. Four nonsimple polygons. In (c) the cut is intended to have zero width.

(c)   (d)

like "cheating," relying on a face that is not simply connected,[5] and leads us into the thicket of the precise definition of what constitutes a polyhedron (Grünbaum 2003). Figure 22.4(b) resolves this tension, for each of its faces is homeomorphic to a disk, and yet it still has no edge unfolding (much less obviously). Still, one feels that the essence is reached only with convex faces. That there exists polyhedra with convex faces that cannot be unfolded without overlap was established in 1999, and will be described in Section 22.4. Thus, under every reasonable interpretation, the open problem dissolves without the convexity restriction on the polyhedra.

Simple polygon. The problem statement demands that the unfolding be a single nonoverlapping *simple polygon*. A simple polygon cannot overlap itself, as does (a) of Figure 22.5, and it must be simply connected, ruling out holes as in (d). More subtly, even its boundary should not self-touch, as in (b) and (c). However, given the paucity of results in this area, an algorithm that unfolded a convex polyhedron to polygons whose interior is simply connected would be equally interesting, for such polyhedra could be folded from a paper cutout. We will return to the possibility of a hole in a single piece below.

---

[5] A shape is *simply connected* if every cycle can be contracted to a point, or equivalently, if it is homeomorphic to a disk. Thus, a simply connected region has no holes.

22.1.3   Spanning Trees

We now resume discussion of the unaltered edge-unfolding problem for convex polyhedra. The collection of edges that constitute an edge unfolding must *span* the vertices,
in the sense that every vertex of the polyhedron must have at least one incident cut edge,
for otherwise the vertex retains its 3D structure and cannot be flattened. (We insist that
polyhedron vertices be "real" in the sense that they have nonzero curvature; without
this definition, zero-curvature points would not need incident cut edges.) Thus the cut
edges form a *spanning graph* of the polyhedron *1-skeleton*, the graph formed by its vertices and edges. To achieve a single piece, the edges cannot include a cycle, for a cycle
would disconnect the unfolding. Thus, the cut edges form a *forest*. Finally, to achieve a
simple polygon, the set of cut edges must be connected, because the cut edges unfold
to the boundary of the polygon and the boundary is connected. So, the cut edges form
a *spanning tree*. We record this easy result in a lemma for later reference:

Lemma 22.1.1. *The cut edges of an edge unfolding of a convex polyhedron to a simple
polygon form a spanning tree of the 1-skeleton of the polyhedron.*

If we demand a single, connected piece, but do not insist that it be simply connected,
then it is conceivable that this is achievable by a spanning forest. Indeed, this possibility
can be realized for nonconvex polyhedra, as was first noticed by Andrea Mantler (Bern
et al. 2003). Figure 22.6 shows her example. The unfolding shown in Figure 22.6(e) is
obtained by two disjoint cut trees: one a path of two edges connecting the mountain
peaks, one an н to unfold the base.

Seeing this surprising example raises the question of whether a similar phenomenon
can occur for convex polyhedra. The answer NO is provided by the following lemma
(Bern et al. 2003, Lem. 4):

Lemma 22.1.2. *The cut edges of an edge unfolding of a convex polyhedron to a single,
connected piece form a spanning tree of the 1-skeleton of the polyhedron.*

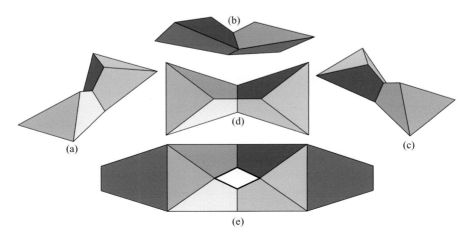

Figure 22.6. (a–c) Three views of polyhedron; (d) overhead view; (e) unfolding. [Based on Figure 2
of Bern et al. 2003.]

Table 22.1: Status of main questions concerning nonoverlapping unfoldings

Shapes	Edge unfolding?	General unfolding?
Convex polyhedra	Open	YES
Nonconvex polyhedra	NO	Open

Proof: We know the cut edges must form a forest; suppose in contradiction to the claim of the lemma that the forest has more than one connected component. Then there is a closed path p on the surface of the polyhedron that avoids all cuts and strictly encloses a connected component of the cutting. In particular, p avoids all vertices of the polyhedron. Let τ denote the total turn angle along the path p. Because p avoids vertices, it unfolds to a connected (uncut) closed path in the assumed planar layout. Because the path is closed, $\tau = 2\pi$. Now we apply the Gauss-Bonnet theorem, specifically, Equation (21.3) (p. 388), $\tau + \gamma = 2\pi$, to conclude that the curvature enclosed on the surface is $\gamma = 0$. (Note: the layout of p does not alter the turn angles, so τ is the same in the layout as on the surface.) But p encloses a component of the cutting, which must span at least one vertex, which, by our definition of a vertex, implies that $\gamma > 0$. We have reached a contradiction: p cannot in fact be flattened.   □

Table 22.1 summarizes the status of the main questions on unfolding. Having now detailed the contours of these problems, we present the (spare) evidence for and against a positive answer to Open Problem 21.1, the upper left entry in the table.

22.2   EVIDENCE FOR EDGE UNFOLDINGS

The strongest evidence for a positive answer to the edge-unfolding problem is that no one has yet found a convex polyhedron that cannot be edge unfolded without overlap. Indeed this led Grünbaum to conjecture the answer YES to Open Problem 21.1 in Grünbaum (1991). In the hundreds of years since Dürer's drawings, presenting a net (by definition, without overlap) has become a standard way to convey the structure of a polyhedron. Nets have been constructed for all the most interesting regular polyhedral shapes, both by hand, and more recently, by a variety of programs. For example, Figure 22.7 displays nets for the Archimedean solids constructed in Mathematica (compare Dürer's slightly different net in Figure 21.3 for the central truncated icosahedron). The program HyperGami (and its successor JavaGami)[6] has allowed users to unfold polyhedra without encountering (to our knowledge) an "ununfoldable" convex polyhedron. A more systematic exploration (Schlickenrieder 1997), to which we will return in a moment, also failed to find an ununfoldable convex polyhedron.

Of course, none of this is substantive evidence, especially because there is much more interest in highly regular polyhedra, which seem to be relatively easy to unfold without overlap. Stronger evidence would be broad classes of polyhedra all of which are known to unfold without overlap. However, the results here are again thin. A case-analysis proof established that all polyhedra of no more than six vertices unfold without overlap (DiBiase 1990), but the proof seems to give little insight into the general

---

[6] http://www.cs.colorado.edu/~ctg/projects/hypergami/.

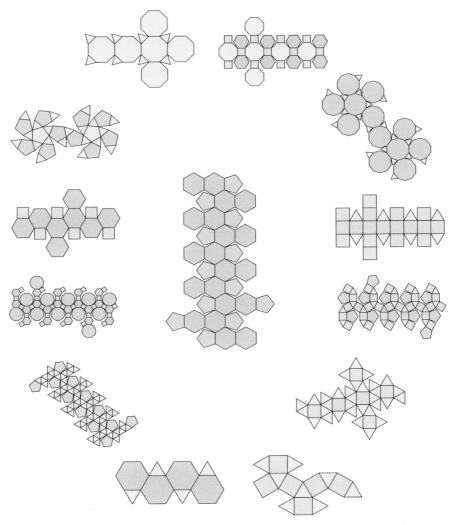

Figure 22.7. Unfoldings of the 13 Archimedean solids. Clockwise, starting at 1 o'clock: great rhombicuboctahedron, truncated dodecahedron, small rhombicuboctahedron, small rhombicosidodecahedron, snub cube, cuboctahedron, truncated tetrahedron, snub dodecahedron, great rhombicosidodecahedron, truncated octahedron, icosidodecahedron, truncated cube; center: truncated icosahedron. [Computed by Eric Weisstein's code Archimedean.m, and PolyhedronOperations.m available at http://www.mathworld.wolfram.com/packages/.]

case. We will describe one highly specialized class that unfolds without overlap in Section 22.5.1.

Most people who try to construct nets for polyhedra rather quickly develop an intuition that there must always be a way to do it for any convex polyhedron, but it has to be admitted that there is not much reason to trust this intuition.

22.3   EVIDENCE AGAINST EDGE UNFOLDINGS

Hand exploration can lead one to wonder if perhaps every edge unfolding avoids overlap, for it is not easy to create overlap. However, even a tetrahedron might have an edge

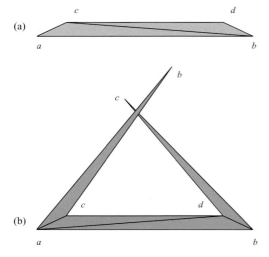

(a)

(b)

Figure 22.8. All faces are colored blue on the outside and red on the inside. (a) A flat doubly covered quadrilateral; (b) unfolding from cutting path (a, b, c, d); (c) unfolding from cutting (a, c, d, b).

(c)

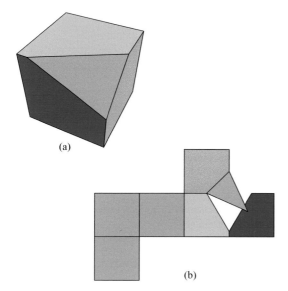

(a)

(b)

Figure 22.9. (a) Cube with corner truncated; (b) unfolding. [Based on Namiki and Fukuda 1993.]

unfolding that overlaps. Figure 22.8 shows an example of a thin, nearly flat tetrahedron that has an overlapping unfolding. The example also works if it is exactly flat, in which case it is a double-sided trapezoid $abcd$. Although it is easy to unfold without overlap Figure 22.8(c), there is an unfolding Figure 22.8(b) that causes two faces to cross. Another example, perhaps easier to grasp immediately, is shown in Figure 22.9: an unfolding of a cube with one corner truncated.

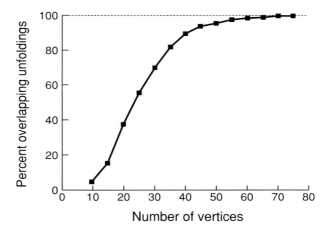

Figure 22.10. Each point in this graph represents an average of five randomly generated polyhedra whose vertices lie on a sphere, and the percentage of 1000 randomly selected unfoldings of each which overlaps. [After Figure 2.12 of Schevon 1989, p. 30.]

So there certainly are overlapping unfoldings, even if not easily found via hand exploration. In curious contrast, it was noticed early on that "most" unfoldings lead to overlap. Schevon explored all unfoldings of random polyhedra in Schevon and O'Rourke (1987) and Schevon (1989). The unfoldings can be tested by methodically generating spanning trees of the polyhedron graph, unfolding, and checking for overlap. There are an exponential number of spanning trees for any graph; in our case, $2^{\Omega(\sqrt{F})}$ for a polyhedron of F faces (Read and Tarjan 1975).[7] Thus it is not feasible to check them all, but it is possible to sample from them randomly (Schevon and O'Rourke 1987). Typical data is shown in Figure 22.10 for polyhedra constructed from random points on a sphere. For a sufficiently large number of vertices n, overlap is common. One can see in the figure that 99% of unfoldings of polyhedra with $n \geq 70$ vertices led to overlap. This led to the conjecture that as $n \to \infty$, the probability of overlap in a random unfolding of a random polyhedron of n vertices approaches 1, under any reasonable definition of "random."

However, a negative answer to Open Problem 21.1 requires a polyhedron for which exactly 100% of its unfoldings overlap.

22.3.1 Schlickenrieder's Thesis

Some of the strongest negative evidence is from the same source as some of the strongest positive evidence: the thesis of Wolfram Schlickenrieder (1997). Following a decade after (Schevon and O'Rourke 1987), this work tested potential unfolding algorithms on a test suite of tens of thousands of polyhedra. For each of the 34 distinct algorithms he implemented, he was able to find polyhedra that thwarted that algorithm.

Figure 22.11 shows a typical nonoverlapping unfolding for one particular algorithm, while Figure 22.12 shows overlap using a different algorithm on a "turtle" polyhedron, a class of nearly flat polyhedra defined by Schlickenrieder that were among the most difficult for the algorithms to unfold without overlap.

---

[7] More precisely, at least 2^k, where $k = \lceil \frac{1}{2}[\sqrt{8(F-2)+1} - 1] \rceil$.

Figure 22.11. RIGHTMOST-ASCENDING-EDGE-UNFOLD. [Figure 40c of Schlickenrieder 1997.]

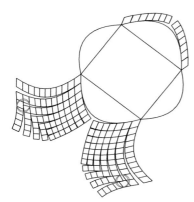

Figure 22.12. NORMAL-ORDER-UNFOLD. [Figure 42b of Schlickenrieder 1997.] Overlap circled.

One result of Schlickenrieder's investigations was a counterexample to a conjecture of Komei Fukuda, posted on his influential Web pages[8] exploring unfolding: Cutting the boundary of a polyhedron along the edges of a shortest-path tree from a fixed "source" vertex leads to nonoverlapping unfolding. The inspiration for this algorithm, SHORTEST-PATHS-UNFOLD, is the nonoverlapping star unfolding (which is not an edge unfolding)—Figure 24.11—to be discussed in Section 24.3. Although he found counterexamples to Fukuda's conjecture, he did not find any polyhedron that could not be unfolded without overlap from *some* source vertex. (A related conjecture of Fukuda's—that a minimum-length spanning tree leads to nonoverlap—was earlier shown false by Günter Rote.)

The best algorithm he found is what he called STEEPEST-EDGE-UNFOLD, the only algorithm we describe here. First, a (unit) direction vector c is selected so that there is a unique highest vertex v_+ and a unique lowest vertex v_-. Then, for every vertex v except v_+, the steepest ascending edge is selected as that edge $e = (v, w)$ that minimizes the angle θ between c and $w - v$, that is, which maximizes $c \cdot u$, where u is the unit vector $(w - v)/|w - v|$. The "objective function" c is chosen to be in general position in the sense that there are no two vertices at the same "height," and no ties for steepest.

[8] "Strange Unfoldings of Convex Polytopes," http://www.ifor.math.ethz.ch/~fukuda/unfold_home/unfold_open.html, 1997.

The cut tree is then the union of all the steepest ascending edges. The reason this is a spanning tree can be seen as follows. Every vertex except v_+ has an incident steepest out-edge. So if there are n vertices, the graph has $n-1$ edges. There can be no cycle, because a cycle would have a unique lowest vertex, which would then have two tied-for-steepest ascending edges, violating the general position of c. An acyclic graph on n vertices with $n-1$ edges must be a tree. The intuition behind this algorithm is that it should produce relatively straight paths, which spread out the unfolding. Figure 22.13 shows a net produced by this algorithm for a 400-vertex spherical polyhedron. Although this intuition is natural, the dual notion of not cutting the flattest edges, which seems equally natural, was found to be less effective (Figure 22.14).

Figure 22.13. STEEPEST-EDGE-UNFOLD. [Figure 35 of Schlickenrieder 1997.]

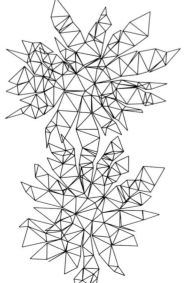

Figure 22.14. FLAT-SPANNING-TREE-UNFOLD. [Figure 38 of Schlickenrieder 1997.]

Although STEEPEST-EDGE-UNFOLD was effective, it failed to avoid overlap for all explored definite methods of selecting c. However, similar to the situation with SHORTEST-PATHS-UNFOLD, no polyhedron was found that completely thwarted STEEPEST-EDGE-UNFOLD: each of the 60,000 polyhedra in his test suite could be unfolded without overlap for *some* objective function c. Indeed, the simple strategy of choosing c randomly and retrying when the unfolding overlapped always succeeded (indeed, in no more than six iterations). This led Schlickenrieder to conjecture that there is always a c for any polyhedron that leads to overlap. However, recently this was shown to fail on a particular constructed counterexample (Lucier 2004).

We are left, then, with several procedures for finding a nonoverlapping unfolding, but none are definitive algorithms, guaranteed to succeed in all cases.

22.4   UNFOLDABLE POLYHEDRA

As mentioned earlier (p. 309), dropping the stipulation that the polyhedron is convex leads to examples that have no nonoverlapping edge unfolding to a single piece (see Figure 22.4). However, the examples in these figures use nonconvex faces. The question of whether there is a (genus-zero) polyhedron, all of whose faces are convex polygons, was raised by Schevon (1987) and by Dobilin (Grünbaum 2001). A positive answer was provided by several researchers independently at about the same time (Bern et al. 1999, 2003; Tarasov 1999), and improved shortly thereafter in various ways (Grünbaum 2001). In particular, Grünbaum found a very simple example of only 13 faces (Grünbaum 2002). Here we describe the more complex ununfoldable "spiked tetrahedron" of Bern et al. (2003), which employs a similar basic shape—a star-shaped polyhedron with sharp spikes—but is unique among the examples in having all (36) faces triangles.

The basic unit of the example is an open polyhedral surface that cannot be edge unfolded. This unit, called a *hat*, can be seen to be related to the box-on-a-box example from Figure 22.4. It has one interior vertex x of positive curvature (the peak of the hat), and three interior vertices $\{a, b, c\}$ of negative curvature. There are three "hat corners" $\{A, B, C\}$ on the hat boundary. It is formed by nine isosceles triangles: three with base angle α forming the spike, three with base angle β, and three with apex angle γ, arranged as in Figure 22.15.

The total face angle incident to a vertex a at the base of the spike is

$$2\alpha + (\pi - 2\beta) + 2(\pi - \gamma)/2 = 2\pi + 2[\alpha - (\beta + \gamma/2)].$$

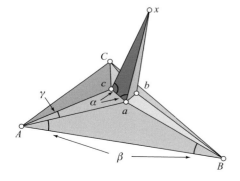

Figure 22.15. An open triangulated hat that cannot be edge-unfolded.

This exceeds 2π (and so a has negative curvature) whenever $\alpha > \beta + \gamma/2$. If α is even larger, satisfying $\alpha > 2\beta + \gamma$, then a has negative curvature even when one spike triangle is removed. For example, for $\alpha = 82.3°$, $\beta = 25.7°$, and $\gamma = 13.5°$ (the angles actually used in the figure), we have $\alpha = 82.3° > 2\beta + \gamma = 64.9°$. And the curvature at a with one spike triangle removed is

$$2\pi - [\alpha + (\pi - 2\beta) + 2(\pi - \gamma)/2] = 360° - 377.4° = -17.4°.$$

We now argue that a hat cannot be cut along edges to unfold to a single nonoverlapping piece. The cut edges must form a forest that spans the interior vertices. Consider one tree in this forest. Because the spike base vertices have negative curvature, they cannot be leaves of the tree. If the tree had two boundary corners as leaves, it would disconnect the surface into more than one piece. Because a tree must have two leaves, it must therefore be a single path from the hat tip x to a boundary vertex.

There are only two combinatorially distinct possibilities for this path, and, as Figure 22.16 shows, both leave one spike base vertex with only one spike face removed. Because we have just shown this vertex has negative curvature, the unfolding overlaps there.

Finally, we glue four hats together, as shown in Figure 22.17, forming a "spiked tetrahedron." We now argue that this polyhedron has no edge unfolding.

Consider a cut tree T for this object that avoids overlap. T restricted to one hat cannot look like the paths shown in Figure 22.16. We now consider the possibilities inside one hat, using the labeling shown in Figure 22.18, to reach the conclusion that there is a path of T in the hat connecting two hat corners $\{A, B, C\}$.

The hat peak x must connect via T to outside the hat, and so there must be a path from x through a hat corner, say B. If this path passes through all three spike base

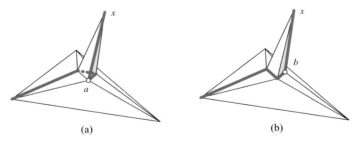

(a)   (b)

Figure 22.16. Both possible cut trees leave one spike base vertex (marked) with only one spike face (shaded) removed.

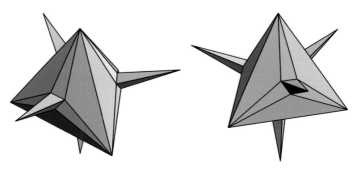

Figure 22.17. Two views of an ununfoldable triangulated polyhedron.

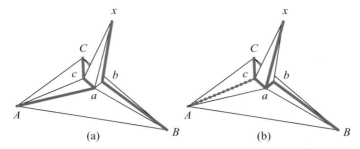

Figure 22.18. Possible restrictions of a cut tree to one hat.

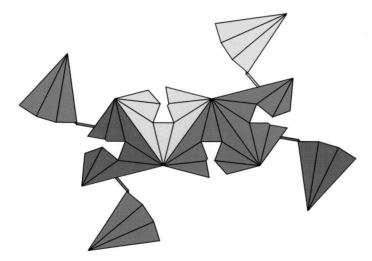

Figure 22.19. A general unfolding of a spiked tetrahedron.

vertices $\{a, b, c\}$, then any additional edge of T (and there must be at least one to avoid Figure 22.16) will connect two corners. So suppose that the path from x to B passes through just one (b: Figure 22.18(a)) or two (a and b: Figure 22.18(b)) base vertices. In the former case, a and c need to be spanned by T, but neither can be a leaf (because both have negative curvature). So there must be a path from A to C as shown. In the latter case, c needs to be spanned by T, but again it cannot be a leaf. The path through c could go either to C or to A (or to both), and in either case, we have a path between two corners.

Now the hat corners are the four corners of the tetrahedron. So we have these four corners connected by four corner-to-corner paths inside the hats. Viewing each of these four paths as an arc connecting four corner nodes, we can see that it is impossible to avoid a cycle, for a tree on n nodes can have only $n-1$ arcs. This violates Lemma 22.1.2, and establishes that the spiked tetrahedron has no edge unfolding to a net.

One might wonder if the spiked tetrahedron can be unfolded without restricting the cuts to edges. Figure 22.19 illustrates a strategy (for a differently parametrized version of the polyhedron) which throws the four spikes safely out to the boundary of the tetrahedron unfolding.

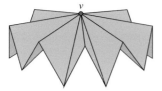

Figure 22.20. An open surface that has no general unfolding to a net.

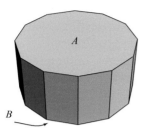

Figure 22.21. A prism.

In fact, no example is known that cannot be unfolded without overlap:

Open Problem 22.3: General Nonoverlapping Unfolding of Polyhedra.[a]
Does every closed polyhedron have a general unfolding to a nonoverlapping polygon?

---

[a] Bern et al. (2003).

However, it is easy to construct open polyhedral surfaces that have no net. Figure 22.20 is a simple example. Its one interior vertex v has negative curvature and so cannot be a leaf of a cut tree. So the leaves of the cut tree must lie on the boundary. Because a tree has at least two leaves, the surface must be disconnected by the cuts.

22.5 SPECIAL CLASSES OF EDGE-UNFOLDABLE POLYHEDRA

One might think that, although no one has been able to prove that every convex polyhedron is edge-unfoldable to a net, surely many special infinite classes of polyhedra have been established as edge-unfoldable. Such is not the case. Results here are sparse. An undergraduate thesis (DiBiase 1990) established that all polyhedra of 4, 5, or 6 vertices may be edge-unfolded to a net, but, as mentioned earlier, the proof does not seem to give insight for larger n. For some classes of polyhedra, a nonoverlapping unfolding is obvious and therefore uninteresting. For example, all *prisms*—formed by two congruent, parallel convex polygons (the *base* and *top*) and rectangle side faces (see Figure 22.21)—can be unfolded by cutting all side edges and all but one top edge. Let us call this a *volcano* unfolding; it blows out the side faces around the base and flips out the top.

But even generalizing to prismatoids has not be accomplished. A *prismatoid* is the convex hull of parallel base and top convex polygons. Its lateral faces are triangles or trapezoids. To give some sense of why this is not straightforward, we outline a proof that a subclass of prismatoids, the prismoids, may always be edge-unfolded without overlap, and indicate where the generalization to prismatoids fails. After sketching a proof that "dome" polyhedra can be edge-unfolded without overlap, we look at convex unfoldings, orthogonal polyhedra, and conclude with open problems in Section 22.5.5.

22.5.1   Prismoids

A *prismoid* is a prismatoid with the base and top angularly similar convex polygons, oriented so that corresponding edges are parallel. (The base and top do not have to be similar, just equiangular.) All lateral faces are trapezoids (see Figure 22.22).

Label the vertices of the top A as $a_0, a_1, \ldots, a_{n-1}$ and the corresponding vertices of the base B as $b_0, b_1, \ldots, b_{n-1}$. The side quadrilateral faces, which we call *flaps*, have vertices $(a_i, a_{i+1}, b_{i+1}, b_i)$.

There are at least two natural edge unfoldings of a prismoid. The first cuts only one side edge (a_i, b_i), unfolding all side faces into a strip, and attaches the top and base to opposite sides of this strip. This is related to "band unfoldings" discussed in Section 24.5.3. Whether this strategy can avoid overlap for all prismoids remains unclear. The second is the volcano unfolding, which cuts every side edge (a_i, b_i), does not cut any bottom edge (b_i, b_{i+1}), and cuts all but one top edge (a_j, a_{j+1}). Thus all flaps attach to the bottom, with the top attached to the top of one flap (Figure 22.23).

Although it may seem obvious that volcano unfoldings avoid overlap, there is an issue: where to place the top, that is, which top edge $a_j a_{j+1}$ should be left intact? Not every choice leads to nonoverlap, as demonstrated in Figure 22.24. The failure in this example is caused by attaching A to a short flap, rather to one that throws A further away. This leads to a natural rule, which we now describe.

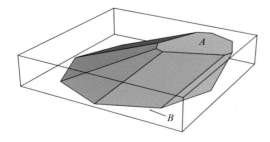

Figure 22.22. A prismoid. The base B is a regular octagon.

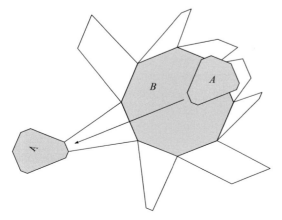

Figure 22.23. A nonoverlapping volcano unfolding of the prismoid of Figure 22.22.

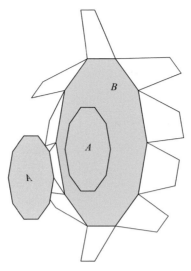

Figure 22.24. A volcano unfolding of a prismoid (not shown) that overlaps.

Let c_i be the unfolded position of a_i in the plane of B, and let a'_i be the projection of a_i onto this same plane. Here is the rule: Attach A to that outer flap edge $c_j c_{j+1}$ whose displacement from $a'_j a'_{j+1}$ is largest over all flaps. This is the choice used in Figure 22.23. It is established by O'Rourke (2001b) that $c_j c_{j+1}$ must be on the (2D) convex hull of all c_i points, and so it is "safe" to place A there. One more fact is needed to complete the proof of nonoverlap:

Lemma 22.5.1. *In any volcano unfolding of a prismoid, no pair of side flaps overlap.*

Of course, it is obvious that adjacent flaps do not overlap, but an argument is needed for nonadjacent flaps. We will not discuss this issue further, but rather revisit it for the proof for domes, which seems more likely to have applicability beyond these special cases.

It would seem that prismatoids should also unfold without overlap via a volcano-like unfolding, but the generality of the side faces makes this unfolding more unruly. Figure 22.25 shows one example, where cutting all top A edges except $a_0 a_2$, and side edges $b_0 a_2$, $b_1 a_0$, and $b_2 a_1$, leads to overlap. Of course it is easy to unfold this polyhedron without overlap (e.g., by retaining the top edge $a_0 a_1$ instead), but no general rule is known.

Curiously, although there is as yet no algorithm for unfolding polyhedral prismatoids, there is a notion of unfolding that can be defined for "smooth prismatoids," a natural limit of polyhedral approximations, and here there is an algorithm for volcano unfolding any smooth prismatoid (Benbernou et al. 2004) (see Figure 22.26 for an example).

22.5.2 Domes

Another narrow class of shapes that are known to be edge-unfoldable is what we call "domes." A *dome* is a polyhedron with a distinguished *base* or ground face G, and the property that every nonbase face shares an edge with G (see Figure 22.27).

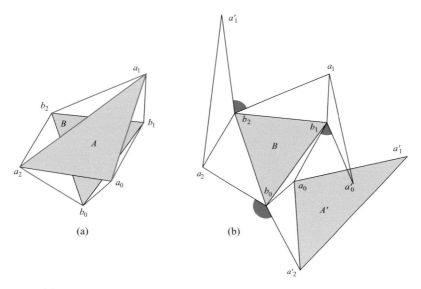

Figure 22.25. (a) A nearly flat prismatoid, viewed from above, cut tree in red; (b) overlapping volcano unfolding. Primes indicate unfolded items.

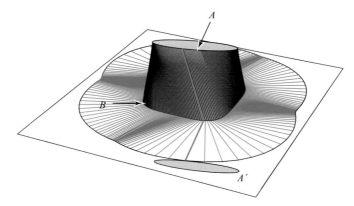

Figure 22.26. Volcano unfolding of a smooth prismatoid, the convex hull of an ellipse top A twisted with respect to a rounded quadrilateral base B.

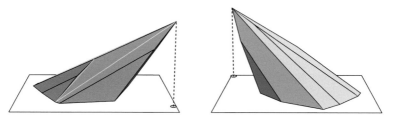

Figure 22.27. Front and back view of the same dome. The projection of the apex onto the base plane is shown for perspective.

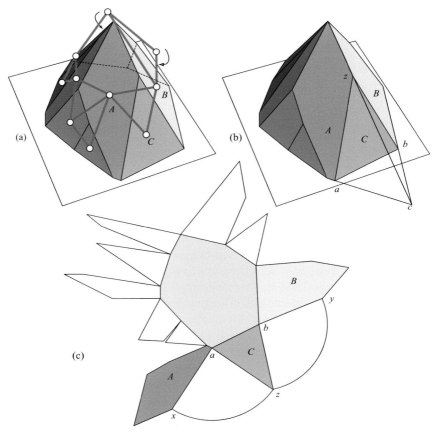

Figure 22.28. (a) Outerplanar dual graph of faces; (b) extension of incident faces A and B to encompass C; (c) proof of Lemma 22.5.4: two empty sectors xz and yz are shown.

The method of unfolding domes is similar to that just sketched for prismoids: It is again a "volcano" unfolding, but there is no top face, and the side flaps are more complex. We provide only a sketch of the main ideas of the inductive proof here.[9]

At the top level, for a given dome P, we select some edge of G to remove, extend adjacent faces to make a dome P' with fewer faces, apply an induction hypothesis to P', and then reestablish it for P. The selection of the edge to be removed is based on the following lemma:

Lemma 22.5.2. *At least two of the nonbase faces of a dome are triangles.*

Proof: Create a dual graph, one node per nonbase face, with two nodes connected by an arc if they share an edge (see Figure 22.28(a)). This graph is planar, and because each face is incident to G, the graph is outerplanar: every node is on the exterior face. It is well known that an outerplanar graph must have at least two degree-2 nodes (e.g., Harary 1972, Cor. 11.9a), a result that can be seen by triangulating the graph, and noting that the dual of this triangulation is a tree and so must have two leaves. A node of degree 2 corresponds to a triangle C surrounded by two faces A and B. In Figure 22.28(a), there are several degree-2 nodes in this graph, including face labeled C. $\square$

---

[9] This is joint work with Patricia Cahn.

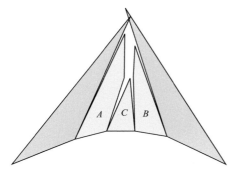

Figure 22.29. An impossible dome overlap in which no pair of adjacent faces overlap.

Let P have n faces. The plan is to select a triangle guaranteed by this lemma, remove it by extending the adjacent faces, apply induction to the resulting dome of $n - 1$ faces, and reestablish the induction hypothesis.

Of course, it may well be that the extension of the adjacent faces do not meet. But, fortunately, the lemma guarantees two degree-2 nodes, and at most one can have adjacent faces diverge (because divergence implies an angle greater than π, and there is only 2π total angle to G.) So let C be the triangle face whose adjacent faces do meet, as in Figure 22.28(b). One possible induction hypothesis is simply what we want ultimately to prove: that the unfolding does not overlap. But it seems one needs some detailed structure in the nonoverlap hypothesis to prove a result. For example, Figure 22.29 shows that merely relying on adjacent faces avoiding overlap is too weak a hypothesis. Certainly if we could ensure that rays between adjacent faces never intersect, then this situation could not hold. But in fact there might be intersections between such rays (see Figure 22.28(c)), so that hypothesis is too strong.

We have employed a hypothesis following an idea present in Aronov and O'Rourke (1992). Let the vertices of the base polygon G be $(v_0, v_1, v_2, \ldots, v_{n-1})$. In the volcano unfolding, v_i is the center of a circle sector determined by the incident cut edge, which splits into two edges and opens by an angle equal to the curvature at v_i. The strong induction hypothesis we employ is that these sectors are empty:

Lemma 22.5.3. *The volcano unfolding of a dome has the property that each sector determined by a base vertex v_i is empty (the closed sector is empty of any unfolded portions of the dome).*

So the essence of the proof is reestablishing this induction hypothesis after extension of face A and B to swallow C (cf. Figure 22.28(c) for notation).

Lemma 22.5.4. *The two n-dome sectors centered at a and at b are nested inside the $(n-1)$-dome sector centered at c.*

Proof: (*Sketch*) Figure 22.30(a) shows the situation both after C has been removed (ab replaced by c), when the $n - 1$ induction hypothesis (IH) holds—which guarantees that the sector (c, a_2, b_2) is empty of the unfolding of other portions—and after returning to n faces. We seek to prove that the sectors centered on $a \in A$ and $b \in B$ lie inside the IH sector centered on c as they clearly do in the figure.

Let z be the tip of the triangle C, and assume for contradiction that z lies on the sector centered at c (or outside) (see Figure 22.30(b)). Because C is cut away from A and

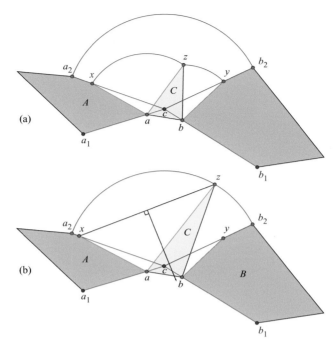

Figure 22.30. (a) The situation after reduction to $n - 1$ faces and return to n faces; (b) perpendicular bisector of xz to the "wrong side" of c.

B, the points x and y that correspond to the points on A and on B that match C's apex z must lie along the segments ca_2 and cb_2 respectively. Let us focus on x. The sector in question extends from x to z; consider the chord xz. It is clear that the perpendicular bisector passes to the "wrong side" of c (a must lie on a_1c), for only if $x = a_2$ would it pass through c. This shows that z is in the sector. Now we need to show that the entire arc xz is inside. But we simply repeat the above argument for any point z' on this arc that is outside or on the arc a_2b_2; the perpendicular bisector argument holds equally well. □

Many details on the exact geometric conditions that hold in this figure are needed to turn this sketch into a proof. This nesting lemma then yields our claim: Each sector is empty of other portions of the unfolding, and so the volcano unfolding of a dome does not overlap. The sector-nesting is illustrated most clearly in Figure 22.31, which is derived from the unfolding in Figure 22.32 (the polyhedron was shown earlier in Figure 22.27).

22.5.3 Convex Unfoldings

As mentioned earlier, the first discussion in print (of which we aware) of Open Problem 21.1 was by Shephard in 1975 (Shephard 1975). His paper focused on a specialization (suggested by Sallee) to convex nets. He defined a polyhedron P to have a *convex net* if there is an edge unfolding of P such that the resulting net is a convex polygon, and a *strictly* convex net if every whole edge of the net is an edge of P. The regular tetrahedron has convex nets, but no strictly convex nets (see Figure 22.33).

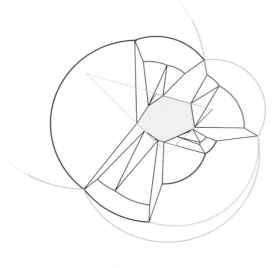

Figure 22.31. Sector nesting (Lemma 22.5.4) for unfolding in Figure 22.32.

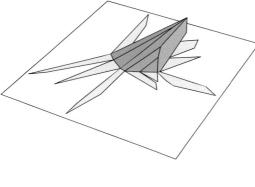

Figure 22.32. Unfolding of dome in Figure 22.27.

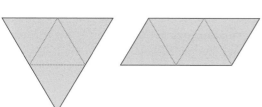

Figure 22.33. Two nonstrictly convex nets of a regular tetrahedron. [After Figure 1 of Shephard 1975.]

Sallee asked for a characterization of polyhedra with convex nets, but Shephard addressed a related question: Which polyhedra P have a combinatorially equivalent polyhedron P' so that P' has a (strictly) convex net? This shift of attention from geometric particulars to classes of combinatorially equivalent polyhedra permitted Shephard to obtain a number of results. For example, the classes of pyramids, bipyramids, prisms, antiprisms, cyclic polytopes, and wedges (duals of cyclic polytopes) all have among them instances of polyhedra with strictly convex nets. We will not present details of his results, but we will employ a key lemma of his when exploring the folding of convex polygons, the reverse of unfolding to a convex net (Section 25.7.1).

We should also mention another fruitful specialization Shephard introduced. He defined a net as *Hamiltonian* if the edge-unfolding cut tree is a Hamiltonian path.[10]

---

[10] A path that visits each vertex of the polyhedron exactly once.

Because it is known that some polyhedra have no Hamiltonian path (e.g., Coxeter 1973, p. 8), there are polyhedra with no Hamiltonian net. Even characterizing those polyhedra that have a convex Hamiltonian net seems challenging. This issue is touched upon in Section 25.10 (Figure 25.59, p. 432) and in Demaine et al. (2000b).

22.5.4 Orthogonal Polyhedra

Another natural restriction is to *orthogonal polyhedra*, those in which each pair of adjacent faces meets orthogonally, at a dihedral angle of either 90° or 270°. Many manufactured objects are orthogonal, so this is a class of objects with some practical applications. Although no nontrivial classes of orthogonal polyhedra are known to be edge-unfoldable, two interesting classes were established as unfoldable with more general cuts in Biedl et al. (1998b). First we define these classes as in the paper, and then modify their definitions so that the modified classes become edge-unfoldable. We will describe the results of this paper, but not include its intricate proofs.

Establish a Cartesian coordinate system with orthogonal x, y, z axes. We will use the term *x-plane* to denote a plane orthogonal to the x-axis and *x-face* for a face lying in an x-plane. An *x-edge* is parallel to the x-axis. The terms are defined analogously for "y-" and "z-."

Orthostacks. The first class of objects that can be unfolded is the class of "orthostacks." An *orthostack* is an orthogonal polyhedron whose every intersection with a z-plane is either empty or a single orthogonal polygon without holes. It can be viewed as stacked from pieces as follows. Let $P_0, \ldots, P_{n-1}$ be simple orthogonal polygons without holes, each lying in a z-plane at height z_i. Let E_i be a vertical *extrusion* of P_i between a range of z values, $z_i \leq z \leq z_{i+1}$. Then $S = \bigcup_i E_i$ is an *orthostack* if, for each i, the intersection of S with z_i is a single orthogonal polygon without holes. An example is shown in Figure 22.34(a).

We now sketch the unfolding algorithm. Let the *band* B_i be all the x- and y-faces of E_i, that is, the faces generated by extrusion. The overall idea is to unfold each band to a single strip and to attach the z-faces to either side of this strip. B_i is then connected to B_{i+1} so that the strips form a staircase pattern. However, we could not make this idea work treating each band as an indivisible unit. So further partition each band into an upper and lower half: $B_i = B_i^+ \cup B_{i+1}^-$. Finally, let Z_i be all the z-faces lying in the plane $z = z_i$; note that some are upward facing and some downward facing. The

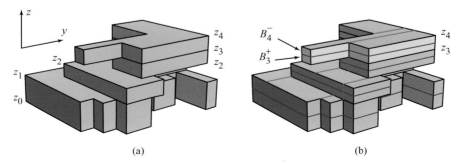

Figure 22.34. (a) A four-piece orthostack: $S = E_0 \cup E_1 \cup E_2 \cup E_3$; (b) refined orthostack, highlighting the band B_3, and showing only the additional edges needed by the algorithm.

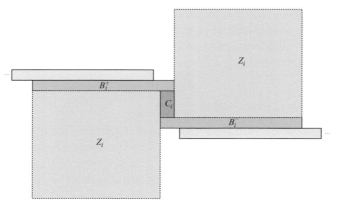

Figure 22.35. One unit of a generic orthostack unfolding. The Z_i faces are distributed into the two regions indicated.

basic component of the unfolding is $B_i^- \cup Z_i \cup B_i^+$, with the two half-bands joined by a rectangle connector $C_i \subset Z_i$, as roughly depicted in Figure 22.35. The details are formidable.

The final unfolding need not only cut each band in half, but also to make additional orthogonal cuts across faces: some z-cuts, some y-cuts. Recently concepts have been developed to capture these "refinements" of the polyhedron. A *grid unfolding* adds edges to the surface by intersecting the polyhedron with planes parallel to coordinate planes through every vertex. These extra edges have been used to achieve grid "vertex unfoldings" of orthogonal polyhedra (Damian et al. 2006; Demaine et al. 2005), a notion to be described in Section 22.6. A $k_1 \times k_2$ *refinement* of a surface (Demaine and O'Rourke 2003) partitions each face further into a $k_1 \times k_2$ grid of faces; thus, a 1×1 refinement is an unrefined grid unfolding. The orthostack algorithm sketched achieves a 2×1 refined grid unfolding, where the 2 derives from the band splitting. See Figure 22.34(b) for a coarser refinement that would suffice (but is difficult to describe). It remains open to achieve a grid unfolding of orthostacks, let alone an edge unfolding (Open Problem 22.4).

Orthotubes. The second class of orthogonal polyhedra that can be unfolded to a net is the "orthotubes." An *orthotube* is composed of rectangular bricks glued face-to-face to form a path or a cycle. Let $b_0, \ldots, b_{n-1}$ be a sequence of bricks. The condition on gluing is that $b_i \cap b_{i+1}$ is a face of each, and for bricks b_i and b_j not adjacent in the sequence, either they do not meet at all, or $b_i \cap b_j$ is an edge or a vertex of $b_i \cap b_{i-1}$ or $b_i \cap b_{i+1}$. This latter condition comes into play at corner turns of the tube. An orthotube is a union of bricks glued this way (see Figure 22.36(a)). Orthotubes can be cycles, and even knots.

We will not even hint at the algorithm for unfolding, but content ourselves with a rephrasing to *gridded orthotubes*, which merely retain every edge of every brick used in the construction (see Figure 22.36(b)). Then the algorithm in Biedl et al. (1998b) establishes that any (1×1) gridded orthotube may be edge-unfolded to a net. A curious property of the unfolding is that the creases are all mountain folds.

Even extending from orthotubes to "orthotrees" (trees built from bricks by face-to-face gluing) is unresolved at this writing (Damian et al. 2005a). Unfolding orthogonal

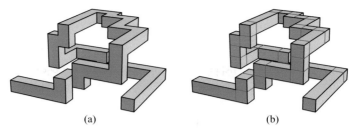

(a)   (b)

Figure 22.36. (a) Orthotube; (b) gridded orthotube.

polyhedra is an area of active research, with several recent results (Damian and Meijer 2004; Damian et al. 2005b, 2007). One illustration of our ignorance in this area is this:

Open Problem 22.4: Edge-Unfolding Polyhedra Built From Cubes. [a] Is there any genus-zero orthogonal polyhedron P built by gluing together cubes face-to-face that cannot be edge unfolded, where all cube edges on the surface of P are considered edges available for cutting?

---

[a] Posed with George Hart (2004).

It is only known that orthotubes built from cubes can be edge unfolded.

22.5.5   Other Classes

A search for unfoldable classes of polyhedra, convex or nonconvex, continues. The most recent advance is on a class named *higher-order deltahedra*. A *deltahedron* is a polyhedron all of whose faces are equilateral triangles. "Higher order" permits faces to be edge-to-edge unions of congruent equilateral triangles. Daniel Bezdek proved that convex higher-order deltahedra do have an edge unfolding and some nonconvex ones do not.[11]

We suggest here several more special classes that might be worth exploring. Because sharp angles are a source for local overlap in edge-unfoldings, it may be easier to solve this problem:

Open Problem 22.5: Edge-Unfolding for Nonacute Faces.   Prove (or disprove) that every convex polyhedron all of whose faces have no acute angles (i.e., all angles are $\geq 90°$) has a nonoverlapping edge-unfolding.

For example, a dodecahedron has all non-acute angles.

---

[11] Personal communication from Karoly Bezdek, February 2005.

Another approach along the same lines is as follows:

Open Problem 22.6: Edge-Unfolding for Nonobtuse Triangulations. Prove (or disprove) that every triangulated convex polyhedron, all of whose angles are non-obtuse, i.e., $\leq 90°$, has a nonoverlapping edge-unfolding. If the polyhedron has F faces, what are the fewest nets into which it may be cut along edges? (Cf. Open Problem 22.1.)

The point here is that if all triangles are nonobtuse, then any adjacent pair form a convex quadrilateral, whereas without the nonobtuse condition, they might form a nonconvex quadrilateral.

Rather than restrict the shape of the polyhedra, one could instead explore polyhedra with many edges from which to choose to achieve an edge unfolding. The "grid refinement" for orthogonal polyhedra used in the previous section (p. 330) leads to this attractive problem:

Open Problem 22.7: Vertex Grid Refinement for Orthogonal Polyhedra. [a]
Is there some fixed, finite degree $k_1 \times k_2$ of refinement that guarantees edge-unfolding of all orthogonal polyhedra?

---

[a] Demaine and O'Rourke (2005).

For example, there is an algorithm to edge-unfold a 4×5 vertex grid refinement of a certain class of polyhedra, Manhattan towers (Damian et al. 2005b). A recent algorithm unfolds all orthogonal polyhedra (Damian et al. 2007), but does not achieve a fixed degree of refinement.

One can extend the notion of refinement to nonorthogonal polyhedra:

Open Problem 22.8: Refinement for Convex Polyhedra. [a] Using any successive, regular refinement process, how many times must a convex polyhedron's faces be refined before it becomes edge-unfoldable?

---

[a] Demaine and O'Rourke (2005).

One particular process, shown in Figure 22.37, is to first triangulate the surface, and then at each successive refinement level, partition each triangle into four subtriangles by dividing at the midpoint of each edge (*1-to-4 refinement*). This has the effect of dividing edges, but not face angles. Barycentric subdivision—connecting each vertex of a face to the face's centroid—divides angles but not edges. An alternating mixture of the two seems a good candidate refinement process.

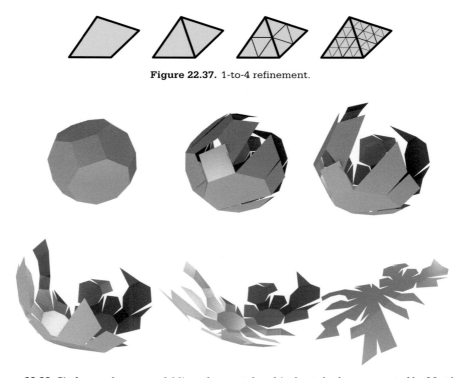

Figure 22.37. 1-to-4 refinement.

Figure 22.38. Six frames from an unfolding of a great rhombicuboctahedron computed by Matthew Chadwick. (Scale varies frame to frame.) Note that two faces interpenetrate in the second and third frames. http://celeriac.net/unfolder/.

As mentioned earlier (p. 307), there are two general methods of unfolding convex polyhedra that cut through the interior of faces. For both of these, adding $O(n^2)$ additional edges to the polyhedron that constitute these cuts suffice to make it edge-unfoldable. Refinement would permit approximating these unfoldings. So the problem can be viewed as asking whether some finite approximation suffices to avoid overlap.

Finally, we have concentrated solely on achieving nonoverlap in the final, planar unfolding, but little is known about avoiding intersections throughout the unfolding process. Figure 22.38 shows an unfolding that self-intersects in 3D but ultimately avoids overlap in the plane. This unfolding linearly interpolates dihedral angles; locally varying the rate of unfolding could avoid the intersection. This issue will be revisited when we discuss "continuous blooming" in Section 24.1.1 (p. 362).

22.6   VERTEX UNFOLDINGS

We end this section of negative results on edge unfolding with a positive result obtained by loosening the definition of edge unfolding (p. 299), changing the requirement that the net be a single, nonoverlapping, simply connected piece, by dropping the "simply connected" qualification. The polyhedron is still to be cut along edges, but unfolded to a nonoverlapping connected region that (in general) does not form a simple polygon, because its interior might be disconnected. Thus faces of the polyhedron are connected in the unfolding at common vertices, but not necessarily along edges. Such an unfolding is called a *vertex unfolding*. With this easier goal, the following result was obtained in Demaine et al. (2003a):

Theorem 22.6.1. *Every connected triangulated* 2-*manifold (possibly with boundary) has a vertex unfolding, which can be computed in linear time.*

This result includes simplicial (triangulated) polyhedra of any genus, manifolds with any number of boundary components, and even manifolds like the Klein bottle that cannot be topologically embedded in 3-space. In particular, both the edge-ununfoldable polyhedron in Figure 22.17 and the ununfoldable open manifold in Figure 22.20 have vertex unfoldings. However, despite the generality of Theorem 22.6.1, the result relies very much on the restriction that every face is a triangle: it remains open whether every nonsimplicial convex polyhedron has a vertex unfolding.

In this section we sketch some of the ideas of the proof of the above theorem, and in Section 26.3.3 we mention extensions to higher dimensions.

Figure 22.39 shows a number of vertex unfoldings of convex polyhedra, which immediately hints at the main algorithmic idea: to find a path from triangle to triangle on the surface, connecting through common vertices, and to lay out the triangles roughly along a (horizontal) line. We now explain this idea more precisely.

Let $\mathcal{M}$ be a triangulated 2-manifold, possibly with boundary. The *vertex-face incidence graph* of $\mathcal{M}$ is the bipartite graph whose nodes are the faces (triangles) and vertices of $\mathcal{M}$, with an arc (v, f) whenever v is a vertex of face f. We define a *face path* as an alternating series of nodes and arcs

$$(v_0, f_1, v_1, f_2, v_2, \ldots, f_k, v_k)$$

with each arc incident to the surrounding nodes, no arc appearing twice,[12] and each face appearing exactly once (but vertex nodes may be repeated). In any face path, v_{i-1} and v_i are distinct vertices of face f_i for all i (else an arc would be repeated).

The definition of a face path is designed to satisfy the following lemma:

Lemma 22.6.2. *If $\mathcal{M}$ has a face path, then $\mathcal{M}$ has a vertex unfolding in which each triangle of the path occupies an otherwise empty vertical strip of the plane.*

Proof: Let p be a face path of $\mathcal{M}$. Suppose inductively that a face path p has been laid out in strips up to face f_{i-1}, with all triangles left of vertex v_i, the rightmost vertex of f_{i-1}. Let (v_i, f_i, v_{i+1}) be the next few nodes in p; recall that $v_i \neq v_{i+1}$. Rotate face f_i about vertex v_i so that v_i is leftmost and v_{i+1} rightmost, and the third vertex of f_i lies horizontally between (or at the same horizontal coordinate as v_i or v_{i+1}). Such rotations exist because f_i is a triangle: if the triangle angle θ at v_{i+1} is acute, $v_i v_{i+1}$ can be horizontal; if $\theta > \pi/2$, rotation at v_i by $\theta - \pi/2$ suffices. Place f_i in a vertical strip with v_i and v_{i+1} on its left and right boundaries. Repeating this process for all faces in p produces a nonoverlapping vertex unfolding. $\square$

The idea is indicated more clearly in Figure 22.40: (a) shows a vertex unfolding of the triangulated surface of a cube, obtained from a face path; (b) repeats the first row of Figure 22.39. Note that the vertices do not necessarily lie on a line.

---

[12] In graph terminology, face paths are therefore *trails, walks* in which no arc is repeated.

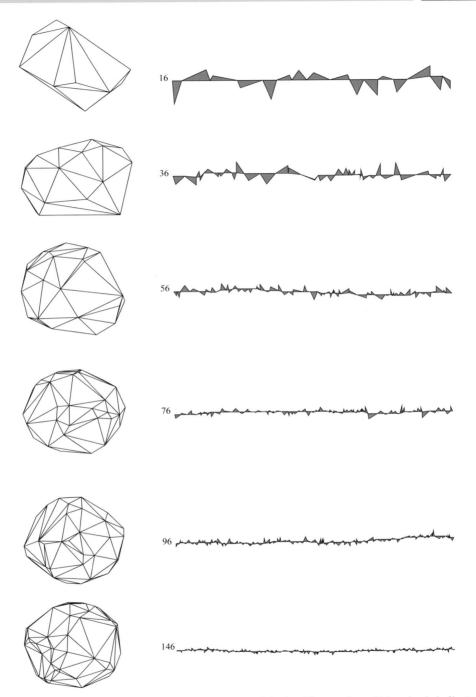

Figure 22.39. Vertex unfoldings of random convex polyhedra. The number of triangles is indicated between the polyhedron and the unfolding (which are not shown to the same scale). The unfoldings were constructed using an earlier, less general algorithm (Demaine et al. 2001b) [implemented with Dessislava Michaylova].

(a)

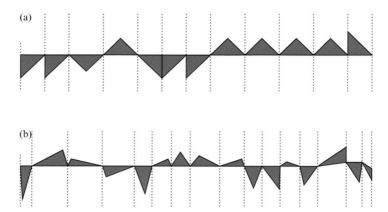

(b)

Figure 22.40. Laying out face paths in vertical strips. (a) Cube; (b) 16-face convex polyhedron.

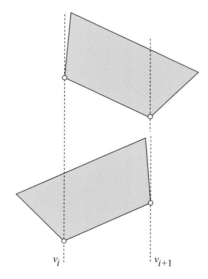

v_i v_{i+1}

Figure 22.41. This trapezoid cannot be laid out in a strip with v_i and v_{i+1} horizontally extreme.

It is crucial in the proof of Lemma 22.6.2 that each face be a triangle. Already with a quadrilateral face, the step of arranging v_i and v_{i+1} on the strip boundaries might necessarily leave part of the face exterior to the strip (see Figure 22.41).

Although Lemma 22.6.2 guarantees that a face path leads to a vertex unfolding, it could well be that the face path crosses itself, in the sense that it contains the pattern $(\ldots, A, v, C, \ldots, B, v, D, \ldots)$ with the faces incident to the vertex v appearing in the cyclic order A, B, C, D. This leads to a nonphysical situation that runs against the grain of the idea behind unfoldings. So it is natural to seek a *noncrossing face path*. Fortunately, any face path can be converted to a noncrossing face path, employing a result in Biedl et al. (1998c, Lem. 1), and illustrated in Figure 22.42. A noncrossing face path can be viewed as a noncrossing thread running over the surface, connecting faces through shared vertices.

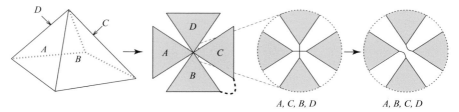

Figure 22.42. Vertex-unfolding the top four triangles of the regular octahedron and making the connections planar.

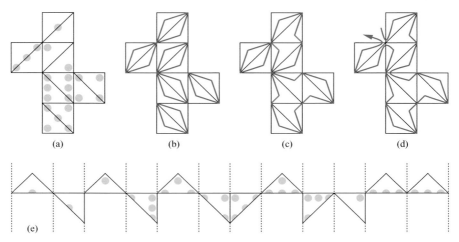

Figure 22.43. Tree manifold (a) to scaffold (b) to connected scaffold (c) to face cycle (d) to strip layout (e). The layout is based on a different face path than used in Figure 24.22(a). [Figure by Jeff Erickson, Figure 4 of Demaine et al. 2003a, by permission, Marcel Dekker.]

The proof of the existence of a face path for any triangulated 2-manifold is largely combinatorial,[13] and will only be hinted here, illustrated with a triangulated cube. The first step is to cut open the given manifold to a "tree manifold" via an arbitrary spanning tree (see (a) of Figure 22.43). Of course, this would not, in general, unfold without overlap, but in any case we treat it entirely as a combinatorial object. Second, a "scaffold," which necessarily exists, is found: a subgraph of the incidence graph in which every face appears and has degree 2, and at most two vertices have odd degree. In (b) of the figure, no vertices have odd degree. Step 3 is to convert the scaffold into a connected scaffold, (c). Finally, any Eulerian walk (a path that traverses each arc exactly once and passes through every node) through a connected scaffold is a face path (d), which can then be laid out as per Lemma 22.6.2, (e).

Theorem 22.6.1 generalizes to higher dimensions, a point to which we will return in Section 26.3.3.

Although the strip-layout algorithm could fail for nonsimplicial polyhedra, some other layout algorithm might succeed. However, there are polyhedra with no face paths at all. For example, the truncated cube, Figure 22.44(a), has no face path: no pair of its

---

[13] The existence of a face path was established independently in Bartholdi and Goldsman (2004), but in a context (TINs—"triangulated irregular networks") in which the restriction that no arc appears twice was irrelevant.

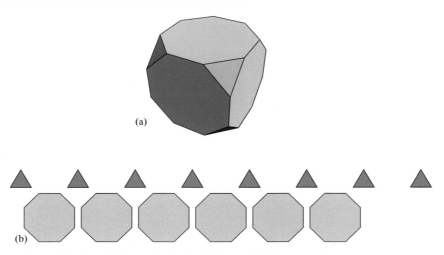

Figure 22.44. (a) A truncated cube; (b) its eight-triangle and six-octagon faces.

eight triangles can be adjacent in a path (because they are pairwise disjoint), but its six octagons are not enough to separate the triangles in a hypothetical face path.

Thus solving this open problem requires a new idea:

Open Problem 22.9: Vertex Unfolding. [a] Does every convex polyhedron have a nonoverlapping vertex unfolding?

---

[a] Demaine et al. (2003a).

Because every edge unfolding is a vertex unfolding, this might be easier than Open Problem 21.1.

Most recently, there has started to be exploration of vertex unfoldings of orthogonal polyhedra (Demaine et al. 2005). The strongest result to date is that every genus-zero orthogonal polyhedron, refined by the vertex grid, has a nonoverlapping (and non-crossing) vertex unfolding (Damian et al. 2006).

23 Reconstruction of Polyhedra

In this chapter we present three fundamental theorems on the structure of convex polyhedra that are essential to understanding folding and unfolding: Steinitz's theorem (below), Cauchy's rigidity theorem (Section 23.1), and Alexandrov's theorem (Section 23.3). It will be useful to view these results as members of a broad class of geometric "realization" problems: reconstructing or "realizing" a geometric object from partial information. Often the information is combinatorial, or combinatorics supplemented by some geometric (metric) information. A classic example here is "geometric tomography" (Gardner 1995b), one version of which is to reconstruct a 3D shape from several X-ray projections of it. When we fold a polygon to form a convex polyhedron, we have partial information from the polygon and seek to find a polyhedron that is compatible with, that is, which realizes, that information.

Steinitz's theorem. The constraints on realization available from purely combinatorial information are settled by Steinitz's theorem. The graph of edges and vertices of a convex polyhedron form a graph (known as the *1-skeleton* of the polyhedron).

Theorem 23.0.3. *G is the graph of a convex polyhedron if and only if it is simple, planar, and 3-connected.*

It is not so difficult to see that the three properties of G are necessary: G is simple because it has no loops or multiple edges; it is planar by stereographic projection, and it is at least plausible that the removal of no pair of vertices can disconnect G. The proof that any graph possessing these three properties is realized by some convex polyhedron is not straightforward. As this will be peripheral to our main thrust, we will not discuss it (see, e.g., Ziegler 1994, Lec. 4).

Pieces of information. Consider all the "pieces" of information inherent in a polyhedron.

- *Graph*: the graph of the edges and vertices on the surface of the polyhedron. We will always consider the graph to have a particular planar embedding so that the face structure is determined.
- *Face angles*: the angles between edges adjacent on a face of the polyhedron, measured within the plane of the face.
- *Edge lengths*: the Euclidean lengths of the edges.
- *Face areas*

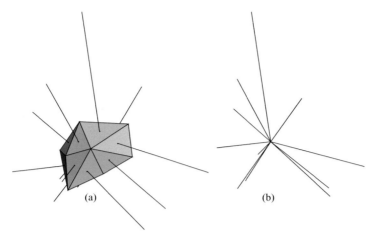

(a)   (b)

Figure 23.1. (a) A polyhedron with area-weighted face normals; (b) normals translated to origin.

- *Face normals*: unit vectors normal to the faces, pointing toward the exterior.
- *Dihedral angles*: the angles between faces adjacent across an edge; measured in 3-space, through the interior of the polyhedron (so always no larger than π).
- *Inscribed* or *circumscribed*: whether the polyhedron is inscribed or circumscribed within or around a sphere, respectively.

One can now consider mixtures of these sources of information and see which combinations determine others. For example, Stoker (1968) proved that *graph + face angles* ⇒ *dihedral angles*. Thus two polyhedra with the same graph and face angles are *isogonal*, a concept weaker than congruence. Karcher (1968) proved that *graph + dihedral angles + circumscribable* ⇒ *unique polyhedron*. One of the most beautiful reconstruction results is Minkowski's theorem.

Minkowski's theorem.

Theorem 23.0.4 (Minkowski). *face areas + face normals* ⇒ *unique polyhedron (up to translation)*.

A precursor, easier result is that the face normals of a convex polyhedron, each scaled by the area of the face, sum to zero (see Figure 23.1). The more substantive direction is the reverse: Any set of nonzero, nonparallel vectors that sum to zero (*equilibrated* vectors) is the area-weighted normal vectors of a unique polyhedron. Both results hold in arbitrary dimension d, with "face normal" replaced by "facet normal" and "face area" by "facet $(d-1)$-dimensional volume." For a proof, see Grünbaum (1967, p. 339).

A computational solution to this problem is of interest in computer vision, for several methods are available for finding surface orientation, that is, normal vectors. This led James Little to develop a convex optimization approximation scheme for reconstructing a convex polyhedron in $\mathbb{R}^3$ from its "extended Gaussian image" (Little 1983, 1985). Later, the computational complexity of the reconstruction problem in $\mathbb{R}^d$ was settled by Gritzmann and Hufnagel (1995), who showed that computing n vertices for

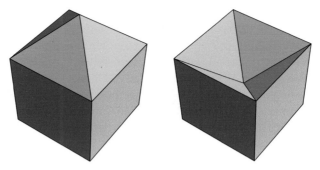

Figure 23.2. Two different polyhedra with the same facial structure.

a d-polytope whose facet normals are within ε of given facet volumes can be solved in polynomial time in n and d.

Although there has been work on discrete analogs to Minkowski's theorem (Auren-hammer 1998), there is a sense in which numerical approximations cannot be avoided, for even with rational input, there may be no polytope with rational vertex coordinates that reconstructs the input exactly.

23.1 CAUCHY'S RIGIDITY THEOREM

In this section we describe and prove Cauchy's rigidity theorem, drawing on the presentations in Aigner and Ziegler (1999) and Cromwell (1997). We then turn to the algorithmic side of Cauchy rigidity in Section 23.4, an important unsolved problem.

As Cromwell reports, the problem solved by Cauchy goes back to Euclid, who defined equality between polyhedra as follows (Euclid 1956, Bk. XI, Def. 10):

> Equal and similar solid figures are those contained by similar planes equal in multitude and magnitude.

In modern terminology, two polyhedra are congruent if they have congruent faces similarly surrounded (or similarly arranged about each vertex). This is certainly false for nonconvex polyhedra, as the example in Figure 23.2[1] demonstrates: the roof over the cube can dent inward as well as protrude outward, while retaining the congruency of the faces and their arrangements about each vertex. But giving Euclid the benefit of the doubt, he was likely thinking only of convex polyhedra. Still, as Legendre realized,

> Definition 10 is not a proper definition but a theorem which it is necessary to prove... (Quoted in Cromwell 1997, p. 224).

It was this necessity that was filled by Cauchy's 1813 proof (Cauchy 1813). One phrasing of his theorem is as follows (following Aigner and Ziegler 1999):

Theorem 23.1.1. *If two closed, convex polyhedra are combinatorially equivalent, with corresponding facets congruent, then the polyhedra are congruent; in particular, the dihedral angles at each edge are the same.*

---

[1] After Abb. 61 in Steinitz and Rademacher (1934, p. 58).

Over a century later, Steinitz discovered and repaired (Steinitz and Rademacher 1934) a flaw in a key lemma that is part of Cauchy's proof, what we earlier called Cauchy's arm lemma (Lemma 8.2.1, p. 143). Although that lemma often bears both their names as a result, the rigidity theorem remains recognized as a great achievement of Cauchy.

The term *stereoisomer* is used in chemistry to mean "combinatorially equivalent, with corresponding facets congruent." So this theorem is sometimes stated as follows: convex stereoisomers are congruent. It is called Cauchy's "rigidity theorem" because it immediately implies that a convex polyhedron is rigid: it cannot flex. In fact, Alexandrov phrases it this way:

> If the faces of a nondegenerate closed convex polyhedron are rigid then the polyhedron itself is rigid (Alexandrov 2006, p. 207).

Cauchy's theorem is, however, strictly stronger than claiming that convex polyhedra are rigid for it is conceivable that they are rigid but there are several isolated, incongruent convex shapes realized by stereo-isomers. In the terminology of Chapter 4, Cauchy proved that convex polyhedra are globally rigid (p. 48). Cauchy's theorem was strengthened (originally by Max Dehn in 1916) to the notion of infinitesimal rigidity (p. 49); see, e.g., Connelly 1993; Pak 2006.

We now turn to a proof of the theorem.

23.1.1 Proof of Cauchy's Theorem

The proof consists of three main steps, each highly original and much used subsequently:

1. Connection to dihedral angle sign changes of spherical polygons,
2. application of Cauchy's arm lemma, and
3. application of Euler's formula to obtain a contradiction.

The proof is by contradiction. Let P and P' be two incongruent convex polyhedra that are nevertheless stereo-isomers. Because their combinatorial structure is the same, their vertices, edges, and faces can be matched one-to-one.

Let $v \in P$ and $v' \in P'$ be corresponding vertices. Intersect P with a small sphere S_v centered on v. Each face F incident to v intersects S_v in an arc of a great circle. For the plane containing F passes through the center of S_v and so intersects S_v in a great circle, and restricting to F selects an arc of this circle whose length is proportional to the face angle of F at v. Therefore, the collection of faces of P incident to v intersects S_v in a convex spherical polygon $Q = P \cap S_v$ (see Figure 23.3). The angle at a vertex p of Q is equal to the dihedral angle of the edge $e \in P$ that penetrates S_v at p.

Now we compare the corresponding spherical polygon $Q' = P' \cap S_{v'}$. Because the faces of P and P' are congruent, the arcs of Q and Q' have the same length. But, because by hypothesis P and P' are not themselves congruent, there must be some v such that Q and Q' differ in at least one dihedral angle. Mark each angle of Q with a symbol from $\{+, 0, -\}$ depending on whether the angle is larger, equal, or smaller in Q' than the corresponding angle in Q.

Now we will prove the lemma we employed in Part I without proof, Lemma 5.3.1:

Lemma 23.1.2. *The total number σ of alternations in sign around the boundary of Q is at least 4.*

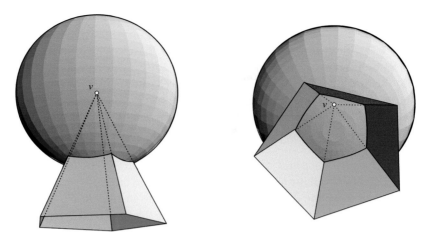

Figure 23.3. Two views of the spherical polygon Q produced by intersecting a polyhedron in the neighborhood of a vertex v with a sphere S_v centered on v.

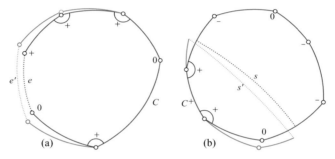

Figure 23.4. The spherical polygon Q. (a) All angles $\{+, 0\}$: the subchain C opens; (b) angles partition into $\{+, 0\}$ and $\{-, 0\}$: the subchain C^+ opens.

An "alternation in sign" is a $+/-$ transition with any number of intervening 0's.

Proof: There must be an even number of sign alternations, for an odd number of transitions from a $+$ would leave a $-$ to cycle back to the starting $+$, which would be an additional sign change. So the claim reduces to showing that σ cannot be 0 or 2.

1. $\sigma = 0$. There are no sign changes. Because Q and Q' are not identical, there must be at least one vertex of Q labeled with $+$ or $-$, say the former without loss of generality. Therefore, all vertices of Q are labeled either $+$ or 0 (see Figure 23.4(a)). Now apply Cauchy's arm lemma, Lemma 8.2.1 in Part I. We proved this in Section 8.2 for plane convex polygons, but it holds as well (and with a nearly identical proof) for spherical polygons. We restate it here in the form we need:

Lemma 23.1.3. *If a spherical convex chain C is opened by increasing some or all of its internal angles, but not beyond π, then the distance between the endpoints of the chain (strictly) increases.*

Let C be Q with any edge e deleted. The lemma applies to C and shows that $|e'| > |e|$: the length of the deleted edge increases (see again Figure 23.4(a)). But

e' is the edge of Q' corresponding to e, and we saw earlier that these lengths are fixed by the equal face angles incident to v from which they derive.

2. $\sigma = 2$. If there are exactly two sign changes, then we can draw a chord s across Q separating the $+$ and $-$ signs (see Figure 23.4(b)). Applying Cauchy's arm lemma to the positive chain $C^+ \subset Q$ delimited by s shows that s must increase in length to s' in Q': $|s'| > |s|$. Switching viewpoint to Q', it will be labeled with opposite signs, and its positive chain $C^{+'} \subset Q'$ shows that s' must increase in length to s in Q: $|s| > |s'|$. We have again reached a contradiction.   □

The final step of the proof is to show that Lemma 23.1.2 is incompatible with Euler's formula. Let G be the plane graph of vertices, edges, and faces of P, with its edges labeled by $\{+, 0, -\}$ to mark the dihedral angle changes with respect to P'. Let the number of vertices, edges, and faces of G be V, E, F. Let Σ be the total sum of the number of sign alternations around all vertices of G. $\Sigma = \sum_{v \in P} \sigma(v)$. Lemma 23.1.2 immediately yields a lower bound: $\Sigma \geq 4V$. We now show that this is too large by counting the sign changes another way.

First we simplify matters by reducing the three labels $\{+, 0, -\}$ to two, $\{+, -\}$, simply by mapping every 0 label to $+$. This mapping is entirely combinatorial; there is no geometric significance to these additional $+$'s. Let Σ^+ be the total number of sign alternations around all vertices of G with these new labels. We claim that $\Sigma^+ \geq \Sigma$. For the mappings

$$(+, 0, \ldots, 0, +) \rightarrow (+, +, \ldots, +, +)$$

and

$$(+, 0, \ldots, 0, -) \rightarrow (+, +, \ldots, +, -)$$

leave the number of sign changes unaltered, while

$$(-, 0, \ldots, 0, -) \rightarrow (-, +, \ldots, +, -)$$

increases the number by 2.

Now we count Σ^+ by traversing around the boundary of each face of G. Note that every sign alternation around v is reflected in a sign alternation around the edges of some face f incident to v (see Figure 23.5). So this second accounting is a legitimate method for computing Σ^+.

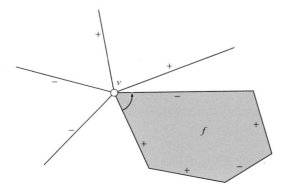

Figure 23.5. A $+/-$ sign change around v is also a sign change in a traversal of the boundary of f.

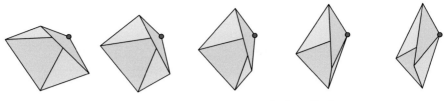

Figure 23.6. Half of a regular octahedron, flexing.

Next, note that if a face f has $2k$ or $2k + 1$ edges, then it has at most $2k$ sign changes. Thus a pentagon can have four sign alternations (as does f in the figure) but not five. Letting f_k be the number of faces of G with k edges, we then know

$$\Sigma^+ \leq 2\,f_3 + 4\,f_4 + 4\,f_5 + 6\,f_6 + 6\,f_7 + \cdots. \tag{23.1}$$

Now we compute a similar sum. If we count edges by $k f_k$ terms, we would double count each edge. So

$$2E = 3\,f_3 + 4\,f_4 + 5\,f_5 + 6\,f_6 + 7\,f_7 + \cdots,$$

$$E = \frac{1}{2}(3\,f_3 + 4\,f_4 + 5\,f_5 + 6\,f_6 + 7\,f_7 + \cdots).$$

Substituting this into Euler's formula $V = 2 + E - F$ yields

$$V = 2 + \frac{1}{2}(3\,f_3 + 4\,f_4 + 5\,f_5 + 6\,f_6 + 7\,f_7 + \cdots),$$

$$- (f_3 + f_4 + f_5 + f_6 + f_7 + \cdots).$$

$$V = 2 + \frac{1}{2}(f_3 + 2\,f_4 + 3\,f_5 + 4\,f_6 + 5\,f_7 + \cdots),$$

$$4V = 8 + 2\,f_3 + 4\,f_4 + 6\,f_5 + 8\,f_6 + 10\,f_7 + \cdots.$$

Comparing this last expression for $4V$ with Equation (23.1), we see that it is strictly greater than Σ^+. So we have concluded that $4V > \Sigma^+ \geq \Sigma \geq 4V$, a contradiction. This completes the proof of Cauchy's rigidity theorem.

The use of Euler's formula as a key step in the proof hints that it is false if the polyhedron is not closed. And indeed, in general a convex polyhedral manifold with boundary is not rigid, that is, it can flex.[2] Figure 23.6 illustrates this flexing around a degree-4 vertex. Note that $link(v)$, the polygon composed of the opposite edges of the facets incident to v, flexes with the dihedral angle sign pattern $(+, -, +, -)$, satisfying Lemma 23.1.2, as it must.

23.2 FLEXIBLE POLYHEDRA

Although Cauchy's rigidity theorem shows that every convex polyhedron with rigid faces is rigid, it leaves open the possibility that a nonconvex polyhedron might not be rigid,

---

[2] In his 1950 book (Alexandrov 2005, Chap. 5), Alexandrov initiated the study of which convex polyhedra, modified by cutting out a single polygonal hole, can flex, and a precise characterization was found in 1958 by Shor (Alexandrov 2005, Sec. 12.3).

Table 23.1: Coordinates of vertices in Figure 23.7

Vertex	Coordinates
a	$(2, 0, 0)$
b	$(1, \sqrt{3}, 0)$
c	$(-2, 0, 0)$
d	$(-1, \sqrt{3}, 0)$
e	$(0, 0, 1)$
f	$(0, 0, -1)$

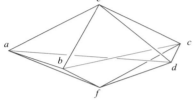

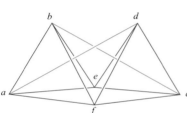

Figure 23.7. Two views of one of Bricard's flexible octahedra. See http://www.gfalop.org/III for a (rigid) 3D Java applet model.

that is, might be *flexible*. In this section we briefly recount the remarkable positive resolution of this issue, relying heavily on the detailed presentation by Cromwell (1997).

23.2.1 Bricard's Flexible Octahedra

Some 85 years after Cauchy's 1813 proof, a French engineer named Raoul Bricard discovered several flexing octahedra, but they all self-intersect. His octahedra are now called *Bricard's flexible octahedra*. They are best thought of as flexible 3D linkages rather than as self-intersecting polyhedra. We illustrate one in Figure 23.7; its vertex coordinates are listed in Table 23.1. We explain its flexing motion informally. The two red diagonals ad and bc form a contraparallelogram (cf. p. 32) with the edges ab and cd, lying in a common plane (the xy-plane in the coordinate system used in Table 23.1). As this contraparallelogram flexes according to its single degree of freedom,[3] it pulls the two triangles attached to the "outer" edges ab and cd ($\triangle abe$, $\triangle abf$; $\triangle cde$, $\triangle cdf$), which varies the separation between e and f. The resulting motion is pleasingly symmetric.

---

[3] See http://www.gfalop.org/I for a Java applet illustrating this motion.

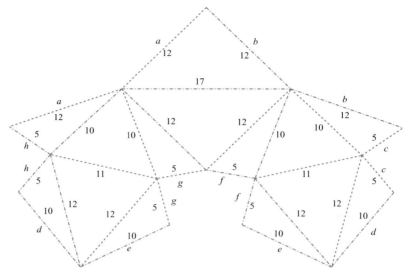

Figure 23.8. Unfolding of Steffen's flexible polyhedron. Dashes are valley folds, dash-dots represent mountain folds. Numbers indicate (integer!) edge lengths, letters gluing instructions.

23.2.2 Connelly's Flexible Polyhedron

Euler's 1766 conjecture that all polyhedra are rigid was provided a second strong piece of evidence when Herman Gluck proved that almost all triangulated polyhedra are infinitesimally rigid and (by the results of Part I, Chapter 4) therefore rigid. More formally, the infinitesimally rigid polyhedra form an open, dense set in the space of all polyhedra (Gluck 1976).

This result made it all the more surprising when Robert Connelly constructed a flexible polyhedron in 1978 (Connelly 1978, 1981). He started from Bricard's flexible octahedron shown in Figure 23.7 as a framework, but then added "crinkled" surfaces to avoid self-intersection. This enabled him to reduce the self-intersection to be edge-to-edge (rather than two properly interpenetrating faces). Finally, he used another of Bricard's flexing octahedra as the basis for replacing those crossing edges by faces that completely avoid self-intersection. The result is a somewhat intricate (30-face (not all triangles), 50-edge, and 22-vertex) polyhedron (of genus zero) that flexes in a small range.

23.2.3 Steffen's Flexible Polyhedron

Subsequently there have been several simplifications of Connelly's original flexible polyhedron, whose endpoint may be Klaus Steffen's 14-triangle, 9-vertex example. As it was later proven that all triangulated polyhedra of eight vertices are rigid, Steffen's example is minimal in this sense. The polyhedron can be seen as highly symmetric when unfolded (see Figure 23.8). Despite this symmetry and the small combinatorial complexity, the folded polyhedron is not easily comprehended. See Figure 23.9 for two representations. The flexing motion is more intricate still.[4]

---

[4] See http://www.mathematik.com/Steffen/ for a carefully computed flexing motion represented in an animated GIF.

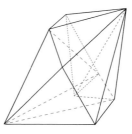

Figure 23.9. Steffen's flexible polyhedron from one point of view: hidden surfaces removed (left); hidden lines shown dashed (right).

23.2.4   The Bellows Conjecture

It was soon noticed that the volume of all the known flexible polyhedra remains constant throughout the flexing motion. Dennis Sullivan conjectured (Cromwell 1997, p. 245) that this was always the case: no polyhedron could form a "bellows"; if the volume did change, then poking a hole at a vertex would permit its flexing to suck in and force out air. This "Bellows conjecture" was settled positively in 1997 (Connelly et al. 1997). As Mackenzie put it (Mackenzie 1998), polyhedra can bend but not breathe!

The key to the proof is a generalization of Heron's formula, which expresses the area A of a triangle in terms of its three side lengths a, b, c:

$$s = (a + b + c)/2, \tag{23.2}$$
$$A^2 = s(s - a)(s - b)(s - c). \tag{23.3}$$

Heron's formula was generalized by the painter Piero della Francesca in the fifteenth century to express the volume of a tetrahedron as a complicated polynomial of its six-edge lengths (discussed further in Section 23.4, p. 354). Sabitov extended this in a remarkable way: he proved that the volume of a polyhedron satisfies a grand generalization of Francesca's formula, again a polynomial only of the edge lengths (Astrelin and Sabitov 1999; Sabitov 1996b). (For an octahedron, this polynomial is already degree 16.) Because the edge lengths of a flexing polyhedron are constant, the polynomial is fixed and the volume could only change discretely by jumping from one root of this polynomial to another. But clearly the volume changes continuously with flexing; and so it must remain constant.

23.3   ALEXANDROV'S THEOREM

A. D. Alexandrov proved a significant generalization of Cauchy's rigidity theorem in 1941, which he phrased at that time as follows (Alexandrov 1996):

> For every convex polyhedral metric, there exists a unique polyhedron (up to a translation or a translation with a symmetry) realizing this metric.

In this section we reformulate this theorem, crudely sketch its proof, and discuss some of its subsequent generalizations.

Convex polyhedral metrics. A manifold M is "given an intrinsic metric" by specifying a distance function (a *metric*) between any two points as the length of the shortest path on the manifold between them.[5] Such a metric is *intrinsic* because it is determined by lengths of paths in the manifold. In a *convex polyhedral metric*, each point of the manifold has a neighborhood that is isometric to a circular cone (which may degenerate to a plane). The metric determined by great circular arcs on a sphere is not polyhedral, but the metric determined by shortest paths on a convex polyhedron is a convex polyhedral metric.

Alexandrov's theorem combines both an existence claim and a uniqueness claim. Pogorelov states the Alexandrov's existence claim in the following more explicit form (Pogorelov 1973, p. 20):

> Any convex polyhedral metric given ... on a manifold homeomorphic to a sphere is realizable as a closed convex polyhedron (possibly degenerating into a doubly covered plane polygon).

Alexandrov gluing. In this form it may seem obscure, or perhaps even tautological. We turn then to a rephrasing in terms of an important technique used by Alexandrov: *gluing* of polygons, by matching or identifying a finite number of equal-length subsections of the boundaries of the polygons with one another (Alexandrov 2005, p. 52). The results hold for gluing just one polygon; then portions of its boundary are glued to other portions of its boundary. In terms of gluings, we may state Alexandrov's theorem as follows:

Theorem 23.3.1. *Any such gluing of polygons that*

   (a) uses up the perimeter of all the polygons with boundary matches;
   (b) glues no more than 2π total angle at any point; and
   (c) results in a complex homeomorphic to a sphere,

any such gluing corresponds to a unique convex polyhedron (where a doubly covered polygon is considered a polyhedron).

We will call a gluing with these three properties an *Alexandrov gluing.* Condition (a) makes the complex a manifold, condition (c) ensures it is homeomorphic to a sphere, and condition (b) establishes that each point has a neighborhood that is isometric to a circular cone, and so the metric is polyhedral. Specializing these conditions to a single polygon implies that any closing up of the perimeter of a polygon that constitutes an Alexandrov gluing yields a unique convex polyhedron. This forms the basis of determining when a polygon can fold to a polyhedron, and to which polyhedra it might fold, explored at length in Chapter 25.

Relation to Cauchy rigidity. There are two primary ways in which Alexandrov's theorem (Theorem 23.3.1) is a significant extension of Cauchy's rigidity theorem (Theorem 23.1.1). First, Alexandrov's theorem includes both existence and uniqueness, whereas Cauchy's theorem is concerned only with uniqueness: Cauchy's theorem does not tell us when a collection of faces might form a convex polyhedron, only that when they do, they form only one. Second, Alexandrov generalizes the preconditions in two ways. First, he does not require the joining of flat, rigid polygonal faces. One can imagine gluing together curved, bent polygonal surfaces (e.g., polygonal pieces of paper).

---

[5] More precisely, the infimum of the lengths of curves joining the points.

Second, the boundaries of the given gluing polygons are (potentially) unrelated to the edges of the resulting polyhedron. In other words, Alexandrov's theorem does not require specification of the creases in the paper, whereas in Cauchy's theorem, the faces and therefore edges of the polyhedron are prespecified. The combination of these generalizations make it one of the most powerful tools in this area.

23.3.1   Uniqueness

As mentioned in Part I, Chapter 4, uniqueness (global rigidity) is stronger than rigidity, for it might have been that there could be several realizations, each of which is rigid. Nevertheless, sometimes uniqueness is informally equated with rigidity in the literature. Cauchy's rigidity theorem (Theorem 23.1.1) is typical, for it says that the realization within the specified class is unique. Alexandrov phrases his uniqueness in the following lemma (Alexandrov 1996):

Lemma 23.3.2. *Two isometric polyhedra are congruent.*

   Proof: *(Sketch)* Let P_1 and P_2 be two isometric convex polyhedra, but with different combinatorial structures. Their vertices must be in one-to-one correspondence, for their angle sum of less than 2π is preserved by isometry. So only their edges differ. On P_1 draw shortest paths corresponding to the edges of P_2, and on P_2 draw shortest paths for the edges of P_1. Now each is subdivided into complexes with identical structures. Introduce pseudovertices wherever shortest paths and edges cross, and obtain a finer face structure for each polyhedron. By construction, these face structures are identical, and Cauchy's theorem applies, showing that P_1 and P_2 must be congruent.   □

Cauchy to Alexandrov. The reduction to Cauchy's theorem here suggests a connection that can be exploited to find creases. Suppose we had a robust algorithm for Cauchy rigidity: given the face polygons and combinatorial structure of a polyhedron, it would construct the 3D coordinates of the vertices of the unique realizing polyhedron guaranteed by Cauchy's theorem. By "robust" we mean it would also work if some faces are coplanar, for example, if a cube face is given as four triangles meeting at a "flat vertex" rather than as a square. We now argue that this algorithm could be used to find the unique realizing polyhedron corresponding to an Alexandrov gluing, that is, it could be used to find the creases. The argument is due to Boris Aronov,[6] although its essence appears already in a footnote of Alexandrov's (Alexandrov 2005, p. 100, fn. 13).

   An Alexandrov gluing yields immediately the points of the complex that become the vertices of the polyhedron P: exactly those points whose "total angle" glued there is strictly less than 2π. We know that every edge of P is the unique shortest path between its endpoint vertices, because an edge of a polyhedron is a straight segment in 3D, and any other path between those vertices is not straight in 3D, and so is longer. In the glued complex, construct the shortest path between all pairs of vertices (see Figure 23.10).

   The edges of P are a subset of these paths. Introduce pseudovertices wherever shortest paths cross, producing a partition of the complex into convex polygons. Now apply the presumed Cauchy-rigidity algorithm to this complex. It would effectively crease the

---

[6]   Personal communication, June 1998.

Figure 23.10. Shortest paths between all pairs of vertices on a convex polyhedron [Figure 1 of Kaneva and O'Rourke 2000].

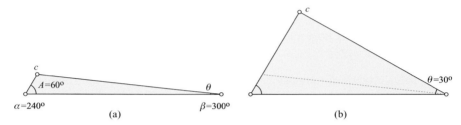

Figure 23.11. (a) Two copies of $A = 60°$ flatten α; (b) two copies of $\theta = 30°$ flatten β.

complex exactly where the edges of P must lie, and leave the other segments of the shortest-paths arrangement with dihedral angle π, flat on the surface of P.

Sketch of existence proof. We now give a rough sketch of Alexandrov's existence proof for Theorem 23.3.1, to give a flavor of some of the techniques used. The overall structure of the proof is inductive, based on the number of vertices of the metric. (Recall from above that the vertices are easily identified in the gluing.) He moves from a realization of a metric of $n - 1$ vertices to one of n vertices by a deformation of metrics tracked continuously by polyhedron realizations of those metrics.

Let $n \geq 5$; we leave aside complexes with fewer vertices. By Theorem 21.3.1, the total curvature of the n vertices is 4π. This implies that there are at least two vertices a and b whose curvatures are less than π. Suppose otherwise. Then at least four vertices have curvature $\geq \pi$, and the total curvature would exceed 4π.

Because a and b have angle deficits less than π, they each have total angle more than π; call these angles α and β respectively. Let γ be the shortest path connecting a and b on the manifold. Cut the manifold along γ and insert two congruent triangles there, back to back, both with base γ, and base angles $A = \pi - \alpha/2$ at a and θ at b. Here θ is a variable angle to be set later. The apex of each triangle then has angle $C = \alpha/2 - \theta$ (see Figure 23.11(a) for an example). Note that the new complex has total angle $\alpha + 2(\pi - \alpha/2) = 2\pi$ at a, so a ceases to be a vertex. However, it has not eliminated the vertex at b, which now has total angle $\beta + 2\theta$, and a new vertex c has been introduced with total angle $2C = \alpha - 2\theta$. (Note that the total curvature has not been altered.) Thus the new complex also has n vertices.

Now vary θ from 0 to $\pi - \beta/2$, keeping A fixed, and note that when θ reaches $\pi - \beta/2$, the vertex b has total angle 2π and so is eliminated (Figure 23.11(b)). At this end of the θ range, then, both a and b are eliminated, replaced by c, and so the resulting complex has $n - 1$ vertices. Thus we have a continuous variation of the metrics from the given complex of n vertices to one with $n - 1$ vertices. This latter complex is realizable by induction, say with polyhedron P'. Adjusting pseudovertex a of P' slightly creates a "nearby" polyhedron with n vertices. Then the metric for P' can be connected continuously to that for the original given metric, also of n vertices.

We do not include here the considerable portions of the proof devoted to establishing properties of the manifold of metrics realizable as a polyhedron. Alexandrov was unable to ascertain whether the "manifold of developments of strictly positive curvature, with fixed structure" is connected, and this complicated this aspect of the proof (cf. Alexandrov 2005, p. 200).

23.3.2 Extension to Smooth Surfaces

There have been several extensions of Alexandrov's theorem, both by him and by his students, particularly Pogorelov. The latter phrases one existence extension as follows (Pogorelov 1973, p. 23):

> An intrinsic metric of nonnegative curvature, given [...] on a manifold homeomorphic to a sphere, is realizable as a closed convex surface (possibly degenerating into a doubly covered convex plane domain).

One can similarly extend the gluing technique to gluing together closed regions, each with intrinsic metric and nonnegative curvature, identifying their boundary curves in such a way that the "integral geodesic curvature" is nonnegative.

Extensions of the uniqueness (or "monotypy") results have been more technically difficult. Cauchy was extended by Alexandrov, as we have seen, and a series of results followed on surfaces: the sphere is rigid (Minding), the sphere is unique given its metric (Liebmann and Minkowski), closed, "regular" surfaces are rigid (Liebmann, Blaschke, Weyl), uniqueness of these surfaces within a certain class (Cohn-Vossen), and other results we will not recount.[7] The endpoint of this work was obtained by Pogorelov in 1952, and may be simply stated (Pogorelov 1973, p. 167):

> Isometric closed convex surfaces are congruent.

23.3.3 D-Forms

We now apply these extensions of Alexandrov's theorem to a question raised by Helmut Pottmann and Johannes Wallner in Pottmann and Wallner (2001, p. 418) concerning objects they and their originator Tony Wills call "D-forms." Let c_1 and c_2 be two smooth, closed, convex curves of the same length drawn on separate pieces of paper. Choose a point $p_1 \in c_1$ and $p_2 \in c_2$, and glue the two curves to each other starting with p_1 glued to p_2. The paper forms an elegant convex shape in space called a *D-form*.[8] The curves c_1 and c_2 join to form a space curve **c** bounding two developable surfaces S_1 and S_2. Figure 23.12 shows an example from Pottman and Wallner. Figure 23.13(b) shows a simpler example constructed from two identical "racetracks," rectangles capped by semicircles. In this example, all the 4π of curvature is concentrated along the sections of **c** where a circle arc of one racetrack is glued to a straight side of the other. Thus, if

---

[7] See Pogorelev (1973, p. 121ff).
[8] Although the two curves could be polygons and the shape nonconvex, the term "D-form" is used for the convex shape produced from smooth curves.

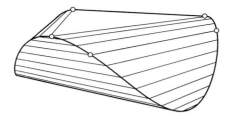

Figure 23.12. Based on Figure 6.49 of Pottmann and Johannes (2001, p. 401), by permission.

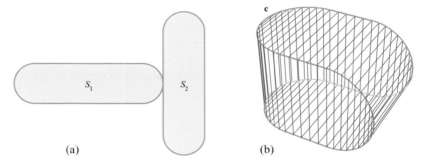

(a)   (b)

Figure 23.13. D-form (b) constructed from two "racetracks" (a).

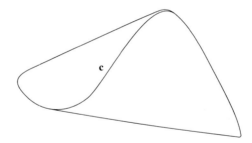

Figure 23.14. A pita-form.

the circle radius is 1, the semicircle arc length is π, and the curvature $\gamma(p)$ is 1 at each of the points p of these four sections of $\mathbf{c}$, so that $\int_{p \in \mathbf{c}} \gamma(p) = 4\pi$.

There is an analogous and perhaps simpler shape that can be formed by gluing the perimeter of one smooth convex disk to itself. We call these "pita-forms," and will study their polygonal counterparts in Section 25.7.2 (p. 412). Figure 23.14 shows a crude line drawing of a pita-form.

Pottman and Wallner ask two questions about their D-forms:

1. "It is not clear under what conditions a D-form is the convex hull of a space curve."
2. "After some experiments we found that, surprisingly, both S_1 and S_2 were free of creases, but we do not know whether this will be so in all cases."

Answering these questions formally seems to require mathematical machinery beyond that developed in this book. We therefore only suggest informal reasoning in support of this claim, answering their first question: a D-form is always the convex hull of $\mathbf{c}$.

The two convex regions satisfy the conditions of the gluing theorem of Alexandrov and Pogorelov (Pogorelov 1973, p. 33), and so we know the resulting metric is itself intrinsic and of positive curvature. Then Alexandrov's Theorem 2 from Alexandrov (1996) yields existence of a closed convex surface realizing the metric, and Pogorelov's theorem just quoted above yields uniqueness of this convex surface, the D-form D.

It is useful to imagine the original planar convex regions as bounded by convex polygons rather than smooth curves. In fact, the extensions of Alexandrov's theorem are established by increasingly accurate polygonal approximations. With polygons, it is clear that all vertices of the resulting polyhedron D_P lie along (the now polygonal) $\mathbf{c}_P$, or, in other words, no polyhedron vertices are created in the interior of the paper regions. So the vertices of D_P are precisely the vertices along $\mathbf{c}_P$. Therefore D_P is the convex hull of the vertices of $\mathbf{c}_P$. Taking the limit, we see that D is the convex hull of $\mathbf{c}$.

Their question about D-form creases remains open. Pita-forms necessarily have two "kinks" in $\mathbf{c}$ at the perimeter-halving (p. 382) points (one is visible in Figure 23.14). But our experiments suggest that, in spite of these kinks, pita-form surfaces also have no creases, except in a degenerate flat folding obtainable from a shape with a line of reflection symmetry.

Open Problem 23.1: D-Forms and Pita-Forms. [a] Might a D-form, or a (non-degenerate) pita-form, have a crease?

---

[a] Pottmann and Wallner (2001, p. 418).

The lack of an effective method for reconstructing polyhedra from an Alexandrov gluing extends to these forms, which is why our examples are either approximate drawings (Figures 23.12 and 23.14) or symmetric enough for an ad hoc computation (Figure 23.13). We turn now to this reconstruction issue.

23.4   SABITOV'S ALGORITHM

Alexandrov's assessment of the problem of constructing the unique convex polyhedron guaranteed by Cauchy's Theorem 23.1.1 is captured in these two quotes:

> ... if we begin to glue this polyhedron [from polygonal pieces], it will take its shape in a sense automatically. However, we can say nothing other than make a few obvious but completely insufficient remarks about this shape (Alexandrov 2006, p. 21).

> The general solution to the problem of determining the structure of a polyhedron from its development, i.e., finding the actual polyhedron edges, seems hopelessly difficult. Only for the tetrahedron can a complete answer be found (Alexandrov 1958, p. 89).

The reason Alexandrov's existence proof (Section 23.3.1, Theorem 23.3.1) does not lead to a reconstruction algorithm is pinpointed in Fedorchuk and Pak (2005) as the use of the "Implicit Function Theorem" in his proof. Although there remains no known practical reconstruction method, there is at least a finite algorithm, due to Sabitov (1996a), which we sketch in this section.[9]

---

[9] We base our presentation partly on personal communication with Sabitov, August 2002.

As mentioned in Section 23.2.4, Sabitov constructed an expression for the volume of a polyhedron, with given specific combinatorial structure and vector ℓ of squared edge lengths, as a polynomial in one variable V, the volume, with coefficients polynomials in the lengths with rational coefficients (Sabitov 1996b, 1998). Here we will need to study these volume polynomials in more detail. The general form may be written as follows:

$$V^{2N} + a_1(\ell) V^{2(N-1)} + a_2(\ell) V^{2(N-2)} + \cdots + a_N(\ell) V^0 = 0. \tag{23.4}$$

The volume is a root of this polynomial. For a tetrahedron this expression reduces to

$$V^2 + a_1(\ell) = 0, \tag{23.5}$$

where $\ell = (l_1^2, l_2^2, l_3^2, l_4^2, l_5^2, l_6^2)$, l_i are the six edge lengths, and the polynomial $a_1(\ell)$ has the form

$$\frac{1}{144} \sum_{i,j,k} \delta_{ijk} l_i^2 l_j^2 l_k^2, \tag{23.6}$$

where the sum runs over a certain subset of all triples of indices, and $\delta_{ijk} = \pm 1$. Not only is the volume of the unique convex polyhedron with the given combinatorial structure and edge lengths a root of Equation (23.4), the volume of every nonconvex polyhedron with the same structure and lengths is a root of the same equation. The volume polynomial includes more: it represents the volume of a broader class of objects than polyhedra, and this broader class will play a crucial role in Sabitov's algorithm. So we now turn to describing this class of "generalized polyhedra."

A *generalized polyhedron* is, technically, any simplicial 2-complex homeomorphic to an orientable manifold of genus $g \geq 0$, mapped into $\mathbb{R}^3$ by a continuous function linear on each simplex; we will consider only closed manifolds. A simplicial complex is an object constructed by attaching simplices to one another at their facets. So a simplicial 2-complex is created by gluing triangles edge-to-edge and/or vertex-to-vertex. The orientable requirement rules out, for example, the Klein bottle. Without detailing every technical nuance of this definition, let us mention the most important generalization for our present purposes: the map need not be an embedding into $\mathbb{R}^3$, which means that the surface may self-intersect. So we need a corresponding generalization of the notion of "volume." This is relatively straightforward, following the definition used to compute the area of a polygon by summing signed triangle areas.[10] The volume of a generalized polyhedron is obtained by choosing an arbitrary point p (say, the origin), and summing the signed volumes of the tetrahedra determined by p and each triangle face of the complex. For true polyhedra (of any genus), this corresponds to the conventional volume. The volume polynomial Equation (23.4) encodes as roots the volumes of all the generalized polyhedra with the given structure and lengths. (For the tetrahedron, the two roots represent positive and negative orientations of the surface.)

It is easy to see that the degree $2N$ of the volume polynomial can sometimes be exponential in n, the number of vertices of the polyhedron. Consider a "row of houses" polyhedron composed of k copies of the "house" polyhedron shown earlier in Fig. 23.2. A 2D depiction of its structure is shown in Figure 23.15(a). The roof of each house can stick up or dent down into the cube, as in 23.15(b). If we arrange the roof heights

[10] See, e.g., O'Rourke (1998, p. 21).

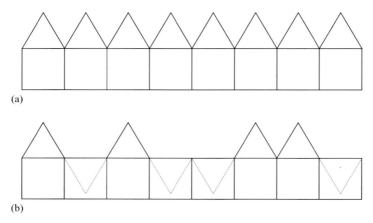

(a)

(b)

Figure 23.15. A row of "house" polyhedra. (a) 11111111; (b) 10100110.

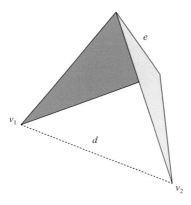

Figure 23.16. Tetrahedron T determining the unknown diagonal d.

to be slightly different for each house, we can easily ensure that all the 2^k volumes achieved by the binary up/down choices are distinct. This yields a lower bound of $2^{\Omega(n)}$ on $2N$. Recently a tight upper bound of $2N \leq 2^m$, where m is the number of edges of the polyhedron P, was established in Fedorchuk and Pak (2005, Cor. 6.2). Thus, for a generic icosahedron, the polynomial degree is bounded by 2^{30}, or about 10^9. Of course, the polynomial equation (23.4) has at most N real roots.

We can now describe Sabitov's algorithm for solving the Cauchy rigidity problem. Let e be any edge of a triangulated polyhedron P whose combinatorial structure and edge lengths are known. Let $\triangle_1$ and $\triangle_2$ be the two triangles sharing e, and let d be the length of the diagonal connecting the two vertices v_1 and v_2 of $\triangle_1$ and $\triangle_2$ not on e (see Figure 23.16). Although d may be determined (e.g., if $v_1 v_2$ is an edge of P), in general, d is unknown, being either an internal or an external diagonal. Let T be the tetrahedron that is the hull of $\{e, v_1, v_2\}$. The volume of T is a polynomial of form Equation (23.5) in its lengths, whose only unknown is d. Although we know neither d nor whether it is internal or external to P, we can express the volume of P as

$$\text{vol}(P) = \text{vol}(P') - \delta \text{vol}(T), \qquad (23.7)$$

where P' is the generalized polyhedron in which e is replaced by $v_1 v_2$, and δ is ± 1, depending on whether d is internal $(+)$ or external $(-)$ to P. Note that in general this removal of T could cause self-intersection of the surface, so P' is not necessarily a poly-hedron; but it is a generalized polyhedron, because the surface remains a 2-complex homeomorphic to a closed, orientable manifold. It will be convenient to rearrange our equation to this form:

$$\text{vol}(P') = \text{vol}(P) + \delta \text{vol}(T). \tag{23.8}$$

Our goal is now to establish this claim: Equation (23.8) is equivalent to a polynomial in one unknown, d. In particular, the unknown δ can be eliminated. We will only sketch the proof. First, all the terms for $\text{vol}(P)$ are known, so we can find the roots and try each one. So we can treat $\text{vol}(P)$ as a known number V. $\text{vol}(T)$, however, contains the unknown d.

Consider the volume polynomial for $\text{vol}(P')$, which is a polynomial in V'. Equation (23.8) can be viewed as $V' = V + \delta \text{vol}(T)$. So if we substitute $V + \delta \text{vol}(T)$ for V' throughout, the polynomial in V' becomes a polynomial in V, the lengths, and the unknowns d and δ. Sabitov shows that "routine" (but clever) algebraic manipulation permits isolating δ as a factor on one side of the equation, which squaring then eliminates (because $\delta = \pm 1$). The result is a polynomial whose only unknown is d:

$$c_0 d^{2k} + c_1 d^{2(k-1)} + \cdots + c_k d^0 = 0, \tag{23.9}$$

where the coefficients c_i are polynomials in the known lengths and V, which is known as a root of $\text{vol}(P)$. Sabitov proved that the c_i's are all identically zero only in the case of flexible polyhedra so that situation does not apply here. The roots of Equation (23.9) then yield candidates for d.

So, by trying each root for V, and each root for d, for each edge e, we reach a set of all the possible d's. The recent extension of Sabitov's work (Fedorchuk and Pak 2005) es-tablishes that every diagonal length $d_{ij} = |v_i v_j|$ is in fact a root of a polynomial of degree at most 4^m. Now, once we "guess" the d_{ij}'s from all these possibilities, construction of the polyhedron is immediate: start with one triangle face $\triangle v_1 v_2 v_3$, and use $\{d_{1i}, d_{2i}, d_{3i}\}$ to compute the coordinates of v_i (as described in Section 25.6.2, p. 403). When a set of guesses produces a legal convex polyhedron, we are finished by the uniqueness guaranteed by Cauchy's rigidity theorem 23.1.1.

The evident impracticality of Sabitov's algorithm leaves us with this important un-solved problem:

Open Problem 23.2: Practical Algorithm for Cauchy Rigidity. [a] Find either a polynomial-time algorithm or even a numerical approximation procedure that takes as input the combinatorial structure and edge lengths of a triangulated convex polyhedron, and outputs coordinates for its vertices.

---

[a] O'Rourke (2000b).

We will see in Section 25.1 examples of reconstruction of polyhedra with few faces (up to octahedra) by ad hoc methods.

24 Shortest Paths and Geodesics

INTRODUCTION

Both shortest paths and geodesics are paths on a surface which "unfold" or "develop" on a plane to straight lines. This gives them a special role in unfolding, and underlies the two known (general) unfoldings of a convex polyhedron to a nonoverlapping polygon, the source and the star unfolding (mentioned previously on p. 306). In this chapter we describe these two unfoldings, as well as investigate shortest paths and geodesics for their own sake and for their possible future application to unfolding.

Let P be a polyhedral surface embedded in $\mathbb{R}^3$, composed of flat faces. We will insist that each vertex be a "true," nonflat vertex, one with incident face angle $\neq 2\pi$. A *shortest path* on P between two points x and y on P is a curve connecting x and y whose length, measured on the surface, is shortest among all curves connecting the points on P. Several important properties of shortest paths are as follows:

1. There always exists a shortest path between any two points, but it may not be unique: several distinct but equally shortest curves may connect the points.
2. Shortest paths are simple in that they never self-cross (otherwise they could be shortcut).
3. A shortest path never passes through a vertex of positive curvature[1] (although it may begin or end at a vertex) (Alexandrov 2005, p. 72). So on a convex polyhedron, shortest paths do not pass through vertices (Sharir and Schorr 1986, Lem. 4.1), and on a nonconvex polyhedral surface, they only may pass through a vertex of negative curvature (Mitchell et al. 1987, Lem. 3.4).
4. A shortest path p meets a convex face f of P at most once: $p \cap f$ is empty, a point, or a line segment (Mitchell et al. 1987, Lem. 3.2).
5. If a shortest path p passes through an interior point of an edge e, the planar unfolding of the two faces sharing e unfolds the two segments of p on the faces to a single straight segment.
6. Thus, the planar unfolding of a shortest path is a straight-line segment.

We will detail further properties as needed below.

A *geodesic curve* on a surface is a curve γ with the property that for any two sufficiently close points $x, y \in \gamma$, the portion of γ between x and y is the shortest path on the surface connecting x and y. Thus geodesics are locally shortest paths, that is, locally length minimizing. Every shortest path is a geodesic, but geodesics are often not shortest

---

[1] See p. 301 for a definition.

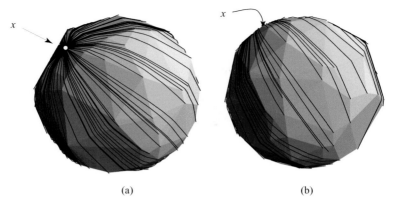

(a)   (b)

Figure 24.1. Two views of the shortest paths from a source point x to all $n = 100$ vertices of a convex polyhedron; x is obscured in the "back view" (b). These shortest paths are shown lifted slightly from the surface to ensure their visibility. [Computations from Xu 1996.]

paths. An important property of geodesics (of course inherited by shortest paths) is that, from every point x, there is a geodesic issuing in each direction (Hilbert and Cohn-Vossen 1952, p. 221). We will save further exploration of the properties of geodesics for Section 24.4.

24.1.1 Source Unfolding

The source unfolding for convex polyhedra was introduced to the computational geometry community in the late 1980s (Mount 1985; Sharir and Schorr 1986), but had been studied for (smooth) Riemannian manifolds earlier (Kobayashi 1967; Volkov and Podgornova 1971).

The construction relies crucially on this lemma (Alexandrov 2005, p. 75; Sharir and Schorr 1986, Lem. 4.2):

Lemma 24.1.1. *Let p_1 and p_2 be distinct shortest paths emanating from x, which meet again at a point $y \in p_1 \cap p_2$ distinct from x. Then either one of the paths is a subpath of the other, or neither p_1 nor p_2 can be extended past y while remaining a shortest path.*

An immediate corollary is that two shortest paths cross at most once.

We now fix one "source" point x on a convex polyhedron surface P and consider the shortest paths from x to all other points on P. For most points y on P, the shortest path from x to y is unique. The set of all those points y to which is there is more than one shortest path from x is called the *ridge tree* in Sharir and Schorr (1986) and the *cut locus* in Kobayashi (1967). We will follow the latter terminology, using the symbol T_x. The shortest paths from x naturally cover the surface of P, as depicted in Figure 24.1. This figure shows only the shortest paths to the vertices, but by the noncrossing property, the shortest paths to other points must lie between the vertex paths. Along any geodesic γ from x, the point at which γ ceases to be a shortest path is a cut locus point. The back view (Figure 24.1(b)) hints at these cut locus points without explicitly showing them. For that, we turn to a simpler example, the pyramid shown, in Figure 24.2.

Here the cut locus T_x is the dashed tree seen most clearly in Figure 24.2(b). For example, each point of the vertical edge of the cut locus on face C, incident to the apex v_0, has equal-length shortest paths from x, wrapping around the pyramid left and right.

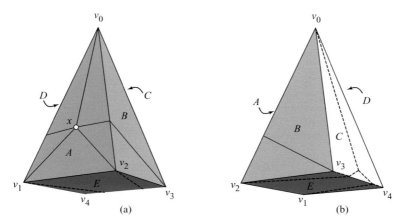

Figure 24.2. Pyramid, front (a) and side (b) view. Shortest paths to five vertices from source x are shown solid; the cut locus is dashed. Coordinates of vertices are $(\pm 1, \pm 1, 0)$ for v_1, v_2, v_3, v_4, and $v_0 = (0, 0, 4)$; $x = (0, 3/4, 1)$. [Based on Figure 1 of Agarwal et al. 1997.]

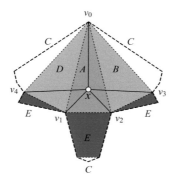

Figure 24.3. Source unfolding for the example in Figure 24.2. Shortest paths to vertices are solid, polyhedron edges are lightly dashed and the cut locus is the dashed outer boundary. [Based on Figure 3 of Agarwal et al. 1997.]

The degree-3 point of the cut locus on face C has three equal-length shortest paths from x, two left and right around the sides, one underneath, crossing face E.

Finally we may define the *source unfolding*[2] with respect to a nonvertex point x as the unfolding produced by using the cut locus T_x as the cut tree and unfolding to a planar layout as usual. Figure 24.3 shows the source unfolding for the pyramid.

That the source unfolding is nonoverlapping is in some sense obvious, for it merely arrays the shortest paths from x in $360°$ about x, and each shortest path unfolds straight. So the source unfolding is a star-shaped region composed of rays from x. But it is important to understand the source unfolding's relationship to the cut locus, and toward that end we list a number of properties. Let us define T_x to be the closure of the cut locus, for as we will see, on a polyhedron it is generally an open set.

1. T_x is a tree whose leaves are the vertices of P (Sharir and Schorr 1986, Lem. 4.5).
2. The edges of T_x are geodesics (Sharir and Schorr 1986, Lem. 4.3).
3. The edges of T_x are in fact shortest paths (Agarwal et al. 1997, Lem. 2.4).

---

[2] Called the "planar layout" in Mount (1985) and Sharir and Schorr (1986), the "outward layout" in Chen and Han (1996) and the "source foldout" in Miller and Pak (to appear).

By (1), T_x is a spanning tree, so it is a valid cut tree, and will unfold the surface flat. By (2) (or (3)), the boundary of the unfolding is polygonal. And the observation that it is composed of shortest-path rays about x, none of which cross one another, guarantees that the unfolding is nonoverlapping. These claims lead to the following theorem:

Theorem 24.1.2. *The source unfolding of a convex polyhedron with respect to any non-vertex point x is a nonoverlapping polygon.*

To provide some intuition on the crucial spanning tree claim (1), we now prove the following lemma (Alexandrov 1958, p. 72; Sharir and Schorr 1986, Lem. 4.1).

Lemma 24.1.3. *A shortest path on a convex polyhedron P does not pass through any vertex of P.*

Proof: Suppose to the contrary that a shortest path p from x to y passes through a vertex v. Lay out the faces of P incident to v in the plane so that $p = xy$ is either straight or as straight as possible. If xy cannot lay out straight, it must be because the face angle θ incident to v to one side (say, the left) is greater than π (see Figure 24.4(a)). There must be some angle deficit α to the right side of xy, because a true vertex has some positive curvature. Then rotating the incident faces to close up the α gap to the right permits short-cutting xy in a neighborhood of v (see Figure 24.4(b)), contradicting the assumption that p is a shortest path. If xy can be laid out straight (see Figure 24.4(c)), then there must be some angle deficit α to one side or the other (or both). Again rotating faces to close the α gap leads to a shortcut; see Figure 24.4(d). □

It should now be plausible that to every vertex of P is incident a segment of the cut locus T_x. Consider a shortest path from x to a vertex v. As we have just seen, this path

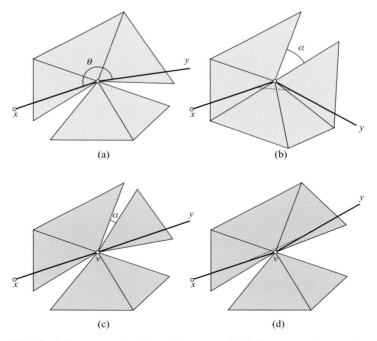

Figure 24.4. (a,c) Shortest paths through a vertex; (b,d) Shortcuts after rotating faces.

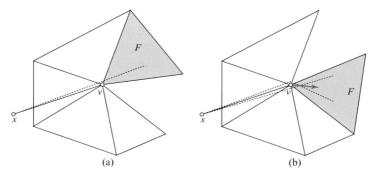

(a)   (b)

Figure 24.5. Shortest paths just left (a) and right (b) of v meet at an edge of T_x incident to v.

does not pass through v. What about shortest paths from x to points just "beyond" v? Those are reached by paths that pass slightly left or slightly right of v and which "wrap around" v. As the unfoldings in Figure 24.5 make clear, there must be an edge of T_x emerging from v representing those points reached equally well to the left and right of v. In general, v itself will not be part of the cut locus: there is a unique path to vertices of P if they are in "general position." This is why we defined T_x to be the closure of the cut locus, to add in the vertex to the incident cut locus segment.

We have restricted x to be a nonvertex not because shortest paths cannot be defined when the source is a vertex—they certainly can—but because T_v for a vertex v is not a spanning tree: it does not span v itself. Thus the source unfolding upon cutting T_v will not flatten in the vicinity of v.

Although the nonoverlap of the source unfolding is almost self-evident, it is less clear whether self-intersection can be avoided during the unfolding process, from 3D all the way down to 2D. Connelly, as reported in Miller and Pak (to appear), asked whether the source unfolding can be *continuously bloomed*, that is, unfolded without self intersection so that all dihedral angles increase monotonically. Although we conjecture this holds, it remains much less clear whether every general unfolding can be executed continuously (cf. again Figure 22.38).

Before turning to the second general nonoverlapping unfolding, we make a brief detour through algorithms for finding shortest paths.

24.2   SHORTEST PATHS ALGORITHMS

In a sense to be made more clear in the next section, finding shortest paths from a source x to any other point on a polyhedron P can be accomplished once the shortest path from x to all the vertices of P have been computed. Thus algorithms concentrate on the latter task. Curiously, the algorithms work equally well on nonconvex polyhedra, or polyhedral manifolds with boundary, as they do on convex polyhedra. ("Equally well" in asymptotics; implementations handling nonconvexities are more complicated.)

24.2.1   The Continuous Dijkstra Approach

After early $O(n^5)$ and $O(n^3 \log n)$ algorithms, an important $O(n^2 \log n)$ algorithm was developed by Mitchell et al. (1987). This work introduced the "continuous Dijkstra" approach, paralleling Dijkstra's famous algorithm (Dijkstra 1959) for finding

Figure 24.6. A polyhedron whose $n = 99$ vertices lie on the surface of an ellipsoid. [Figure 7 of Kaneva and O'Rourke 2000.]

shortest paths in a discrete graph, a technique which has found many subsequent applications.

Dijkstra's algorithm finds the shortest paths from a node x to all other nodes in a graph whose arcs are weighted by their length. It can be viewed as simulating the spreading of paint originating from x and traveling down the incident arcs at a uniform rate. When paint first reaches a node y, its travel time t must accurately represent the shortest path from x to y. The spreading can be simulated in discrete time steps, where the next node to be "expanded" as a further source of spreading paint is the one labeled with the minimum t.

The continuous Dijkstra algorithm is exactly analogous: one can view it as spreading paint from the source x at a uniform rate (Figure 24.6 may provide some intuition), but over the whole surface rather than just down arcs. The key to simulating this process in discrete time steps is the notion of "propagating an interval." Let $p = xy$ be a shortest path from x on the polyhedron P. It crosses a particular set of edges of P in a particular order, which constitute the *edge sequence* of p. The algorithm of Mitchell et al. (1987) bundles together all those shortest paths with the same edge sequence, and propagates them as a unit. An *interval* I is a subsegment of a polyhedron edge e such that the shortest paths xy to every point $y \in I$ have the same edge sequence. They prove that the set of points on any e whose shortest paths from x have the same edge sequence is in fact a subsegment of e. Whereas in Dijkstra's graph algorithm, the next node to be expanded is that with the smallest t, in the continuous algorithm, the next interval to be propagated is the one whose closest point $z \in I$ to x is smallest. As with Dijkstra's algorithm, it is certain that the shortest path to at least this one point has already been found by the algorithm.

The situation prior to propagation is as illustrated in Figure 24.7. The set of paths to points on I are angularly contiguous at x, forming a cone apexed at x. I is then "projected" across the next face f. If this encompasses one or more vertices (such as v in the figure), the cone is split, for the edge sequences to either side of a vertex will be different.

At any one time, an edge of P has already been reached by several intervals, and before newly projected interval "children" (I_1 and I_2 in the figure) are inserted into the interval data structure, there must be adjustments made: a new interval might dominate an old interval entirely, or new and old intervals might trim one other, joining at a point tied in distance via two different edge sequences, a point potentially on the cut locus.

Naively, the bifurcation evident in Figure 24.7 might lead one to anticipate exponential complexity, but the domination and trimming of intervals were proved in Mitchell et al. (1987) to lead to only $O(n^2)$ intervals, resulting in an $O(n^2 \log n)$ algorithm.

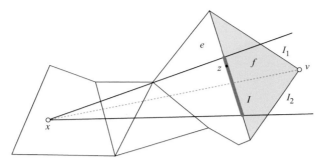

Figure 24.7. An unfolding of paths from x to an interval I; z is the closest point to x among all the points of I.

24.2.2 Chen and Han's Algorithm

Nearly a decade later, a second, slightly faster algorithm was developed by Chen and Han (1996). Their improvement to $O(n^2)$ is probably less important than its simplicity, which has invited at least two implementations. Here we report on one (Kaneva and O'Rourke 2000);[3] the other is proprietary (Lanthier et al. 1997).

Chen and Han's algorithm is perhaps most notable for *not* following the Dijkstra paradigm. Assume that the surface is triangulated. At the top level is a loop over all F faces. The algorithm builds a *cone tree* to store all shortest paths sharing a common edge sequence, as in the continuous Dijkstra approach. But the intervals are propagated without regard to whether they contain a true shortest path. Rather, each iteration of the loop selects a leaf of the cone tree at level ℓ for propagating across one face f, crossing the last edge e of its edge sequence. When the extended cone encompasses the vertex v of f opposite e, the cone splits to two children. In this algorithm, cones always include a vertex on each of its two bounding rays; trimming in Mitchell et al. (1987) does not retain this property. Now the potential exponential explosion is avoided by Chen and Han's key "one-split" lemma: If two cones "occupy" v from the same edge e, then at most one cone has two children that include shortest paths. After the main loop has iterated F times, the best occupier of each vertex represents the shortest path to that vertex. The paths do not grow in shortest-first fashion; rather the frontier represents all cones that have crossed ℓ faces, regardless of their length.

Extension to nonconvex surfaces is accomplished by treating each negative-curvature vertex as a new "pseudosource" point. Although this is conceptually simple, it represents a serious implementation complication, for the one-split rule does not hold at negative-curvature vertex: in fact, because a negative vertex can have arbitrarily large total incident face angle, there is no limit to the number of children it may spawn. Nevertheless, the algorithm maintains its $O(n^2)$ complexity.

Figure 24.8 shows two views of a terrain (a polyhedral manifold with boundary) of more than 2,000 vertices, and all the shortest paths from a source point near a corner. The pseudosource points are especially evident in the overhead view.

Finally we remark that there is no restriction that the polyhedron have genus zero. Figure 24.9 shows an example of genus 5. Indeed, because the implementation is based entirely on a user-supplied 2D complex, the input need not represent a surface

---

[3] http://cs.smith.edu/~orourke/ShortestPaths/.

Figure 24.8. Terrain of $n = 2{,}140$ vertices and $F = 4{,}274$ faces. Top: oblique view; bottom: overhead view. [Based on Figure 11 of Kaneva and O'Rourke 2000.]

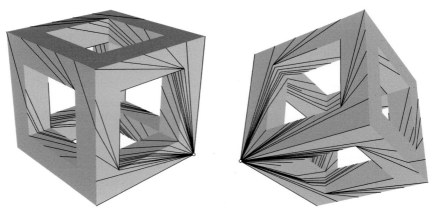

Figure 24.9. Two views of a wireframe cube: $n = 160$ ($F = 336$) [Based on Figure 14 of Kaneva and O'Rourke 2000.]

embeddable in 3-space without self-intersections; nor need it be orientable—it works on a Klein bottle as well as on a polyhedron.

It has been a long-standing open problem in the community to break the quadratic barrier for computing the shortest paths from a source point to all vertices on a convex polyhedron of n vertices. Because it is easily seen (e.g., in Figure 24.6) that there can be $\Omega(n^2)$ crossings between polyhedron edges and shortest paths to vertices, a subquadratic algorithm must avoid treating each such crossing as an "event," unlike the two algorithms just described. A decade after Chen and Han's work, finally this problem has been solved with an intricate, optimal $O(n \log n)$ algorithm (Schreiber and Sharir 2006).

24.3 STAR UNFOLDING

The second general unfolding to a nonoverlapping net is the *star unfolding*. It derives from the observation that the shortest paths from a source x to all vertices of a convex polyhedron P form a spanning tree of the vertices, as is evident in Figures 24.1 and 24.6. We must restrict x to be a point such that the shortest path to each vertex is unique, for otherwise they do not form a tree. Call such a point x a *generic* point.[4] The star unfolding is produced by cutting along all these shortest paths. As Alexandrov was the first to realize,[5] this must lead to what he called a "geodesic polygon" that can be unfolded flat. Figure 24.10 shows his drawing of the process. However, he only used this unfolding to prove that the polygon had a triangulation with certain properties, and in a footnote allowed for the possibility that it might self-overlap.[6] That the star unfolding in fact does not overlap was established in Aronov and O'Rourke (1992). Although the cut tree is more straightforward than that for the source unfolding, the star unfolding itself is less easily comprehended. We now take some time to explore this without, however, proving nonoverlap, as the only proof known is not straightforward.

We return to the pyramid in Figure 24.2. Cutting the shortest paths from x to the five vertices produces the star unfolding shown in Figure 24.11(a). Each cut xv_i maps to two edges of the boundary of the star unfolding. The source point x has as many "images" in the unfolding as there are vertices of the polyhedron P. The vertices of the boundary of the star unfolding alternate as deriving from x, or deriving from a polyhedron vertex:

$$(x_0, v_1, x_1, v_4, x_2, v_0, x_3, v_3, x_4, v_2)$$

in the figure. One can view the star unfolding as composed of triangles $\triangle x_i, v_j, v_k$ glued to a *core*,[7] a polygon connecting the vertices of the polyhedron in the same order as is their shortest paths from x: $(v_1, v_4, v_0, v_3, v_2)$ in the figure (see also Figure 24.2(a)).

Although the star unfolding produces shapes that remind one of star bursts, in fact they are not necessarily *star-shaped* in the technical sense of having a point inside that can see the entire interior. For example, Figure 24.12 displays four star unfoldings for

---

[4] Note that x could be a vertex. But x cannot lie on the cut locus of any vertex v treated as source, for then it would have more than one shortest path to v.

[5] For this reason, the star unfolding is called the "Alexandrov unfolding" in Miller and Pak (to appear). In Chen and Han (1996) it is called the "inward layout."

[6] A translation from Alexandrov (1955, p. 226) is, "Of course the polygon Q [the star unfolding] may overlap itself when unfolded." See also Alexandrov (2006, p. 198).

[7] Called the *kernel* in Aronov and O'Rourke (1992).

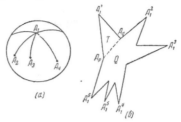

Так как кратчайшие, исходящие из одной точки, не могут пересекаться, то в результате такого разрезывания получим (геодезический) многоугольник Q, гомеоморфный кругу. Он не содержит вершин внутри, а потому согласно теореме 5 § 8 главы 1 его можно разбить диагоналями на треугольники, каждый из которых развёртывается на плоскость.

Однако мы хотим доказать, что многоугольник Q можно разбить диагоналями на треугольники так, что в полученной развёртке ни у какого треугольника две вершины не будут склеиваться друг с другом, т. е. не будут соответствовать одной вершине развёртки.

У многоугольника Q имеются вершины двух типов: $e-1$ вершин отвечают вершинам A_2, A_3, ..., A_e развёртки и другие $e-1$ вершин отвечают одной вершине A_1 (черт. 88, б). Эти вершины «второго рода» A_1^1, A_1^2, ..., A_1^{e-1} расположены на периметре многоугольника Q через одну. Так как одной и той же вершине развёртки отвечают только вершины второго рода, то нам нужно доказать следующее:

Черт. 88.

Figure 24.10. Alexandrov's figure of the star unfolding (1950, p. 181).

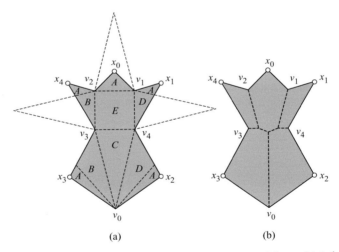

(a)   (b)

Figure 24.11. Construction of the star unfolding corresponding to Figure 24.2. (a) The superimposed dashed edges show the "natural" unfolding obtained by cutting along the four edges incident to v_0. The A, B, C, D, and E labels indicate portions of the star unfolding derived from those faces. (b) The cut locus displayed inside the star unfolding. [Based on Figures 5 and 6 of Agarwal et al. 1997.]

randomly generated convex polyhedra and none of them are star-shaped. (Of course, the source unfolding is always star-shaped, with x the point of central visibility.)

24.3.1 Nonoverlap of the Star Unfolding

The proof of nonoverlap (Aronov and O'Rourke 1992) is by induction. The reduction of the polyhedron P from n to $n-1$ vertices is accomplished selecting two vertices forming the base of one of the triangles $\triangle x_i$, v_j, v_k mentioned above, with a special relationship to the cut locus. A local modification is made, replacing v_j and v_k by one new vertex, producing a star unfolding of a polyhedron P' of $n-1$ vertices. Crucial here is

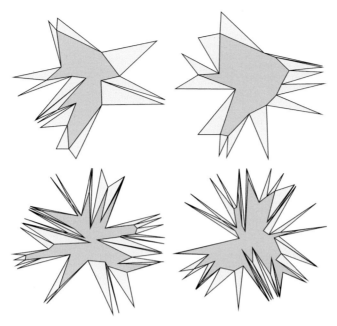

Figure 24.12. Four star unfoldings: $n = 13, 13, 36, 42$ vertices, left to right, top to bottom. The core is shaded darker in each figure. [Code written by Julie DiBiase and Stacia Wyman.]

applying Alexandrov's theorem (Theorem 23.3.1, p. 349), to guarantee that the reduced unfolding does fold back to a convex polyhedron. Nonoverlap of the star unfolding of P' then yields, employing a nesting similar to that used for dome unfolding (Figure 22.32, p. 328) but more delicate, nonoverlap of the star unfolding of P. This establishes the following:

Theorem 24.3.1. *The star unfolding of a convex polyhedron with respect to any generic point x produces a nonoverlapping polygon.*

We now show that Theorem 24.3.1 is tight in one sense: the star unfolding might overlap even for a polyhedron with just one vertex v of negative curvature. Recall (p. 303) that the curvature at v is 2π minus the sum of the face angle incident to v. So a vertex of negative curvature of a paper polyhedron has an "excess" of paper glued to v. If the source x is at v itself, then this excess of paper is split apart by the cuts to the other vertices, and it is at least plausible that the star unfolding then avoids overlap. However, this does not hold for the example shown in Figure 24.13.

The 9-vertex heart-shaped polyhedron has one vertex x of negative curvature. Four vertices, a, b, c, and d, form a 2×2 square in a vertical plane in space, with the corresponding vertices with primed labels forming a parallel 3×3 square, $\frac{1}{4}$ back; x is midway between the square planes. a' is displaced $(-\frac{1}{2}, -\frac{1}{2}, -\frac{1}{4})$ from a and so separated by a distance of $\frac{3}{4}$, and similarly for the other square vertices. Vertex x has curvature $-194.9°$. All other vertices have positive curvature, two just barely (b and c have curvature $5.9°$). The shortest paths from x to the other 8 vertices each follow an edge of the polyhedron.

The star unfolding from x is therefore easily constructed by laying out the faces of the polyhedron appropriately, as shown in Figure 24.13(b). The shapes of the side trapezoids—(a, b, b', a'), etc.—cause the layout to curl in such a way that the two

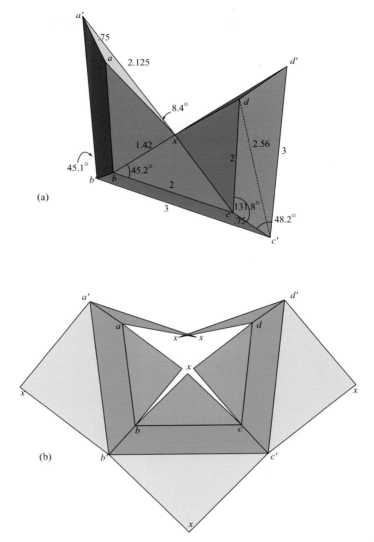

Figure 24.13. A polyhedron (a) whose star unfolding (b) from x overlaps. Several edge lengths and face angles are detailed, as well as a few critical computations.

"spikes," $\triangle xaa'$ and $\triangle xdd'$, cross in the vicinity of x. It is, however, easy to find a nonoverlapping unfolding of this polyhedron. For example, just leave the a-spike attached at its xa' edge to the back triangle $\triangle xa'b'$, cutting edges $a'a$ and ax.

As far as we know, the general unfolding problem (Open Problem 22.3) is open even for polyhedra with just one vertex of negative curvature (see Figure 24.14).

24.3.2 Cut Locus Is a Voronoi Diagram

A second theorem established by Aronov and O'Rourke (1992) leads to an easy computation of the cut locus via the Voronoi diagram (see Box 24.1):

Theorem 24.3.2. *The cut locus is the portion of the Voronoi diagram of the source images x_i that lies inside the star unfolding (in fact, inside the core of the star unfolding).*

Box 24.1: Voronoi Diagrams

Let S be a set of n distinct points (or *sites*) $x_0, \ldots, x_{n-1}$ in the plane. The Voronoi diagram of S is a partition of the plane into convex regions (*Voronoi regions*), one per site x_i, of all the points of the plane closer to x_i than to any other site (see Figure 24.15). As defined, with "closer" meaning "strictly closer," the regions are open sets, whose boundaries are straight segments (*Voronoi edges*), whose interior points are tied for closeness to the two sites of the two regions sharing that segment, and whose endpoints (*Voronoi vertices*) are tied for closeness to the three or more sites that meet at that point. All the regions of sites on the convex hull of S are unbounded; all the others are bounded, convex polygons.

Figure 24.14. A polyhedron with one vertex of negative curvature. The dimensions have been chosen so that the vertices at the small base regular octagon have curvature exactly zero.

This theorem is illustrated in Figure 24.11(b), and should at least be plausible from the properties of Voronoi diagram edges, for the cut locus edges are points equidistant from two source images. A nonobvious corollary is that the source images are the vertices of the convex hull of the star unfolding. Again we will not prove this theorem or its corollaries, but just draw on its computational consequences.

The star unfolding is easily computed from the shortest paths from x to each vertex v_j: The internal angle at a source image x_i is the angle at x on P between two adjacent shortest paths, and the internal angle at v_j is the total face angle incident to v_j on P. Computing a layout of the star unfolding then yields locations for the source images x_i. Computing the Voronoi diagram of these images is a thoroughly understood problem, with several $O(n \log n)$ algorithms available. Finally, clipping the diagram to the core, by truncating the edges at the v_j, is also easily accomplished. Thus the cut locus can be computed within the same time bound as computing the shortest paths, $O(n \log n)$.

The star unfolding has a number of other computational consequences detailed in Agarwal et al. (1997).

Imagining polyhedra approaching in the limit a surface with smoothly curved sections (e.g., Figure 24.16), it is natural to wonder if Theorem 24.3.2 applies to such surfaces. The natural analog of the Voronoi diagram is the "medial" or "symmetric" axis of a Jordan curve (Blum 1967; Lee 1982), which is the locus of centers of interior

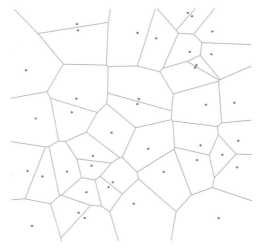

Figure 24.15. A Voronoi diagram of 40 sites, clipped to a rectangular region.

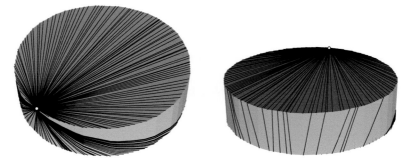

Figure 24.16. Shortest paths to all vertices of a polyhedral cylinder. [Figure 6 of Kaneva and O'Rourke 2000.]

disks that meet the curve in more than one point. This leads to the following open problem:

Open Problem 24.1: Star Unfolding of Smooth Surfaces. [a] Prove or disprove the conjecture that the cut locus for a point x on a smooth convex surface develops (without overlap) to the medial axis of the boundary of the star unfolding, which is composed of the images of the source x.

---

[a] Aronov and O'Rourke (1992).

24.3.3 Core/Anticore

Finally we mention an interesting consequence of the nonoverlap of the star unfolding and its core. Let x be the source point on the polyhedron P and $v_0, v_1, v_2, \ldots, v_{n-1}$ the vertices of P sorted angularly by the shortest paths from x. Recall that the boundary of the core is then $(v_0, v_1, v_2, \ldots, v_{n-1})$. Viewed on the surface of P, this core boundary is a geodesic polygon, a sort of crooked belt around P, with the core to one side. Call this geodesic polygon the *ss-divide* (for source/star-unfolding divide) and call the other side of P, the one including x, the *anticore*. The anticore is entirely contained in the source unfolding and so also does not overlap (cf. Figure 24.3). Thus, for any nonvertex x, the ss-divide connecting the vertices in a cycle partitions P into two halves each of which unfolds without overlap.

One might hope that the core and anticore can be joined along a shared edge to form a single, nonoverlapping polygon, but there are polyhedra for which, for a particular source x, every such joining leads to overlap, and it seems plausible that examples might exist where the same holds for every x. Nevertheless, several other questions suggest themselves concerning the core and anticore, all unexplored as of this writing. Let P be a convex polyhedron and x a source point.

1. For which P and x can the core and anticore be joined to form a polygon?
2. For which P and x is the core star-shaped?
3. For which P and x is the core convex?
4. For which P and x are the core and anticore congruent polygons?
5. Is there a polynomial-time algorithm for determining if a given polyhedron may be dissected into two congruent (in 3D) halves via a geodesic polygonal "belt" all of whose vertices are polyhedron vertices?

The last question is only loosely suggested by the others, but is an intriguing "dissection" question (Frederickson 1997); for another, see Open Problem 25.6.

24.4   GEODESICS: LYUSTERNIK–SCHNIRELMANN

We describe in this section an old theorem: every closed, convex surface has at least three distinct, closed geodesics. Aside from the natural beauty of this theorem, it is connected to shortest paths (Chapter 24), and to curve development (Section 24.5). And it may play a role in establishing polyhedron unfolding results.

Geodesics. Recall (from p. 358) that geodesics are locally shortest paths. Because of this, they share with shortest paths characteristics 5 and 6 in our earlier list (p. 358) in that they unfold, or develop onto a plane, as straight lines.

We will focus primarily on convex surfaces, and on *simple* geodesics, those that do not self-intersect, that is, they are *embedded* on the surface (in the sense that no two distinct points of the geodesic map to the same point of the surface). Finally, we are especially interested in simple, closed geodesics, which come in two varieties:

1. "We call a geodesic curve *closed* if it returns to its starting point without having a corner and without intersecting itself before it reaches the starting point" (Hilbert and Cohn-Vossen 1952, p. 222).
2. A *geodesic loop* is a simple geodesic that returns to its starting point but forms a "corner" there in the sense that the tangents do not match, and so the curve does not rejoin itself smoothly.

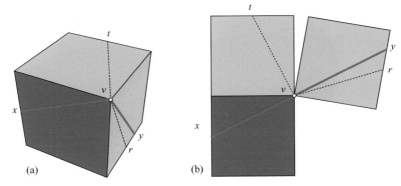

Figure 24.17. (a) Any continuation of the segment xv angularly between vr and vt is a quasi-geodesic; (b) rotation of faces unfolds vy collinear with xv.

Note that the term "closed geodesic" is used in the literature to imply both simplicity (non-self-intersection) and smooth joining; we will follow this established usage.[8]

Quasigeodesics. Although most of the literature on geodesics on surfaces comes out of the discipline of differential geometry and assumes that the surfaces are smooth, every result we will need applies equally well to the surface of a polyhedron. The extension to "polyhedral metrics" (which we encountered earlier, p. 348) was carried out by A. D. Alexandrov and his students starting in 1948,[9] via the notion of a "quasigeodesic," the natural extension of geodesics to nonsmooth surfaces. We will define this notion only for a polyhedral surface, although the definition encompasses a wider class of surfaces. There are two *surface angles* defined at each point p on a directed curve Γ on a surface: $\rho(p)$, the total incident surface (face) angle at p to the right at p, and $\lambda(p)$, the angle to the left. Define the left and right *turn angles* at p to be $\tau_L(p) = \pi - \lambda(p)$ and $\tau_R(p) = \pi - \rho(p)$ respectively, which can be viewed as left and right deviations from straightness. A *quasigeodesic* is a curve on a surface such that both τ_L and τ_R are nonnegative at each point (Alexandrov and Zalgaller 1967, p. 16; Pogorelov 1973, p. 28).

Every geodesic is a quasigeodesic, because $\tau_L = \tau_R = 0$ at every point of a geodesic: they are straight curves and do not turn. Geodesics on polyhedral surfaces enjoy this same property at every point not coincident with a polyhedron vertex, for there is π face angle on either side. This is true even on interior points of edges. When a geodesic passes through a polyhedron vertex v, however, there are many continuations that could be unfolded straight, as illustrated in Figure 24.17. A curve that makes a negative left turn at v could not be straightened, for $\tau_L(v) < 0$ implies that $\lambda(v) > \pi$, which means there is too much surface angle to the left to permit it to unfold straight. So Alexandrov's definition is indeed the natural definition from the point of view of unfolding to a straight line: quasigeodsics are exactly those curves that can unfold/develop to straight lines.[10] They can also be viewed as the natural limit of geodesics on smooth surfaces: Pogorelov proved that a curve is a quasigeodesic if and only if it is "the limit of geodesic

---

[8] The term *periodic geodesic* is sometimes used to mean closed but perhaps self-intersecting (Berger 2003, p. 44).

[9] See the citations in Alexandrov and Zalgaller (1967).

[10] In Polthier and Schmies (1998) a "straightest geodesic" is defined as the continuation that splits the surface angle so that $\lambda(v) = \rho(v)$.

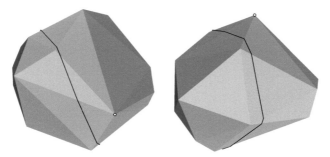

Figure 24.18. Two views of a closed quasigeodesic on a convex polyhedron of 21 vertices, 38 faces, and 57 edges. (The marked vertex is the same in both views.)

curves that lie on convex surfaces converging to the given surface" (Alexandrov and Zalgaller 1967, p. 16).

The Lyusternik–Schnirelmann theorem. That there are at least three closed geodesics on every closed convex surface had been "conjectured for a long time" (Hilbert and Cohn-Vossen 1952, p. 222) before proved by Lyusternik and Schnirelmann in 1929 (Lyusternik and Schnirelmann 1929). Their proof relied on an earlier proof by Birkhoff (1927) that there is at least one closed geodesic. The Lyusternik–Schnirelmann proof "contains some gaps" which were not filled in until 1978 (Ballmann et al. 1983). Pogorelov extended the theorem by establishing that polyhedral surfaces have three closed quasigeodesics (Pogorelov 1949). A single closed quasigeodesic is shown in Figure 24.18. Most recently, after a long pursuit it has been established that every surface homeomorphic to a sphere has an infinite number of self-crossing closed geodesics (Cipra 1993).

Closed quasigeodesics on polyhedra. Aside from numerical simulations (e.g., Lysyanskaya 1997), there is no algorithm for actually finding the three quasigeodesics guaranteed by Pogorelov's extension of the Lyusternik–Schnirelmann theorem:

Open Problem 24.2: Closed Quasigeodesics. [a] Find a polynomial-time algorithm to find a closed quasigeodesic on a convex polyhedron.

---

[a] Posed with Stacia Wyman (1990).

One impediment here is this negative result,[11] which shows that any algorithm that lists all closed quasigeodesics cannot be polynomial-time:

Theorem 24.4.1. *A polyhedron of n vertices might have $2^{\Omega(n)}$ distinct, closed, simple quasigeodesics.*

Although it seems natural to build up a closed geodesic from shortest paths, this potential approach is seemingly blocked by the fact that for any k, there is a convex body whose shortest geodesic is not composed of k shortest paths. However, Croke proved

---

[11] With Boris Aronov, unpublished.

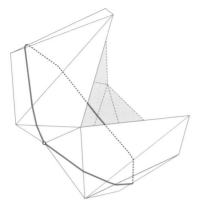

Figure 24.19. The faces F crossed by the closed quasi-geodesic in Figure 24.18 (from a different viewpoint). The quasigeodesic passes through the marked vertex.

that the shortest curve that partitions the surface into two pieces each of curvature 2π is a closed geodesic (Croke 1982).

24.4.1 Relation to Edge-Unfolding

One reason to study quasigeodesics is that they unfold or develop as straight lines. A closed quasigeodesic seems a particularly attractive "backbone" for a non-overlapping unfolding. This leads to another unsolved problem, which we explain in this section.

Let γ be a closed quasigeodesic on a convex polyhedron P, and let F be the set of faces of P whose interior is touched by γ. F forms an irregular "band" on P surrounding γ (see Figure 24.19). There is a natural edge-unfolding of the faces in F, arranged around the straight-line development of γ. Although it is clear that this unfolding avoids overlap locally, and it is natural to hope that it avoids overlap everywhere, this is not clear:

> **Open Problem 24.3: Closed Quasigeodesic Edge-Unfolding.** For a given closed quasigeodesic γ, is it true that the set of faces whose interior is touched by γ unfolds along γ without overlap? Is this at least true for the shortest closed quasigeodesic?

A positive answer would provide a natural way to partition the edge-unfolding problem into the two convex "caps" that remain on either side of the F-band.

One reason to anticipate a negative answer to this question is that the set of faces crossed by an open geodesic, or even a shortest path, can unfold with overlap. This issue was first raised in Schevon (1989, Chap. 5), which included an example equivalent to Figure 24.20 (a repeat of Figure 22.8(b)) to show that the claim is FALSE for geodesics. Recently Günter Rote[12] showed that essentially the same example can be achieved with a shortest path crossing the faces in the same order as the horizontal geodesic.

24.5 CURVE DEVELOPMENT

Continuing on the theme of developing curves without overlap, we describe two further results in this section, without formally proving either. Both involve developing curves

[12] Personal communication, August 2005.

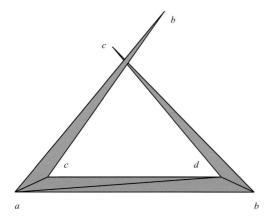

Figure 24.20. Faces cut by a horizontal geodesic unfold with overlap.

that are only piecewise geodesic. Nonoverlap of a developed geodesic is trivial, as it develops to a straight-line segment, but nonoverlap of any nongeodesic curve requires analysis.

Let $c_0, c_1, \ldots, c_n$ be the *corners* of curve C, the points at which it turns: either by crossing an edge of P with a dihedral angle different from π, or by passing through a vertex of P, or by turning on the interior of a face of P. Orient C so each of its edges has a direction. Recall (p. 373) the two surface angles defined at each point $p \in C$: $\rho(p)$, the total incident face angle at p to the right of the directed curve C at p, and $\lambda(p)$, the angle to the left. Only at a corner c_i of C is either surface angle different from π. If c_i is not a vertex of P, then $\rho(p) + \lambda(p) = 2\pi$; at a vertex the two angles sum to less than 2π by the angle deficit (p. 303). Define the *right development* of C to be a planar drawing of the polygonal chain C as the chain $B = (b_0, b_1, \ldots, b_n)$ with the same edge lengths, $|b_i b_{i+1}| = |c_i c_{i+1}|$ for $i = 0, \ldots, n-1$, and with exterior angle ρ_i to the right of b_i the same as the right surface angle $\rho(c_i)$, for all i. Define *left development* similarly. If C avoids vertices of P, the right and left developments are congruent, and are called *the* development of C. If C includes one or more vertices, then any equal-lengths polygonal chain that at b_i turns an angle between $\rho(c_i)$ to the right and $\lambda(c_i)$ to the left is called *a* development of C. Although unnecessary, for simplicity we will assume throughout this section that our curves include no vertices, so there is a unique development.

The two results are that both closed convex curves, and "slice curves," develop without self-intersection.

24.5.1 Closed Convex Curves

A closed convex curve C on the surface of a convex polyhedron P is a closed polygonal curve with $\lambda(c_i) \le \pi$ for all i. It was proved in O'Rourke and Schevon (1989)[13] that such curves always develop without self-intersection, starting from any point on C. We now illustrate the proof via an example. Figure 24.21(a) shows a curve $C = (c_0, \ldots, c_5, c_0)$ on the surface of a cube P. If C encloses no curvature, that is, encloses no vertices—for example, if it lies entirely within one face—then it develops to a closed convex polygon, with planar internal angles at each b_i equal to $\lambda(c_i)$. If it does enclose curvature, remove all vertices inside C, and retriangulate the interior by taking the convex hull of the c_i.

[13]   William Thurston first suggested that this might be true.

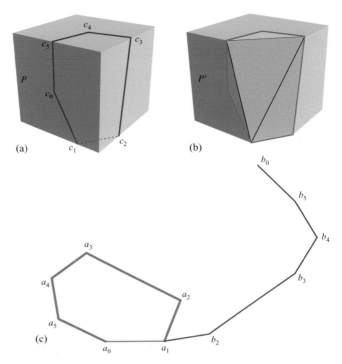

Figure 24.21. (a) A convex polygonal curve C on a polyhedron P; (b) P': after removal of interior vertices and retriangulation; (c) development of C from P' (closed) and from P (open).

The resulting modified cube is shown in Figure 24.21(b). Now C encloses no curvature on P' by construction, so it develops to a convex polygon $A = (a_0, \ldots, a_5, a_0)$.

Let θ_i be the internal angle at the vertex a_i of this planar polygon. A lemma in O'Rourke and Schevon (1989) establishes the plausible claim that $\theta_i \le \lambda(c_i)$ for all i: cutting away material to form P' only can reduce the angle. Because by assumption we have $\lambda(c_i) \le \pi$, we may apply Cauchy's arm lemma, Lemma 8.2.1 (p. 143). Cutting B at any point $p \in A$, and then "opening" the internal angles from θ_i to $\lambda(c_i)$, ensures that the two images of p increase in separation. See Figure 24.21(c), where $p = a_0$. A further implication, which we will prove in a more general form below, is that the opened curve $B = (b_0, \ldots)$ does not self-intersect.

24.5.2 Slice Curves

Let Π be a (2D) plane, and P a convex polyhedron. The intersection $C = P \cap \Pi$ is either a vertex, edge, or face of P, or it is a closed curve called a *slice curve*. Viewed in $\mathbb{R}^3$, a slice curve is a convex polygon, but viewed as a curve on P it is more complex. The second result (O'Rourke 2003) is that a slice curve develops to a non-self-intersecting curve in the plane.

That the development of a slice curve is not, in general, convex, is illustrated in Figure 24.22, which shows two different slices through a cube.

The proof is however very similar to that used in the preceding section. Let Q be the convex polygon in the slice plane Π whose boundary is C (see Figure 24.23). Let $A = (a_0, a_1, \ldots, a_n)$ be a polygonal chain representing the development of convex polygon Q, with a_i corresponding to c_i. Rather than look at the internal angle θ_i, we examine the

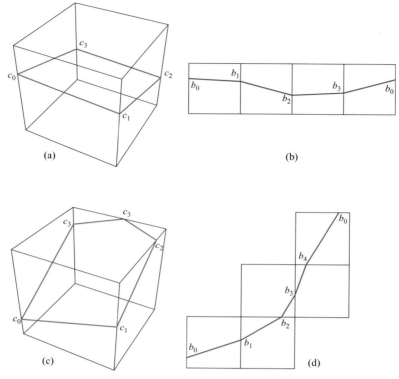

Figure 24.22. (a) Slice curve $C = (c_0, c_1, c_2, c_3)$; (b) its development B; (c) slice curve $(c_0, c_1, c_2, c_3, c_4)$; (d) its development. [Based on Figure 1 of O'Rourke 2003, by permission, Elsevier.]

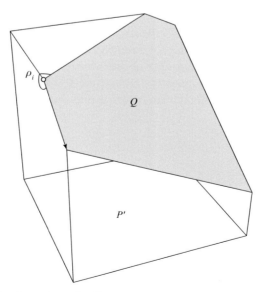

Figure 24.23. Q is a slice through a cube P. Angles are shown at corner c_i of $\Gamma = \partial Q$, the boundary of Q. [Based on Figure 4 of O'Rourke 2003, by permission, Elsevier.]

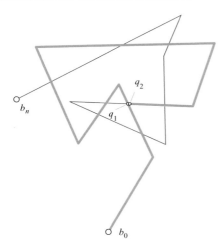

Figure 24.24. Violation of Theorem 24.5.1. $q_1 = q_2$ is the first point of self-contact; the initial portion of B, up to q_2, is highlighted. [Based on Figure 3 of O'Rourke 2003, by permission, Elsevier.]

turn angle α_i at a_i, $\alpha_i = \pi - \theta_i$. It is not difficult to see (O'Rourke 2003) that if β_i is the turn angle at b_i for the development $B = (b_0, b_1, \ldots, b_n)$ of C the plane, then (ignoring the easily handled degenerate intersections of Π with P) $-\alpha_i < \beta_i < \alpha_i$, that is, $|\beta_i| < \alpha_i$. This condition satisfies the generalization of Cauchy's arm lemma, Lemma 8.2.2 (p. 146): If a convex curve is straightened, with the new turn angles no larger (in absolute value) than the old, then distances increase. As with Cauchy's theorem, this additionally implies non-self-intersection of the opened curve, less obvious in this case (O'Rourke 2003, Thm. 4):

Theorem 24.5.1. *If $A = (a_0, \ldots, a_n)$ is a closed convex chain (closed in the sense that $a_n = a_0$), then any reconfiguration to $B = (b_0, b_1, \ldots, b_n)$, with turn angles satisfying $|\beta_i| < \alpha_i$ (i.e., Lemma 8.2.2), is a noncrossing chain.*

   Proof: Suppose to the contrary that B self-intersects. Let q_2 be the first point of B, measured by distance along the chain from the shoulder b_0, that coincides with an earlier point $q_1 \in B$. Thus q_1 and q_2 represent the same point of the plane, but different points along B (see Figure 24.24). Because B crosses itself, these "first touching points" exist, and we do not have both $q_1 = b_0$ and $q_2 = b_n$ (because that would make B a noncrossing closed chain). Let p_1 and p_2 be the points of A corresponding to q_1 and q_2.
   An easy corollary to Lemma 8.2.2 guarantees that $|q_1 q_2| \geq |p_1 p_2|$, that is, every pair of corresponding points (not just the chain endpoints) separates. But $|q_1 q_2| = 0$, and because the q's do not coincide with the original b_0 and b_n, $|p_1 p_2| > 0$. This contradiction establishes the claim. □

24.5.3 Band Unfolding

Surrounding C by a narrow enough band on the surface of P shows that this band unfolds without overlap. It is then natural to wonder if a vertex-free strip sandwiched between two parallel slicing planes—a potentially thick band—may be unfolded without overlap. That this question is not settled by Lemma 8.2.2 above is illustrated by the example in Figure 24.25,[14] which demonstrates that not every edge unfolding of such

---

[14] Found in joint work with Anna Lubiw (1998).

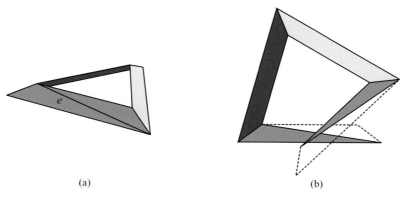

(a)   (b)

Figure 24.25. (a) A truncated tetrahedron; top and bottom faces removed; (b) an edge-unfolding of the band of side faces obtained by cutting e. [Based on Figure 5 of O'Rourke 2003, by permission, Elsevier.]

a strip avoids overlap. However, this example left open whether *some* edge unfolding always avoids overlap, or whether some nonedge cut avoids overlap. For example, in O'Rourke (2003) it was conjectured that cutting the shortest path between the two slice curves (which path is not, in general, an edge of P) develops the strip without overlap.

Recent progress has settled these questions affirmatively. First, the existence of a nonoverlapping edge unfolding was established in a special case (Aloupis et al. 2004): if the top polygon A is nested inside the base polygon B, in the sense that orthogonal projection to the plane of B places A inside B. Paeng[15] established that there is always a geodesic cut that leads to nonoverlap. Aloupis (2005) removed the nested restriction of Aloupis et al. (2004), thus establishing that there is an edge to cut that unfolds any band without overlap.

Returning to the question of edge-unfolding prismatoids (Section 22.5.1, p. 323), the existence of a nonoverlapping edge-unfolding of a band shows that the side faces of a prismatoid can be unfolded as a unit without overlap. It is natural to hope that the top and bottom faces A and B can be affixed to opposite sides of the band, resulting in a nonoverlapping edge unfolding of a prismatoid, but this has not yet been established.

[15] S.-G. Paeng, personal communication, November 2004.

25 Folding Polygons to Polyhedra

FOLDING POLYGONS: PRELIMINARIES

The question

When can a polygon fold to a convex polyhedron?

was first explicitly posed in 1996 (Lubiw and O'Rourke 1996). Here we mean fold without overlap (in constrast to the wrapping permitted in Theorem 15.2.1, p. 236) and of course without leaving gaps. We have seen in Section 23.3 (p. 348) that Alexandrov's Theorem provides an answer: whenever a polygon has an Alexandrov gluing. (Recall that, in this context, a "polyhedron" could be a flat, doubly covered polygon.) This then reorients the question to

When does a polygon have an Alexandrov gluing?

Before turning to an algorithmic answer to this question, we first show that not all polygons have an Alexandrov gluing, that is, not all polygons are *foldable*, and indeed the foldable polygons are rare.

25.1.1 Not-Foldable Polygons

Lemma 25.1.1. *Some polygons cannot be folded to any convex polyhedron.*

Proof: Consider the polygon P shown in Figure 25.1. P has three consecutive reflex vertices (a, b, c), with the exterior angle β at b small. All other vertices are convex, with interior angles strictly larger than β.

Either the gluing *zips* at b, gluing edge ba to edge bc, or some other point(s) of ∂P glue to b. The first possibility forces a to glue to c, exceeding 2π there; so this gluing is not Alexandrov. The second possibility cannot occur with P, because no point of ∂P has small enough internal angle to fit at b. Thus there is no Alexandrov gluing of P. □

It is natural to wonder what the chances are that a random polygon could fold to a convex polyhedron. This is difficult to answer without a precise definition of "random,"

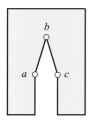

Figure 25.1. A not-foldable polygon.

but we feel any reasonable definition would lead to the same answer. Let us assume that random polygons on n vertices satisfy two properties:

1. The polygon edge lengths are drawn from some continuous density distribution.
2. As $n \to \infty$, the polygon has at least two reflex vertices. (Of course one would expect the number of reflex and convex vertices to approach balance, but we only need this weaker assumption.)

Under these assumptions, we can prove this result:

Theorem 25.1.2. *The probability that a random polygon of n vertices can fold to a polyhedron approaches 0 as $n \to \infty$.*

We defer a proof to Section 25.1.3 below, using technical machinery developed in the next section.

25.1.2   Perimeter Halving

In contrast to these results, we show here that every convex polygon folds to a convex polyhedron, and indeed we shall claim to infinitely many geometrically distinct (incongruent) polyhedra.

For two points x, $y \in \partial P$, let $P(x, y)$ be the open interval of ∂P counterclockwise from x to y and let $|x, y|$ be its length. Define a *perimeter-halving gluing* as one that glues $P(x, y)$ to $P(y, x)$, where $|x, y| = |y, x|$.

Theorem 25.1.3. *Every perimeter-halving gluing of a convex polygon folds to a convex polyhedron.*

Proof: Let the perimeter of a convex polygon P be L. Let $x \in \partial P$ be an arbitrary point on the boundary of P, and let $y \in \partial P$ be the midpoint of perimeter around ∂P measured from x, that is, y is the unique point satisfying $|x, y| = |y, x| = L/2$. See Figure 25.2 for an example.

Now glue $P(x, y)$ to $P(y, x)$ in the natural way, mapping each point z with $|x, z| = d$ to the point z' the same distance from x in the other direction: $|z', x| = d$. We claim this is an Alexandrov gluing. Because P is convex, each point along the gluing path has $\leq 2\pi$ angle incident to it: the gluing of two nonvertex points results in exactly 2π, and if either point is a vertex, the total angle is strictly less than 2π. The resulting surface is clearly homeomorphic to a sphere. By Alexandrov's Theorem 23.3.1, this gluing corresponds to a unique convex polyhedron Q_x.   □

We call a point in the interior of a polygon edge that glues only to itself, that is, where a crease folds the edge in two, a *fold point*. A fold point becomes a vertex of the polyhedron with curvature π. Points x and y in Figure 25.2 are fold points. One

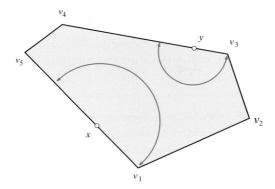

Figure 25.2. A perimeter-halving fold of a pentagon. The gluing mappings of vertices v_1 and v_3 are shown.

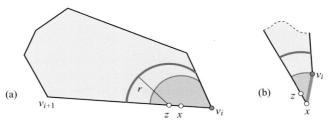

Figure 25.3. (a) x varies over (z, v_i) on ∂P; (b) xv_i is the shortest path between those vertices on $\mathcal{P}$.

can imagine sliding x around ∂P, moving y correspondingly, achieving a continuum of different polyhedra Q_x. We now show that indeed this process produces an infinite number of incongruent polyhedra.[1]

Theorem 25.1.4. *For any convex polygon P, perimeter-halving folds produce an uncountably infinite number of incongruent polyhedra.*

Proof: The idea of the proof is to vary the perimeter-halving fold point x near any vertex v_i of P, and show that an uncountable number of vertex–vertex shortest path lengths are produced. Identify a point z on $v_i v_{i+1}$ so that no vertices of P lie within a radius of $r = |zv_i|$ of z. Clearly such a z exists. Now let x vary over the open interval (z, v_i). Then, as Figure 25.3 shows, when folded at x, the open "geodesic disk" (p. 303) of radius $|xv_i| < r$, centered at x on the resulting polyhedron surface $\mathcal{P}$, contains no vertices. Both x and v_i are vertices of $\mathcal{P}$, and the empty geodesic disk ensures that the segment xv_i is indeed the shortest path between these two vertices.

Now, varying x produces an uncountably infinite number of distinct shortest path lengths $|xv_i|$. If it were the case that perimeter halving produces only a countable number of distinct polyhedra, then, because each has a finite number of vertex–vertex shortest paths, there would only be a countable number of shortest path lengths. This contradiction establishes the claim of the theorem. □

An interesting question raised by this proof is whether an infinite number of the polyhedra resulting from perimeter-halving foldings of a given convex polygon can be congruent.

---

[1] This proof improves on that in Demaine et al. (2000b).

Perimeter halving will serve as the model of "rolling belts," to be introduced in Section 25.5.

We mention here that the corresponding question for nonconvex polyhedra is unsolved, that is, whether an analog of Lemma 25.1.1 holds:

Open Problem 25.1: Folding Polygons to (Nonconvex) Polyhedra. Does every simple polygon fold (by perimeter gluing) to some simple polyhedron? In particular, does some perimeter halving always lead to a polyhedron?

Permitting nonconvex polyhedra alters the problem substantially, and as of this writing, the tools needed to answer these questions seem not yet developed.

25.1.3   Random Polygons Do Not Fold

We return now to proving Theorem 25.1.2. Define a *subpolygon* as a closed chain of vertices and edges left unglued after partially gluing a polygon perimeter. For example, in a perimeter halving with fold points x and y, if just the middle third of $P(x, y)$ is glued to its counterpart, this creates two subpolygons constituted by the first and last thirds. We know from Theorem 25.1.3 that a subpolygon all of whose vertices are convex can fold via perimeter halving, for the proof of that theorem holds for a proper subset of the vertices of a convex polygon just as it does for the full polygon. We now show that (sub)polygons with one reflex vertex can fold; but random polygons with two or more reflex vertices cannot.

Lemma 25.1.5. *A polygon or subpolygon with one reflex vertex r_1 folds by zipping at r_1, that is, performing a perimeter-halving fold with one endpoint r_1.*

Proof: The described gluing satisfies Alexandrov's conditions: the zip of the reflex vertex creates a point with less than 2π around r_1, and the convex vertex gluings are also Alexandrov by the same reasoning used in the proof of Theorem 25.1.3.   □

This lemma can be viewed as a direct extension of Theorem 25.1.3. (Note that the polygon in Figure 25.1 has three reflex vertices, and so this lemma does not apply.) Next we show that the step up to two reflex vertices fails for random polygons.[2]

Lemma 25.1.6. *A subpolygon with two or more reflex vertices, and random edge lengths, folds with probability zero.*

Proof: We first prove the lemma for exactly two reflex vertices, r_1 and r_2. Let $c_1, \ldots, c_k$ be the convex vertices. The proof is by induction on k. If $k = 0$, then r_1 and r_2 are connected by edges of random length, which have zero probability of matching by Assumption (1) of our definition of randomness (p. 382). They could glue only by pinching the longer edge, creating a glued point of angle 3π, violating the Alexandrov gluing conditions (see Figure 25.4(a)).

---

[2]   The argument below improves upon that originally presented in Demaine et al. (2000b).

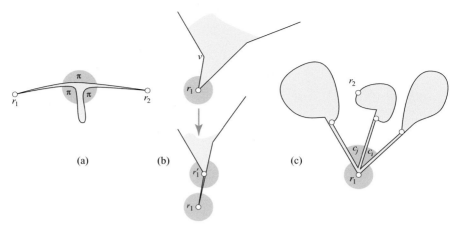

Figure 25.4. Foldings illustrating aspects of the proof of Lemma 25.1.6. (Paper is exterior to curves shown.) (a) Base case not Alexandrov; (b) zipping at r_1 produces a new reflex vertex r_1'; (c) gluing several convex vertices into r_1.

Suppose now $k > 0$. We focus on the gluing at r_1, and show that neither of the two options eliminates the reflexivity:

1. r_1 is zipped by gluing its two incident edges together, becoming a leaf in the gluing tree. If the other endpoint v of the shorter of these edges is r_2, then the gluing is not Alexandrov. Otherwise, we produce a new reflex vertex r_1' formed by the convex vertex v glued to an edge interior point (see Figure 25.4(b)). The new subpolygon has $k - 1$ convex vertices, still at least two reflex vertices, and cannot fold by induction.
2. One or more (say $m > 0$) convex vertices are glued into r_1. This produces $m + 1$ subpolygons, each with a new reflex vertex at its "mouth" (see Figure 25.4(c)). One of these subpolygons contains r_2 as well, and it cannot fold by induction.

This establishes the lemma for two reflex vertices. Suppose the subpolygon has more than two reflex vertices. The argument above holds unchanged, except for the base case, which now consists of r_1 connected to r_2 by chains that contain only reflex vertices, which clearly cannot fold without gluing an edge interior point to a reflex vertex, which is non-Alexandrov. □

This lemma then immediately implies Theorem 25.1.2, under our assumptions on randomness. In essence, without matching edge lengths, reflexivities cannot be eliminated.

The reliance of the proof on Assumption (1), which implies the unlikeliness of matching edge lengths, makes it natural to wonder if the same result holds for polygons all of whose edge lengths are the same. Again we believe it does (Demaine et al. 2000b):

Conjecture 25.1.1. *The probability that a random polygon of n vertices, all of whose edges have unit length, can fold to a convex polyhedron approaches* 0 *as n* → ∞.

The upshot of all this is that foldability is rare, which leads us to the question of deciding when it is possible.

Figure 25.5. An unfolding of a randomly generated convex polyhedron. The polyhedron was generated as in O'Rourke (1998) and the unfolding produced according to Namiki and Fukuda (1993).

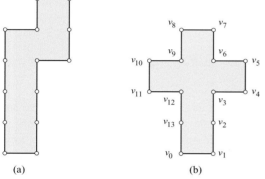

(a)   (b)

Figure 25.6. (a) A polygon that cannot fold to a cube; (b) an unfolding of a cube.

25.2   EDGE-TO-EDGE GLUINGS

Having seen now that foldability is in general rare, albeit guaranteed for convex polygons, we turn to the question of deciding when folding is possible.[3]

We approach this decision problem in this section with a natural restriction: each whole polygon edge must glue to another whole edge of the polygon. We call this an *edge-to-edge* gluing. For polygons created via a "generic" polyhedron unfolding, this restriction renders our question easy to answer. Consider the polygon in Figure 25.5. It was produced by unfolding the convex hull of random points in a sphere. The random generation produced distinct polyhedron edge lengths (cf. Assumption (1), p. 382). Thus each edge of the polygon has only one possible mate: the only other one with the same length. Checking Alexandrov's conditions then amounts to checking the angle sum at the vertices of the unique gluing. The problem is more difficult when many edges have the same length (perhaps all are identical), and so there are many choices for edge matches. For example, the polygon in Figure 25.6(a) does not fold to a cube, but that in (b) of course does. (The shape in (b) is called the *Latin cross*, and will be scrutinized closely in the sequel (Section 25.6).) However, the polygon in (a) is foldable to other polyhedra, and the Latin cross folds to polyhedra other than just the cube. This is just to hint that matters are not straightforward.

---

[3] The material in this section is largely drawn from Lubiw and O'Rourke (1996).

Naive search of all possible edge-to-edge gluings and testing each one for satisfaction of the Alexandrov conditions would explore an exponential number of possible matches. In the edge-to-edge case, we can achieve polynomial time: we next describe an $O(n^3)$ decision algorithm.

25.2.1 Dynamic Programming Formulation

Let the polygon P have vertices $(v_0, v_1, \ldots, v_{n-1})$ connected by edges $e_i = v_i v_{i+1}$ of length $\ell_i = |e_i|$, with indices increasing in counterclockwise order. We do not need our polygons to be simple in this section (although they may not have holes). All index arithmetic is mod n. The (closed) polygonal chain from v_i counterclockwise to v_j is denoted by $P[i, j]$. In anticipation of generalizations in subsequent sections, we use the notation $\{e_i, e_j\}$ to indicate that e_i is matched, or identified with, or glued to e_j (all synonymous phrases), and $\{v_i, v_j\}$ if v_i is glued to v_j.

The basic insight of the algorithm is that if e_i is matched with e_j, then we have two smaller decision problems: folding $P[i + 1, j]$ and $P[j + 1, i]$. These two subproblems are isolated from one another by the two matched edges. For if an edge in $P[i + 1, j]$ is glued to an edge in $P[j + 1, i]$, any resulting surface will not be homeomorphic to a sphere: the "cross identification" of edges creates a torus hole. Certain choices of initial edge matches lead to subproblems with no solution. Others lead to a subproblem that can be solved but that is incompatible with the angle conditions at the endpoints of e_i and e_j, the "mouth" of the subproblem. And others lead to compatible, solvable subproblems, which result in a folding. We design an algorithm based on dynamic programming, formulated as follows:

The key quantity associated with the matching $\{v_i, v_j\}$ is $\alpha_{\min}(i, j)$. We first provide an informal definition. Define $\alpha_{\min}(i, j) = \infty$ if there is no "legal" folding of the chain $P[i, j]$. If there is at least one legal folding of the chain, $\alpha_{\min}(i, j)$ is the minimum face angle incident to $v_i = v_j$ contributed by the chain, minimum over all legal foldings of $P[i, j]$. A formal definition is as follows:

D1. For $|j - i|$ odd, $\alpha_{\min}(i, j) = \infty$.

D2. For $i = j$, $\alpha_{\min}(i, j) = 0$.

D3. For $|j - i| \geq 2$ and even, let k have different parity from i, and $i < k \leq j - 1$. $\alpha_{\min}(i, j)$ is the minimum over all such k of the "extra" angle Δ_k at $v_i = v_j$ resulting from a folding that matches $\{e_i, e_k\}$. If $\ell_i \neq \ell_k$, then $\Delta_k = \infty$. Otherwise this match creates two subproblems (see Figure 25.7):

1. $\{v_{i+1}, v_k\}$. If $i + 1 = k$, this subproblem is vacuous and Δ_k is determined by the second subproblem. Assume then that $i + 1 \neq k$. If $\alpha_{i+1} + \alpha_k + \alpha_{\min}(i + 1, k) \geq 2\pi$, the gluing $\{e_i, e_k\}$ is not Alexandrov, so $\Delta_k = \infty$. Otherwise Δ_k is determined by the second subproblem.

2. $\{v_{k+1}, v_j\}$. The extra angle is 0 if $k + 1 = j$, and $\Delta_k = \alpha_{k+1} + \alpha_{\min}(k + 1, j)$ otherwise.

D1 of this definition simply reflects the fact that between any two matched vertices must lie an even number of edges, because the edges are matched in pairs. D2 says that a vertex can always match with itself (when a cut on the polyhedron surface terminates at the vertex), resulting in no extra angle glued to $v_i = v_j$.

The heart of the procedure is D3, which we illustrate with the cube unfolding in Figure 25.6(b), computing $\alpha_{\min}(1, 5)$, corresponding to matching $\{v_1, v_5\}$. There

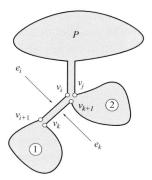

Figure 25.7. The match $\{v_i, v_j\}$ creates two subproblems for every possible gluing $\{e_i, e_k\}$.

are two values of k within the relevant range: $k = 2$ and $k = 4$. The first corresponds to matching $\{e_1, e_2\}$. This creates two subproblems, $\{v_2, v_2\}$ and $\{v_3, v_5\}$. The first is possible (D1) and so we turn to the second, which yields an extra angle of $\Delta_2 = \alpha_3 + \alpha_{\min}(3, 5) = 270° + 0° = 270°$. The $k = 4$ case corresponds to the match $\{e_1, e_4\}$, which creates the two subproblems $\{v_2, v_4\}$ and $\{v_5, v_5\}$. The first is legal since $\alpha_2 + \alpha_4 + \alpha_{\min}(2, 4) = 180° + 90° + 0° = 270° \leq 360°$. The second yields $\Delta_4 = 0$ (by D1). Finally, $\alpha_{\min}(1, 5) = \min\{\Delta_2, \Delta_4\} = 0°$, meaning that there is a way to fold the chain $P[1, 5]$ so that no extra angle is matched to $v_1 = v_5$.

It is clear that $\alpha_{\min}(i, j)$ can be computed for all i and j by dynamic programming: values for $|j - i| = d$ depend on values of $\alpha_{\min}$ for strictly smaller index separation. Thus the entire table of $\alpha_{\min}$ values can be computed in the order: $d = 0, 2, 4, \ldots, n - 2$.

With this table in hand, the question of whether P can fold may be answered by seeing if e_0 can be glued to some other edge e_m. This is answered by examining the two subproblems exactly as in D3 of the $\alpha_{\min}$ definition,[4] $\{v_1, v_m\}$ and $\{v_{m+1}, v_0\}$, and then verifying that the angle conditions are satisfied at the endpoints of $e_0 = e_m$:

$$\alpha_1 + \alpha_m + \alpha_{\min}(1, m) \leq 2\pi,$$

$$\alpha_{m+1} + \alpha_0 + \alpha_{\min}(m + 1, 0) \leq 2\pi.$$

The entire algorithm takes $O(n^3)$ time.[5] The dynamic programming table has size $O(n^2)$: $\lfloor (n - 1)/2 \rfloor$ rows for the $|j - i| = d$ separations, and n entries for each (i, j) pair with $j = (i + d) \bmod n$. Computing one $\alpha_{\min}$ entry (D3) is an $O(n)$ computation: checking every other k between i and j, computing Δ_k by table lookup, and taking the minimum of these Δ_k values.

25.2.2   Example

We will show the complete dynamic programming table for one example. Consider the polygon in Figure 25.8, an example due to Shephard. It has $n = 10$ vertices, whose angles (in degrees) are as follows:

i	0	1	2	3	4	5	6	7	8	9
α_i	60	108	236	98	218	60	108	236	98	218

---

[4] In fact one could extend the table to $d = n$ to capture this case.
[5] We thank Helen Cameron and Afroza Sultana (personal communication, May 2004) for correcting the erroneous $O(n^2)$ claim in Lubiw and O'Rourke (1996).

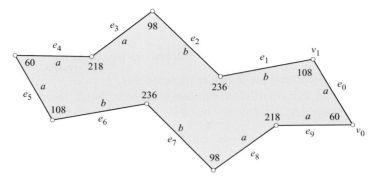

Figure 25.8. A polygon with two distinct edge-to-edge foldings. (Based on Figure 2 in Shephard 1975.)

We have $\sum \alpha_i = 1440° = 8 \cdot 180° = (n-2)\pi$. The edge lengths are $a = 4$ and $b = 5$ as illustrated. We initialize $\alpha_{\min}(i, j)$ to 0 for $i = j$ in the first table, and to either 0 or ∞ for $|i - j| = 2$, depending on whether the adjacent edges match in length:

(i, j)	(0, 2)	(1, 3)	(2, 4)	(3, 5)	(4, 6)	(5, 7)	(6, 8)	(7, 9)	(8, 0)	(9, 1)
$\alpha_{\min}(i, j)$	∞	0	∞	0	0	∞	0	∞	0	0

(i, j)	(0, 4)	(1, 5)	(2, 6)	(3, 7)	(4, 8)	(5, 9)	(6, 0)	(7, 1)	(8, 2)	(9, 3)
$\alpha_{\min}(i, j)$	0	98	∞	∞	108	0	108	∞	∞	108

We will explain the computation of the first entry of this table, $\alpha_{\min}(0, 4) = 0°$. D3 explores two values of k, $k = 1, 3$, corresponding to the two ways to fold the four edges in $P[0, 4]$. The first glues $\{e_0, e_1\}$, an edge-length mismatch. The second glues $\{e_0, e_3\}$, which forces $\{e_1, e_2\}$. Both of these matches are legal foldings. The first subproblem of D3 is $\{v_1, v_3\}$, which results in the angle $\alpha_1 + \alpha_3 + \alpha_{\min}(1, 3) = 108° + 98° + 0° = 206°$, a legal folding. The second subproblem of D3 is $v_4, v_0 = v_4$, a vacuous subproblem, so $\Delta_3 = 0°$. Finally, $\alpha_{\min}(0, 4) = \min\{\Delta_1, \Delta_3\} = 0°$.

The tables for $|j - i| \in \{6, 8\}$ follow:

(i, j)	(0, 6)	(1, 7)	(2, 8)	(3, 9)	(4, 0)	(5, 1)	(6, 2)	(7, 3)	(8, 4)	(9, 5)
$\alpha_{\min}(i, j)$	0	∞	∞	60	0	0	∞	∞	60	0

(i, j)	(0, 8)	(1, 9)	(2, 0)	(3, 1)	(4, 2)	(5, 3)	(6, 4)	(7, 5)	(8, 6)	(9, 7)
$\alpha_{\min}(i, j)$	108	158	∞	0	∞	108	168	∞	0	∞

We come now to the final set of tests: finding a match for e_0. There are five possible matches, only a few of which we detail:

1. $m = 1$: $\{e_0, e_1\}$. This is an immediate length mismatch, gluing a to b.

2. $m = 3$: $\{e_0, e_3\}$. This leads to a legal matching, whose complete set of edge matchings is

$$(\{e_0, e_3\}, \{e_1, e_2\}, \{e_4, e_9\}, \{e_5, e_8\}, \{e_6, e_7\}).$$

The polyhedron is an (irregular) octahedron.

3. $m = 5$: $\{e_0, e_5\}$. We will step through this one in detail. The two subproblems are $\{v_1, v_5\}$ and $\{v_6, v_0\}$. The first results in this angle calculation:

$$\alpha_1 + \alpha_5 + \alpha_{\min}(1, 5) = 108° + 60° + 98° = 266°.$$

The second results in this angle calculation:

$$\alpha_6 + \alpha_0 + \alpha_{\min}(6, 0) = 108° + 60° + 108° = 276°.$$

This represents a second legal folding of the polygon, whose edge matchings are

$$(\{e_0, e_5\}, \{e_1, e_2\}, \{e_3, e_4\}, \{e_6, e_7\}, \{e_8, e_9\}) .$$

This is a so-called "stack polytope" (Brönsted 1983, p. 129), which can be built from a tetrahedron by adding two tetrahedra above two of its facets.

4. $m = 7$: $\{e_0, e_7\}$. This is a length mismatch.

5. $m = 9$: $\{e_0, e_9\}$. The attempt to glue v_1, v_9 eventually leads to the angle computation

$$\alpha_1 + \alpha_9 + \alpha_{\min}(1, 9) = 108° + 218° + 158° = 484°,$$

which exceeds 2π and so violates Alexandrov's conditions.

The two polyhedra produced by the two legal foldings are of different combinatorial types.

25.2.3   Five Edge-to-Edge Foldings of the Latin Cross

Running the dynamic programming algorithm on the Latin cross of Figure 25.6(b) leads to no less than five distinct foldings, which so surprised us at first that we thought the algorithm was in error. The foldings correspond to these five polyhedra (see Figure 25.9):

1. The cube.

2. A flat, doubly covered quadrilateral, a degenerate polyhedron permitted by Alexandrov's theorem:

$$(\{e_0, e_3\}, \{e_1, e_2\}, \{e_4, e_{13}\}, \{e_5, e_6\}, \{e_7, e_{12}\}, \{e_8, e_{11}\}, \{e_9, e_{10}\}).$$

The vertices of the quadrilateral are $\{v_0, v_4\}$, v_2, v_6, and v_{10}.

3. A tetrahedron:

$$(\{e_0, e_1\}, \{e_2, e_3\}, \{e_4, e_{13}\}, \{e_5, e_6\}, \{e_7, e_{12}\}, \{e_8, e_{11}\}, \{e_9, e_{10}\}).$$

In this matching, only four vertices have a complete angle strictly less than 2π: v_1, v_3, v_6, and v_{10}. An animation of the folding to a tetrahedron is shown in Figure 25.10. Note, for example, that the set of vertices $\{v_5, v_7, v_{13}\}$ join in the interior of one of the tetrahedron's faces: $\alpha_5 + \alpha_7 + \alpha_{13} = 90° + 90° + 180° = 360°$.

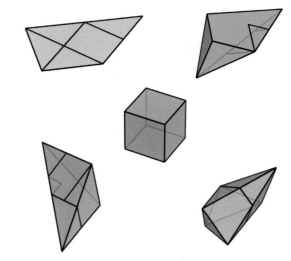

Figure 25.9. The five edge-to-edge foldings of the Latin cross.

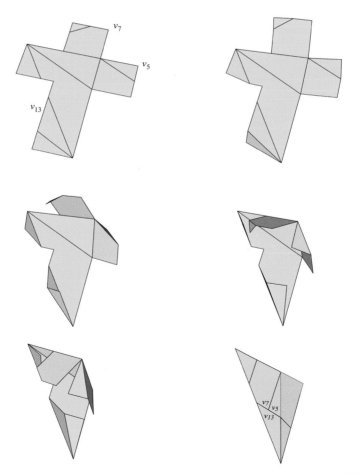

Figure 25.10. Folding the Latin cross to a tetrahedron (Demaine et al. 1999c).

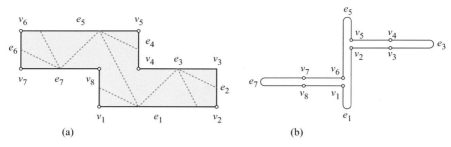

Figure 25.11. (a) A polygon, with fold creases shown dashed; (b) a gluing tree T_G corresponding to the crease pattern.

4. A four-sided pyramid:

$$(\{e_0, e_3\}, \{e_1, e_2\}, \{e_4, e_9\}, \{e_5, e_8\}, \{e_6, e_7\}, \{e_{10}, e_{13}\}, \{e_{11}, e_{12}\}) .$$

5. An octahedron:

$$(\{e_0, e_3\}, \{e_1, e_2\}, \{e_4, e_7\}, \{e_5, e_6\}, \{e_8, e_9\}, \{e_{10}, e_{13}\}, \{e_{11}, e_{12}\}) .$$

How we reconstruct the 3D shape of the polyhedra from the Alexandrov gluing will be described in Section 25.8. These five foldings of the Latin cross are illustrated in a video (Demaine et al. 1999c),[6] from which Figures 25.9 and 25.10 are drawn.

Finally, the polygon of Figure 25.6(a), which cannot fold to a cube, can edge-to-edge fold to several flat, doubly covered polygons, as well as to several nondegenerate polyhedra: two different quadrilaterals, a pentagon, a hexagon, a pentahedron, a hexahedron, and an octahedron.

25.3   GLUING TREES

We define now the central combinatorial object that we use to represent polygon foldings: a gluing tree.[7] Let polygon P have vertices $v_1, \ldots, v_n$, labeled counterclockwise, and edges e_i, $i = 1, \ldots, n$, with each edge the open segment of ∂P following v_i. Recall (p. 349) that a gluing G identifies ∂P with itself in equal-length subsections. The *gluing tree* T_G for a particular gluing is a labeled tree representing the combinatorial structure of G. Any point of ∂P that is identified with more, or fewer, than one other distinct point of ∂P becomes a node of T_G, as does any point to which a vertex is glued. (Note that this means there may be tree nodes of degree 2.) So every vertex of P maps to a node of T_G. Each node is labeled with the set of all the elements (vertices or edges) that are glued together there. Fold points (p. 382) are labeled with the polygon edge that is creased. An example is shown in Figure 25.11. Here and throughout we will use the convention that the polygon is folded toward the viewer, so that the "paper" is exterior to the gluing tree; folding away produces a mirror-image polyhedron. T_G in Figure 25.11(b) has eight nodes, labeled $\{e_1\}$, $\{v_2, v_5, e_5\}$, $\{v_3, v_4\}$, $\{e_3\}$, $\{e_5\}$, $\{v_1, v_6, e_1\}$, $\{v_7, v_8\}$, and $\{e_7\}$, where the notation $\{e_i\}$ indicates a fold point in e_i, that is, e_i glues to itself. Each node of a gluing

---

[6] http://theory.csail.mit.edu/~edemaine/metamorphosis/.
[7] Some of the material in this section is based on Demaine et al. (2000b, 2002a).

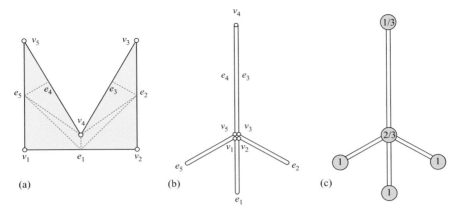

Figure 25.12. (a) A nonconvex pentagon (creases shown dashed); (b) the corresponding gluing tree; (c) gluing tree with curvatures (in units of π) marked at each polyhedron vertex.

tree receives as many labels as its degree. The four leaves of this particular tree are all fold points.

Note that there is not a straightforward correspondence between the nodes of T_G and the vertices of the resulting polyhedron, call it Q. In this case, this 8-node tree corresponds (in a nonobvious way) to a 4-vertex tetrahedron. For example, node $\{v_7, v_8\}$ of T_G has curvature 0 and so is not a vertex of Q. We return to this point below with a second example, which we analyze a bit more closely.

Let P be the nonconvex pentagon shown in Figure 25.12(a), folded along the crease lines as shown. The corresponding gluing tree is shown in (b).[8] Here three leaves are fold points, one leaf is the reflex vertex v_4, and the central degree-4 node of the tree is formed by gluing all the vertices in $\{v_1, v_2, v_3, v_5\}$ together.

Let us verify that this is an Alexandrov gluing by computing the curvature at each node of T_G. An Alexandrov gluing requires at most 2π of paper glued at any point, which is equivalent to requiring nonnegative curvature (p. 303) at each point. Points of the gluing that do not correspond to nodes of T_G all have curvature 0, for two angles of π are glued to each side. Each fold point has a curvature of π. The leaf v_4 has curvature $60° = \pi/3$. Finally, the central node has total angle $90° + 90° + 30° + 30° = 240°$, and so curvature of $120° = 2\pi/3$. Figure 25.12(c) redisplays the gluing tree with all the nonzero curvatures indicated in circles, in units of π. This is a convenient form, for it is exactly these points that correspond to vertices of the polyhedron Q. This also makes it is easy to verify the Gauss–Bonnet theorem (p. 304): $1 + 1 + 1 + \frac{1}{3} + \frac{2}{3} = 4$.

By Alexandrov's theorem (p. 348), then, we know that the indicated gluing leads to a unique polyhedron, a 5-vertex shape because there are 5 points of nonzero curvature. It is shown in Figure 25.13. We defer discussion of how the creases were determined, and the shape constructed, to Section 25.6.2.

25.3.1 Rolling Belts

Rolling belts (Demaine et al. 2002a) are an important structural feature of gluing trees that determine whether the polygon may fold to an infinite number of noncongruent

---

[8]   Note that in contrast to the previous section, the gluing tree does not separately notate the edge-to-edge gluing $\{e_3, e_4\}$, which can easily be inferred from the tree node labels.

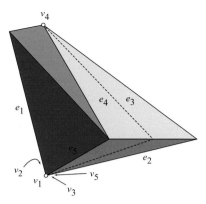

Figure 25.13. The polyhedron Q resulting from the folding in Figure 25.12.

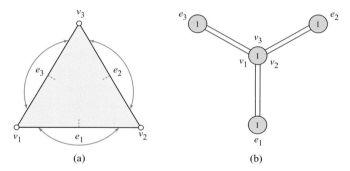

Figure 25.14. (a) An equilateral triangle creased in the interior of each edge; (b) the gluing tree.

shapes. A *rolling belt* is a path in the gluing tree that satisfies two properties:

(a) A rolling belt path connects two leaves of the gluing tree, each of curvature $\leq \pi$, that is, either convex-vertex or fold-point leaves.

(b) The face angle to each side of the path is, at every point, convex, that is, $\leq \pi$.

The simplest example of a rolling belt is provided by a perimeter-halving folding of a convex polygon (Figure 25.2). The name reflects the possibility of "rolling" the path/belt, with the two endpoints sliding around ∂P. In Figure 25.11, the path from $\{e_3\}$ to $\{e_7\}$ is not a rolling belt because, for example, v_8 is reflex, but the path from $\{e_1\}$ to $\{e_5\}$ is, because, for example, v_1 and v_6 together form an angle of π. One can view a rolling belt as arising when a partial folding leaves a subpolygon ring that is structurally a convex polygon, which then admits perimeter-halving folds of that ring/polygon.

Another example is provided by the folding of an equilateral triangle illustrated in Figure 25.14. Here there are three rolling belts: $(\{e_1\}, \{e_2\})$, $(\{e_2\}, \{e_3\})$, and $(\{e_3\}, \{e_1\})$.

25.3.2 Gluing Tree Properties

Gluing trees are fundamentally discrete structures, characterized as follows (Demaine et al. 2002a):

Figure 25.15. A gluing tree with three fold-point leaves, two forming a rolling belt.

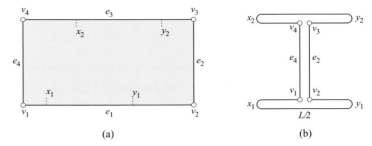

(a)   (b)

Figure 25.16. I gluing tree for an $L \times W$ rectangle.

Lemma 25.3.1. *Gluing trees satisfy these properties:*

1. *Each leaf is either a vertex of the polygon, or a fold point.*
2. *At most one nonvertex may be glued at any gluing-tree junction of degree $d \geq 3$.*
3. *At most four leaves of the gluing tree can be fold points or convex vertices, and therefore a gluing tree may have at most two rolling belts with distinct endpoints.*
4. *The case of four fold-point leaves is possible only under special circumstances.*

Proof: Gluing two interior edge points together already consumes the 2π permitted in an Alexandrov gluing, leaving no slack to achieve $d > 2$. Each fold point consumes π curvature, and four fold points exhaust the 4π of curvature available according to the Gauss–Bonnet Theorem 21.3.1. Folding at a convex vertices consumes more than π of curvature. We will discuss the "special circumstances" below. □

Setting aside the special case of two independent rolling belts, a gluing tree might have the features illustrated in Figure 25.15: a rolling belt, with trees hanging off it, and one of those trees having a fold-point leaf.

The case of four fold-point leaves is possible only when the tree has exactly four leaves. The combinatorial structure of the tree must then be what we shall call either I or +: with two degree-3 junctions, or one degree-4 junction. The former structure permits double-belt trees, which can be seen to arise from folding an $L \times W$ rectangle as follows. Glue the two opposite edges of length W together to form a cylinder. Now glue the bottom rim of the cylinder to itself by creasing at two diametrically opposed points x_1 and y_1. Similarly glue the top rim to itself by creasing at two points x_2 and y_2 (see Figure 25.16). Both (x_1, y_1) and (x_2, y_2) are rolling belts. But note that, for example, (x_1, y_2) is not a rolling belt, because there is $3\pi/2$ face angle to one side of the path at two points, violating (b) of the definition. The gluing is Alexandrov even if the crease points on the top and bottom are not located at corresponding points on their rims, as we will see in Section 25.7.3. When the vertical beam of the I shrinks to zero length, the structure becomes the + shape. These issues will be revisited in Lemma 25.7.1.

25.3.3   Hirata Half-Lengths Theorem

Although there are clearly a finite number of gluing trees (an exponential number, as we will see in Section 25.4), the number of distinct (noncongruent) polyhedra produced by these gluings can be infinite, as we saw in Section 25.1.2. The finite/infinite distinction depends on whether some gluing tree contains a rolling belt, which generates a continuum of polyhedra. Although we will discuss in Section 25.5 an algorithm that can detect, for any given polygon, whether a rolling belt can be produced, no useful characterization is known:

Open Problem 25.2: Finite Number of Foldings. [a] Characterize those polygons that have only a finite number of foldings, that is, for which no folding contains a rolling belt.

---

[a]   Posed by Koichi Hirata, December 2002.

Taking this finite/infinite distinction as a primitive, Koichi Hirata established[9] this beautiful connection between arbitrary gluings and edge-to-edge gluings:

Theorem 25.3.1 (Hirata). *Let P be a polygon for which no Alexandrov gluing contains a rolling belt. If all of P's edge lengths are integers, then all gluing trees for P can be found by partitioning each edge of P into subedges of length $\frac{1}{2}$, forming a "refined polygon" P', and generating all edge-to-edge gluings of P'.*

By scaling P, the same holds true when all edge lengths are integer multiples of any real "unit" length δ.

25.4   EXPONENTIAL NUMBER OF GLUING TREES

Although there are clearly an exponential number of ways one might try to find an Alexandrov gluing of a polygon of n vertices, even for edge-to-edge gluings, it is not immediately evident whether one particular polygon can actually have an exponential number of Alexandrov gluings. Shephard's example (Figure 25.8) has two, and we've seen that the Latin cross has at least five (and more will be detailed in Section 25.6). Here we provide an example that establishes a worst-case exponential lower bound (Demaine et al. 2000b, 2002a).[10]

---

[9]   Personal communication, September 2000.
[10]   Comments by Günter Rote sharpened the proof.

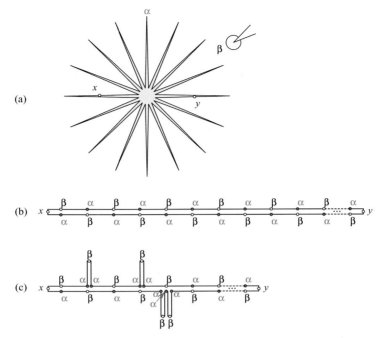

(a)

(b)

(c)

Figure 25.17. (a) Star polygon P, $m = 16$, $n = 34$; (b) base gluing tree; (c) a gluing tree after several contractions.

Theorem 25.4.1. *For any even n, there is a polygon P of n vertices that has $2^{\Omega(n)}$ combinatorially distinct edge-to-edge Alexandrov gluings.*

Proof: The polygon P is illustrated in Figure 25.17(a). It is a centrally symmetric star, with an even number m of vertices with a small convex angle $\alpha \approx 0$, alternating with m vertices with large reflex angle $\beta < 2\pi$. (The choice of even m is for illustration convenience; the proof is no different for odd m.) All edges have the same (say, unit) length. We call this polygon an m-star. We choose α small enough so that m copies of α can join with one of β and still be less than 2π:

$$m\alpha + \beta < 2\pi. \tag{25.1}$$

Because P has $2m$ vertices (ignoring flat vertices x and y, to be described shortly), we have $m(\alpha + \beta) = (2m - 2)\pi$, which together with Equation (25.1) leads to

$$\alpha < \frac{2\pi}{m(m-1)}. \tag{25.2}$$

Now we add two vertices x and y at the midpoints of edges, symmetrically placed so that y is half the perimeter around ∂P from x. Let $n = 2m + 2$ be the total number of vertices of P.

The "base" gluing tree is illustrated in Figure 25.17(b). In this gluing, x and y are fold vertices, and each α is matched with a β. Because all edge lengths are the same, and because $\alpha + \beta < 2\pi$, this path is an Alexandrov gluing. We label it $T_{00\cdots 0,00\cdots 0}$, where $m/2$ zeros $00\cdots 0$ represent the top chain, and another $m/2$ zeros represent the bottom chain.

The other gluing trees are obtained via "contractions" of the base tree. A *contraction* makes any particular β-vertex, not adjacent to x or to y, a leaf of the tree by gluing its two adjacent α-vertices together. Label a β-vertex 0 or 1 depending on whether it is uncontracted or contracted respectively. Then a series of contractions can be identified with a binary string. For example, Figure 25.17(c) displays the tree $T_{010100\cdots,00110\cdots0}$. Note that k adjacent contractions result in $k+1$ α-vertices glued together.

We now claim that if the number of contractions in the top chain is the same as the number in the bottom chain (call such a series of contractions *balanced*), then the resulting tree represents an Alexandrov gluing. Fix the position of x to the left, and contract toward the left, as in Figure 25.17(c). Then it is evident that the alternating "parity" pattern of α's and β's is not changed by contractions. Each contraction shortens the path by 2 units, and an even shortening does not affect the parity pattern. If the top and bottom chains are contracted the same number of times (twice each in Figure 25.17(c)), then their lengths are the same.

Because the parity pattern remains the same, no two β-vertices are glued together. The β-leaves are legal gluings because $\beta < 2\pi$. And the clusters of α-vertices glued to one β-vertex have total angle $<2\pi$ by Equation (25.1). Thus the gluing is Alexandrov.

Finally, we count the number of balanced bit patterns. Suppose k ones are set among the $m/2 - 1$ bits in the top chain. Then k must be selected in the bottom chain for balance. If we imagine flipping all the bits in the bottom chain, then the total number of ones selected is $k + (m/2 - 1) - k = m/2 - 1$. So it is clear that there are $\binom{m-2}{m/2-1}$ patterns: of the total (top and bottom) $m - 2$ bits, we select $m/2 - 1$. (For example, $m = 16$ (Figure 25.17(a)) leads to $\binom{14}{7} = 3432$ balanced patterns.)

It is a standard bound on binomial coefficients that $\binom{a}{b} \geq (a/b)^b$, so $2^{m/2-1}$ is a lower bound. Because P has $n = 2m + 2$ vertices, $\Omega(2^{m/2-1}) = \Omega(2^{(n-6)/4}) = 2^{\Omega(n)}$.   □

Figure 25.18 shows six gluings of a 4-star. The first two in the top row correspond to the perimeter-halving construction used in the proof. Hand-folding exploration suggests that all six of these gluings fold to noncongruent polyhedra, each with the combinatorial structure of the regular octahedron. Two of our conjectured crease patterns are shown in Figure 25.19.

Finally, we argue that Theorem 25.4.1 is tight, in the sense that any polygon can have at most $2^{O(n)}$ Alexandrov gluings. There are at most $2^{O(N)}$ unlabeled rooted plane trees

Figure 25.18. Six gluing patterns for a 4-star. Matched edges are indicated by arcs.

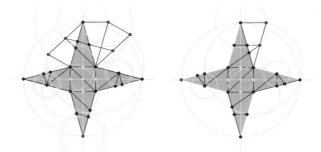

Figure 25.19. Conjectured crease patterns for the first two gluing patterns in the top row of Figure 25.18. [Constructions in Cinderella.]

on N nodes.[11] Let us assign a node of a gluing tree the bit 1 if its set of labels includes some vertex v_i, and the bit 0 if its label is $\{e_j\}$, that is, it is a fold point. There are 2^N bit patterns that may be assigned to the N nodes of a particular gluing tree. Now select any node with bit 1 and place label v_1 there. Because the node labels must occur in the order $(v_1, e_1, v_2, e_2, v_3, \ldots)$ in a traversal around the gluing tree, the choice of where to place v_1, together with the bit pattern, determines the full labeling. Thus, the total number of gluings is $2^{O(N)}$ trees times 2^N bit patterns times n choices for v_1. This leads to $n2^{O(N)}$. Because $N = O(n)$, this upper bound is $2^{O(n)}$, as claimed.

25.5 GENERAL GLUING ALGORITHM

We now turn to a general gluing algorithm, one that does not make the assumption of edge-to-edge gluing. We follow an algorithm implemented by Anna Lubiw,[12] and later (p. 402) mention an alternative algorithm due to Koichi Hirata. From one point of view, the algorithm is a direct generalization of the dynamic programming algorithm presented in Section 25.1. We describe only an enumeration algorithm, which, by the results of the previous section, could require exponential time just to list all gluings. Finding a polynomial-time decision algorithm is an open problem, mentioned at the end of this section.

We first need to develop notation for gluings of part of the polygon boundary. Let P be the polygon, with vertices v_i and edges e_i as in the previous section. A *partial gluing tree* is a gluing tree for just some contiguous section of the polygon boundary, supplemented by information concerning its interface to the remainder of the polygonal chain. These data are the *mouth angle* μ and the *mouth interval* M. μ is the total angle glued at the mouth. For gluings of a vertex v_i into the interior of an edge e_j, M is the interval of lengths along e_j for which the gluing tree is valid, as depicted in Figure 25.20. For vertex–vertex gluings, M is [0, 0]. Note that M refers only to the interval at the mouth, and implies only that the combinatorial structure is valid throughout that interval. M does not reflect dependencies between intervals. For example, in Figure 25.21, the position of v_4 on e_2 determines the position of v_5 and M, but this relationship is not recorded in the partial tree.

Our goal is to compute all the partial gluing trees corresponding to the counterclockwise polygonal chain between two gluing points, either vertices or edges. We represent

---

[11] More precisely, there are $\binom{2N}{N}/(N+1) = O(4^N) = 2^{O(N)}$ such trees.
[12] Personal communication, Fall 2000.

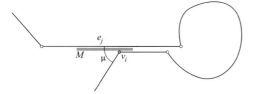

Figure 25.20. Mouth angle μ and mouth interval M.

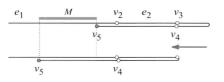

Figure 25.21. Interval M for v_5 implies sliding of v_4 on e_2.

each tree by a list of its node labels, from which it is easy to construct the determined tree structure.

Let $G(v_i, v_j)$ be the set of partial gluing trees possible when gluing v_i to v_j, and let $G(v_i, e_j)$ be the set when gluing v_i to some interior point of e_j. $G(e_i, v_j)$ is defined similarly, and $G(e_i, e_j)$ is not needed, as this is subsumed by gluing a vertex to a vertex or to an edge.

We illustrate with three increasingly complex partial gluings of the pentagon of Figure 25.12 (p. 393):

$$G(e_3, v_4) = \emptyset, \tag{25.3}$$

$$G(e_4, v_5) = (\mu = 7\pi/6, M = [0, 1], (\{e_4, v_5\}, \{e_4\})), \tag{25.4}$$

$$G(v_2, e_3) = \Big\{ (\mu = 3\pi/2, M = [0, 1], (\{v_2, e_3\}, \{e_2, v_3\}, \{e_2\})), \tag{25.5}$$

$$(\mu = 5\pi/3, M = [0, 1], (\{v_2, v_3, e_3\}, \{e_2\}, \{e_3\})) \Big\}. \tag{25.6}$$

$G(e_3, v_4)$ is empty because the reflex vertex v_4 cannot glue into the interior of e_3. $G(e_4, v_5)$ places v_5 somewhere along the unit-length e_4. The mouth angle is $\pi + \pi/6$, and any position along e_4 ($M = [0, 1]$) yields the same gluing tree, with an e_4 fold point. $G(v_2, e_3)$ can be folded in two ways: either by closing up at $\{v_2, e_3\}$ with a mouth angle of $3\pi/2$, or by pulling in v_3 to the same point, producing a mouth angle of $(3/2 + 1/6)\pi = 5\pi/3$; both are valid for any point along e_3.

One can see in this last example a hint of the general situation: any gluing $G(\,)$ could either "close up" or "pull in" at the gluing point. We call these two operations *zip* and *tug*. More formally, $\text{zip}(v_i, v_j)$ is the set of partial gluing trees that result from "zipping" together e_i with e_{j-1}—the edges of the chain incident to the mouth. And $\text{tug}(v_i, v_j)$ is the set of partial gluing trees that result from "tugging in" some other vertex or edge from the chain and gluing it to the same point as v_i and v_j. Both functions are similarly defined for gluing v_i to e_j or e_i to v_j.

It is now relatively easy to define $G(\,)$, $\text{zip}(\,)$, and $\text{tug}(\,)$ in terms of one another in a mutually recursive tangle. We will just present enough of this to see the flavor. Full details would require presenting full code. At the top level, we want to compute $G(x_i, x_j)$, where we use x to represent either v or e. By the observations above,

$$G(x_i, x_j) = \text{zip}(x_i, x_j) \cup \text{tug}(x_i, x_j) \tag{25.7}$$

$\text{zip}(v_i, v_j)$:

1. If $\alpha_i + \alpha_j > 2\pi$, then
   - Return $\emptyset$.

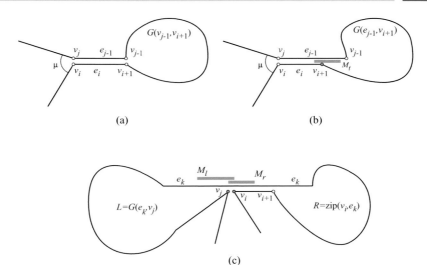

Figure 25.22. zip( ): equal edge lengths (a) and unequal lengths (b); (c) tug( ), pulling in e_k.

2. If $|e_i| = |e_{j-1}|$, then
   - Return $(\mu = \alpha_i + \alpha_j,\ M = [0, 0],\ \{v_i, v_j\}) \oplus G(v_{i+1}, v_{j-1})$.
3. If $|e_i| < |e_{j-1}|$, then
   (a) For each partial tree t in $G(v_{i+1}, e_{j-1})$
       i. If the mouth interval M_t for t includes the position of v_{i+1} (namely, $|e_{j-1}| - |e_i|$), then
          - Return $(\mu = \alpha_i + \alpha_j,\ M = [0, 0],\ \{v_i, v_j\}) \oplus t$.
4. If $|e_i| > |e_{j-1}|$, then [analogous actions]

Step 1 disallows the gluing if it violates Alexandrov's 2π condition. Step 2 says that if the two edges being zipped are the same length, then the set of partial gluing trees is just the set of trees obtained by gluing the endpoints of the edges, each extended by the one tree node $\{v_i, v_j\}$. We use the $\oplus$ notation to capture the extension operation (see Figure 25.22(a)). Step (3) handles the case where the edge lengths place v_{i+1} in the interior of e_{j-1} (see Figure 25.22(b)). Here the relevant recursive call is $G(v_{i+1}, e_{j-1})$, but each partial tree t must be checked to see if its mouth interval covers the spot where v_{i+1} is forced to glue as a result of the zipping. For each such t, the tree is extended by the one new node.

The function tug(x_i, x_j) pulls in either a vertex or an edge into the x_i/x_j junction, recursively constructing the gluings to either side of the element drawn between. We describe just one case of this function, pulling an edge e_k between two vertices v_i and v_j.

tug(v_i, v_j, e_k):
   1. $R \leftarrow$ zip(v_i, e_k).
   2. $L \leftarrow G(e_k, v_j)$.
   3. For each $l \in L$ and $r \in R$
      (a) If $\mu_l + \mu_r + \pi \le 2\pi$ and
      (b) If M_l and M_r overlap at the $\{v_i, v_j\}$ gluing point, then
          - Return $(\mu = \mu_l + \mu_r + \pi,\ M = M_l \cap M_r,\ \{v_i, v_j\}) \oplus (l, r)$.

Step (1) relies on the logic that some element (vertex or edge) must be glued against a portion of e_i next to v_i. The function considers only pulling in e_k adjacent to v_i. If e_k

is not glued adjacent to v_i, then some other element is pulled in between v_i and e_k, and will be captured in another call to tug( ). So we may assume that v_i and e_k zip; Step (1) computes all these partial gluing trees and stores them in R. Step (2) stores all possible gluings between e_k and v_j in L. It should be clear that we can make this zipping assumption only to one side of e_k, so the recursive call here is to the general $G(\ )$ (see Figure 25.22(c)). Step (3) then considers all combinations in L and R (in a double loop similar to a cross product), producing a new partial gluing tree only when (a) the angle sum satisfied Alexandrov's conditions and (b) the mouth intervals to either side of the gluing point cover the gluing point. In these cases, the appropriate new tree is constructed (again indicated by the $\oplus$).

Although we are leaving out many details, all other cases are analogous to the two just described. We complete the description by explaining how a complete set of gluings for the entire polygon boundary is obtained from these functions: by invoking $G(v_1, v_1)$. Although there are a few details specific to this special starting case, the idea is the same: v_1 must be glued somewhere, which means either its incident edges are zipped, or some other elements are tugged in to join v_1. The result is a listing of the set of all Alexandrov gluing trees for the whole polygon. When applied to the pentagon of Figure 25.12, the algorithm finds 26 gluings. The one illustrated in Figure 25.12(b) is

$$G(v_1, v_1) = (\mu = 4\pi/3, M = [0, 0], \{v_1, v_2, v_3, v_5\}, \{e_1\}, \{e_2\}, \{v_4\}, \{e_5\}).$$

This is not the only possible gluing tree enumeration procedure. A substantively different algorithm was developed by Koichi Hirata.[13] His pays more heed to the underlying geometric structure of the foldings during the enumeration. For example, he explicitly identifies the rolling belts. As mentioned earlier (p. 393), one can view a rolling belt as arising when a partial folding leaves a ring that is structurally a convex polygon, which then admits perimeter-halving folds. Hirata's code identifies such situations, leaving the set of combinatorially distinct foldings within each rolling belt implicit. It then proceeds to explore other ways to fold the belt, by gathering in three or more points forming a junction.

Both of these algorithms are worst-case exponential time, and necessarily so, as we have seen from Theorem 25.4.1. However, it remains possible that if one simply desires a decision algorithm, like that presented in Section 25.2 for edge-to-edge gluings, it could be accomplished in polynomial time. This remains an intriguing possibility:

Open Problem 25.3: Polynomial-time Folding Decision Algorithm. [a] Given a polygon P of n vertices, determine in time polynomial in n if P has an Alexandrov folding, and so can fold to some convex polyhedron.

---

[a] Demaine et al. (2002a) and Lubiw and O'Rourke (1996).

25.6 THE FOLDINGS OF THE LATIN CROSS

With a gluing algorithm in hand, it is theoretically possible to explore the convex polyhedra foldable from any given polygon. But the difficulty of reconstructing the 3D

---

[13] Personal communication, June 2000. http://weyl.ed.ehime-u.ac.jp/cgi-bin/WebObjects/Polytope2.

shape from an Alexandrov gluing makes this a problematic enterprise. It can, however, be carried out for polygons with few vertices. With collaborators,[14] we have chosen to concentrate on two representative polygons, one nonconvex and one convex: the Latin cross (Figure 25.6(b)) and the square. Although our choice here was somewhat arbitrary, aside from the cube/square symmetry, there is some logic in that the nonconvex polygon folds to a finite set of shapes (as we shall see, 23), whereas the convex polygon folds to an infinite number of shapes.

25.6.1   The 85 Foldings

We have already seen in Figure 25.9 that the Latin cross has five edge-to-edge foldings. Running the general algorithms of Lubiw and of Hirata without edge-gluing restrictions leads to 85 combinatorially distinct foldings (including the five edge-to-edge gluings). Not all of these foldings result in different 3D shapes, however. Via ad hoc reconstruction techniques (to be described in Section 25.6.2), we determined that just 23 noncongruent polyhedra result from the 85 gluings. We next present an inventory of representative gluings[15] leading to each of these 23 polyhedra, and then describe the shapes themselves.

We label the foldings $F_1, \ldots, F_{85}$, in the order output by Lubiw's algorithm, and the incongruent polyhedra $P_1, \ldots, P_{23}$. The five edge-to-edge foldings of the Latin cross (p. 390) are identified as follows according to these labelings:

1. the cube: $F_1 \to P_1$;
2. the double-quadrilateral: $F_{60} \to P_2$;
3. the tetrahedron: $F_{34} \to P_4$;
4. the pentahedron: $F_{12} \to P_{12}$;
5. the octahedron: $F_{15} \to P_{23}$.

We display representative foldings that lead to the 23 polyhedra in Figures 25.23–25.26, after which we describe the polyhedra in more detail.

25.6.2   Shape Reconstruction

As discussed in Sections 23.1 and 23.4, there is no known practical algorithm for reconstructing the 3D shape of a triangulated polyhedron given it edge lengths, despite Cauchy's Theorem 23.1.1 guaranteeing unique existence. In this section we describe a collection of ad hoc techniques that enable reconstruction of polyhedra of few vertices, and permitted us to reconstruct the shapes of the foldings of the Latin cross and of the square.

25.6.2.1 Tetrahedron

Just as the lengths of the three edges of a triangle determine the shape of the triangle, so do the lengths of the six edges of a tetrahedron determine the 3D shape of the tetrahedron.[16] We describe the method we used to compute this shape. Let p_0, p_1, and p_2 be the vertices of the "base" of a tetrahedron set on the xy-plane, and let p_3 be the apex

---

[14]   Rebecca Alexander, Heather Dyson, Martin Demaine, Anna Lubiw, and Koichi Hirata.
[15]   See http://theory.csail.mit.edu/~edemaine/aleksandrov/cross/photos.html for the full list.
[16]   And the shape determines the volume (cf. Equation (23.5)).

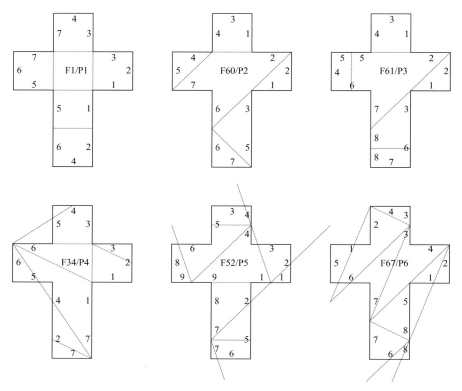

Figure 25.23. Foldings to $P_1, \ldots, P_6$: cube, two quadrilaterals, three tetrahedra.

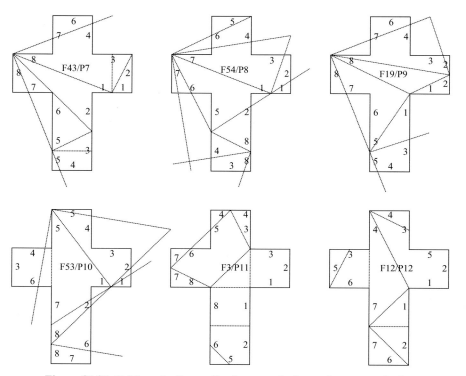

Figure 25.24. Foldings to $P_7, \ldots, P_{12}$: four tetrahedra and two pentahedra.

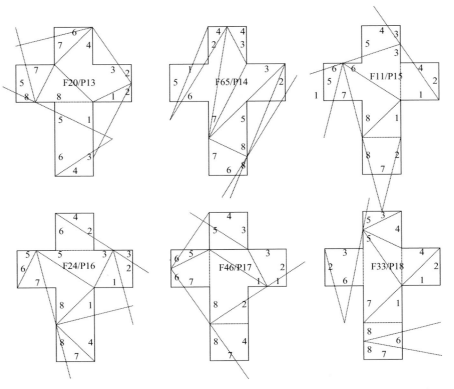

Figure 25.25. Foldings to $P_{13}, \ldots, P_{18}$: a pentahedron, four hexahedra, and an octahedron.

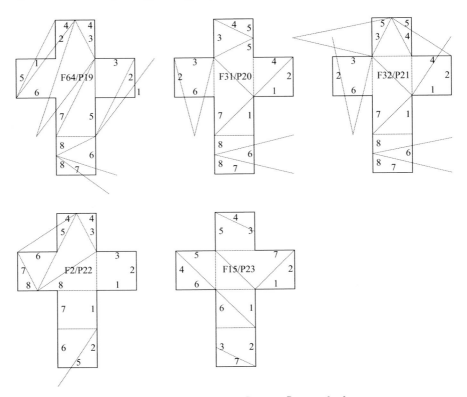

Figure 25.26. Foldings to $P_{19}, \ldots, P_{23}$: octahedra.

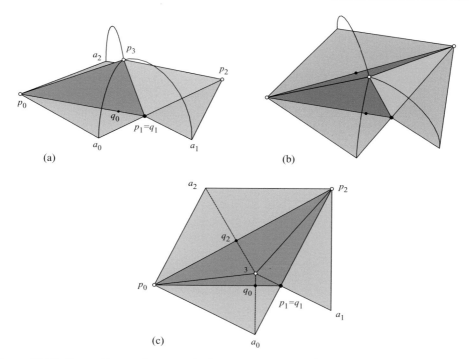

Figure 25.27. Computing p_3 for a tetrahedron, in this case P_4 (Figure 25.33), the edge-to-edge tetrahedral folding of the Latin cross (Figure 25.9).

with $z > 0$. Coordinates for the base vertices can be computed from the three base edge lengths in the standard manner. This leaves only computation of coordinates for p_3.

Think of unfolding the three faces incident to p_3 and rotating about their base edges until each lies flat in the xy-plane. Call the images of p_3 in the plane a_0, a_1, a_2, with each a_i opposite $p_i p_{i+1}$. Then, as is clear from Figure 25.27(a–b), each a_i travels on a circle centered on a point q_i on an axis through $p_i p_{i+1}$, with radius $a_i q_i$. So we can construct the unfolding in the xy-plane, and find the common intersection of the three circles above the plane.

One need not directly perform an intersection of the 3D circles to compute coordinates for p_3. Let p'_3 be the projection of p_3 onto the xy-plane. Then p'_3 is determined by the intersection of the three altitudes $a_i q_i$ extended, as shown in Figure 25.27(c). Any pair then yields the x- and y-coordinates of p_3, leaving only its z-coordinate unknown. There are many ways to compute this. For each $i = 0, 1, 2$, the tetrahedron edge length $|p_3 - p_i|$ is the hypotenuse of a right triangle whose base is $|p'_3 - p_i|$ and whose altitude is our unknown z. This gives three independent computations for z, which we average for numerical accuracy. This is the method we used to compute 3D coordinates for all the tetrahedra reconstructions.

25.6.2.2 Five-Vertex Polyhedra: Hexahedra

For a polyhedron all of whose faces are triangles (*simplicial* polyhedra), Euler's formula implies that $F = 2V - 4$.[17] For $V = 5$ we must have $F = 6$, a hexahedron. If two triangles are coplanar, then we obtain a pentahedron with one quadrilateral face. The number of

---

[17] $V - E + F = 2$ and $3F = 2E$ lead to $F = 2V - 4$.

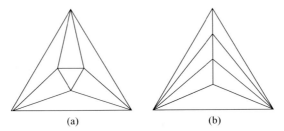

Figure 25.28. The two combinatorial types of simplicial octahedra.

combinatorially distinct "triangulations of the sphere" is a well-studied quantity (Bowen and Fisk 1967). For $V = 5$, there is just one combinatorial type, which can be viewed geometrically as gluing two tetrahedra that have congruent base triangles base to base. As this gluing places all nine edges of the resulting hexahedron on the surface (i.e., none become internal diagonals), the shape of a hexahedron can be determined by reconstructing the two tetrahedra. This is the method we used. Detecting quadrilateral faces is accomplished by computing the dihedral angle at each edge, from the dot product of unit face normals.

25.6.2.3 Six-Vertex Polyhedra: Octahedra

For $V = 6$, we must have $F = 8$, an octahedron. There are two combinatorially distinct octahedra: one equivalent to the regular octahedron, with all six vertices of degree 4, and one with two vertices each of degrees 3, 4, and 5 (see Figure 25.28). The geometry of the latter is that of three stacked tetrahedra, that is, gluing a tetrahedron to a face of a hexahedron. Thus these latter polyhedra have all 12 surface edges deriving from the tetrahedra (again, none become internal diagonals), and their shape can be reconstructed by reconstructing each tetrahedron and performing the 3D rotations necessary to glue the common faces to one another.

The structure of a degree-4 octahedron is more challenging, the first instance where we get a hint of why Cauchy's rigidity problem has no easy solution. For here there is an internal diagonal of unknown length d. This unknown length can be written as a polynomial in the 12 edge lengths, and determined as a root. This is Sabitov's algorithm in another guise, and is already approaching impractability for $n = 6$. We implemented instead a different approach.

Recall from Figure 23.6 (p. 345) that each "half" of an octahedron can flex freely. What varies in this flexing is exactly the unknown internal diagonal length d. This suggests that we view an octahedron as the joining together of two hexahedra, glued at two common faces that share the unknown diagonal. For a given d, the shape of each hexahedron is determined, and can be reconstructed as described above. Let e_1 and e_2 be the hexahedra edges corresponding to this internal diagonal. The dihedral angles of e_1 and e_2 can be computed to see if they match, that is, sum to 2π (as they must to join seamlessly in 3D). Define $f(d)$ to be the difference between the sum of these dihedral angles and 2π. We seek the roots of $f(d)$. Although if written out explicitly, this function would be a polynomial of some high degree (perhaps 8), we found in practice that there is a unique root within the relevant range for d, so that it could be found by a straightforward numerical search. With d determined, it only remains to perform the appropriate 3D rotation of one hexahedron to glue properly to the other. This is what

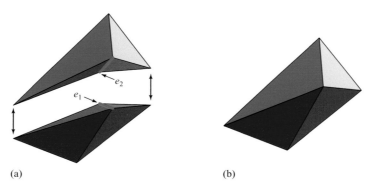

(a) (b)

Figure 25.29. (a) Two hexahedra with dihedral angles 157.8° and 202.2° along edges e_1 and e_2; (b) the joining of the two to form octahedron P_{23} of Figure 25.36.

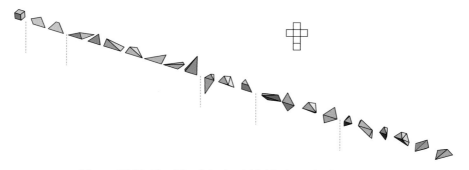

Figure 25.30. The 23 polyhedra foldable from the Latin cross.

we implemented and used to reconstruct all the degree-4 octahedra. An example for P_{23} is shown in Figure 25.29.

It is clear that these ad hoc techniques cannot extend much beyond octahedra. For $V = 7, 8, 9$ there are 5, 14, and 50 combinatorially distinct simplicial polyhedra respectively, and only those with at most one unknown internal diagonal could be reconstructed by these naive methods.

25.6.3   The 23 Shapes

The Latin cross unfolding of the cube can refold (in the Alexandrov sense) into 23 distinct convex polyhedra, as displayed in Figures 25.30 and 25.31: the cube, two flat quadrilaterals (Figure 25.32), seven tetrahedra (Figure 25.33), three pentahedra, each with one or more quadrilateral faces (Figure 25.34), four hexahedra (Figure 25.35), and six octahedra (Figure 25.36). More details about the combinatorial structures are included in the captions to these figures.

25.6.4   No Rolling Belts

We now sketch a proof due to Koichi Hirata[18] that there are only a finite number of foldings of the Latin cross, so that the 85 detailed above are in fact a complete list. Recall the definition of a rolling belt from Section 25.3.1 (p. 393): a belt can roll only with $\leq \pi$ to each side at every point. It is only rolling belts that permit an infinite number

---

[18]   Personal communication, December 2000.

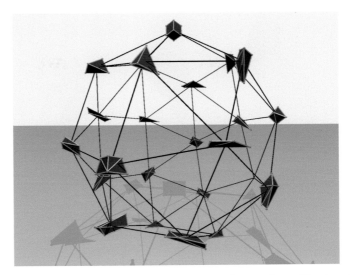

Figure 25.31. The 23 polyhedra from Figure 25.30 arranged on surface of (an imaginary) sphere. (The connecting lines have no significance; they are included only to enhance the 3D effect.) [Image by Alexandra Berkoff and Sonya Nikolova.]

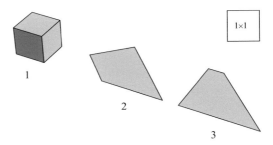

Figure 25.32. The cube and two flat quadrilaterals. (P_2 is the edge-to-edge quadrilateral of Figure 25.9.)

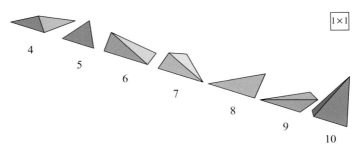

Figure 25.33. Latin cross tetrahedra. (P_4 is the edge-to-edge tetrahedron of Figure 25.9.)

of foldings. Hirata proved that the Latin cross does not admit any rolling belts. He starts with this lemma, which holds in general.

Lemma 25.6.1. *If a subsegment ab of $e_i = v_i v_{i+1}$ of ∂P is part of a rolling belt B, then the whole edge e_i is in B.*

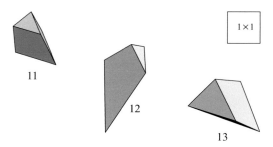

Figure 25.34. Latin cross pentahedra. P_{11} has six vertices and is composed of two triangles and three quadrilaterals. P_{12} and P_{13} have five vertices and one quadrilateral face. (P_{12} is the edge-to-edge pentahedron of Figure 25.9.)

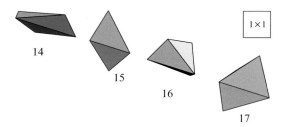

Figure 25.35. Latin cross hexahedra, all with five vertices, two of degree 3 and three of degree 4.

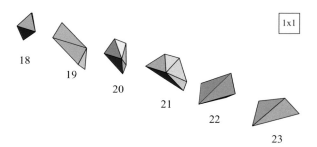

Figure 25.36. Latin cross octahedra, each of six vertices. P_{18}, P_{19}, P_{20}, and P_{21} have two vertices each of degrees 2, 3, and 4, and are composed of three stacked tetrahedra. P_{22} and P_{23} have all vertices of degree 4, and are combinatorially equivalent to a regular octahedron. (P_{23} is the edge-to-edge octahedron of Figure 25.9.)

Proof: One of a or b must be interior to e_i; assume a is. Assume for contradiction that a marks the end of e_i's participation in B. Then the angle to one side of B at a is π from e_i, plus some positive additional angle from the path of B touching e_i at a. Thus the angle to one side of B exceeds π, a contradiction to the definition of a rolling belt.   □

We now consider the Latin cross, as labeled in Figure 25.6(b) (p. 386). None of the edges incident to the four reflex angles could be part of a rolling belt, for that would immediately place more than π to one side. So the only candidates are e_0, e_4, e_7, and e_{10} symmetrical with e_4. Let the cube folded from the Latin cross be a unit cube, so all edge lengths are 1. Any rolling belt of length 1 would glue two reflex vertices to one another, as can be easily checked. So any rolling belt must include more than one edge. Edge e_0 cannot be part of a rolling belt, for that would place its adjacent length-2 edge outside the belt in contact with a reflex vertex.

Now the ends of a rolling belt must be vertices by Lemma 25.6.1, and they must be convex vertices by the definition of a rolling belt. An exhaustive review of the limited options for rolling belts of length 2 or 3 (the only remaining possibilities) shows that each fails to satisfy Alexandrov's gluing conditions in exceeding 2π glued at some point. This leads to Hirata's result:

Theorem 25.6.2. *The Latin cross foldings do not include any rolling belt.*

25.7 THE FOLDINGS OF A SQUARE TO CONVEX POLYHEDRA

In contrast to the situation just detailed for one nonconvex polygon, we know from Theorem 25.1.4 (p. 383) that every convex polygon folds to an infinite variety of convex polyhedra. It remains an incompleted project to understand the 3D structure of all these polyhedra for any given convex polygon. In this section, we explore as an initial case study all the convex polyhedra that may be folded from a square. Before that, we detail some constraints that apply, first, to all convex polygons, and second, to the regular polygons, before plunging into the details on squares in Section 25.7.3.

25.7.1 Foldings of Convex Polygons

We start with a lemma that limits the combinatorial structure of gluing trees for convex polygons, due originally to Shephard (1975). This version, from Demaine et al. (2002a) starts with slightly different assumptions, and follows a different proof.

Lemma 25.7.1. *The possible gluing trees for a convex polygon of n vertices fall into four combinatorial types:*

1. *|: a tree of two leaves, that is, a path.*
2. *Y: a tree of three leaves and one internal degree-3 node.*
3. *I: a tree of four leaves and two internal degree-3 nodes.*
4. *+: a tree of four leaves and one internal degree-4 node.*

Proof: Each leaf must have curvature $\geq \pi$, because it is the folding of a boundary point of internal angle $\leq \pi$: either $= \pi$ if a fold point (p. 382), or $< \pi$ if a vertex of the polygon. The Gauss–Bonnet Theorem 21.3.1 (p. 304) says the total curvature is 4π. So the gluing tree T_G cannot have more than four leaves. If T_G has just two or three leaves, then the only possible combinatorial structures are the two claimed: a path, and a Y.

If T_G has exactly four leaves, then each must have curvature exactly π; so each is a fold point. The only additional possible combinatorial structures for a tree with four leaves are the two claimed: a + and a I. □

The I structure, discussed earlier (Figure 25.16), can be realized by folding a hexagon to a rectangle as shown in Figure 25.37. If the rectangle is modified to become a square, the I becomes a +.

Corollary 25.7.2. *Any polyhedron Q folded from a convex n-gon P has at most $n + 2$ vertices.*

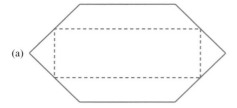

(a)

Figure 25.37. A hexagon (a) folds via an I gluing tree to a doubly-covered rectangle (b).

(b)

Proof: Let us call the vertices of Q *corners* to distinguish them from the vertices of P. The two four-leaf gluing trees of Lemma 25.7.1 lead to Q being a tetrahedron, for there is no curvature left over from the four fold points to concentrate at other corners of Q. Because each corner of Q must either be a fold-point leaf of T_G or involve one or more vertices of P, we have a limit of $n + 3$ corners: three fold-point leaves and one corner of Q deriving from each vertex of P. The I-path leads to $n + 2$ corners by the same logic; and we will see that this is achievable. However, the Y-tree cannot lead to $n + 3$ corners. Consider the junction of the Y. It cannot be constituted by one vertex of P gluing to an edge of P, for then it would be degree-2 rather than 3. So it must involve two or more vertices of P, and the claim follows.　　　□

In particular, this corollary implies that a polyhedron folded from a square has at most six vertices.

25.7.2　Foldings of Regular Polygons

We now specialize the discussion to regular polygons. These are especially simple for large n; for $n \le 6$ there are more combinatorial possibilities. We will not provide the details reported in Demaine et al. (2000b), but rather just cite one particularly simple lemma:

Lemma 25.7.3. *For all $n \ge 3$, regular n-gons fold via perimeter halving, using path gluing trees, to two classes of convex polyhedron:*

1. *A continuum of "pita" polyhedra of $n + 2$ vertices.*
2. *One or two flat, "half-n-gons":*
    (a)　n even: Two flat polyhedra, of $\frac{n}{2} + 2$ and $\frac{n}{2} + 1$ vertices.
    (b)　n odd: One flat polyhedron, of $\frac{n+1}{2} + 1$ vertices.

For $n > 6$, these are the only foldings possible of a regular n-gon.

The flat foldings are illustrated in Figure 25.38.

The *pita polyhedra* are the generic nonflat foldings of regular n-gons. Figure 25.39 shows an example polygon with $n = 12$. Perimeter halving as shown leads to a 3D shape something like Figure 25.40.[19] As $n \to \infty$, the polyhedra approach a doubly covered flat

---

[19]　We were unable to reconstruct these shapes exactly.

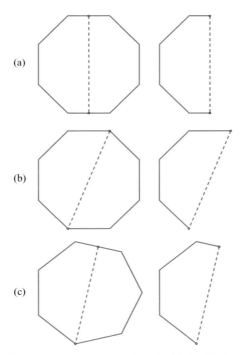

Figure 25.38. (a–b) Flat foldings of an n-gon, n even ($n = 8$); (c) flat folding of an n-gon, n odd ($n = 7$).

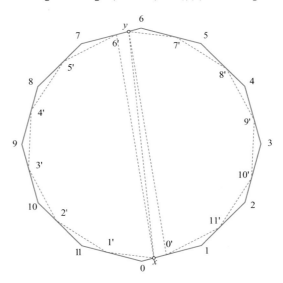

Figure 25.39. Duodecagon, $n = 12$, x and y are perimeter-halving fold vertices. The dashed lines are (largely) conjectured creases.

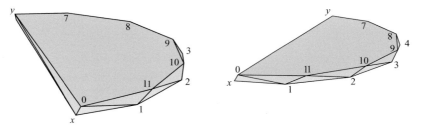

Figure 25.40. Two views of the approximate 3D shape of pita polyhedron folded as per Figure 25.39.

semicircle. The "pita-forms" discussed in Section 23.3.3 (p. 353) are a generalization to arbitrary smooth convex shapes.

25.7.3   The Foldings of a Square

Finally, we turn to one detailed study, drawing from Alexander et al. (2003), which studied the "space" of convex polyhedra foldable from a (unit) square. Even this simple case provides a rich collection of polyhedra: an infinite number of incongruent polyhedra falling into nine classes of distinct combinatorial 3D structure:

1. Five nondegenerate polyhedra:
    (a) Tetrahedra.
    (b) Two pentahedra: one of five vertices and a single quadrilateral face, and one of six vertices and three quadrilateral faces (and all other faces triangles).
    (c) Hexahedra: five-vertex, six-triangle polyhedra with vertex degrees $(3, 3, 4, 4, 4)$.
    (d) Octahedra: six-vertex, eight-triangle polyhedra with all vertices of degree 4.
2. Four flat polyhedra:
    (a) A right triangle.
    (b) A square.
    (c) A $1 \times \frac{1}{2}$ rectangle.
    (d) A pentagon with a line of symmetry.

The combinatorially possible foldings of a square were first enumerated by the two computer programs discussed in Section 25.5. They neither reconstruct the shape of the polyhedra, nor detail the topology of the space of all foldings. These are the aspects on which we concentrate in this section.

In stark contrast to the foldings of the Latin cross, each shape achievable from a square may be deformed continuously into any other through intermediate foldings of the square, that is, no shape is isolated. An illustration of the foldings for a small portion of a continuum is shown in Figure 25.41.

The bulk of Alexander et al. (2003) is a (long) proof detailing all the possible gluing trees, showing that the inventory sketched above is complete, and organizing the continua of all shapes foldable from a square. Here we opt not to discuss the proof, but

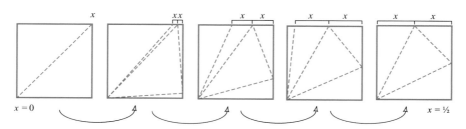

Figure 25.41. Creases for a section of a continuum of foldings: as x varies in $[0, \frac{1}{2}]$, the polyhedra vary between a flat triangle and a symmetric tetrahedron. (This is half of loop B in Figure 25.43.)

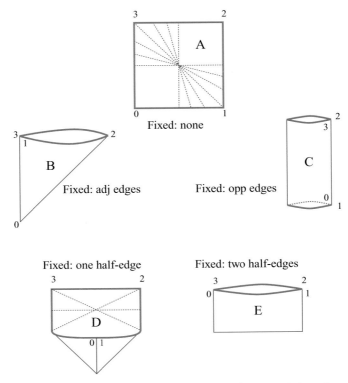

Figure 25.42. The five rolling belts among foldings of a (unit) square. Selected perimeter-halving lines for A and D are shown.

content ourselves with describing the continua (and organizing them differently than Alexander et al. (2003)).[20]

We classify all the foldings of a square according to the type of rolling belt(s) obtained. Recall that the rolling of a belt can be viewed as perimeter halving of a convex loop. We claim that all the foldings of the square can be realized as perimeter halvings of five distinct rolling belts, which we will call A, B, C, D, and E. We can classify these foldings according to which parts of the perimeter are glued to each other in fixed positions (see Figure 25.42):

1. A: No part of the boundary is fixed—the entire boundary is a rolling belt of length 4.
2. B: Two adjacent edges are glued, leaving a belt of length 2.
3. C: Two opposite edges are glued, leaving two belts, each of length 1.
4. D: One edge is creased at its midpoint and the two half-edges glued together, leaving a belt of length 3.
5. E: Two opposite half-edges are glued together, leaving a belt of length 2.

Perimeter halving of each belt creates a continuum of polyhedra. This continuum is naturally viewed as a cycle or "loop" of polyhedra: Starting at halving points x and y,

---

[20] We thank Don Shimamoto for suggesting this reorganization (personal communication, October 2004). In Alexander et al. (2003), some (incongruent) mirror images were not included in the accounting.

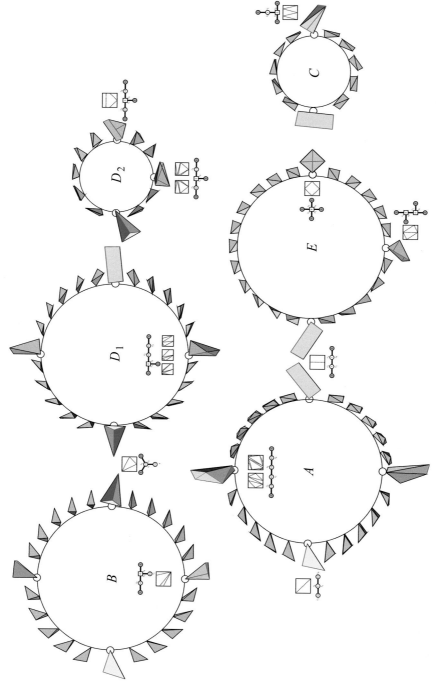

Figure 25.43. Details of all six continuua. Selected crease diagrams and gluing trees are shown.

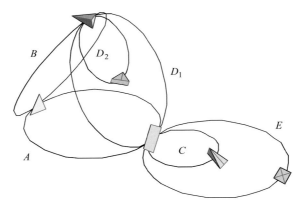

Figure 25.44. The six loops arranged in space, sharing common shapes.

and walking x around the belt perimeter to y's original position, returns the shape to repeat the initial polyhedron. Thus, viewing the cycle as recording this walk of x, the circumference of the loop should be half the length of the rolling belt. However, it is easy to see that in loop A the shape returns to its starting point at the midpoint of that walk, due to the symmetry of the square. We therefore close a cycle at the point at which it first returns to a congruent polyhedron, which reduces the circumference of the A-loop to 1. The B-loop also has circumference 1. Although there are two independent belts in C, it is clear that all that matters in determining the shape of the folding is the relative positions of the two perimeter halvings. So there is just one C-loop of incongruent polyhedra, this time of circumference $\frac{1}{2}$. The E-loop is similar to the B-loop, again of circumference 1. The D-loop is more complicated. Start x at vertex 1 (Figure 25.42), at which point the perimeter-halving fold yields the $1 \times \frac{1}{2}$ rectangle. After walking a distance $\frac{1}{2}$ counterclockwise around the boundary, we reach a symmetric tetrahedron, call it T_s. Walking another $\frac{1}{4}$ brings us to the flat pentagon. Then the shapes repeat in mirror image, which are incongruent to their counterparts except for the symmetric T_s. So one can view the D-loop as pinching at T_s into two loops: D_1 of circumference 1, and D_2 of circumference $\frac{1}{2}$. This yields a total of six loops, displayed in Figure 25.43 (the reason for the layout chosen will be evident shortly).

Examination of this figure shows that the loops share certain polyhedra: the $1 \times \frac{1}{2}$ rectangle appears in the A-, C-, D_1, and E-loops; T_s is in the B-loop, and is shared by D_1 and D_2, and the right triangle is common to A and B. This suggests arranging the loops as depicted in Figure 25.44. Although both the circumferences of the loops, and their points of contact, have significance, the arrangement in 3-space does not.

This diagram represents the topology of the space of incongruent polyhedra foldable from a square. Because the space is a connected 1D network, one can imagine an animation of the polyhedra encountered in a traversal of the network.[21] Establishing a general theory concerning such topology diagrams is an active area of current research. We mention two further results along these lines in Sections 25.8.1 and 25.8.2.

---

[21] See http://cs.smith.edu/~orourke/Square/animation.html for such an animation (following, however, the organization of Alexander et al. (2003)).

25.7.4   Reconstructing the 3D Shapes

We have seen that Corollary 25.7.2 limits the polyhedra foldable from a square to have at most six vertices, and indeed the possible structures are the same as those encountered with the Latin square foldings. Thus the techniques described in Section 25.6.2 apply here as well.

Although we have no systematic method for identifying the creases, in practice, some of the creases are obvious from physical models, leaving only a few uncertainties. These were resolved by "trying" each, and relying on Alexandrov's theorem guaranteeing a unique reconstruction: the wrong choices failed to reconstruct, and the correct choice led to a valid polyhedron.

All the nonflat polyhedra foldable from a square have four, five, or six vertices. Aside from isolated special cases caused by coplanar triangles merging into quadrilaterals, all faces are triangles. In fact, only three distinct combinatorial types are realized: tetrahedra, hexahedra equivalent to two tetrahedra glued base-to-base (i.e., a "triangular double pyramid"), and octahedra combinatorially equivalent to the regular octahedron. Reconstruction then proceeds exactly as described previously.

25.8   CONSEQUENCES AND CONJECTURES

Constructing the entire set of polyhedra permits us to answer a special case of a question posed by Joseph Malkevitch[22]: What is the maximum volume polyhedron foldable from a given polygon of unit area? We found (by numerical search) that the maximum is achieved for a 1×1 square by an octahedron along the A-ring (at about—but apparently not exactly—the 12 o'clock position in Figure 25.43), with a volume of approximately 0.056.[23] It is a perimeter halving folding at a point on one side at about $x = 0.254$. We expected to observe symmetry here but did not (see Figure 25.45).

In light of this example, and the "pita forms" described earlier (p. 353), we can reformulate Malkevitch's question as follows:

Open Problem 25.4: Volume Maximizing Convex Shape. [a] Among all planar convex shapes of unit area (smooth shapes or polygons), which folds, via perimeter halving, to the maximum volume (convex) 3D shape?

---

[a] Demaine and O'Rourke (2006); http://cs.smith.edu/~orourke/TOPP/P62.html.

The restriction to perimeter halving is natural, as smooth shapes may fold only by perimeter halving. And non-perimeter halving foldings do not achieve the maximum for the square.

A related problem is known as the "teabag problem":[24] Glue two unit squares together along their perimeters to form a double-sided square. What is the maximum

---

[22] Personal communication, February 2002.
[23] Although this volume may seem small, it is 60% of the volume of a unit-area sphere: $(4/3)\pi\sqrt{1/(4\pi)}^3 = 1/(6\sqrt{\pi}) \approx 0.094$.
[24] http://www.ics.uci.edu/~eppstein/junkyard/teabag.html.

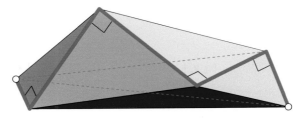

Figure 25.45. The maximum volume convex polyhedron foldable from a square. The red line marks the boundary of the square; the two fold points are circled.

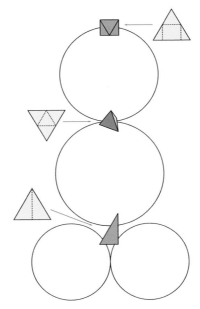

Figure 25.46. The topological structure of the polyhedra foldable from an equilateral triangle. Selected foldings are shown.

volume that can be contained inside? Here there is no restriction to convex shapes; otherwise the volume is zero, for Alexandrov's theorem guarantees that the two-sided square is the unique convex D-form. However, by shaping the corners into four cone shapes that meet along a series of "pleats," a surprisingly large volume can be attained. This problem remains open. It may be that the maximum volume shape is nonpolyhedral, perhaps smooth.

25.8.1 The Foldings of an Equilateral Triangle

Recent work (Akiyama and Nakamura 2005) has detailed the space of polyhedra foldable from an equilateral triangle (without explicitly reconstructing the 3D shapes). The space consists of four rings, with the topological structure roughly depicted in Figure 25.46. An equilateral triangle has four flat foldings (two of which are indicated in the figure), and folds to continua of hexahedra, pentahedra, and tetrahedra.

Although the goal of understanding the structure of all the convex polyhedra foldable from an arbitrary polygon does not appear close with our current understanding, we can enumerate the possibilities for the space of all polyhedra foldable from convex polygons, to which we now turn.

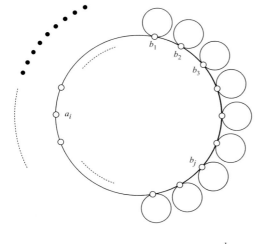

Figure 25.47. The space of foldings of the polygon in Figure 25.48.

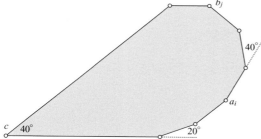

Figure 25.48. $n = 8$, $m = 3$, $\varepsilon = 20°$.

25.8.2   The Space of Foldings of a Convex Polygon

Having studied the space of foldings of a square in the previous section, and uncovered an intricate structure, it is natural to ask for a similar analysis for any convex polygon. Although a detailed understanding of the structure of all the convex polyhedra foldable from an arbitrary convex polygon seems infeasible with our current understanding, we can at least address some high-level questions. Is the space always connected (as it is for the square and equilateral triangle)? Are there at most a constant number of rings for a polygon with n vertices? The answer to both questions is NO.[25] We start by describing a type of "worst-case" example, one whose space of foldings contains $\Omega(n^2)$ isolated shapes, and $\Omega(n)$ rings surrounding the perimeter-halving ring. Its topology is shown in Figure 25.47.

Here and throughout this section we will express angles in units of π in calculations, so that $1 \equiv \pi$, but freely change back to degrees for examples and figures.

The polygon consists of the following vertices:

a_i: m vertices a_i of angle $1 - \varepsilon$ (total turn angle $m\varepsilon$).
b_j: $m+1$ vertices b_j of angle $1 - 2\varepsilon$ (total turn angle $2m\varepsilon + 2\varepsilon$).
c: One vertex c of angle 2ε (turn angle $1 - 2\varepsilon$).

There are a total of $n = 2m + 2$ vertices; $m = n/2 - 1$. The total turn angle is $3m\varepsilon + 1$. Because this turn must equal 2, we choose $\varepsilon = 1/(3m)$; so $m = 1/(3\varepsilon)$. Figure 25.48 shows an example with $m = 3$, and so $\varepsilon = 1/9 = 20°$.

---

[25]   Incisive questions of John Sullivan, August 2003, led to the analysis in this section.

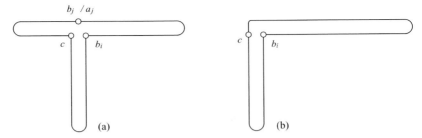

Figure 25.49. (a) Two positions of the T rolling belt; (b) corresponds to the I gluing tree.

In order to connect this specific example to the space of its foldings, we need to introduce a bit more notation. We established earlier (Lemma 25.7.1) that the gluing trees for a convex polygon fall into one of four combinatorial classes: I, Y, I, and +. We now refine this combinatorial classification with geometric distinctions. In particular, we define two geometric classes corresponding to Y: Y and T:

1. If no two angles glued at the junction sum to less than 1, then it is a (strict) Y-junction: no sliding is possible.
2. If at least two of the angles glued at the junction sum to less than 1, then it is a T-junction, admitting a rolling belt.

For the polygon in Figure 25.48, Y-junctions can be formed as follows:

1. Y: $\{a_i, c, a_j\}$ $(i \neq j)$ produces a flat vertex of angle $(1 - \varepsilon) + 2\varepsilon + (1 - \varepsilon) = 2$, with any pair of the three summing to more than 1. So these $\binom{m}{2} = \Omega(n^2)$ matchings are each isolated shapes.
2. T: $\{b_i, c, a_j$ or $b_j\}$ $(i \neq j)$ produces a vertex with one pair of angles (for b_i and c) summing to 1. The $m + 1 = \Omega(n)$ choices for b_i each leads to a rolling belt that encompasses all the third-vertex choices. One position of this rolling belt corresponds to the path folding, as illustrated in Figure 25.49. Thus the ring for the rolling belt attaches to a central path ring.

Thus the space of foldings of this polygon is as depicted in Figure 25.47.

There is a sense in which this example already encompasses the generic situation, in that all other possibilities require special shape conditions. For example, the I gluing tree could have one or both degree-3 nodes of geometric types Y or T, each combination of which forces the polygon to have few vertices (six or fewer) and perhaps a pair of parallel edges. Rather than describe each of these special possibilities, we gather the conditions in Table 25.1 and use this table to indicate why the square has the space of foldings previously detailed in Figure 25.43 (p. 416).

A square is a quadrilateral and so permits the + combinatorial type, which, as the last line of the table indicates, is generically isolated. However, when the angles are all $\frac{1}{2}$, this gluing is achieved in a rolling belt, in ring A in Figure 25.43. The square also admits two T versions of the Y junction and TT versions of the I. Thus, the connectedness of the space of foldings of the square can be seen to be a consequence of various special properties of the square, rather than a reflection of the general situation.

Although it is almost certain that all these gluings lead to distinct polyhedra, it seems difficult to establish this property without a method for reconstructing the 3D structure.

Table 25.1: Inventory of possible foldings of a convex polygon

Combinatorial type	Geometric type	Rolling belts	Angle condition	Number of Realizations	Isolated/ connected	Shape
|		1		1	Ring	
Y	Y	0	3 angles ≤ 2; no 2 angles ≤ 1	n^2	Isolated	
	T	1	2 angles ≤ 1	n	Connected	
I	YY	0	Two pairs 3 angles ≤ 2; no 2 angles ≤ 1	1	Isolated	Parallel hexagon
	YT	1	3 angles ≤ 2; 2 angles ≤ 1	1	Isolated	Parallel pentagon
	T/T	1	Two pairs 2 angles ≤ 1	1	Connected	Parallelogram
	TT	2 = 1	Two pairs 2 angles ≤ 1	1	Connected	Parallelogram
+		0		1	Isolated	Quadrilateral

25.8.3   Dissection-Related Open Problems

In this section we quickly sketch a number of frontier research problems loosely connected to dissections. A *dissection* of a pair of polygons A and B is a partition of each into a finite number of congruent pieces so that the pieces of A can be rearranged to form B and vice versa. Dissections have a long and rich history, as detailed in Frederickson's monograph (Frederickson 1997), concentrating mostly on dissecting 2D shapes, and sometimes or 3D solids. Frederickson followed this book with another (Frederickson 2002) focused on *hinged dissections*, which restrict the rearrangement of the pieces to rotations of the pieces about hinge points. This type of dissection is closer to our concerns, but it is a third type of *piano-hinged dissections*, which fold a polygon along fixed "piano hinges" (creases) into two different doubly covered shapes, that most closely connects dissections to polygon folding. Piano-hinged dissections are just beginning to be explored (Frederickson 2006), and there is little to say yet, except to mention the connection.

These dissections all relate to flat foldings, whose inclusion in Alexandrov's Theorem 23.3.1 raises this natural question:

Open Problem 25.5: Flat Foldings. [a] Characterize those polygons that have a flat folding, that is, which fold to a doubly covered convex polygon.

---

[a] Posed with Koichi Hirata, December 2002.

Another related direction for research is the problem posed by Ferran Hurtado[26] of deciding which convex polygons P can *wrap* which convex polygons Q. Here the notion of wrap is a specialized version of flat folding, where $Q \subset P$ and the "base" face of the wrapping is congruent to Q. This problem can be decided in polynomial time: given P, the Q can be listed, and given Q, the P can be listed, both in polynomial time.

Finally, we discuss a fascinating open problem, which was first posed to us by Martin Demaine, and independently by Ferran Hurtado and Ed Pegg, and natural enough likely to have occurred to many others.[27] One way to phrase the question is this: Can any one of the Platonic solids be cut open, unfolded to a polygon, and refolded to form another Platonic solid? One could of course ask the same of any particular pair of shapes. One could view this as asking for a common "net" for the two shapes (but a net not restricted to edge unfoldings, cf. p. 299), or as a dissection between the surfaces. Frederickson's monograph includes just one surface dissection, a beautiful two-piece dissection of a cube to a regular tetrahedron discovered by Gavin Theobald (Frederickson 1997, p. 246). However, a one-piece dissection seems more challenging.

---

[26] Personal communication, February 2004.
[27] Personal communication from M. Demaine, August 1998; Hurtado asked the question at the Japan Conference on Discrete and Computational Geometry, December 2000. Pegg posed the problem at http://www.mathpuzzle.com/Solution.htm, December 2000.

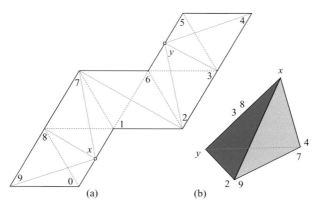

Figure 25.50. A polygon (a) that folds to a regular octahedron and to a tetramonohedron (b), the latter using creases x9, x8, x7, 72, y4, y3, y2; x and y are edge midpoints.

Open Problem 25.6: Fold/Refold Dissections. [a] Can any Platonic solid be cut open and unfolded to a polygon that may be refolded to a different Platonic solid? For example, may a cube be so dissected to a tetrahedron?

---

[a] M. Demaine, F. Hurtado, E. Pegg.

Although it is easy to quickly become convinced that the answer here is likely NO, the following two "close calls" should give pause.

Figure 25.50 shows a polygon that folds to a regular octahedron (in the obvious way), and to a *tetramonohedron*: a tetrahedron all of whose faces are congruent. All four vertices have curvature π, and the common face is an isosceles triangle, but, alas, not an equilateral triangle.

More impressive is the polygon, shown in Figure 25.51, discovered by Koichi Hirata,[28] which folds to both a $1 \times 1 \times 1.232$ rectangular box and a regular tetrahedron (see Figure 25.52).

A related fold/refold dissection question was raised and answered by a problem group at the University of Waterloo[29]: Is there a polygon that folds to two incongruent orthogonal boxes? They found the two examples illustrated in Figure 25.53.

Bipartite space of foldings and unfoldings. The high-level view of the questions we have been considering can be captured in a "bipartite space" graph: on one side are the unfoldings–polygons–and on the other side are the foldings—convex polyhedra. Connect a polygon to a polyhedron by an arc if the polygon can fold to that polyhedron, or equivalently, the polyhedron can be cut open and unfolded to the polygon. Then Open Problem 25.6 can be phrased as follows: Is there a path in this graph (of length 2) between any distinct pair of Platonic solids? We seem far from understanding this bipartite space.

---

[28] Personal communication, December 2000.
[29] Therese Biedl, Timothy Chan, Erik Demaine, Martin Demaine, Anna Lubiw, Ian Munro, and Jeffrey Shallit, September 1999.

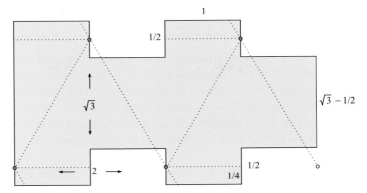

Figure 25.51. A polygon that folds to a regular tetrahedron and to a rectangular box, shown in Figure 25.52.

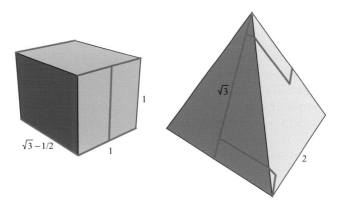

Figure 25.52. The box has dimensions $1 \times 1 \times \sqrt{3} - \frac{1}{2}$; the tetrahedron edge length is 2.

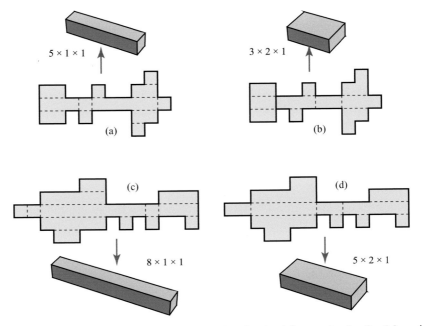

Figure 25.53. (a–b) One polygon that folds to both a $5 \times 1 \times 1$ box and a $3 \times 2 \times 1$ box; (c–d) A different polygon that folds to both an $8 \times 1 \times 1$ box and a $5 \times 2 \times 1$ box.

25.9   ENUMERATIONS OF FOLDINGS

We saw in Section 25.4 that the number of distinct gluing trees for a fixed polygon of n vertices has a tight worst-case bound of $2^{\Theta(n)}$ (Theorem 25.4.1, p. 397). Although this settles a main question concerning enumeration, several interesting issues remain, which we explore in this and the succeeding section. Here we derive a more nuanced bound, still exponential, but expressed in terms of the number of leaves of the gluing tree. This leads to a polynomial bound for convex polygons. The following section enumerates unfoldings.

As mentioned earlier (p. 400), the combinatorial information we use to represent a gluing tree determines the tree.[30] Call the list of node labels the *combinatorial type* of the gluing tree. So the tree illustrated in Figure 25.11(b) (p. 392) has combinatorial type

$$(\{e_1\}, \{v_2, v_5, e_5\}, \{v_3, v_4\}, \{e_3\}, \{e_5\}, \{v_1, v_6, e_1\}, \{v_7, v_8\}, \{e_7\}).$$

Because this list determines the tree, we can count gluing trees by counting combinatorial types. Counting gluing trees is what we mean by "counting foldings," treating all combinatorially equivalent positions of a rolling belt as the same folding. We will not prove every claim in this section, but show the most interesting parts of the reasoning, leaving some details to Demaine et al. (2000b), from which this section is derived.

Let $\lambda \geq 2$ be the number of leaves of the gluing tree. We start with three lemmas, one each for $\lambda = 2, 3, 4$.

Lemma 25.9.1. *A polygon P of n vertices has $O(n^2)$ different gluing trees of two leaves, that is, paths, and in the worst case may have $\Omega(n^2)$ such trees.*

Proof: View ∂P as rolling continuously between the two leaves x and y, like a rolling belt or tank tread. Each specific position corresponds to a perimeter-halving gluing G (Figure 25.2). The combinatorial type T_G changes each time a vertex v_i either passes another vertex v_j or becomes the leaf x or y. Each such event corresponds to two distinct types: the type at the event, and the type just beyond it: for example, (v_i, v_j) and (v_i, e_j). So counting events undercounts by half. If we count the possible pairs (v_i, v_j) for all $i \neq j$, we will double count each type: the event (v_i, v_j) leads to the same type as (v_j, v_i). The undercount by half and overcount by double cancel; thus $n(n-1)$ is the number of types without a vertex at a leaf. Adding in the n possible (v_i, v_i) events, each of which leads to two types, yields an upper bound of $n(n-1) + 2n = O(n^2)$ on the number of combinatorial types.

A worst-case lower bound of $\Omega(n^2)$ is achieved by the example illustrated in Figure 25.54(a). Here $n/2$ vertices of P are closely spaced within a length L of ∂P, and $n/2$ vertices are spread out by more than L between each adjacent pair. Then each of the latter vertices (on the lower belt in the figure) can be placed between each pair of the former vertices (on the upper belt), yielding $n^2/4$ distinct types. This example can be realized geometrically by making the internal angle at each vertex nearly π, that is, by a convex polygon that approximates a circle. □

---

[30]   This is formally proved as Lemma 5.3 of Demaine et al. (2000b).

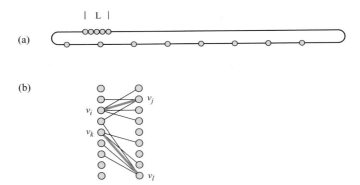

Figure 25.54. (a) $\Omega(n^2)$ combinatorial types achieved by rolling the perimeter belt; (b) only $O(n)$ possible disjoint vv-pairings.

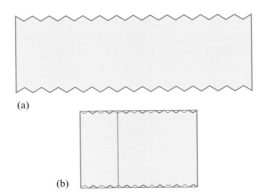

Figure 25.55. (a) Polygon P; (b) one four fold-point gluing; the fold points are midpoints of edges. The dashed lines indicate tips of front teeth bent over and glued behind.

We will leave the next two lemmas (proved by Demaine et al. (2000b)) as claims, as both proofs are relatively unrevealing case analyses:

Lemma 25.9.2. *A polygon P of n vertices folds to at most $O(n^4)$ different gluing trees of three leaves, that is, Y's.*

We do not know whether this bound is tight.

Moving to $\lambda = 4$, the four fold-point gluing trees play a special role: those of exactly four leaves, each of which is a fold point. Recall we encountered these earlier in Figure 25.16.

Lemma 25.9.3. *The number of four fold-point gluing trees for a polygon of n vertices can be $\Omega(n^2)$ and is at most $O(n^4)$.*

The lower bound is established by a variation on the foldings of a rectangle to tetrahedra, examples of which can be seen in ring D of the square continuum in Figure 25.43 (p. 416). The idea is to make each of the two rolling belts in the I structure realize $\Omega(n)$ gluings independently. This can be accomplished by alternating supplementary angles along the belt at equal intervals, illustrated in Figure 25.55 with angles $\pi/2$ and $3\pi/2$. The tetrahedra produced are the same as that obtained by folding a rectangle: the "teeth" mesh seamlessly.

We are now positioned to establish an upper bound on the number of gluing trees, as a function of the number leaves.

Theorem 25.9.4. *The number of gluing trees with λ leaves for a polygon P with n vertices is $O(n^{2\lambda-2})$.*

Proof: Let $g(n, \lambda)$ be the number of gluing trees for P that have λ leaves. The proof is by induction on λ. We know from Lemma 25.3.1 that at most four leaves can be fold points. We assume for the general step of the induction that $\lambda > 4$, and so there is at least one non-fold-point leaf. The base cases for $\lambda \leq 4$ will be considered later.

The bound will use one consequence of the angles or curvature of a gluing (described in this paragraph) and one consequence of the matching edge lengths of a gluing (described in the next paragraph). Because a point interior to an edge of P has angle π, a node of degree d of a gluing tree ($d = 1, 2, \ldots$) glues together d vertices of P or $d - 1$ vertices and one edge of P. Apart from this, we will use nothing else about the angles of the polygon, and in fact, our argument will hold more generally for a closed chain of n vertices, with specified edge lengths.

Given a tree T_G that is not a path, and a leaf l, define the *source* of l as the first node of degree more than 2 along the (unique) path from l into T. The path in T_G from l to its source is called the *branch* of l. For a tree T_G and a non-fold-point leaf corresponding to polygon vertex l, let $s(l)$ be a vertex of P closest to l glued at the source of the leaf. Note that there must be such a vertex, since we cannot glue together two points interior to polygon edges at the source of the leaf. Note—this is the single consequence of matching edge lengths referred to above—that the pair $(l, s(l))$ determines the portion of P's boundary that is glued together to form the branch of l. We can simplify T by cutting off l's branch, resulting in a tree with $\lambda - 1$ leaves. The corresponding simplification of ∂P is to excise the portion of its chain of length $2d$ $(l, s(l))$ centered at l, resulting in a closed chain on at most $n - 1$ vertices. Since there are n choices for l and at most n choices for $s(l)$ we obtain $g(n, \lambda) \leq n^2 g(n - 1, \lambda - 1)$. For the general case there are at most three fold-point leaves, hence $g(n, \lambda) \leq n^{2(\lambda-3)} g(n - (\lambda - 3), 3)$.

Lemmas 25.9.1 and 25.9.2 established the base cases $g(n, 2) = O(n^2)$ and $g(n, 3) = O(n^4)$. Substituting, this yields

$$\begin{aligned} g(n, \lambda) &\leq n^{2(\lambda-3)} O([n - (\lambda - 3)]^4) \\ &= n^{2(\lambda-3)} O(n^4) \\ &= O(n^{2\lambda-2}). \end{aligned}$$

It remains to handle the case of $\lambda = 4$ leaves. We separate into the cases when at least one of these leaves is not a fold-point leaf, where arguments as above (but not detailed here) yield $O(n^6)$, and the case when all four vertices are fold-point leaves. Here Lemma 25.9.3 establishes a bound of $O(n^4)$, smaller than that claimed by the theorem. $\square$

Thus, although the algorithms presented in Section 25.5 (cf. p. 402) definitely are worst-case exponential time, they are polynomial when the number of leaves of the gluing tree is fixed. There are two interesting corollaries of this result:

1. Recall that Lemma 25.7.1 restricts the number of leaves for convex polygons. This leads, after some analysis, to a polynomial bound: A convex polygon P of

n vertices folds to at most $O(n^4)$ different gluing trees. We do not know if this bound is tight.

2. Each leaf consumes some of the 4π curvature available from the Gauss–Bonnet Theorem 21.3.1. If a polygon's reflex vertices have "bounded sharpness" in the sense that there is a $\delta > 0$ so that no (internal) reflex angles exceeds $2\pi - \delta$, then every leaf consumes at least δ, and so the number of leaves is bounded: $\lambda \leq 4\pi/\delta$. Thus, Theorem 25.9.4 implies a polynomial bound for such polygons.

25.10   ENUMERATIONS OF CUTTINGS

We now turn briefly to the complementary view of cutting a polyhedron surface and unfolding to a polygon. Our goal is to count such cuttings.

It will be useful in what follows to distinguish between a *geometric* tree $\mathcal{T}$ composed of a union of line segments, and the *combinatorial* tree T of nodes and arcs. A *geometric cut tree* $\mathcal{T}_C$ for a polyhedron Q is a tree drawn on ∂Q, with each arc a polygonal path, which leads to a polygon unfolding when the surface is cut along $\mathcal{T}$, that is, flattening $Q \setminus \mathcal{T}$ to a plane. A *geometric gluing tree* $\mathcal{T}_G$ specifies how ∂P is glued to itself to fold to a polyhedron. There is clearly a close correspondence between $\mathcal{T}_C$ and $\mathcal{T}_G$, which are in some sense the same object, but it will nevertheless be useful to retain a distinction between them.

It seems difficult to find a completely satisfactory definition of when two cut trees should be counted as distinct. Here we opt to define the *combinatorial cut tree* T_C corresponding to a geometric cut tree $\mathcal{T}_C$ as the labeled graph with a node (not necessarily labeled) for each point of $\mathcal{T}_C$ with degree not equal to 2, and a labeled node for each point of $\mathcal{T}_C$ that corresponds to a vertex of Q (labeled by the vertex label); arcs are determined by the polygonal paths of $\mathcal{T}_C$ connecting these nodes. An example is shown in Figure 25.56. Note that not every node of the tree is labeled, but every polyhedron vertex label is used at some node. All degree-2 nodes are labeled.

This definition has the consequence of counting different geodesics on ∂Q between two polyhedron vertices as the same arc of T_C. Thus the two unfoldings shown in Figure 25.57 have the same combinatorial cut tree, even though the geodesic in (c) spirals twice around compared to once around in (a).

It is no surprise that there can be an exponential number of cutting trees for a given polyhedron. We now describe the examples and sketch the arguments that establish this.

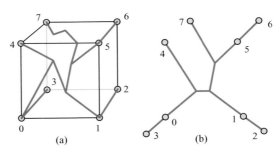

Figure 25.56. (a) Geometric cut tree $\mathcal{T}_C$ on the surface of a cube; (b) the corresponding combinatorial cut tree T_C.

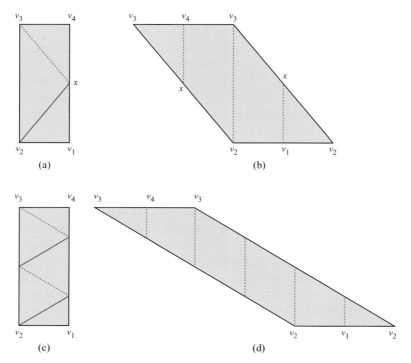

Figure 25.57. (a) Doubly covered rectangle with cut path: x is the midpoint of edge $v_1 v_4$; (b) unfolding; (c–d) another cut path and its unfolding.

Theorem 25.10.1. *There is a polyhedron Q of n vertices that may be cut open with exponentially many $(2^{\Omega(n)})$ combinatorially distinct cut trees, which unfold to exponentially many geometrically distinct simple polygons.*

Proof (*Sketch*)**:** Q is a truncated cone, as illustrated in Figure 25.58(a): the hull of two regular, even n-gons of different radii lying in parallel planes and similarly oriented. Label the vertices on the top face $a_0, \ldots, a_{n-1}$ and $b_0, \ldots, b_{n-1}$ correspondingly on the bottom face. The "base" cut tree, which we notate as $T_{0000000}$, unfolds Q by flipping out each trapezoid $(b_i, b_{i+1}, a_{i+1}, a_i)$, with the top face attached to $a_{n-1}a_0$.

We then define a cut tree based on the digits m_i of a binary number, as an alteration of the base tree $T_{0\cdots0}$. The cut tree $T_{1001101}$ shown in Figure 25.58(b) replaces (a_1, b_1) with (a_0, b_1) because $m_0 = 1$, (a_5, b_5) with (a_4, b_5) because $m_2 = 1$, and so on.

There are $2^{n/2-1} = 2^{\Omega(n)}$ cut trees, and all lead to simple, distinct (incongruent) polygons when slight asymmetries are introduced into the structure of Q. □

Even restricting the cut tree to a path permits an exponential number of unfoldings, as Figure 25.59 hints at. Whether the exponential lower bound holds for convex unfoldings (a convex Hamiltonian unfolding in the notation of Section 22.5.3, p. 327), remains open.

The bound derived is tight, in the following sense:

Theorem 25.10.2. *The maximum number of edge-unfolding cut trees of a polyhedron of n vertices is $2^{O(n)}$, and the maximum number of arbitrary cut trees $2^{O(n^2)}$.*

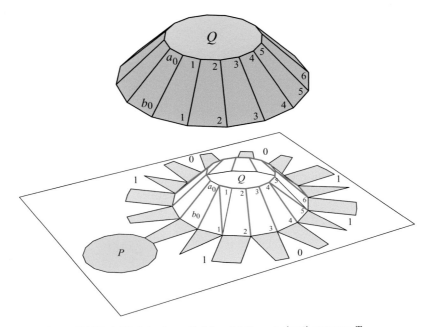

Figure 25.58. (a) Polyhedron Q; (b) unfolding via (red) cut tree $T_{1001101}$.

Proof: For edge unfoldings, the bound depends on the number of spanning trees of a polyhedron graph. We may obtain a bound here as follows.[31] First, triangulating a planar graph only increases the number of spanning trees, so we may restrict attention to triangulated planar graphs. Second, it is well known that the number of spanning trees of a connected planar graph is the same as the number of spanning trees of its dual. So we focus just on 3-regular (cubic) planar graphs. Finally, a result of McKay (1983) proves an upper bound of $O((16/3)^n/n)$ on the number of spanning trees for cubic graphs. This bound is $2^{O(n)}$.

For arbitrary cut trees, the underlying graph might conceivably have a quadratic number of edges, which leads to the bound $2^{O(n^2)}$. □

25.11 ORTHOGONAL POLYHEDRA

The difficulty of reaching definitive conclusions concerning folding polygons to polyhedra has led to exploration of restricted classes of polygons, such as convex polygons (Section 25.7.1). Another common restriction is to *orthogonal polygons*, those in which adjacent edges meet orthogonally, with an internal angle of either 90° or 270°. Although, as we have seen with the Latin cross (Section 25.6), nonconvex orthogonal polygons might fold to convex polyhedra, it is natural to explore foldings of orthogonal polygons to *orthogonal polyhedra*, those in which each pair of adjacent faces meet orthogonally, at a dihedral angle of either 90° or 270°. This area was first studied by Biedl et al. (2005), whose results we next describe.

---

[31] We thank B. McKay (personal communication, January 2000) for guidance here.

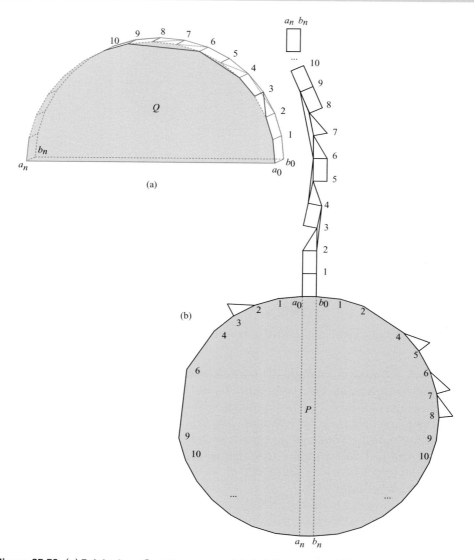

Figure 25.59. (a) Polyhedron Q, with a cut tree labeled $T_{...0022020100}$; (b) unfolding to polygon P.

25.11.1   Orthogonal Nets

Define a *creased net* as a polygon with internal crease chords specified, and an *orthogonal creased net* as an orthogonal polygon with creases specified, and all creases parallel to a polygon edge. To specify the creases is not to specify the gluing, or the dihedral angles. The main result of Biedl et al. (2005) is as follows:

Theorem 25.11.1. *Deciding whether a given orthogonal creased net folds to some orthogonal polyhedron is NP-complete.*

We will not prove this result, but give some sense of why it holds.

2D. First consider the equivalent problem in one dimension lower: Given a polygonal chain, decide if it folds to some orthogonal polygon. They proved this is NP-complete

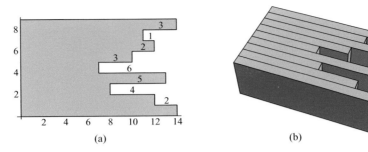

Figure 25.60. (a) Orthogonal polygon corresponding to set $S = \{2, 4, 5, 6, 3, 2, 1, 3\}$ and partition $2 + 4 + 6 + 1 = 5 + 3 + 2 + 3$; (b) polygon extruded to orthogonal polyhedron.

by reduction from PARTITION (p. 25). The proof is nearly identical to that for ruler folding (Theorem 2.2.1, p. 25) and for minimum flat span (p. 133). Given an instance of PARTITION, with S a set of n integers, one forms a chain C so that the edge lengths alternate between 1 and the integers in S, plus three more edges of lengths $(L, n + 1, L)$, where L is 1 more than half the sum of the numbers in S. Then C folds to an orthogonal polygon P if and only if there is a partition of S, which can be read off from the directions of the edges of P. Figure 25.60(a) illustrates the construction on a set of eight numbers. As with the Ruler and Span proofs, to achieve (in this case) closure with the C-shape $(L, n + 1, L)$ edges, the edges must balance heading right and left. Note that L is chosen long enough so that the central edge of the C cannot be reached by the edges to the right.

3D. It is natural to assume that the 3D problem is no easier than the 2D problem and to hope that 2D proof can be extended. Consider the naive extension achieved by extruding the orthogonal polygon into the third dimension, as illustrated in Figure 25.60(b). If one retains the top and bottom faces as congruent to the 2D orthogonal polygon, then the given net has these faces outlined by creases, and so already encodes the partition. This undermines the reduction necessary to achieve the NP-completeness proof, for the idea is to use the net to solve the partition problem. If, on the other hand, the top and bottom faces are partitioned into rectangles as illustrated, then again, in a slightly less direct way, the net encodes the partition in the horizontal lengths of these rectangles.

The challenge to arrange a net that does not encode the partition, but can be folded to an orthogonal polyhedron if and only if the set can be partitioned, was met in Biedl et al. (2005) by a delicate construction employing notches and a frame to force various alignments. We leave the details of this construction to the original paper.

25.11.2 Nonorthogonal Polyhedra

Theorem 25.11.1 holds only with the restriction that the folded polyhedron should be an orthogonal polyhedron. The authors therefore posed the question of whether an orthogonal creased net could ever fold to a nonorthogonal polyhedron. Note that this question is not answered by the various foldings of the Latin cross, for the creases in those examples are not themselves orthogonal.

This question was answered in Donoso and O'Rourke (2001, 2002): NO, an orthogonal creased net can fold only to an orthogonal polyhedron. This extends the range of the NP-completeness result: deciding whether an orthogonally creased net can fold to

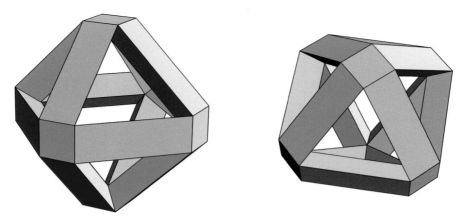

Figure 25.61. Two views of an nonorthogonal polyhedron composed of rectangular faces. [Figure 1 of Donoso and O'Rourke 2002.]

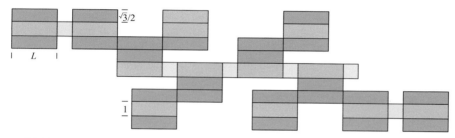

Figure 25.62. An orthogonal creased net for the nonorthogonal polyhedron in Figure 25.61. [Based on Figure 3 of Biedl et al. 2002a.]

a polyhedron is NP-complete. However, the result depends crucially on the genus of the polyhedron. So far we have only discussed polyhedra of genus zero, those homeomorphic to a sphere. The genus counts the number of holes of a polyhedron; for example, a torus has genus one. Surprisingly, for sufficiently high genus, an orthogonal creased net can fold to a nonorthogonal polyhedron. A genus-7 example is shown in Figure 25.61.

An orthogonal creased net for this polyhedron was found in Biedl et al. (2002a); (see Figure 25.62).

The question of Biedl et al. (2005) can be divorced from foldability issues (Donoso and O'Rourke (2002)) by this phrasing:

> If a polyhedron's faces are all rectangles, must all dihedral angles be a multiple of $\pi/2$?

(Note this permits two rectangular faces to be coplanar, which therefore permits a face to be an orthogonal polygon, partitioned into rectangles.) The status of this question is shown in Table 25.2. Both the genus 2 and the genus 6 results are quite intricate. The higher-dimensional analog is completely unresolved (Demaine and O'Rourke 2003).

25.11.3 "Rigid Nets"

We have touched upon the little-explored topic of avoiding self-intersection during a continuous unfolding process, rather than only in the final flat state, only sporadically—

Table 25.2: Do rectangle faces imply 3D orthogonality?

Genus	Orthogonal?	Refrences
0	YES	Donoso and O'Rourke 2002
1	YES	Donoso and O'Rourke 2002
2	YES	Biedl et al. 2002a
3, 4, 5	?	
6	NO	Biedl et al. 2002a
≥ 7	NO	Donoso and O'Rourke 2002

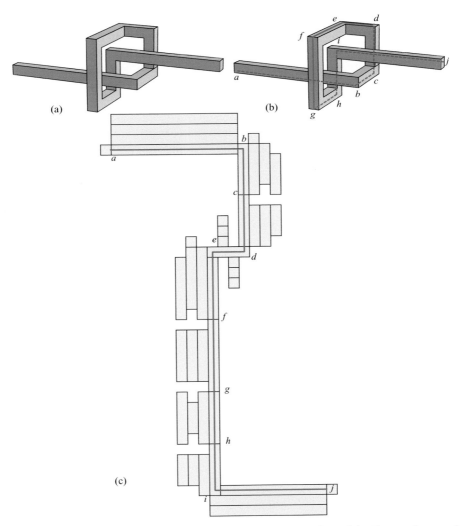

Figure 25.63. (a) Orthogonal polyhedron; (b) "backbone" chain of net; (c) orthogonal creased net for polyhedron. [Based on Figures 6 and 7 in Biedl et al. 2005.]

when discussing a single vertex of a flat folding (Theorem 12.3.1, p. 212), in mentioning "continuous blooming" (p. 362) (cf. Figure 22.38, p. 333), and Theorem 11.6.2 in Part II (p. 190) concerning the ability to continuously move between any two folded states of a polygonal piece of paper. One clean albeit negative result emerged from the same work Biedl et al. (2005), on *rigid nets*: There is an orthogonal creased net that can fold to an orthogonal polyhedron (of genus zero), but only if faces are permitted to pass through one another. Here, in contrast to origami assumptions, the model is that each face is a rigid plate, that is, "rigid origami" in the notation of Chapter 20 (p. 292). Thus, the final folded state is unreachable.

The example is based on the knitting needles locked chain in Figure 6.3 (p. 89). The orthogonal polyhedron is shown in Figure 25.63(a). As in the knitting needles, the two end tubes should be long in comparison to the other links. There is a clever unfolding of the surface to a nonoverlapping net, shown in (c) of the figure. The central "backbone" of the net is marked, and the corresponding path on the surface is illustrated in (b). With the backbone unfolded without overlap, it is not difficult to arrange the other three sides of the tubes on either side of the backbone to also avoid overlap.

Now view the backbone on the net as a polygonal chain in the plane. A continuous folding of the net to the polyhedron that avoided self-intersection throughout would carry along the chain, showing that the chain could be folded without self-intersection to the knitting needles configuration. But we know from Theorem 6.3.1 that this is impossible. This establishes the claim.

Recall that the continuous foldability result of Theorem 11.6.2 is achieved by making many additional creases during the motion than are present in the final folded states. A model intermediate between these arbitrary creasings and rigid faces would permit the faces to bend but disallow the introduction of new creases.

26 Higher Dimensions

Higher dimensions are just beginning to be explored. Here we touch on extensions to higher dimensions in all three parts of the book.

26.1 PART I

1D (one-dimensional) linkages in higher dimensions have been explored for certain problems. For example, many linkage results that permit crossings generalize to higher dimensions, such as the annulus reachability Lemma 5.1.1 (p. 59) and the results on turning a polygon inside-out (Section 5.1.2, p. 63). Many of the generalizations are straightforward, employing nearly identical proofs. Disallowing crossings can lead to fundamentally different situations, however, as we saw with the lack of locked 4D chains and trees (Section 6.4, p. 92).

What remains largely unexplored here are 2D "linkages" in 4D—and higher-dimensional analogs. One model is 2D polygons hinged together at their edges, which have fewer degrees of freedom than 1D linkages in 3D. For example, hinged polygons can be forced to fold like a planar linkage by extruding the linkage orthogonal to the plane (see Figure 26.1). As we have just seen, Biedl et al. (2005) showed that even hinged chains of rectangles do not have connected configuration spaces, with their orthogonal version of the knitting needles (Figure 25.63). It would be interesting to explore these chains of rectangles in 4D. Another connection is to Frederickson's hinged piano dissections (p. 423), which are also just beginning to be explored.

26.2 PART II

Turning to the origami context, the definitions of what constitute a paper, folded state, and folding motions from Section 11.6 have been extended to higher-dimensional paper in Demaine et al. (2006a). For example, for 3D paper, creases are planar polygons, so a crease pattern is a 3D polyhedral cell complex.

Kawasaki (1989a) studied flat origami in $\mathbb{R}^d$, providing a necessary condition that any flat origami must satisfy. Specialized to 3D paper folding in 4D, his condition says that a 3D polyhedral cell complex has a flat folding only if for any closed curve γ that avoids all vertices and edges and passes through faces of the complex transversally in the order $f_1, f_2, \ldots, f_n$, the composition of the reflections through those faces,

$$R(f_1) \circ R(f_2) \circ \cdots \circ R(f_n),$$

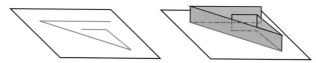

Figure 26.1. Extruding a linkage into an equivalent collection of polygons (rectangles) hinged together at their edges.

is the identity transformation I. This is a natural condition, similar to Theorem 12.2.14 (p. 211), because walking along γ starting from a point p on γ must eventually return to p, and the foldings are all reflections $R(f_i)$. We are not aware of a generalization of Kawasaki's theorem 12.2.1 (p. 199) for a single vertex of such a flat origami.

The natural generalization of the fold-and-cut Theorem 17.2.2 remains open. Recall that this theorem states that any planar graph G drawn with straight lines on 2D paper may be folded flat to a common line L so that one cut along L cuts out precisely G. The analog of the plane graph G in the higher-dimensional context is a *polyhedral complex* of faces: each face is a polyhedron in some dimension, and two i-dimensional faces meet at zero or one $(i-1)$-dimensional faces.

Open Problem 26.1: Higher-Dimensional Fold-and-Cut. Given an arbitrary polyhedral complex drawn on a d-dimensional piece of paper, is it always possible to fold (through $\mathbb{R}^{d+1}$) the paper flat (into d-space) so that the facets of the complex all map to a common hyperplane?

For example, for $d = 3$, imagine one or more polyhedra in 3-space. The challenge is to fold 3-space through 4D so that all the faces of the polyhedra end up lying in a common plane. Slicing along this plane through the 3-space constituted by this folded 4-space would cause the polyhedra to "fall out." The even harder version of the fold-and-cut problem that demands folding the k-dimensional faces of the complex, and no more, to a common k-flat, for all $k = 0, 1, \ldots, d$, is also open, even for $d = 2$ as we mentioned in Part II (p. 278).

The analogous generalization of flattening to higher dimensions also remains unexplored:

Open Problem 26.2: Flattening Complexes. [a] Which polyhedral complexes in $d \geq 3$ dimensions can be flattened (i.e., have flat folded states)?

---

[a] Demaine et al. (2001a).

26.3   **PART III**

26.3.1   Tesseract

Imagine an orthographic projection of the unfolding of a cube to a Latin cross onto the plane containing the bottom face, the one none of whose edges are cut. The 2D

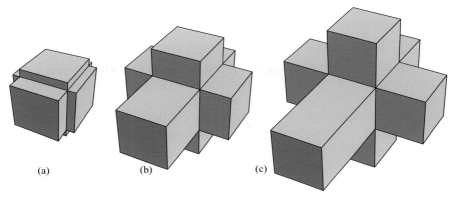

Figure 26.2. Three snapshots of a hypercube unfolding: (a) after rotation by 10°; (b) after 44°; (c) after 90°. [Computations by Patricia Cahn.]

Figure 26.3. "Corpus Hypercubus". [Salvador Dali, 1955. ©2006 Salvador Dali, Gala-Salvador Dali Foundation / Artists Rights Society, New York.]

projection of this unfolding starts as a square (the shadow of the cube, coincident with the bottom face), becomes an orthogonal cross (four rectangles surrounding the bottom square face), and eventually takes the shape of the familiar Latin cross. The analogous process of unfolding a hypercube (a "tesseract") and projecting it orthogonally to the 3-flat containing the "bottom" cubical facet follows a similar pattern, as illustrated in Figure 26.2. Initially the shadow is a cube, which morphs to a cube surrounded by six growing rectangular boxes and ends with the hypercube unfolding, a shape familiar to many from the famous Dali painting (Figure 26.3).

26.3.2 3-Manifolds Built of Boxes

The result mentioned in Section 25.11.2 (p. 433) that polyhedra built from rectangles must be orthogonal can be rephrased this way: For any polyhedral 2-manifold homeomorphic to a sphere $\mathbb{S}^2 \subset \mathbb{R}^3$, all of whose facets are rectangles, adjacent facets must either meet orthogonally or are coplanar. The analogous question one dimension higher was raised in Demaine and O'Rourke (2003): For any polyhedral 3-manifold homeomorphic to a sphere $\mathbb{S}^3 \subset \mathbb{R}^4$, all of whose facets are rectangular boxes, is it true that adjacent facets lie either in orthogonal 3-flats or within the same 3-flat? Very roughly, must a 3-manifold built from boxes be itself orthogonal? Of course the question generalizes to arbitrary dimensions.

26.3.3 Vertex Unfolding

One of the few unfolding results in higher dimensions is the extension of the results on "vertex unfolding" discussed in Section 22.6 (p. 333). In particular, the main theorems of that section extend directly:

Theorem 26.3.1. *Every connected simplicial d-manifold (possibly with boundary) has a facet cycle, and therefore a vertex unfolding.*

A *simplicial* manifold is one composed of simplices: the generalization of triangles to higher dimensions. Thus a 4D manifold has tetrahedral facets. The strip construction becomes a slab construction, placing the simplex facets between parallel hyperplanes.

26.3.4 Source Unfolding and Star Unfolding

It has recently been established by Miller and Pak (to appear) that of the two known general unfoldings of convex polyhedra to nets—the source unfolding (Section 24.1.1, p. 359) and the star unfolding (Section 24.3, p. 366)—only the former generalizes to higher dimensions. In particular, they prove that the source unfolding, based on the shortest paths from a source point x, leads to a polyhedral nonoverlapping "foldout," whose boundary is the cut locus with respect to x, exactly analogous to the 3D-to-2D situation (and they provide a wavefront algorithm for constructing the unfolding). Surprisingly, they show that the obvious analog to the star unfolding does not even exist. Recall the star unfolding cuts the shortest paths from x to the vertices of the polyhedron P. The analog is to cut the shortest paths from x to the points of nonzero curvature ("warped points" in their notation) of the polyhedral complex P. Although these shortest paths themselves form a polyhedral complex, they do not form a "cut set," which roughly means that the set does not cut enough points to unfold P. It is left an open problem in Miller and Pak (to appear) to augment the shortest paths to nonzero curvature points in a "canonical" way and obtain a generalized star unfolding.

It is fitting to end the book with the natural analog of the unsolved edge-unfolding problem (Open Problem 21.1, p. 300) which—hardly surprisingly—remains wide open:

Open Problem 26.3: Ridge Unfolding. [a] Does every convex d-polytope have a *ridge unfolding*, a cutting of $(d-2)$-dimensional facets that unfolds the polytope into $\mathbb{R}^{d-1}$ as a single, simply connected object, without overlap?

_____

[a] Jeff Erickson, personal communication, 2001.

For $d = 3$, this asks if every convex polyhedron (3-polytope) has a cutting of its edges (1-dimensional facets) that unfolds to a plane ($\mathbb{R}^2$) without overlap, that is, Open Problem 21.1.

Bibliography

Tim Abbott and Reid Barton. Generalizations of Kempe's universality theorem, Manuscript, December 2004. Cited on 37, 39.

Hisashi Abe. Possibility of trisection of arbitrary angle by paper folding. In SUGAKU Seminar, July 1980. Cited on 285.

Aaron Abrams and Robert Ghrist. Finding topology in a factory: Configuration spaces. *Am. Math. Mon.*, 109(2):140–150, 2002. Cited on 18.

Colin C. Adams. *The Knot Book*. W. H. Freeman, New York, 1994. Cited on 89, 92, 128.

Colin Adams, Bevin Brennan, Deborah Greilsheimer, and Alexander Woo. Stick numbers and composition of knots and links. *J. Knot Theory Ramifications*, 6(2):149–161, 1997. Cited on 90.

Pankaj K. Agarwal, Boris Aronov, Joseph O'Rourke, and Catherine A. Schevon. Star unfolding of a polytope with applications. *SIAM J. Comput.*, 26:1689–1713, 1997. Cited on 360, 367, 370.

Richa Agarwala, Serafim Batzoglou, Vlado Dancik, Scott E. Decatur, Martin Farach, Sridhar Hannenhalli, S. Muthukrishnan, and Steven Skiena. Local rules for protein folding on a triangular lattice and generalized hydrophobicity in the HP model. *J. Comput. Biol.*, 4(2):275–296, 1997. Cited on 160.

Hee-Kap Ahn, Prosenjit Bose, Jurek Czyzowicz, Nicolas Hanusse, Evangelos Kranakis, and Pat Morin. Flipping your lid. *Geombinatorics*, 10(2):57–63, 2000. Cited on 80.

Oswin Aichholzer, David Alberts, Franz Aurenhammer, and Bernd Gärtner. A novel type of skeleton for polygons. *J. Universal Comput. Sci.*, 1(12):752–761, 1995. Cited on 256.

Oswin Aichholzer and Franz Aurenhammer. Straight skeletons for general polygonal figures in the plane. In *Proc. 2nd Annu. Int. Conf. Comput. Combin.*, volume 1090 of *Lecture Notes in Computer Science*, pages 117–126. Springer-Verlag, Berlin, 1996. Cited on 256, 257.

Oswin Aichholzer, David Bremner, Erik D. Demaine, Henk Meijer, Vera Sacristán, and Michael Soss. Long proteins with unique optimal foldings in the H-P model. *Comput. Geom. Theory Appl.*, 25(1–2):139–159, May 2003. Cited on 161, 162, 164.

Oswin Aichholzer, Carmen Cortés, Erik D. Demaine, Vida Dujmović, Jeff Erickson, Henk Meijer, Mark Overmars, Belén Palop, Suneeta Ramaswami, and Godfried T. Toussaint. Flipturning polygons. *Discrete Comput. Geom.*, 28:231–253, 2002. Cited on 76, 80.

Oswin Aichholzer, Erik D. Demaine, Jeff Erickson, Ferran Hurtado, Mark Overmars, Michael Soss, and Godfried T. Toussaint. Reconfiguring convex polygons. *Comput. Geom. Theory Appl.*, 20:85–95, 2001. Cited on 72, 73.

Martin Aigner and Günter M. Ziegler. *Proofs from THE BOOK*. Springer-Verlag, Berlin, 1999. 2nd corrected printing. Cited on 144, 341.

Jin Akiyama and Gisaku Nakamura. Foldings of regular polygons to convex polyhedra I: Equilateral triangles. In *Proc. 2003 Indonesia–Japan Conf. Combinatorial Geom. Graph Theory*, volume 3330 of *Lecture Notes in Computer Science*, pages 34–43, Springer-Verlag, Berlin, 2005. Cited on 419.

Jin Akiyama, Takemasa Ooya, and Yuko Segawa. Wrapping a cube. *Teach. Math. Appl.*, 16(3):95–100, 1997. Cited on 238.

Rebecca Alexander, Heather Dyson, and Joseph O'Rourke. The convex polyhedra foldable from a square. In *Proc. 2002 Japan Conf. Discrete Comput. Geom.*, volume 2866 of *Lecture Notes in Computer Science*, pages 38–50, Springer-Verlag, Berlin, 2003. Cited on 414, 415, 417.

Alexandr D. Alexandrov. *Vupyklue Mnogogranniki*. Gosydarstvennoe Izdatelstvo Tehno-Teoreticheskoi Literaturu, 1950 [in Russian]. See Alexandrov (1958) for German translation and Alexandrov (2005) for English translation. Cited on 367.

Alexandr D. Alexandrov. *Die Innere Geometrie der Konvexen Flächen.* Math. Lehrbucher und Mono-graphien. Akademie-Verlag, Berlin, 1955. Cited on 366.

Alexandr D. Alexandrov. *Konvexe Polyeder*. Math. Lehrbucher und Monographien. Akademie-Verlag, Berlin, 1958. Translation of the 1950 Russian edition. Cited on 299, 354.

Alexandr D. Alexandrov. Existence of a convex polyhedron and a convex surface with a given metric. In Yu. G. Reshetnyak and S. S. Kutateladze, editors, *A. D. Alexandrov: Selected Works: Part I*, pages 169–173. Gordon and Breach, Australia, 1996. Translation of *Doklady Akad. Nauk SSSR, Matematika*, volume 30, No. 2, 103–106, 1941. Cited on 348, 350, 354.

Alexandr D. Alexandrov. *Convex Polyhedra*. Springer-Verlag, Berlin, 2005. Monographs in Mathematics. Translation of the 1950 Russian edition by N. S. Dairbekov, S. S. Kutateladze, and A. B. Sossinsky. Cited on 299, 345, 349, 350, 352, 358, 359, 361.

Alexandr D. Alexandrov. Intrinsic geometry of convex surfaces. In S. S. Kutateladze, editor, *A. D. Alexandrov: Selected Works: Part II*, pages 1–426. Chapman & Hall, Boca Raton, FL, 2006. Cited on 342, 354, 366.

Alexandr D. Alexandrov and Victor A. Zalgaller. *Intrinsic Geometry of Surfaces*. American Mathematical Society, Providence, RI, 1967. Cited on 373, 374.

Greg Aloupis. *Reconfigurations of Polygonal Structures*. PhD thesis, McGill University, School of Computer Science, 2005. Cited on 380.

Greg Aloupis, Erik D. Demaine, Vida Dujmović, Jeff Erickson, Stefan Langerman, Henk Meijer, Ileana Streinu, Joseph O'Rourke, Mark Overmars, Michael Soss, and Godfried T. Toussaint. Flat-state connectivity of linkages under dihedral motions. In *Proc. 13th Annu. Int. Symp. Algorithm Comput.*, volume 2518 of *Lecture Notes in Computer Science*, pages 369–380. Springer-Verlag, Berlin, 2002a. Cited on 137, 138, 142, 152.

Greg Aloupis, Erik D. Demaine, Stefan Langerman, Pat Morin, Joseph O'Rourke, Ileana Streinu, and Godfried Toussaint. Unfolding polyhedral bands. In *Proc. 16th Can. Conf. Comput. Geom.*, pages 60–63, 2004. Cited on 380.

Greg Aloupis, Erik D. Demaine, Henk Meijer, Joseph O'Rourke, Ileana Streinu, and Godfried T. Toussaint. Flat-state connectedness of fixed-angle chains: Special acute chains. In *Proc. 14th Can. Conf. Comput. Geom.*, pages 27–30, 2002b. Cited on 138.

Greg Aloupis and Henk Meijer. Reconfiguring planar dihedral chains. In *Proc. 22nd Eur. Workshop Comput. Geom.*, 2006. Cited on 138.

Roger C. Alperin. A mathematical theory of origami constructions and numbers. *New York J. Math.*, 6:119–133, 2000. Cited on 287, 289.

Roger C. Alperin. A grand tour of pedals of conics. *Forum Geometricorum*, 4:143–151, 2004. Cited on 285, 287.

Helmut Alt, Christian Knauer, Günter Rote, and Sue Whitesides. On the complexity of the linkage reconfiguration problem. In J. Pach, editor, *Towards a Theory of Geometric Graphs*, volume 342 of *Contemporary Mathematics*, pages 1–14. American Mathematical Society, New York, 2004. Cited on 18, 20, 22, 23, 24, 114.

Nancy M. Amato and Guang Song. Using motion planning to map protein folding landscapes and analyze protein folding kinetics of known native structures. Technical Report TR01-001, Depertment of Computer Science, Texas A&M University, 2001. Cited on 157.

Kenneth Appel and Wolfgang Haken. Every planar map is four colourable, Part I: discharging. *Illinois J. Math.*, 21:429–490, 1977. Cited on 31.

Esther M. Arkin, Michael A. Bender, Erik D. Demaine, Martin L. Demaine, Joseph S. B. Mitchell, Saurabh Sethia, and Steven S. Skiena. When can you fold a map? *Comput. Geom. Theory Appl.*, 29(1):166–195, 2004. Cited on 204, 224, 226, 230, 231.

Esther M. Arkin, Sándor P. Fekete, and Joseph S. B. Mitchell. An algorithmic study of manufacturing paperclips and other folded structures. *Comput. Geom. Theory Appl.*, 25:117–138, 2003. Cited on 14.

Esther M. Arkin, Martin Held, Joseph S. B. Mitchell, and Steven S. Skiena. Hamiltonian triangulations for fast rendering. *Visual Comput.*, 12(9):429–444, 1996. Cited on 235.

Boris Aronov, Jacob E. Goodman, and Richard Pollack. Convexification of planar polygons in $\mathbb{R}^3$. Manuscript, June 2002, http://math.nyu.edu/faculty/pollack/convexifyingapolygon10-27-99.ps. Cited on 83.

Boris Aronov and Joseph O'Rourke. Nonoverlap of the star unfolding. *Discrete Comput. Geom.*, 8:219–250, 1992. Cited on 326, 366, 367, 369, 371.

Andrej V. Astrelin and Idzhad Sabitov. A canonical polynomial for the volume of a polyhedron. *Russian Math. Surv.*, 54(2):430–431, 1999. Cited on 348.

Franz Aurenhammer, Friedrich Hoffmann, and Boris Aronov. Minkowski-type theorems and least-squares clustering. *Algorithmica*, 20:61–76, 1998. Cited on 341.

Mihai Bădoiu, Erik D. Demaine, MohammadTaghi Hajiaghayi, and Piotr Indyk. Low-dimensional embedding with extra information. *Discrete Comput. Geom.*, 36(4):609–632, 2006. Cited on 48, 49.

Devin J. Balkcom. *Robotic Origami Folding*. PhD thesis, Carnegie Mellon University, August 2004. Cited on 292.

Devin J. Balkcom, Erik D. Demaine, and Martin L. Demaine. Folding paper shopping bags. In *Proc. 14th Annu. Fall Workshop Comput. Geom.*, pages 14–15. MIT, Cambridge, MA, November 2004. Cited on 292, 293.

Devin J. Balkcom and Matthew T. Mason. Introducing robotic origami folding. In *IEEE Int. Conf. Robot. Autom.*, pages 3245–3250, 2004. Cited on 292.

Werner Ballmann, Gudlaugur Thorbergsson, and Wolfgang Ziller. On the existence of short closed geodesics and their stability properties. In *Seminar on Minimal Submanifolds*, pages 53–63. Princeton University Press, Princeton, NJ. 1983. Cited on 374.

Jérôme Barraquand, Lydia Kavraki, Jean-Claude Latombe, Tsai-Yen Li, Rajeev Motwani, and Prabhakar Raghavan. A random sampling scheme for path planning in large-dimensional configuration spaces. *Int. J. Robot. Res.*, 16(6):759–774, 1997. Cited on 157.

Jérôme Barraquand and Jean-Claude Latombe. A Monte-Carlo algorithm for path planning with many degrees of freedom. In *Proc. IEEE Int. Conf. Robot. Autom.*, pages 1712–1717, May 1990. Cited on 12.

John J. Bartholdi, III and Paul Goldsman. The vertex-adjacency dual of a triangulated irregular network has a Hamiltonian cycle. *Oper. Res. Lett.*, 32:304–308, 2004. Cited on 337.

Saugata Basu, Richard Pollack, and Marie-Françoise Roy. Computing roadmaps of semi-algebraic sets on a variety. *J. Am. Math. Soc.*, 13:55–82, 2000. Cited on 18, 19.

Alex Bateman. Computer tools and algorithms for origami tessellation design. In *Origami³: Proc. 3rd Int. Meeting Origami Sci., Math, Educ. (2001)*, pages 39–51. A K Peters, Wellesley MA., 2002. Cited on 293.

sarah-marie belcastro and Thomas C. Hull. A mathematical model for non-flat origami. In *Origami³: Proc. 3rd Int. Meeting Origami Sci., Math, Educ. (2001)*, pages 39–51, A K Peters, Wellesley, MA, 2002a. Cited on 210, 211, 212.

sarah-marie belcastro and Thomas C. Hull. Modelling the folding of paper into three dimensions using affine transformations. *Linear Algebra Appl.*, 348:273–282, June 2002b. Cited on 210, 211, 212.

Nadia Benbernou, Patricia Cahn, and Joseph O'Rourke. Unfolding smooth prismatoids. In *Proc. 14th Annu. Fall Workshop Comput. Geom.*, pages 12–13, November 2004, arXiv:cs.CG/0407063. Cited on 323.

Nadia Benbernou and Joseph O'Rourke. On the maximum span of fixed-angle chains. In *Proc. 18th Can. Conf. Comput. Geom.*, 2006. Cited on 135.

Bonnie Berger, Jon Kleinberg, and Tom Leighton. Reconstructing a three-dimensional model with arbitrary errors. *J. Assoc. Comput. Mach.*, 46(2):212–235, 1999. Cited on 48.

Bonnie Berger and Tom Leighton. Protein folding in the hydrophobic–hydrophilic (*HP*) model is NP-complete. *J. Comput. Biol.*, 5(1):27–40, 1998. Cited on 160.

Marcel Berger. *A Panoramic View of Riemannian Geometry*. Springer-Verlag, Berlin, 2003. Cited on 303, 373.

Marshall Bern, Erik D. Demaine, David Eppstein, and Barry Hayes. A disk-packing algorithm for an origami magic trick. In *Proc. Int. Conf. Fun Algorithms*, pages 32–42, June 1998. See also Bern et al. (2002). Cited on 255, 263, 266.

Marshall Bern, Erik D. Demaine, David Eppstein, and Barry Hayes. A disk-packing algorithm for an origami magic trick. In *Origami³: Proc. 3rd Int. Meeting Origami Sci., Math, Educ. (2001)*, pages 17–28, A K Peters, Wellesley, MA, 2002. Cited on 255, 263, 266.

Marshall Bern, Erik D. Demaine, David Eppstein, and Eric Kuo. Ununfoldable polyhedra. In *Proc. 11th Can. Conf. Comput. Geom.*, pages 13–16, 1999. See Bern et al. (2003). Cited on 318.

Marshall Bern, Erik D. Demaine, David Eppstein, Eric Kuo, Andrea Mantler, and Jack Snoeyink. Ununfoldable polyhedra with convex faces. *Comput. Geom. Theory Appl.*, 24(2):51–62, 2003. Cited on 311, 318, 321.

Marshall W. Bern and David Eppstein. Quadrilateral meshing by circle packing. *Int. J. Comput. Geom. Appl.*, 10(4):347–360, 2000. Cited on 266.

Marshall Bern and Barry Hayes. The complexity of flat origami. In *Proc. 7th ACM-SIAM Symp. Discrete Algorithms*, ACM/SIAM, New York/Philadelphia, pages 175–183, 1996. Cited on 169, 199, 207, 214, 216, 217, 222, 232.

Marshall W. Bern, Scott Mitchell, and Jim Ruppert. Linear-size nonobtuse triangulation of polygons. *Discrete Comput. Geom.*, 14:411–428, 1995. Cited on 266.

Dimitris Bertsimas and Santosh Vempala. Solving convex programs by random walks. *J. Assoc. Comput. Mach.*, 54(4):540–556, 2004. Cited on 105.

Therese Biedl. Polygons needing many flipturns. *Discrete Comput. Geom.*, 35(1):131–141, 2006. Cited on 81.

Therese Biedl, Timothy M. Chan, Erik D. Demaine, Martin L. Demaine, Paul Nijjar, Ryuhei Uehara, and Ming-wei Wang. Tighter bounds on the genus of nonorthogonal polyhedra built from rectangles. In *Proc. 14th Can. Conf. Comput. Geom.*, pages 105–108, August 2002a. Cited on 434, 435.

Therese Biedl, Erik D. Demaine, Martin L. Demaine, Sylvain Lazard, Anna Lubiw, Joseph O'Rourke, Mark Overmars, Steve Robbins, Ileana Streinu, Godfried T. Toussaint, and Sue Whitesides. Locked and unlocked polygonal chains in 3D. In *Proc. 10th ACM-SIAM Symp. Discrete Algorithms*, ACM/SIAM, New York/Philadelphia, pages 866–867, January 1999. See Biedl et al. (2001a). Cited on 82, 84, 88, 89, 90, 113.

Therese Biedl, Erik D. Demaine, Martin L. Demaine, Sylvain Lazard, Anna Lubiw, Joseph O'Rourke, Mark Overmars, Steve Robbins, Ileana Streinu, Godfried T. Toussaint, and Sue Whitesides. Locked and unlocked polygonal chains in three dimensions. *Discrete Comput. Geom.*, 26(3):269–282, 2001a. Cited on 74, 82, 84.

Therese Biedl, Erik D. Demaine, Martin L. Demaine, Sylvain Lazard, Anna Lubiw, Joseph O'Rourke, Steve Robbins, Ileana Streinu, Godfried T. Toussaint, and Sue Whitesides. On reconfiguring tree linkages: Trees can lock. *Discrete Appl. Math.*, 117(1–3):293–297, 2002b. Cited on 94, 114.

Therese Biedl, Erik D. Demaine, Martin L. Demaine, Anna Lubiw, Joseph O'Rourke, Mark Overmars, Steve Robbins, Ileana Streinu, Godfried T. Toussaint, and Sue Whitesides. On reconfiguring tree linkages: Trees can lock. In *Proc. 10th Can. Conf. Comput. Geom.*, pages 4–5, 1998a. See Biedl et al. (2002b). Cited on 88, 94, 95, 96.

Therese Biedl, Erik D. Demaine, Martin L. Demaine, Anna Lubiw, Joseph O'Rourke, Mark Overmars, Steve Robbins, and Sue Whitesides. Unfolding some classes of orthogonal polyhedra. In *Proc. 10th Can. Conf. Comput. Geom.*, pages 70–71, 1998b. Full version available at: http://cgm.cs.mcgill.ca/cccg98/proceedings/cccg98-biedl-unfolding.ps.gz. Cited on 309, 310, 329, 330.

Therese Biedl, Erik D. Demaine, Christian A. Duncan, Rudolf Fleischer, and Stephen G. Kobourov. Tight bounds on maximal and maximum matchings. In *Proc. 12th Annu. Int. Symp. Algorithms Comput.*, volume 2223 of *Lecture Notes in Computer Science*, pages 308–319. Springer-Verlag, Berlin, December 2001b. Cited on 308.

Therese Biedl, Michael Kaufmann, and Petra Mutzel. Drawing planar partitions II: HH-drawings. In *Proc. 24th Int. Workshop Graph-Theoret. Concepts Comput. Sci.*, volume 1517 of *Lecture Notes in Computer Science*, pages 124–136. Springer-Verlag, Berlin, 1998c. Cited on 336.

Therese Biedl, Anna Lubiw, and Julie Sun. When can a net fold to a polyhedron? *Comput. Geom. Theory Appl.*, 31(3):207–218, June 2005. Cited on 431, 432, 433, 434, 435, 436, 437.

George D. Birkhoff. *Dynamical Systems*, volume IX of *AMS Colloquium Publications*. American Mathematical Society, New York, 1927. Cited on 374.

Harry Blum. A transformation for extracting new descriptors of shape. In W. Wathen-Dunn, editor, *Models for the Perception of Speech and Visual Form*, pages 362–380. MIT Press, Cambridge, 1967. Cited on 370.

Lenore Blum, Felipe Cucker, Michael Shub, and Steve Smale. *Complexity and Real Computation*. Springer-Verlag, New York, 1998. Cited on 44.

Robert Bohlin and Lydia Kavraki. Path planning using lazy PRM. In *Proc. Int. Conf. Robot. Autom.*, volume 1, pages 521–528, 2000. Cited on 157.

Valérie Boor, Mark Overmars, and A. Frank van der Stappen. The Gaussian sampling strategy for probabilistic roadmap planners. In *Proc. IEEE Int. Conf. Robot. Autom.*, pages 1018–1023, 1999. Cited on 156.

Prosenjit Bose, Francisco Gómez, Pedro Ramos, and Godfried T. Toussaint. Drawing nice projections of objects in space. In *Proc. 3rd Int. Symp. Graph Drawing (1995)*, volume 1027 of *Lecture Notes in Computer Science*, pages 52–63. Springer-Verlag, Berlin, 1996. Cited on 122.

Robert Bowen and Stephen Fisk. Generations of triangulations of the sphere. *Math. Comput.*, 21:250–252, 1967. Cited on 407.

David Brill. Justin's origami trisection. *Br. Origami*, 107:14, 1984. Cited on 285.

Aren Brönsted. *An Introduction to Convex Polytopes*. Springer-Verlag, New York, 1983. Cited on 390.

Nicolaas G. de Bruijn. Problems 17 and 18. *Nieuw Arch. Wiskunde*, 2:67, 1954. Answers in *Wiskundige Opgaven met de oplossingen*, 20:19–20, 1955. Cited on 125.

Sergio Cabello, Erik D. Demaine, and Günter Rote. Planar embeddings of graphs with specified edge lengths. In *Proc. 11th Int. Symp. Graph Drawing (2003)*, volume 2912 of *Lecture Notes in Computer Science*, pages 283–294, Springer-Verlag, Berlin, 2004. Cited on 48.

Julio Gonzalez Cabillon. Re: [HM] Peaucellier. *Historia Matematica Mailing List Archive*, 1999. 4 February 1999, jgc@adinet.com.uy. Cited on 30.

Gruia Călinescu and Adrian Dumitrescu. The carpenter's ruler folding problem. In J. E. Goodman, J. Pach, and E. Welzl, editors, *Combinatorial and Computational Geometry*, volume 52 of *Mathematics Sciences Research Institute Publications*, pages 155–166. Cambridge University Press, Cambridge, UK, 2005. Cited on 26.

Jorge Alberto Calvo, Danny Krizanc, Pat Morin, Michael A. Soss, and Godfried T. Toussaint. Convexifying polygons with simple projections. *Inform. Process. Lett.*, 80(2):81–86, 2001. Cited on 119, 122.

John Canny. *The Complexity of Robot Motion Planning*. ACM–MIT Press Doctoral Dissertation Award Series. MIT Press, Cambridge, MA, 1987. Cited on 18, 19.

John Canny. Some algebraic and geometric computations in PSPACE. In *Proc. 20th Annu. ACM Symp. Theory Comput.*, ACM, New York, pages 460–467, 1988. Cited on 18, 19, 20.

Jason H. Cantarella, Erik D. Demaine, Hayley N. Iben, and James F. O'Brien. An energy-driven approach to linkage unfolding. In *Proc. 20th Annu. ACM Symp. Comput. Geom.*, ACM, New York, pages 134–143, June 2004. Cited on 88, 111, 112.

Jason Cantarella and Heather Johnston. Nontrivial embeddings of polygonal intervals and unknots in 3-space. *J. Knot Theory Ramifications*, 7:1027–1039, 1998. Cited on 88, 89, 90, 91, 154.

Steven Casey. Chessboard. In R. J. Lang, editor, *Proc. West Coast Origami Guild*, number 19, pages 3–12, August 1989. Cited on 238.

Michael L. Catalano-Johnson and Daniel Loeb. Problem 10716: A cubical gift. *Amer. Math. Mon.*, 108(1):81–82, January 2001. Posed in volume 106, 1999, page 167. Cited on 237, 238.

Augustin L. Cauchy. Sur les polygones et les polyèdres, seconde mémoire. *J. École Polytechnique*, XVIe Cahier, Tome IX:113–148, 1813. OEuvres Complètes, IIe Série, volume 1, pages 26–38. Paris, 1905. Cited on 341.

Hue Sun Chan and Ken A. Dill. The protein folding problem. *Physics Today*, pages 24–32, February 1993. Cited on 158, 159, 160.

Bernard Chazelle, Herbert Edelsbrunner, Michelangelo Grigni, Leonidas J. Guibas, John Hershberger, Micha Sharir, and Jack Snoeyink. Ray shooting in polygons using geodesic triangulations. *Algorithmica*, 12:54–68, 1994. Cited on 108.

Jindong Chen and Yijie Han. Shortest paths on a polyhedron. *Int. J. Comput. Geom. Appl.*, 6:127–144, 1996. Cited on 360, 364, 366.

Shiing-Shen Chern. Curves and surfaces in Euclidean space. In S. S. Chern, editor, *Studies in Global Geometry and Analysis*, volume 4 of *Studies in Mathematics*, pages 16–56. Mathematical Association of America, Washington, DC, 1967. Cited on 147.

Barry Cipra. Collaboration closes in on closed geodesics. *What's Happening in the Mathematical Sciences*, 1:27–30, 1993. Cited on 374.

Barry Cipra. In the fold: Origami meets mathematics. *SIAM News*, 34(8):200–201, October 2001. Cited on 169.

Roxana Cocan and Joseph O'Rourke. Polygonal chains cannot lock in 4D. In *Proc. 11th Can. Conf. Comput. Geom.*, pages 5–8, 1999. See Cocan and O'Rourke (2001). Cited on 88, 92.

Roxana Cocan and Joseph O'Rourke. Polygonal chains cannot lock in 4D. *Comput. Geom. Theory Appl.*, 20:105–129, 2001. Cited on 88, 92, 93, 94.

Robert Connelly. A flexible sphere. *Math. Intelligencer*, 1:130–131, 1978. Cited on 347.

Robert Connelly. Flexing surfaces. In D. A. Klarner, editor, *The Mathematical Gardner*, pages 79–89. Wadsworth, 1981. Cited on 347.

Robert Connelly. Rigidity and energy. *Invent. Math.*, 66:11–33, 1982. Cited on 116.

Robert Connelly. On generic global rigidity. In P. Gritzman and B. Sturmfels, editors, *Applied Geometry and Discrete Mathematics: The Victor Klee Festschrift*, volume 4 of *DIMACS Series in Discrete Mathematics and Theoretical Computer Science*, pages 147–155, AMS Press, 1991. Cited on 49.

Robert Connelly. Rigidity. In *Handbook of Convex Geometry*, volume A, pages 223–271. North-Holland, Amsterdam, 1993. Cited on 43, 292, 342.

Robert Connelly. Generic global rigidity. *Discrete Comput. Geom.*, 33:549–563, 2005. Cited on 48, 49.

Robert Connelly and Erik D. Demaine. Geometry and topology of polygonal linkages. In J. E. Goodman and J. O'Rourke, editors, *Handbook of Discrete and Computational Geometry*, 2nd edition, chapter 9, pages 197–218. CRC Press LLC, Boca Raton, FL, 2004. Cited on 18, 28, 31, 113.

Robert Connelly, Erik D. Demaine, Martin L. Demaine, Sándor Fekete, Stefan Langerman, Joseph S. B. Mitchell, Ares Ribó, and Günter Rote. Locked and unlocked chains of planar shapes. In *Proc. 22nd Annu. ACM Symp. Comput. Geom.*, ACM, New York, pages 61–70, 2006. Cited on 119.

Robert Connelly, Erik D. Demaine, and Günter Rote. Every polygon can be untangled. In *Proc. 16th Eur. Workshop Comput. Geom.*, pages 62–65. Ben-Gurion University, the Negev, Israel, January 2000a. Cited on 88.

Robert Connelly, Erik D. Demaine, and Günter Rote. Straightening polygonal arcs and convexifying polygonal cycles. In *Proc. 41st Annu. IEEE Symp. Found. Comput. Sci.*, IEEE, Los Alamitos, pages 432–442, November 2000b. Cited on 88, 96.

Robert Connelly, Erik D. Demaine, and Günter Rote. Infinitesimally locked self-touching linkages with applications to locked trees. In J. Calvo, K. Millett, and E. Rawdon, editors, *Physical Knots: Knotting, Linking, and Folding of Geometric Objects in* $\mathbb{R}^3$, pages 287–311. American Mathematical Society, Providence, RI, 2002a. Cited on 95, 113, 114, 115, 116, 117, 118.

Robert Connelly, Erik D. Demaine, and Günter Rote. Straightening polygonal arcs and convexifying polygonal cycles. Technical Report B02-02, Freie Universität Berlin, 2002b. Cited on 58, 104, 109.

Robert Connelly, Erik D. Demaine, and Günter Rote. Straightening polygonal arcs and convexifying polygonal cycles. *Discrete Comput. Geom.*, 30(2):205–239, September 2003. Cited on 50, 87, 96, 97, 98, 99, 100, 104, 105, 106, 112.

Robert Connelly, Idzhad Sabitov, and Anke Walz. The bellows conjecture. *Beiträge Algebra Geom.*, 38:1–10, 1997. Cited on 348.

Robert Connelly and Herman Servatius. Higher-order rigidity – What is the proper definition? *Discrete Comput. Geom.*, 11(2):193–200, 1994. Cited on 52.

Robert Connelly and Walter Whiteley. Second-order rigidity and pre-stress stability for tensegrity frameworks. *SIAM J. Discrete Math.*, 9(3):453–491, August 1996. Cited on 52, 53, 56.

H. S. M. Coxeter. *Regular Polytopes*, 2nd edition. Dover, New York, 1973. Cited on 329.

H. S. M. Coxeter and Samuel L. Greitzer. *Geometry Revisited*. Mathematical Association of America, Washington, DC, 1967. Cited on 30.

Thomas E. Creighton, editor. *Protein Folding*. W. H. Freeman & Co., June 1992. Cited on 148.

Luigi Cremona. *Graphical Statics (English translation)*. Oxford University Press, Oxford, UK, 1890. Cited on 58.

Pierluigi Crescenzi, Deborah Goldman, Christos H. Papadimitriou, Antonio Piccolboni, and Mihalis Yannakakis. On the complexity of protein folding. *J. Comput. Biol.*, 5(3):423–466, 1998. Cited on 159, 160.

Christopher Croke. Poincaré's problem and the length of the shortest closed geodesic on a convex hypersurface. *J. Diff. Geom.*, 17(4):595–634, 1982. Cited on 375.

Peter Cromwell. *Polyhedra*. Cambridge University Press, Cambridge, UK, 1997. Cited on 143, 341, 346, 348.

Artur Czumaj and Willy-B. Strothmann. Bounded degree spanning trees. In *Proc. 5th Annu. Eur. Symp. Algorithms*, volume 1284 of *Lecture Notes in Computer Science*, pages 104–117, Springer-Verlag, Berlin, 1997. Cited on 308.

Mirela Damian, Robin Flatland, Henk Meijer, and Joseph O'Rourke. Unfolding well-separated orthotrees. In *15th Annu. Fall Workshop Comput. Geom.*, pages 23–25, November 2005a. Cited on 330.

Mirela Damian, Robin Flatland, and Joseph O'Rourke. Unfolding Manhattan towers. In *Proc. 17th Can. Conf. Comput. Geom.*, pages 204–207, 2005b. Cited on 331, 332.

Mirela Damian, Robin Flatland, and Joseph O'Rourke. Grid vertex-unfolding orthogonal polyhedra. In *Proc. 23rd Symp. Theoret. Aspects Comput. Sci.*, volume 3884 of *Lecture Notes in Computer Science*, pages 264–276. Springer-Verlag, Berlin, 2006. Cited on 330, 338.

Mirela Damian, Robin Flatland, and Joseph O'Rourke. Epsilon-unfolding orthogonal polyhedra. *Graphs Combinatorics*, to appear, 2007. Cited on 331, 332.

Mirela Damian and Henk Meijer. Grid edge-unfolding orthostacks with orthogonally convex slabs. In *14th Annu. Fall Workshop Comput. Geom.*, pages 20–21, November 2004. Cited on 331.

Robert J. Dawson. On removing a ball without disturbing the others. *Math. Mag.*, 57:27–30, 1984. Cited on 125.

Erik D. Demaine. *Folding and Unfolding*. PhD thesis, Department of Computer Science, University of Waterloo, 2001. Cited on 154.

Erik D. Demaine and Martin L. Demaine. Computing extreme origami bases. Technical Report CS-97-22, Department of Computer Science, University of Waterloo, May 1997. Cited on 94.

Erik D. Demaine and Martin L. Demaine. The power of paper folding. Unpublished manuscript, 2000. Cited on 289.

Erik D. Demaine and Martin L. Demaine. Recent results in computational origami. In *Origami³: Proc. 3rd Int. Meeting Origami Science, Math, Educ. (2001)*, pages 3–16, A K Peters, Wellesley, MA, 2002. Cited on 169.

Erik D. Demaine, Martin L. Demaine, and Anna Lubiw. Folding and one straight cut suffice. In *Proc. 10th Annu. ACM-SIAM Symp. Discrete Algorithm*, ACM/SIAM, New York/Philadelphia, pages 891–892, January 1999a. Cited on 255.

Erik D. Demaine, Martin L. Demaine, and Anna Lubiw. Polyhedral sculptures with hyperbolic paraboloids. In *Proc. 2nd Annu. Conf. BRIDGES: Math. Connections Art, Music, Sci.*, Winfield, KS, pages 91–100, 1999b. Cited on 294.

Erik D. Demaine, Martin L. Demaine, and Anna Lubiw. Folding and cutting paper. In *Proc. 1998 Japan Conf. Discrete Comput. Geom.*, volume 1763 of *Lecture Notes in Computer Science*, pages 104–117. Springer-Verlag, Berlin, 2000a. Cited on 255, 257, 259.

Erik D. Demaine, Martin L. Demaine, and Anna Lubiw. Flattening polyhedra. Unpublished manuscript, 2001a. Cited on 279, 282, 283, 438.

Erik D. Demaine, Martin L. Demaine, Anna Lubiw, and Joseph O'Rourke. Examples, counterexamples, and enumeration results for foldings and unfoldings between polygons and polytopes. Technical Report 069, Smith College, Northampton, MA, July 2000b. arXiv:cs.CG/0007019. Cited on 329, 383, 384, 385, 412, 426, 427.

Erik D. Demaine, Martin L. Demaine, Anna Lubiw, and Joseph O'Rourke. Enumerating foldings and unfoldings between polygons and polytopes. *Graphs Comb.*, 18(1):93–104, 2002a. See also Demaine et al. (2000b). Cited on 402.

Erik D. Demaine, Martin L. Demaine, Anna Lubiw, Joseph O'Rourke, and Irena Pashchenko. Metamorphosis of the cube. In *Proc. 15th Annu. ACM Symp. Comput. Geom.*, ACM, New York, pages 409–410, 1999c. Video and abstract. Cited on 391, 392.

Erik D. Demaine, Martin L. Demaine, and Joseph S. B. Mitchell. Folding flat silhouettes and wrapping polyhedral packages: New results in computational origami. *Comput. Geom. Theory Appl.*, 16(1):3–21, 2000c. Cited on 169, 232, 236, 237.

Erik D. Demaine, Satyan L. Devadoss, Joseph S. B. Mitchell, and Joseph O'Rourke. Continuous foldability of polygonal paper. In *Proc. 16th Can. Conf. Comput. Geom.*, pages 64–67, August 2004. Cited on 172.

Erik D. Demaine, Satyan L. Devadoss, Joseph S. B. Mitchell, and Joseph O'Rourke. Continuous foldability of polygonal paper. Unpublished manuscript, 2006a. Cited on 172, 187, 437.

Erik D. Demaine, David Eppstein, Jeff Erickson, George W. Hart, and Joseph O'Rourke. Vertex-unfoldings of simplicial manifolds. Technical Report 072, Department of Computer Science, Smith College, Northampton, MA, October 2001b. arXiv:cs.CG/0110054. Cited on 335.

Erik D. Demaine, David Eppstein, Jeff Erickson, George W. Hart, and Joseph O'Rourke. Vertex-unfoldings of simplicial manifolds. In A. Bezdek, editor, *Discrete Geometry*, pages 215–228. Marcel Dekker, 2003a. Cited on 334, 337, 338.

Erik D. Demaine, Blaise Gassend, Joseph O'Rourke, and Godfried. T. Toussaint. Polygons flip finitely: Flaws and a fix. In *Proc. 18th Can. Conf. Comput. Geom.*, 2006b. Cited on 76.

Erik D. Demaine, John Iacono, and Stefan Langerman. Grid vertex-unfolding of orthostacks. In *Proc. 2004 Japan Conf. Discrete Comput. Geom.*, volume 3742 of *Lecture Notes in Computer Science*, pages 76–82. Springer-Verlag, Berlin, November 2005. Cited on 330, 338.

Erik D. Demaine, Stefan Langerman, and Joseph O'Rourke. Short interlocked linkages. In *Proc. 13th Can. Conf. Comput. Geom.*, pages 69–72, August 2001c. Cited on 123.

Erik D. Demaine, Stefan Langerman, and Joseph O'Rourke. Geometric restrictions on polygonal protein chain production. *Algorithmica*, 44(2):167–181, February 2006c. Cited on 148, 150.

Erik D. Demaine, Stefan Langerman, Joseph O'Rourke, and Jack Snoeyink. Interlocked open linkages with few joints. In *Proc. 18th Annu. ACM Symp. Comput. Geom.*, ACM, New York, pages 189–198, June 2002b. Cited on 123, 125, 126, 129, 130.

Erik D. Demaine, Stefan Langerman, Joseph O'Rourke, and Jack Snoeyink. Interlocked closed and open linkages with few joints. *Comp. Geom. Theory Appl.*, 26(1):37–45, 2003b. Cited on 123, 124, 126, 127, 129.

Erik D. Demaine and Joseph O'Rourke. Open problems from CCCG'99. In *Proc. 12th Can. Conf. Comput. Geom.*, pages 269–272, August 2000. Cited on 123.

Erik D. Demaine and Joseph O'Rourke. Open problems from CCCG 2002. In *Proc. 15th Can. Conf. Comput. Geom.*, pages 178–181, 2003. arXiv:cs/0212050. Cited on 330, 434, 440.

Erik D. Demaine and Joseph O'Rourke. Open problems from CCCG 2003. In *Proc. 16th Can. Conf. Comput. Geom.*, pages 209–211, 2004. Cited on 308.

Erik D. Demaine and Joseph O'Rourke. Open problems from CCCG 2004. In *Proc. 17th Can. Conf. Comput. Geom.*, pages 303–306, 2005. Cited on 332.

Erik D. Demaine and Joseph O'Rourke. Open problems from CCCG 2005. In *Proc. 18th Can. Conf. Comput. Geom.*, pages 75–80, 2006. Cited on 418.

Julie DiBiase. *Polytope Unfolding*. Undergraduate thesis, Smith College, 1990. Cited on 312, 321.

Edgar W. Dijkstra. A note on two problems in connexion with graphs. *Numer. Math.*, 1:269–271, 1959. Cited on 362.

Ken A. Dill. Dominant forces in protein folding. *Biochemistry*, 29(31):7133–7155, August 1990. Cited on 158.

Nikolai P. Dolbilin. Rigidity of convex polyhedrons. *Quantum*, 9(1):8–13, 1998. Cited on 299.

Melody Donoso and Joseph O'Rourke. Nonorthogonal polyhedra built from rectangles. Technical Report 073, Department of Computer Science, Smith College, Northampton, MA, October 2001. arXiv:cs/0110059. Cited on 433.

Melody Donoso and Joseph O'Rourke. Nonorthogonal polyhedra built from rectangles. In *Proc. 14th Can. Conf. Comput. Geom.*, pages 101–104, August 2002. Cited on 433, 434, 435.

David Dureisseix. Chessboard. *Br. Origami*, 201:20–24, April 2000. Cited on 238, 239.

Albrecht Dürer. *The Painter's Manual: A Manual of Measurement of Lines, Areas, and Solids by Means of Compass and Ruler Assembled by Albrecht Dürer for the Use of all Lovers of Art with Appropriate Illustrations Arranged to be Printed in the Year MDXXV.* Abaris Books, New York, 1977 (1525). English translation by Walter L. Strauss of "Unterweysung der Messung mit dem Zirkel un Richtscheyt in Linien Ebnen uhnd Gantzen Corporen." Cited on 300.

John W. Emert, Kay I. Meeks, and Roger B. Nelsen. Reflections on a Mira. *Amer. Math. Mon.*, 101(6):544–549, June–July 1994. Cited on 287, 289, 290.

David Eppstein. Faster circle packing with application to non-obtuse triangulation. *Int. J. Comput. Geom. Appl.*, 7:485–492, 1997. Cited on 266.

Paul Erdős. Problem 3763. *Am. Math. Mon.*, 42:627, 1935. Cited on 74.

Euclid. *Elements*. Dover, 1956. Translated by Sir Thomas L. Heath. Cited on 341.

Hazel Everett, Sylvain Lazard, Steve Robbins, Heiko Schröder, and Sue Whitesides. Convexifying star-shaped polygons. In *Proc. 10th Can. Conf. Comput. Geom.*, pages 2–3, 1998. Cited on 88.

Michael Farber. Instabilities of robot motion, arXiv:cs/0205015, 2002. Cited on 63.

Maksym Fedorchuk and Igor Pak. Rigidity and polynomial invariants of convex polytopes. *Duke Math. J.*, 129:371–404, 2005. Cited on 354, 356, 357.

Eugene S. Fergusson. Kinematics of mechanisms from the time of Watt. In *U.S. National Museum Bulletin*, volume 228, pages 185–230. Smithsonian Institute, 1962. Cited on 30.

Thomas Fevens, Antonio Hernandez, Antonio Mesa, Patrick Morin, Michael Soss, and Godfried T. Toussaint. Simple polygons with an infinite sequence of deflations. *Beiträge Algebra Geom.*, 42(2):307–311, 2001. Cited on 78.

Maxim D. Frank-Kamenetskii. *Unravelling DNA*. Addison-Wesley, Reading, MA, 1997. Cited on 131.

Greg N. Frederickson. *Dissections: Plane and Fancy*. Cambridge University Press, Cambridge, UK, 1997. Cited on 372, 423.

Greg N. Frederickson. *Hinged Dissections: Swinging and Twisting*. Cambridge University Press, Cambridge, UK, 2002. Cited on 423.

Greg N. Frederickson. *Piano-Hinged Dissections: Time to Fold*. A K Peters, Wellesley, MA, 2006. Cited on 423.

Erich Friedman. Circles in squares. http://www.stetson.edu/~efriedma/cirinsqu/, 2002. Cited on 246.

Hal Gabow and Robert Tarjan. A linear time algorithm for a special case of disjoint set union. *J. Comput. Syst. Sci.*, 30:209–221, 1985. Cited on 216.

Harold N. Gabow and Herbert H. Westermann. Forests, frames, and games: Algorithms for matroid sums and applications. *Algorithmica*, 7(5–6):465–497, 1992. Cited on 46.

Xiao-Shan Gao, Chang-Cai Zhu, Shang-Ching Chou, and Jian-Xin Ge. Automated generation of Kempe linkages for algebraic curves and surfaces. *Mech. Mach. Theory*, 36(9):1019–1033, 2002. Cited on 39.

Martin Gardner. The cocktail cherry and other problems. In *Mathematical Magic Show*, chapter 5, pages 66–81. Mathematical Association of America, Washington, DC, 1990. Cited on 232, 237.

Martin Gardner. Paper cutting. In *New Mathematical Diversions (Revised Edition)*, chapter 5, pages 58–69. Mathematical Association of America, Washington, DC, 1995a. Appeared in *Sci. Am.*, 1960. Cited on 232, 254.

Richard J. Gardner. *Geometric Tomography*. Cambridge University Press, Cambridge, UK, 1995b. Cited on 339.

Robert Geretschläger. Euclidean constructions and the geometry of origami. *Math. Mag.*, 68(5):357–371, 1995. Cited on 285, 287, 289.

Julie Glass, Stefan Langerman, Joseph O'Rourke, Jack Snoeyink, and Jianyuan K. Zhong. A 2-chain can interlock with a k-chain. In *Proc. 14th Annu. Fall Workshop Comput. Geom.*, pages 18–19, Cambridge, MA, 2004. Cited on 126.

Julie Glass, Bin Lu, Joseph O'Rourke, and Jianyuan K. Zhong. A 2-chain can interlock with an open 11-chain. *Geombinatorics*, 15(4):166–176, 2006. Cited on 126.

Andrew M. Gleason. Angle trisection, the heptagon, and the triskaidecagon. *Am. Math. Mon.*, 95(3):185–194, March 1988. Cited on 289, 290.

Herman Gluck. Almost all simply-connected closed surfaces are rigid. In *Geometric Topology*, volume 438 of *Lecture Notes in Mathematics*, pages 225–239. Springer-Verlag, Berlin, 1976. Cited on 347.

Jack E. Graver. *Counting on Frameworks: Mathematics to Aid the Design of Rigid Structures*. Mathematical Association of America, Washington, DC, 2001. Cited on 29, 43, 45, 48, 51, 52.

Jack Graver, Brigitte Servatius, and Herman Servatius. *Combinatorial Rigidity*. American Mathematical Society, Providence, RI, 1993. Cited on 43.

Peter Gritzmann and Alexander Hufnagel. A polynomial time algorithm for Minkowski reconstruction. In *Proc. 11th Annu. ACM Symp. Comput. Geom.*, ACM, New York, pages 1–9, 1995. Cited on 340.

Martin Grötschel, Lalzo Lovász, and Alexander Schrijver. *Geometric Algorithms and Combinatorial Optimization*, volume 2 of *Algorithms and Combinatorics*, 2nd edition. Springer-Verlag, Berlin, 1993. Cited on 105.

Branko Grünbaum. *Convex Polytopes*. John Wiley & Sons, New York, 1967. See Grünbaum (2003b). Cited on 340.

Branko Grünbaum. Nets of polyhedra II. *Geombinatorics*, 1:5–10, 1991. Cited on 312.

Branko Grünbaum. How to convexify a polygon. *Geombinatorics*, 5:24–30, July 1995. Cited on 75, 76.

Branko Grünbaum. A starshaped polyhedron with no net. *Geombinatorics*, 11:43–48, 2001. Cited on 318.

Branko Grünbaum. No-net polyhedra. *Geombinatorics*, 12:111–114, 2002. Cited on 318.

Branko Grünbaum. Are your polyhedra the same as my polyhedra? In B. Aronov, S. Basu, J. Pach, and M. Sharir, editors, *Discrete and Computational Geometry: The Goodman–Pollack Festschrift*, pages 461–488. Springer, New York, 2003a. Cited on 310.

Branko Grünbaum. *Convex Polytopes*, 2nd edition. Springer, New York, 2003b. Cited on 310.

Branko Grünbaum and Geoffrey C. Shephard. *Tilings and Patterns*. W. H. Freeman, New York, 1987. Cited on 293.

Branko Grünbaum and Joseph Zaks. Convexification of polygons by flips and by flipturns. Technical Report 6/4/98, Department of Mathematics, University of Washington, Seattle, 1998. Cited on 80.

Satyandra K. Gupta. Sheet metal bending operation planning: Using virtual node generation to improve search efficiency. *J. Manuf. Syst.*, 18(2):127–139, 1999. Cited on 14.

Mick Guy and Dave Venables. *The Chess Sets of Martin Wall, Max Hulme and Neal Elias*, Booklet 7. British Origami Society, 1979. 1st revision. Cited on 238, 239.

Ruth Haas, David Orden, Günter Rote, Francisco Santos, Brigitte Servatius, Herman Servatius, Diane Souvaine, Ileana Streinu, and Walter Whiteley. Planar minimally rigid graphs and pseudo-triangulations. *Comput. Geom. Theory Appl.*, 31(1–2):31–61, May 2005. Cited on 46.

Dan Halperin, Jean-Claude Latombe, and Randall H. Wilson. A general framework for assembly planning: The motion space approach. In *Proc. 14th Annu. ACM Symp. Comput. Geom.*, ACM, New York, pages 9–18, 1998. Cited on 20.

Frank Harary. *Graph Theory*. Addison-Wesley, Reading, MA, 1972. Cited on 325.

Harry Hart. On certain conversions of motion. *Cambridge Messenger of Mathematics*, iv:82–88, 116–120, 1875. Cited on 40.

William E. Hart and Sorin Istrail. Fast protein folding in the hydrophobic–hydrophilic model within three-eighths of optimal. *J. Comput. Biol.*, 3(1):53–96, Spring 1996. Cited on 160, 161.

Koshiro Hatori. Origami construction. http://www.jade.dti.ne.jp/~hatori/library/conste.html, 2004. Cited on 288.

Brian Hayes. Proteins. *Am. Scientist*, 86:216–221, 1998. Cited on 158, 162.

Siqian He and Harold A. Scheraga. Macromolecular conformational dynamics in torsional angle space. *J. Chem. Phys.*, 108(1):271–286, 1998. Cited on 155.

Bruce Hendrickson. Conditions for unique graph realizations. *SIAM J. Comput.*, 21(1):65–84, 1992. Cited on 46, 49.

Lebrecht Henneberg. *Die graphische Statik der starren Systeme*. B. G. Teubner, 1911. Cited on 47.

David Hilbert and Stefan Cohn-Vossen. *Geometry and the Imagination*. Chelsea Publishing Company, New York, 1952. Cited on 359, 372, 374.

A. E. Hochstein. Trisection of an angle by optical means. *Math. Teacher*, pages 522–524, January 1963. Cited on 289.

John E. Hopcroft, Deborah A. Joseph, and Sue Whitesides. Movement problems for 2-dimensional linkages. *SIAM J. Comput.*, 13:610–629, 1984. Cited on 22, 23, 35, 36.

John E. Hopcroft, Deborah A. Joseph, and Sue Whitesides. On the movement of robot arms in 2-dimensional bounded regions. *SIAM J. Comput.*, 14:315–333, 1985. Cited on 22, 23, 25, 26, 61, 67, 68.

John E. Hopcroft, Jack T. Schwartz, and Micha Sharir. *Planning, Geometry, and Complexity of Robot Motion*. Ablex Publishing, Norwood, NJ, 1987. Cited on 17.

Harry Houdini. *Paper Magic*, pages 176–177. E. P. Dutton & Company, 1922. Reprinted by Magico Magazine. Cited on 254.

David Hsu, Lydia E. Kavraki, Jean-Claude Latombe, Rajeev Motwani, and Stephen Sorkin. On finding narrow passages with probabilistic roadmap planners. In *Proc. 3rd Workshop Algorithmic Found. Robot.*, pages 141–153. A K Peters, Wellesley, MA, 1998. Cited on 157.

David A. Huffman. Curvature and creases: A primer on paper. *IEEE Trans. Comput.*, C-25:1010–1019, 1976. Cited on 296.

Thomas Hull. On the mathematics of flat origamis. *Congr. Numer.*, 100:215–224, 1994. Cited on 169, 199.

Thomas Hull. A note on "impossible" paper folding. *Am. Math. Mon.*, 103(3):240–241, March 1996. Cited on 285, 289.

Thomas Hull. The combinatorics of flat folds: A survey. In *Origami³: Proc. 3rd Int. Meeting Origami Sci., Math, Educ. (2001)*, pages 29–38, A K Peters, Wellesley, MA, 2002. Cited on 169, 199.

Thomas Hull. Counting mountain–valley assignments for flat folds. *Ars Comb.*, 67:175–187, April 2003a. Cited on 169, 199, 208.

Thomas Hull. Origami and geometric constructions. http://www.merrimack.edu/~thull/omfiles/geoconst.html, 2003b. Cited on 288.

Thomas Hull. *Project Origami: Activities for Exploring Mathematics*. A K Peters, Wellesley, MA, 2006. Cited on 169, 292, 293, 294.

Kenneth H. Hunt. *Kinematic Geometry of Mechanisms*. Oxford University Press, Oxford, UK, 1978. Cited on 12, 30.

Koji M. Husimi. Science of origami. Supplement to *Saiensu* (the Japanese edition of *Sci. Am.*), page 8, October 1980. Cited on 285.

Humiaki Huzita. Right angle billiard games and their solutions by folding paper. In K. Miura, editor, *Proc. 2nd Int. Meeting Origami Sci. Scientific Origami*, Seian University of Art and Design, Otsu, Japan, November–December 1994. Cited on 289.

Humiaki Huzita and Benedetto Scimemi. The algebra of paper folding (origami). In H. Huzita, editor, *Proc. 1st Int. Meeting Origami Sci. Tech.*, pages 215–222, Ferrara, Italy, December 1989. Cited on 285, 289.

Yong K. Hwang and Narendra Ahuja. Gross motion planning – A survey. *ACM Comput. Surv.*, 24(3):219–291, 1992. Cited on 18.

Hayley N. Iben, James F. O'Brien, and Erik D. Demaine. Refolding planar polygons. In *Proc. 22nd Annu. ACM Symp. Comput. Geom.*, ACM, New York, pages 71–79, 2006. Cited on 113.

H. Imai. On combinatorial structures of line drawings of polyhedra. *Discrete Appl. Math.*, 10:79–92, 1985. Cited on 46.

Takayui Ishii. *One Thousand Paper Cranes: The Story of Sadako and the Children's Peace Statu*. Dell, 1997. Cited on 168.

Sorin Istrail, Russell Schwartz, and Jonathan King. Lattice simulations of aggregation funnels for protein folding. *J. Comput. Biol.*, 6(2):143–162, 1999. Cited on 159.

Bill Jackson and Tibor Jordán. Connected rigidity matroids and unique realizations of graphs. *J. Combin. Theory Series B*, 94(1):1–94, 2005. Cited on 48, 49.

Donald J. Jacobs and Bruce Hendrickson. An algorithm for two-dimensional rigidity percolation: The pebble game. *J. Comput. Phys.*, 137(2):346–365, 1997. Cited on 46.

Denis Jordan and Marcel Steiner. Configuration spaces of mechanical linkages. *Discrete Comput. Geom.*, 22:297–315, 1999. Cited on 27.

Deborah A. Joseph and William H. Plantinga. On the complexity of reachability and motion planning questions. In *Proc. 1st Annu. ACM Symp. Comput. Geom.*, ACM, New York, pages 62–66, 1985. Cited on 22, 23, 25.

Jacques Justin. Aspects mathematiques du pliage de papier (Mathematical aspects of paper folding). In H. Huzita, editor, *Proc. 1st Int. Meeting Origami Sci. Tech.*, pages 263–277. Ferrara, Italy, December 1989a. Cited on 199.

Jacques Justin. Resolution par le plaige de l'equation du troisieme degre et applications geometriques. In H. Huzita, editor, *Proc. 1st Int. Meeting Origami Sci. Tech.*, pages 251–261. Ferrara, Italy, December 1989b. Cited on 285, 287, 289.

Jacques Justin. Towards a mathematical theory of origami. In K. Miura, editor, *Proc. 2nd Int. Meeting Origami Sci. Scientific Origami*, pages 15–29, Seian University of Art and Design, Otsu, Japan, November–December 1994. Cited on 173, 199, 204.

Biliana Kaneva and Joseph O'Rourke. An implementation of Chen & Han's shortest paths algorithm. In *Proc. 12th Can. Conf. Comput. Geom.*, pages 139–146, August 2000. Cited on 351, 363, 364, 365, 371.

Vitit Kantabutra. Reaching a point with an unanchored robot arm in a square. *Int. J. Comput. Geom. Appl.*, 7:539–550, 1997. Cited on 69.

Vitit Kantabutra and S. Rao Kosaraju. New algorithms for multilink robot arms. *J. Comput. Syst. Sci.*, 32:136–153, February 1986. Cited on 68.

Michael Kapovich and John J. Millson. On the moduli spaces of polygons in the Euclidean plane. *J. Diff. Geom.*, 42(1):133–164, 1995. Cited on 64.

Michael Kapovich and John J. Millson. Universality theorems for configuration spaces of planar linkages. *Topology*, 41(6):1051–1107, 2002. Cited on 27, 31.

Hermann Karcher. Remarks on polyhedra with given dihedral angles. *Comm. Pure Appl. Math.*, 21:169–174, 1968. Cited on 340.

Kunihiko Kasahara and Toshie Takahama. *Origami for the Connoisseur*. Japan Publications, Inc., 1987. Cited on 199, 203.

Lydia Kavraki, Petr Švestka, Jean-Claude Latombe, and Mark Overmars. Probabilistic roadmaps for path planning in high dimensional configuration spaces. *IEEE Trans. Robot. Autom.*, 12:566–580, 1996. Cited on 155, 156.

Toshikazu Kawasaki. On high dimensional flat origamis. In H. Huzita, editor, *Proc. 1st Int. Meeting Origami Sci. Tech.*, pages 131–142. Ferrara, Italy, December 1989a. Cited on 437.

Toshikazu Kawasaki. On the relation between mountain-creases and valley-creases of a flat origami. In H. Huzita, editor, *Proc. 1st Int. Meeting Origami Sci. Tech.*, pages 229–237. Ferrara, Italy, December 1989b. Unabridged Japanese version in *Sasebo College Tech. Rep.*, 27:153–157, 1990. Cited on 199, 204.

Toshikazu Kawasaki. *Roses, Origami & Math*. Japan Publications Trading Co., 2005. Cited on 199.

Yan Ke and Joseph O'Rourke. Lower bounds on moving a ladder in two and three dimensions. *Discrete Comput. Geom.*, 3:197–217, 1988. Cited on 20.

Alfred Bray Kempe. On a general method of describing plane curves of the nth degree by linkwork. *Proc. London Math. Soc.*, 7:213–216, 1876. Cited on 24, 31.

Alfred Bray Kempe. *How to Draw a Straight Line: A Lecture on Linkages*. Macmillan & Co., 1877. Cited on 29, 30, 33.

Lutz Kettner, David Kirkpatrick, Andrea Mantler, Jack Snoeyink, Bettina Speckmann, and Fumihiko Takeuchi. Tight degree bounds for pseudo-triangulations of points. *Comput. Geom. Theory Appl.*, 25(1–2):3–12, May 2003. Cited on 108.

Kyoung K. Kim, David Bourne, Satyandra Gupta, and S. S. Krishna. Automated process planning for sheet metal bending operations. *J. Manuf. Syst.*, 17(5):338–360, 1998. Cited on 14, 306.

Henry C. King. Planar linkages and algebraic sets. *Turkish J. Math.*, 23(1):33–56, 1999. Cited on 31, 37, 39.

Rob Kirby. Problems in low-dimensional topology. In *Geometric Topology: Proc. 1993 Georgia Int. Topol. Conf.*, volume 2.2 of *AMS/IP Studies in Advanced Mathematics*, pages 35–473, American Mathematical Society and International Press, 1997. Cited on 87.

Marc Kirschenbaum. Chess board. *Paper*, 61:24–30, 1998. Cited on 238.

Shoschichi Kobayashi. On conjugate and cut loci. In S. S. Chern, editor, *Studies in Global Geometry and Analysis*, pages 96–122. Mathematical Association of America, Washington, DC, 1967. Cited on 359.

Jan J. Koenderink. *Solid Shape*. MIT Press, Cambridge, MA, 1989. Cited on 303.

Teun Koetsier. Sylvester's role in the development of the theory of mechanisms. In *Symposium "James Joseph Sylvester (1814–1897)."* Dutch Mathematical Congress, 1999. Cited on 30.

Pascal Koiran. The complexity of local dimensions for constructible sets. *J. Complexity*, 16(1):311–323, 2000. Cited on 44.

Yoshiyuki Kusakari. On reconfiguring radial trees. In *Proc. 2002 Japan Conf. Discrete Comput. Geom.*, volume 2866 of *Lecture Notes in Computer Science*, pages 182–191. Springer-Verlag, Berlin, 2003. Cited on 96.

Yoshiyuki Kusakari, Masaki Sato, and Takao Nishizeki. Planar reconfiguration of monotone trees. *IEEE Trans. Fund.*, E85-A(5):938–943, 2002. Cited on 96.

John Kutcher. *Coordinated Motion Planning of Planar Linkages*. PhD thesis, Johns Hopkins University, 1992. Cited on 62.

Gerard Laman. On graphs and rigidity of plane skeletal structures. *J. Engrg. Math.*, 4:331–340, 1970. Cited on 45.

John D. Lambert. *Numerical Methods for Ordinary Differential Systems: The Initial Value Problem*. John Wiley & Sons, Hoboken, NJ, 1992. Cited on 106.

Robert J. Lang. Mathematical algorithms for origami design. *Symmetry Cult. Sci.*, 5(2):115–152, 1994a. Cited on 240.

Robert J. Lang. The tree method of origami design. In K. Miura, editor, *Proc. 2nd Int. Meeting Origami Sci. Scientific Origami*, Seian University of Art of Design, pages 73–82, Otsu, Japan, November–December 1994b. Cited on 240.

Robert J. Lang. A computational algorithm for origami design. In *Proc. 12th Annu. ACM Symp. Comput. Geom.*, ACM, New York, pages 98–105, 1996. Cited on 240, 242, 243, 244.

Robert J. Lang. *TreeMaker 4.0: A Program for Origami Design*, 1998. http://origami.kvi.nl/programs/treemaker/trmkr40.pdf. Cited on 240, 242, 243, 244, 246, 247, 248, 249, 250, 252.

Robert J. Lang. *Origami Design Secrets: Mathematical Methods for an Ancient Art*. A K Peters, Wellesley, MA, 2003. Cited on 168, 169, 239, 240, 243, 248.

Robert J. Lang. Origami approximate geometric constructions. In B. Cipra, E. D. Demaine, M. L. Demaine, and T. Rogers, editors, *A Tribute to a Mathemagician*, pages 223–239. A K Peters, Wellesley, MA, 2004. Cited on 288.

Mark Lanthier, Anil Maheshwari, and Jörg-Rüdiger Sack. Approximating weighted shortest paths on polyhedral surfaces. In *Proc. 13th Annu. ACM Symp. Comput. Geom.*, ACM, New York, pages 274–283, 1997. Cited on 364.

Rev. Dionysius Lardner. *A Treatise on Geometry and Its Applications to the Arts*. The Cabinet Cyclopædia. Longman, Orme, Brown, Green, & Longmans, and John Taylor, London, 1840. Cited on 168.

Jean-Claude Latombe. *Robot Motion Planning*. Kluwer Academic Publishers, Boston, 1991. Cited on 19.

Jean-Paul Laumond and Thierry Simeon. Notes on visibility roadmaps and path planning. In B. R. Donald, K. M. Lynch, and D. Rus, editors, *Algorithmic and Computational Robotics: New Directions*, pages 317–328. A K Peters, Wellesley, MA, 2001. Cited on 157.

Steven M. LaValle. *Planning Algorithms*. Cambridge University Press, Cambridge, UK, 2006. http://planning.cs.uiuc.edu/. Cited on 19, 155.

Audrey Lee, Ileana Streinu, and Louis Theran. Finding and maintaining rigid components. In *Proc. 17th Can. Conf. Comput. Geom.*, pages 216–219, 2005. Cited on 46.

D. T. Lee. Medial axis transformation of a planar shape. *IEEE Trans. Pattern Anal. Mach. Intell.*, PAMI-4(4):363–369, 1982. Cited on 370.

John M. Lee. *Riemannian Manifolds: An Introduction to Curvature*, volume 176 of *Graduate Texts in Mathematics*. Springer-Verlag, 1997. Cited on 302, 304.

William J. Lenhart and Sue Whitesides. Turning a polygon inside-out. In *Proc. 3rd Can. Conf. Comput. Geom.*, pages 66–69, August 1991. Cited on 62, 64.

William J. Lenhart and Sue Whitesides. Reconfiguring closed polygonal chains in Euclidean *d*-space. *Discrete Comput. Geom.*, 13:123–140, 1995. Cited on 64, 66, 87, 94.

Harry R. Lewis and Christos H. Papadimitriou. *Elements of the Theory of Computation*. Prentice Hall, Saddle River, NJ, 1997. Cited on 23.

David Lister. Two miscellaneous collections of jottings on the history of origami. Unpublished manuscript, 1998. http://www.paperfolding.com/history/. Cited on 167.

James J. Little. An iterative method for reconstructing convex polyhedra from extended Gaussian image. In *Proc. Am. Assoc. Artif. Intell.*, pages 247–250, 1983. Cited on 340.

James J. Little. Extended Gaussian images, mixed volumes, and shape reconstruction. In *Proc. 1st Annu. ACM Symp. Comput. Geom.*, ACM, New York, pages 15–23, 1985. Cited on 340.

Charles Livingston. *Knot Theory*. Mathematical Association of America, Washington, DC, 1993. Cited on 89.

Gerald M. Loe. *Paper Capers*. Magic, Inc., Chicago, 1955. Cited on 254.

Tomas Lozano-Pérez. A simple motion planning algorithm for general robot manipulators. *IEEE Trans. Robot. Autom.*, RA-3(3):224–238, 1987. Cited on 12.

Tomas Lozano-Pérez and Michael A. Wesley. An algorithm for planning collision-free paths among polyhedral obstacles. *Commun. ACM*, 22(10):560–570, 1979. Cited on 12.

Liang Lu and Srinivas Akella. Folding cartons with fixtures: A motion planning approach. *IEEE Trans. Robot. Autom.*, 16(4):346–356, 2000. Cited on 14.

Anna Lubiw and Joseph O'Rourke. When can a polygon fold to a polytope? Technical Report 048, Department of Computer Science, Smith College, Northampton, MA, June 1996. Presented at *Am. Math. Soc.* Conf., 5 October 1996. Cited on 381, 386, 388, 402.

Brendan Lucier. Partial results on convex polyhedron unfoldings, University of Waterloo course *Algorithms for Polyhedra*, 2004. http://www.cs.uwaterloo.ca/~blucier/misc/polyhedra.pdf. Cited on 318.

Vladimir J. Lumelsky. Effect of kinematics on dynamic path planning for planar robot arms moving amidst unknown obstacles. *Int. J. Robot. Autom.*, RA-3(3):207–223, 1987. Cited on 12.

Anna Lysyanskaya. *Constructing a Simple Closed Geodesic on a Polytope*. Undergraduate thesis, Smith College, 1997. Cited on 374.

Lazar Lyusternik and Lev Schnirelmann. Sur le problém de trois géodésiques fermées sur les surfaces de genre 0. *C. R. Acad. Sci. Paris*, 189:269–271, 1929. Cited on 374.

Dana Mackenzie. Polyhedra can bend but not breathe. *Science*, 279:1637, 1998. Cited on 348.

Andrea Mantler and Jack Snoeyink. Banana spiders: A study of connectivity in 3D combinatorial rigidity. In *Proc. 14th Can. Conf. Comput. Geom.*, pages 44–47, 2004. Cited on 47.

Costas D. Maranas, Ioannis P. Androulakis, and Christodoulos A. Floudas. A deterministic global optimization approach for the protein folding problem. In P. M. Pardalos, D. Shalloway, and G. Xue, editors, *Global Minimization of Nonconvex Energy Functions: Molecular Conformation and Protein Folding*, volume 23 of *DIMACS Series Discrete Math. Theoret. Comput. Sci.*, pages 133–150. American Mathematical Society, New York, 1996. Cited on 15.

Giancarlo Mauri, Giulio Pavesi, and Antonio Piccolboni. Approximation algorithms for protein folding prediction. In *Proc. 10th Annu. ACM-SIAM Symp. Discrete Algorithms*, ACM/SIAM, New York/Philadelphia, pages 945–946, January 1999. Cited on 161.

James C. Maxwell. On reciprocal diagrams and diagrams of forces. *Phil. Mag. Ser. 4*, 27:250–261, 1864. Cited on 58.

Brendan D. McKay. Spanning trees in regular graphs. *Eur. J. Combin.*, 4:149–160, 1983. Cited on 431.

Peter Messer. Problem 1054. *Crux Mathematicorum*, 12(10), December 1986. Posed in 1985, page 188. Cited on 289.

Ezra Miller and Igor Pak. Metric combinatorics of convex polyhedra: Cut loci and nonoverlapping unfoldings. *Discrete Comput. Geom.*, to appear. Cited on 360, 362, 366.

Kenneth C. Millett. Knotting of regular polygons in 3-space. *J. Knot Theory Ramifications*, 3(3):263–278, 1994. Cited on 80, 81.

Bhubaneswar Mishra. Computational real algebraic geometry. In J. E. Goodman and J. O'Rourke, editors, *Handbook of Discrete and Computational Geometry*, 2nd edition, chapter 33, pages 743–764. CRC Press LLC, Boca Raton, FL, 2004. Cited on 111.

Jun Mitani and Hiromasa Suzuki. Making papercraft toys from meshes using strip-based approximate unfolding. *ACM Trans. Graph.*, 23(4):259–263, 2004. Proc. ACM Conf. SIGGRAPH 2004. Cited on 308.

Joseph S. B. Mitchell, David M. Mount, and Christos H. Papadimitriou. The discrete geodesic problem. *SIAM J. Comput.*, 16:647–668, 1987. Cited on 358, 362, 363, 364.

John Montroll. *African Animals in Origami*. Dover Publications, 1991. Cited on 233.

Frank Morgan. Riemannian geometry: A beginner's guide. Manuscript, Williams College, 1991. Cited on 304.

Cristian Moukarzel. An efficient algorithm for testing the generic rigidity of graphs in the plane. *J. Phys. A: Math. Gen.*, 29:8079–8098, 1996. Cited on 46.

David M. Mount. On finding shortest paths on convex polyhedra. Technical Report 1495, Department of Computer Science, University of Maryland, 1985. Cited on 359, 360.

Yukio Murakami. Origami geometry (paper folding): A new system making trisection of arbitrary angle possible. In *Proc. Int. IEEE Conf. Syst. Man Cybern.*, pages 1107–1109, 1987. Cited on 285, 289.

Kenneth P. Murphy. *Protein Structure, Stability, and Folding*, volume 168 of *Methods in Molecular Biology*. Humana Press, 2001. Cited on 148.

Béla de Sz. Nagy. Solution to problem 3763. *Am. Math. Monthly*, 46:176–177, 1939. Cited on 74, 75.

Makoto Namiki and Komei Fukuda. Unfolding 3-dimensional convex polytopes: A package for Mathematica 1.2 or 2.0. Mathematica Notebook, University of Tokyo, 1993. Cited on 314, 386.

National standards and emblems. *Harper's New Mon. Mag.*, 47(278):171–181, July 1873. Cited on 254.

Alantha Newman. A new algorithm for protein folding in the HP model. In *Proc. 13th Annu. ACM-SIAM Symp. Discrete Algorithms*, ACM/SIAM, New York/Philadelphia, pages 876–884, January 2002. Cited on 161, 162.

Poul Nissen, Jeffrey Hansen, Nenad Ban, Peter B. Moore, and Thomas A. Steitz. The structural basis of ribosome activity in peptide bond synthesis. *Science*, 289:920–930, 2000. Cited on 148.

Kari J. Nurmela and Patric R. J. Östergøard. Packing up to 50 equal circles in a square. *Discrete Comput. Geom.*, 18:111–120, 1997. Cited on 246.

Jun O'Hara. Morse functions on configuration spaces of planar linkages. arXiv:math.GT/0505462, 2005. Cited on 28.

David Orden, Francisco Santos, Brigitte Servatius, and Herman Servatius. Combinatorial pseudo-triangulations. arXiv:math.CO/0307370. *Discrete Math.*, to appear. Cited on 46.

Joseph O'Rourke. *Computational Geometry in C*, 2nd edition. Cambridge University Press, Cambridge, UK, 1998. Cited on 19, 21, 63, 266, 355, 386.

Joseph O'Rourke. Computational geometry column 39. *Int. J. Comput. Geom. Appl.*, 10(4):441–444, 2000a. Also in *SIGACT News*, 31(3):47–49, 2000a, Issue 116. Cited on 98.

Joseph O'Rourke. Folding and unfolding in computational geometry. In *Proc. 1998 Japan Conf. Discrete Comput. Geom.*, volume 1763 of *Lecture Notes in Computer Science*, pages 258–266. Springer-Verlag, Berlin, 2000b. Cited on 357.

Joseph O'Rourke. On the development of the intersection of a plane with a polytope. Technical Report 068, Department of Computer Science, Smith College, Northampton, MA, June 2000c. arXiv:cs.CG/0006035v3. Cited on 146.

Joseph O'Rourke. An extension of Cauchy's arm lemma with application to curve development. In *Proc. 2000 Japan Conf. Discrete Comput. Geom.*, volume 2098 of *Lecture Notes in Computer Science*, pages 280–291. Springer-Verlag, Berlin, 2001a. Cited on 147.

Joseph O'Rourke. Unfolding prismoids without overlap. Unpublished manuscript, May 2001b. Cited on 323.

Joseph O'Rourke. On the development of the intersection of a plane with a polytope. *Comput. Geom. Theory Appl.*, 24(1):3–10, 2003. Cited on 377, 378, 379, 380.

Joseph O'Rourke and Catherine Schevon. On the development of closed convex curves on 3-polytopes. *J. Geom.*, 13:152–157, 1989. Cited on 376, 377.

Roger Pain. *Mechanisms of Protein Folding.* Oxford University Press, Oxford, UK, 2000. Cited on 148.

Igor Pak. A short proof of rigidity of convex polytopes. *Siberian J. Math.*, 47:859–864, 2006. Cited on 342.

Charles-Nicolas Peaucellier. Lettre de M. Peaucellier. *Nouvelles Ann. Math.*, 3:414, 1864. Letter to the editor ("Lettre de M. Peaucellier, capitaine du Genie, a Nice"). Cited on 30.

Ronald Peikert. Dichteste Packungen von gleichen Kreisen in einem Quadrat (Densest packings of equal circles in a square). *Elem. Math.*, 49(1):16–26, 1994. Cited on 246.

Tony Phillips. Descartes' lost theorem. In *What's New in Math.* American Mathematics Society, 1999. www.ams.org. Feature Column Archive. Cited on 305.

Margherita Piazzolla Beloch. Sulla risoluzione dei problemi di terzo e quarto grado col metodo del ripiegamento della carta. *Scritti Matematici Offerti a Luigi Berzolari, Pavia*, pages 93–96, 1936. Cited on 169, 289.

Michel Pocchiola and Gert Vegter. Computing the visibility graph via pseudo-triangulations. In *Proc. 11th Annu. ACM Symp. Comput. Geom.*, ACM, New York, pages 248–257, 1995. Cited on 108.

Michel Pocchiola and Gert Vegter. Pseudo-triangulations: Theory and applications. In *Proc. 12th Annu. ACM Symp. Comput. Geom.*, ACM, New York, pages 291–300, 1996. Cited on 108.

Aleksei V. Pogorelov. Quasi-geodesic lines on a convex surface. *Mat. Sb.*, 25(62):275–306, 1949. English translation, *Am. Math. Soc. Transl.* 74, 1952. Cited on 374.

Aleksei V. Pogorelov. *Extrinsic Geometry of Convex Surfaces*, volume 35 of *Translations of Mathematical Monographs.* American Mathematical Society, Providence, RI, 1973. Cited on 349, 352, 353, 373.

Konrad Polthier and Markus Schmies. Straightest geodesics on polyhedral surfaces. In H. C. Hege and K. Polthier, editors, *Mathematical Visualization*, page 391. Springer-Verlag, Berlin, 1998. Cited on 373.

Sheung-Hung Poon. On straightening low-diameter unit trees. In P. Healy and N. S. Nikolov, editors, *Proc. 13th Int. Symp. Graph Drawing (2005)*, volume 3843 of *Lecture Notes in Computer Science*, pages 519–521. Springer-Verlag, Berlin, 2006. Cited on 96, 154.

Helmut Pottmann and Johannes Wallner. *Computational Line Geometry.* Springer-Verlag, Berlin, 2001. Cited on 191, 352, 354.

Robert C. Read and Robert E. Tarjan. Bounds on backtrack algorithms for listing cycles, paths and spanning trees. *Networks*, 5:237–252, 1975. Cited on 315.

John H. Reif. Complexity of the mover's problem and generalizations. In *Proc. 20th Annu. IEEE Symp. Found. Comput. Sci.*, IEEE, Los Alamitos, pages 421–427, 1979. Cited on 22, 24.

John H. Reif. Complexity of the generalized movers problem. In J. E. Hopcroft, J. Schwartz, and M. Sharir, editors, *Planning, Geometry and Complexity of Robot Motion*, pages 267–281. Ablex Publishing, Norwood, NJ, 1987. Cited on 22, 23.

Günter Rote, Francisco Santos, and Ileana Streinu. Expansive motions and the polytope of pointed pseudo-triangulations. In B. Aronov, S. Basu, J. Pach, and M. Sharir, editors, *Discrete and Computational Geometry: The Goodman–Pollack Festschrift*, pages 699–736. Springer-Verlag, Berlin, 2003. Cited on 111.

Ben Roth and Walter Whiteley. Tensegrity frameworks. *Trans. Am. Math. Soc.*, 265(2):419–446, 1981. Cited on 56, 57.

T. Sundara Row. *Geometric Exercises in Paper Folding.* Addison & Co., Madras, India, 1893. Edited and revised by W. W. Beman and D. E. Smith, The Open Court Publishing Co., Chicago, 1901. Republished by Dover, New York, 1966. Cited on 168.

Idzhad Sabitov. The volume of a polyhedron as a function of its metric and algorithmical solution of the main problems in the metric theory of polyhedra. In *International School-Seminar Devoted to the N. V. Efimov's Memory*, pages 64–65. Rostov University, 1996a. Cited on 354.

Idzhad Sabitov. The volume of polyhedron as a function of its metric. *Fundam. Prikl. Mat.*, 2(4):1235–1246, 1996b. Cited on 348, 355.

Idzhad Sabitov. The volume as a metric invariant of polyhedra. *Discrete Comput. Geom.*, 20:405–425, 1998. Cited on 355.

G. Thomas Sallee. Stretching chords of space curves. *Geom. Dedicata*, 2:311–315, 1973. Cited on 67.

James B. Saxe. Embeddability of weighted graphs in k-space is strongly NP-hard. In *Proc. 17th Allerton Conf. Commun. Control Comput.*, pages 480–489, 1979. Cited on 48.

Thomas J. Schaefer. The complexity of satisfiability problems. In *Proc. 10th Annu. ACM Symp. Theory Comput.*, ACM, New York, pages 216–226, 1978. Cited on 217.

Catherine Schevon. Unfolding polyhedra, February 1987. sci.math Usenet article. http://www.ics.uci.edu/~eppstein/gina/unfold.html. Cited on 318.

Catherine Schevon. *Algorithms for Geodesics on Polytopes*. PhD thesis, Johns Hopkins University, 1989. Cited on 315, 375.

Catherine Schevon and Joseph O'Rourke. A conjecture on random unfoldings. Technical Report JHU-87/20, Johns Hopkins University, Baltimore, MD, July 1987. Cited on 315.

Wolfram Schlickenrieder. *Nets of Polyhedra*. PhD thesis, Technische Universität Berlin, 1997. Cited on 312, 315, 316, 317.

Isaac J. Schoenberg and Stanislaw K. Zaremba. On Cauchy's lemma concerning convex polygons. *Can. J. Math.*, 19:1062–1077, 1967. Cited on 144.

Yevgeny Schreiber and Micha Sharir. An optimal-time algorithm for shortest paths on a convex polytope in three dimensions. In *Proc. 22nd Annu. ACM Symp. Comput. Geom.*, ACM, New York, pages 30–39, 2006. Cited on 366.

Axel Schur. Über die Schwarzsche Extremaleigenschaft des Kreises unter den Kurven konstantes Krümmung. *Math. Ann.*, 83:143–148, 1921. Cited on 147.

Kan Chu Sen. *Wakoku Chiyekurabe (Mathematical Contests)*, 1721. Excerpts at http://theory.csail.mit.edu/~edemaine/foldcut/sen_book.html. Cited on 254.

Micha Sharir. Algorithmic motion planning. In J. E. Goodman and J. O'Rourke, editors, *Handbook of Discrete and Computational Geometry*, 2nd edition, chapter 47, pages 1037–1064. CRC Press LLC, Boca Raton, FL, 2004. Cited on 19, 173.

Micha Sharir and Amir Schorr. On shortest paths in polyhedral spaces. *SIAM J. Comput.*, 15:193–215, 1986. Cited on 358, 359, 360, 361.

Geoffrey C. Shephard. Convex polytopes with convex nets. *Math. Proc. Camb. Phil. Soc.*, 78:389–403, 1975. Cited on 300, 327, 328, 389, 411.

Don Shimamoto and Catherine Vanderwaart. Spaces of polygons in the plane and Morse theory. *Am. Math. Mon.*, 112(4):289–310, 2005. Cited on 28.

David Singer. *Geometry: Plane and Fancy*. Springer-Verlag, Berlin, 1997. Cited on 144.

John S. Smith. Notes on the history of origami (3rd edition). Unpublished manuscript, 2005. http://www.bitsofsmith.co.uk/history.htm. First edition published as Booklet 1 in a series by the British Origami Society, 1972. Cited on 167.

Jack Snoeyink and Jorge Stolfi. Objects that cannot be taken apart with two hands. *Discrete Comput. Geom.*, 12:367–384, 1994. Cited on 125.

Guang Song and Nancy M. Amato. Using motion planning to study protein folding pathways. In *Proc. 5th Annu. Int. Conf. Comput. Bio.*, ACM, New York, pages 287–296, 2001. Cited on 157.

Guang Song and Nancy M. Amato. A motion planning approach to folding: From paper craft to protein folding. *IEEE Trans. Robot. Autom.*, 20(1):60–71, February 2004. Cited on 157.

Michael Soss. *Geometric and Computational Aspects of Molecular Reconfiguration*. PhD thesis, School of Computer Science, McGill University, 2001. Cited on 132, 135.

Michael Soss and Godfried T. Toussaint. Geometric and computational aspects of polymer reconfiguration. *J. Math. Chem.*, 27(4):303–318, 2000. Cited on 132, 135.

Chris E. Soteros and Stuart G. Whittington. Polygons and stars in a slit geometry. *J. Phys. A: Math. Gen. Phys.*, 21:L857–L861, 1988. Cited on 131.

Michael Spivak. *A Comprehensive Introduction to Differential Geometry*, volume 3. Publish or Perish Press, Berkeley, CA, 1979. Cited on 303.

Ernst Steinitz and Hans Rademacher. *Vorlesungen über die Theorie der Polyeder*. Springer, Berlin, 1934. Cited on 341, 342.

J. J. Stoker. Geometrical problems concerning polyhedra in the large. *Comm. Pure Appl. Math.*, 21:119–168, 1968. Cited on 340.

Ileana Streinu. A combinatorial approach to planar non-colliding robot arm motion planning. In *Proc. 41st Annu. IEEE Symp. Found. Comput. Sci.*, IEEE, Los Alamitos, pp. 443–453. November 2000. Cited on 88, 107, 108, 110.

Ileana Streinu. Combinatorial roadmaps in configuration spaces of simple planar polygons. In S. Basu and L. Gonzalez-Vega, editors, *Proc. DIMACS Workshop Algorithmic Quant. Aspects Real Algebraic Geom. Math. Comput. Sci.*, pages 181–206, 2003. Cited on 111.

Ileana Streinu. Pseudotriangulations, rigidity, and motion planning. *Discrete Comput. Geom.*, 34:587–635, 2005. Cited on 107, 109, 110.

Ileana Streinu and Walter Whiteley. The spherical carpenter's rule problem and conical origami folds. In *Proc. 11th Annu. Fall Workshop Comput. Geom.*, Brooklyn, New York, November 2001. Cited on 212, 213.

Kokichi Sugihara. On redundant bracing in plane skeletal structures. *Bull. Electrotech. Lab.*, 44:78–88, 1980. Cited on 46.

William A. Sutherland. *Introduction to Metric and Topological Spaces*. Oxford University Press, Oxford, UK, 1975. Cited on 174.

Alexei S. Tarasov. Polyhedra with no natural unfoldings. *Russian Math. Surv.*, 54(3):656–657, 1999. Cited on 318.

William P. Thurston and Jeffrey R. Weeks. The mathematics of three-dimensional manifolds. *Sci. Am.*, 251(1):108–120, July 1984. Cited on 61.

Michael J. Todd. Mathematical programming. In J. E. Goodman and J. O'Rourke, editors, *Handbook of Discrete and Computational Geometry*, 2nd edition, chapter 46, pages 1015–1036. CRC Press LLC, Boca Raton, FL, 2004. Cited on 54.

Godfried T. Toussaint. Movable separability of sets. In G. T. Toussaint, editor, *Computational Geometry*, pages 335–375. North-Holland, Amsterdam, 1985. Cited on 125.

Godfried T. Toussaint. A new class of stuck unknots in Pol_6. *Beiträge Algebra Geom.*, 42(2):301–306, 2001. Cited on 90, 91.

Godfried T. Toussaint. Simple proofs of a geometric property of four-bar linkages. *Am. Math. Mon.*, 110(6):482–494, 2003. Cited on 67, 73.

Godfried Toussaint. The Erdős–Nagy theorem and its ramifications. *Comput. Geom. Theory Appl.*, 31(3):219–236, 2005. Cited on 67, 75, 80, 81.

Luca Trevisan. When Hamming meets Euclid: The approximability of geometric TSP and MST. In *Proc. 29th Annu. ACM Symp. Theory Comput.*, ACM, New York, pages 21–29, 1997. Cited on 159.

John M. Troyer, Fred E. Cohen, and David M. Ferguson. Langevin dynamics of simplified protein models. In *Electronic Comput. Chem. Conf.*, 1997. Paper 76; http://www.cmpharm.ucsf.edu/~troyer/eccc/. Cited on 15.

Marc van Kreveld, Jack Snoeyink, and Sue Whitesides. Folding rulers inside triangles. *Discrete Comput. Geom.*, 15:265–285, 1996. Cited on 69.

Ju. A. Volkov and E. G. Podgornova. The cut locus of a polyhedral surface of positive curvature. *Ukrainian Geom. Sb.*, 11:15–25, 1971. In Russian. Cited in Miller and Pak (to appear). Cited on 359.

Cheng-Hua Wang. *Manufacturability-Driven Decomposition of Sheet Metal Products*. PhD thesis, Robotics Institute, Carnegie Mellon University, 1997. Cited on 14, 306.

Pierre-Laurent Wantzel. Recherches sur les moyens de reconnaître si un problème de géométrie peut se résoudre avec la règle et le compas. *Journ. de Math. Pures et Appl.*, 2:366–372, 1837. Cited on 285.

Bernd Wegner. Partial inflation of closed polygons in the plane. *Beiträge Algebra Geom.*, 34(1):77–85, 1993. Cited on 78.

Eric W. Weisstein. *CRC Concise Encylopedia of Mathematics*. CRC Press LLC, Boca Raton, FL, 1999. Cited on 128.

Margaret Wertheim. Cones, curves, shells, towers: He made paper jump to life. *New York Times*, page F2, June 22, 2004. Cited on 296.

Walter Whiteley. Motions and stresses of projected polyhedra. *Structural Topology*, 7:13–38, 1982. Cited on 58.

Walter Whiteley. Rigidity and scene analysis. In J. E. Goodman and J. O'Rourke, editors, *Handbook of Discrete and Computational Geometry*, 2nd edition, chapter 60, pages 1327–1354. CRC Press LLC, Boca Raton, FL, 2004. Cited on 43.

Stacia Wyman. *Geodesics on Polytopes*. Undergraduate thesis, Smith College, 1990. Cited on 374.

Dianna Xu. *Shortest Paths on the Surface of a Polytope*. Undergraduate thesis, Smith College, 1996. Cited on 359.

Yechiam Yemini. Some theoretical aspects of position-location problems. In *Proc. 20th Annu. IEEE Symp. Found. Comput. Sci.*, IEEE, Los Alamitos, pages 1–8, 1979. Cited on 48.

Akira Yoshizawa. *Atarashi Origami Geijutsu (New Origami Art)* [in Japanese], 1954. Cited on 168.

Jianmin Zhao and Norman I. Badler. Inverse kinematics positioning using nonlinear programming for highly articulated figures. *ACM Trans. Graph.*, 13(4):313–336, 1994. Cited on 12.

Günter M. Ziegler. *Lectures on Polytopes*, volume 152 of *Graduate Texts in Mathematics*. Springer-Verlag, Berlin, 1994. Cited on 339.

Konrad Zindler. Eine räumliche Geradführung. *Sitzungsberichte Wien*, 140:399–402, 1931. Cited on 31.

Index

Printed in the United States
By Bookmasters